D1566420

**Nourse, Alan E**
Universe, earth, and atom



A simple but comprehensive explanation of the whole range of physics traces the progress of the science from the early Greek philosopher-scientists to Einstein's startling concepts and the impact of modern physics.

# UNIVERSE, EARTH, AND ATOM

*Books by Dr. Alan E. Nourse*

Universe, Earth, and Atom
Nine Planets
Junior Intern
So You Want to Be a Doctor
So You Want to Be a Physicist
So You Want to Be a Scientist
So You Want to Be a Surgeon
So You Want to Be a Nurse (with Eleanore Halliday, R.N.)
So You Want to Be a Chemist (with James C. Webbert)
So You Want to Be an Engineer (with James C. Webbert)
So You Want to Be a Lawyer (with William B. Nourse)

# Universe, Earth, and Atom

*The Story of Physics*

**by Alan E. Nourse, M. D.**

1817

HARPER & ROW, PUBLISHERS

*New York and Evanston*

For Christopher, Jonathan, Rebecca and Benjamin
and their mother



FIRST EDITION

LIBRARY OF CONGRESS CATALOG CARD NUMBER: 69-13493

# CONTENTS

## Part V: The Universe of the Inconceivably Small



## Part VI: Practicalities and Promises: The Impact of Modern Physics

# *Acknowledgments*

Any book of this nature inevitably is the result of the work, help, and encouragement of many people other than the one whose name is on the manuscript. The author wishes to express his indebtedness in particular to Harold E. Grove, who provided the initial spark for the project; to Professor Louis Trimble, of the University of Washington in Seattle, for his careful reading and annotation of early drafts of the manuscript; to Dr. Hong-Yee Chiu, of the Goddard Institute for Space Science and the State University of New York at Stonybrook, Long Island, for his professional criticism of the advanced draft of the book; to Jerry Jermann for the diagrams and drawings; to George W. Jones and Carl D. Brandt for their perseverance and encouragement in seeing the project through to its present form; and to Elinor Busby, Doris Vinnedge, and Becky Nourse for their help in preparation of the manuscript.

Little wheel spin and spin,
Big wheel turn around and 'round. . . .

*Buffy Sainte-Marie*

# *Introduction*

This book is written about the world of modern physics and the work of physicists today—how they know what they know, what they are still trying to find out, and what their discoveries mean to us all, both now and in the future.

We are living today in an age of research and discovery more intense and exciting than in any comparable period in history. Some of our scientists are probing the mysteries of life in the nuclei of microscopic cells; others are preparing to explore the outer reaches of our solar system. Discovery eclipses discovery as we learn more and more about ourselves, our planet, and our universe. But of all areas of research in science today there is none more fascinating, and none more baffling to the nonscientist, than the work of modern physics—the study of the structure and function of the physical universe of which we are a part.

On first thought it might seem odd that so many people are so completely lost when it comes to understanding the ideas of modern physics. Other areas of science—medicine, space technology, psychology—do not seem so formidable and confusing. Yet when we hear about a new discovery in physics we suddenly find ourselves in deep and murky water. We don't even understand what these discoveries *are,* much less what they may mean to us in our modern technological society. And the knowledge that these discoveries may have profound effects upon our lives—perhaps even on our survival—makes such confusion not merely regrettable but downright dangerous.

Such widespread public confusion—ignorance, if you will—is not the fault of the physicists who are doing the research. Many of them have tried diligently to explain to the nonphysicist just what, exactly, they were doing, in the pages of well-written books—literally hundreds of books. Why, then, do we need another, and one written by a layman in the world of physics at that?

The reason is simple and personal. It is my conviction that existing books, however brilliantly written, have simply not done the job they set out to do. Readers seeking a broad general understanding of what is happening in physics just don't understand what they are reading. They find themselves grappling with baffling or flatly incomprehensible ideas that have no discernible relationship to anything they have ever experienced in their lives. They become trapped in a quagmire of confusing terms, definitions, and abstract concepts, and then feel angry and somehow cheated when they find that these concepts are, in fact, shaking the very ground they walk on, whether they understand what is going on or not.

Obviously something is missing. Some *comprehensible frame of reference* is needed, some way of drawing together a huge and confusing body of information about the laws of nature and the discoveries of modern physics so that they can be readily understood, visualized, and related in *some* way to the everyday world we live in. No one can tackle totally unfamiliar ideas and apparently fantastic and incomprehensible concepts without some familiar frame of reference as a starting place, some place to stand. The major goal of this book is to find such a frame of reference.

A secondary goal is to discuss what has happened throughout past centuries of research in physics, and what is happening today, as clearly as possible, not in terms of exquisite technical detail, but by showing the general direction that scientific thinking has followed in the past, and is still following today. This means choosing repeatedly between what is really significant and what is really not. It means drawing illustrative examples and analogies freely, and using comprehensible generalizations, even at the price of sacrificing some degree of precise scientific accuracy. In this book the crying need to express general ideas clearly must necessarily override the scientist's absolute devotion to accuracy.

What frame of reference do we propose? For one thing, an historical frame of reference, for there is no more fascinating approach to modern concepts in physics than to see how physicists themselves came to their own conclusions, changing and rejecting and modifying the things they once thought to be true, step by painful step throughout the ages. Another part of our frame of reference will be to consider in detail some of the baffling riddles that physicists have struggled to solve—the innumerable perplexing contradictions, for example, between theoretical predictions and actual laboratory observations that have been confounding scientists since the earliest times, forcing them to reappraise what they thought was right before and find out what new concepts may lie closer to the truth. Still another part will be to see how the most complex and fantastic concepts of modern physics actually impinge on the way we live today and will be living tomorrow. Finally, part will be concerned with the future—with the direction that today's search for clearer understanding of the universe is taking us, and with what that direction will mean, or *can* mean, in the world of tomorrow.

To accomplish this, we will first consider a way of looking at the universe around us, or, more accurately, several quite different ways, for this is the only avenue to understanding what is happening in the universe in comprehensible terms. For centuries, one very limited way of regarding the universe completely dominated scientific thinking. Then, within the last century, that one view was found inadequate to explain things which were actually observed to be happening—to the amazement and chagrin of practically all scientists everywhere. Bit by bit in the last few decades that whole "classical" view of the structure and function of the universe, as it always had been understood, began to crumble before the challenge of a few brilliant men, and a different way of regarding the universe—in many ways a wildly incredible way—seemed to be needed.

Other views of the universe *were* needed, and were found. We will see what these different ways of looking at things were, how they evolved in the first place, and why they were so desperately needed to achieve a more complete understanding of how the universe really worked. At first each of these views may seem complete in itself, quite unrelated to any others, as though we lived not in one universe but in several at once, with one set of rules applying in one area and another set in another. But we will see how physicists have discovered, bit by bit, that each of these views fits together with each of the others, that each is no more than a view of the same universe, in all its incredible complexity, regarded from a different angle; and we will see how discoveries made from each of these angles have proven consistent with discoveries from the other angles, and how each has had a profound influence on the everyday lives of all of us.

Finally, we will consider where research in physics has taken us up to the present time, just how it *is* affecting our lives, and where it may be taking us tomorrow. Many physicists today believe that we have reached an all-but-impenetrable barrier to more detailed understanding of the universe . . . that only another totally new way of looking at the physical universe, an approach even more fantastically different than any yet explored, can possibly breach that barrier and lead us on to more detailed knowledge. Others, equally well grounded in modern physics, insist that the complex and esoteric mathematical techniques even now being used by theoretical physicists have already provided that "new view" and that most researchers simply have not yet succeeded in fully understanding these techniques and their long-range implications. At the same time, multitudes of new practical uses have been found for the new knowledge already accumulated, and workers in modern physics have begun approaching goals and achievements that have been dreamed of for centuries. We will discuss these new frontiers as they stand today, and try to foresee some of the directions in which current research may be leading us.

In approaching a book of this sort, certain things must be clearly understood from the beginning. Obviously, this is not a learned scientific treatise, not a textbook of physics. Rather, it is a book of general information written for the intelligent but untrained layman who (like the author) seeks to understand in clear and simple terms what the current work in modern physics is doing, how it got where it is, and where it is now going. Much technical detail and a certain degree of scientific precision will be sacrificed, not because they are unimportant (obviously they are immensely important to the professional physicist) but because they are *unimportant to the goal of this book.* They would hinder more than they would help in presenting a clear, readable discussion of the general principles, exciting ideas, and revolutionary discoveries of modern physics which today are affecting the lives of physicist and layman alike. Too much public comprehension of physics has already been slaughtered on the altar of scientific accuracy; this book, it is hoped, will help revive the victim.

This is not to say that accuracy will simply be thrown to the winds. It does mean that we will walk a razor's edge between "writing down" to more sophisticated readers, on the one hand, and outdistancing others who are attempting once again to grapple with some basic ideas that have eluded them in a dozen previous encounters. Here we must plead for patience; many items that are "common knowledge" to some readers will be new and difficult concepts to others. Certain technical terms *must* be defined and used constantly as we go along; we will seek out the clearest definitions and descriptions that we can find. In general, we will assume that the reader has a certain basic acquaintance with high-school-level mathematics, chemistry, and physics, and is aware of the common terminology associated with such everyday phenomena as electricity, magnetism, radioactivity, and so forth—but when in doubt, we will clarify fundamentals before moving to the more complex. Wherever possible we will avoid mathematical formulas in favor of verbal descriptions; too often, formulas are nothing more than so many blank spaces on the page to many readers—a barrier to understanding rather than a help.

Finally, we will make certain arbitrary choices as to what to discuss and what to pass over. A book ten times longer than this one could not hope to consider all the staggering number of phenomena that modern physics is concerned with, from the interaction of subatomic particles to the cosmic explosion of distant galaxies, and from the behavior of light and other radiant energy permeating the universe to the nature and behavior of time. We can only touch the high spots, selecting topics that will contribute the most to clear general understanding.

In short, we are setting ambitious goals for this book, and must accept certain limitations in order to attain those goals. This is a book about *how things work* and *why they work the way they do,* as far as is known today. If we can describe, in understandable and nontechnical language, some of the great discoveries that have been made in physics over the centuries, and gain some understanding of what those discoveries really mean to us, we will be approaching our goal. And if, in addition, we can convey a sense of the excitement of those discoveries and stir a sense of wonder at the mysteries that human minds are penetrating even today, and at the frightening and promising enigmas that still remain unanswered in the universe around us, then this book will have been worth the writing.

DR. ALAN E. NOURSE
North Bend, Washington

# Part I

## *Physics in Perspective*

# CHAPTER 1

## *The Physics of Common Sense*

On a cool dark night almost four hundred years ago a young man walked up a hill in northern Italy with a lantern in his hand. This man was a scientist whose name and fame would one day be known to every school child; but on this particular night, unknown to him, he was setting out on one of the most celebrated wild-goose chases in all the history of science.

No one knows exactly what night it was, nor whether it was summer or winter. We might imagine that it was a balmy evening with no moon, for this particular man was a very acute thinker and would have picked the time and circumstances best suited to the success of his experiment. At a distance across the valley his laboratory assistant climbed another hill with another lantern, probably quite convinced that his master was afflicted by demons. Yet the assistant knew that his master had performed other experiments with rather surprising results, and—who could say?—perhaps this would be another.

In any case, this event was more than a mere walk in the country. This scientist was intent upon actually measuring the speed with which a beam of light would travel from one point to another. He thought he had devised a way to accomplish this feat. He had observed that other kinds of signals traveled a given distance at a given, measurable speed. He knew, for example, that the disturbance created by a stone dropped in a quiet pond could be followed and the velocity of its movement timed. He had also observed that when a distant woodsman was seen cutting down a tree, the sound of each blow required a measurable period of time to reach the observer after the ax struck—the time necessary for the sound to travel from the tree to the ear. Thus, it seemed logical to our young scientist that a beam of light would take a discrete period of time to travel from its source on one hilltop to an observer on a distant hilltop. His goal was to measure this time interval and then, knowing the distance between hilltops, to calculate the velocity of the light beam.

The procedure planned was very simple. With his assistant watching from a distant hilltop, the scientist would unmask his lantern. The instant his assistant saw the light, he would unmask his lantern in turn. The difference in time between opening the first lantern and observing the answering light from the second should then be equivalent to the time required for a light beam to travel to the distant hilltop and back again.

It was a well-conceived experiment. One might have expected that; this particular man was one of the most acute scientific observers in all history, and a clever experimenter as well. Despite all that, the experiment ended up as a spectacular failure. The answering light from the distant hill was seen to appear at the very same instant that the scientist opened his lantern. There was no time lag observed. What was more, the same thing happened each time, no matter how many times the experiment was repeated.

To Galileo Galilei, his hilltop experiment that night could mean only one thing: that light had *no* measurable velocity, but rather, spread instantaneously to all parts of the universe at the same time. Today, of course, we know that that conclusion was wrong. Light does indeed have a measurable velocity; it requires a definite interval of time to travel from one point to any other point in the universe. But the fault did not lie in Galileo's thinking. He had no way to guess that the distance between the hilltops that he had chosen was so small compared to the enormously swift speed of light that the time lag he was trying to measure was simply not perceptible to the human eye. Galileo's experiment was sound enough. His instruments were simply too crude to measure the time lag; and if anyone had told him that light (which, incidentally, is not a particle but a wave) actually traveled 186,000 *miles* in a single second, he probably would have laughed uproariously. After all, common sense said that *nothing* could travel that fast.

Four centuries later, in the year 1886 in a completely different part of the world, another scientist performed another famous experiment, designed to settle, once and for all, a problem that had been perplexing scientists for generations. As in the case of Galileo, Albert Michelson's experiment was also concerned with measurement of the velocity of light. And, like Galileo's earlier experience, Michelson's experiment turned out to be a spectacular failure. There, however, the resemblance between the two experiments ended.

Michelson's experiment took place, not on a windy hilltop, but in the basement laboratory of Michelson's friend and fellow physicist Edward Morley, near Cleveland, Ohio. The apparatus for the experiment was ponderous. It involved a huge slab of stone five feet square and more than a foot thick, floating in a container of liquid mercury with an elaborate system of mirrors arranged at the four sides of the slab and a semireflecting mirror set in the center. In the darkened room a spotlight sent a beam of light from one side of the slab to the side opposite. The beam was reflected back by the mirror there—but part of the beam was diverted by the semireflecting mirror at the center of the slab, and directed back and forth between the other sides in a direction perpendicular to the original beam. Finally, the two beams of light reflected across the slab in opposite directions were directed to the same white card "target" after each half of the split light beam had traversed the slab many times, one half in one direction, the other in the other (see Fig. 1).

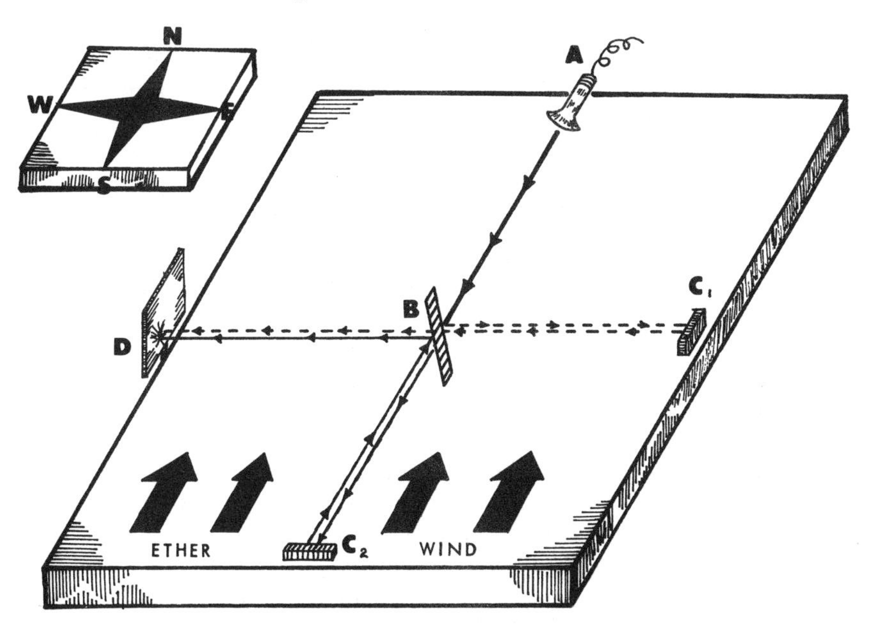


*Fig. 1* A simplified diagram of the Michelson-Morley experiment. A beam of light from source A was divided by semi-reflective mirror B so that part of the beam (unbroken arrows) went straight through to mirror $C_2$ while part (broken arrows) was diverted to mirror $C_1$. Both beams traveled equal distances against the "ether wind" to reach target card D, but one beam (broken arrows) had to travel three times as far *across* the "ether wind" as the other, and was thus expected to reach card D slightly *later* than the other beam, creating an interference pattern.

Michelson was not directly concerned with measuring the speed of light in this experiment. That had long since been accomplished with such a high degree of accuracy that most scientists were in agreement about the measurement. Rather, Michelson was determined to find out to what degree a beam of light would be slowed down as it traveled head-on through a mysterious substance known as the "universal ether," a weightless, invisible substance which virtually all scientists of the day believed permeated all the space surrounding the earth and all the space between the stars.

The notion that all space was filled with this strange invisible ether had a long and respectable history. True, no scientist had ever *seen* the ether, nor was there any direct evidence that it even existed. Yet its existence was widely accepted as necessary in order to explain how certain commonly observed phenomena could possibly occur. Physicists knew, for example, that sound waves and water waves had to travel through some medium. They also knew that light traveled at a great velocity from distant stars to the earth. But there was clear evidence that light was also a form of wave. Thus physicists argued that light also had to travel *through some medium,* whether the medium could actually be detected or not. The universal ether, although never actually detected, was assumed to exist as a necessary medium through which light waves could travel from one place to another.

At best, the idea of the ether had always been a little awkward. For one thing, if the ether was assumed to exist, other things also had to be assumed as a result. It was known, for example, that the earth was moving through space in its orbit around the sun at a velocity of about 18 miles per second. If all space were filled with ether, then it followed that the surface of the earth should be subjected to a constant "wind" of ether passing by as the earth plunged through it. But if so, why had this "ether wind" never been detected? Numerous attempts to demonstrate it had failed. Some scientists were even beginning to doubt that such a thing existed. Others maintained that it *had* to exist, but remained undetected for lack of instruments sufficiently sensitive to measure it.

This was the puzzle that Michelson and Morley had decided to tackle with their elaborate apparatus. They were certain they had devised a way to prove beyond question that an "ether wind" existed on earth. They reasoned that light waves traveling *with* the ether wind ought to be carried along faster than light waves traveling either directly into it or "across-wind," so to speak. By splitting a beam of light in half and sending one half to and fro in the direction of the ether wind by means of mirrors, and sending the other half of the beam to and fro across-wind, they believed the velocity of the crosswind light waves would lag behind slightly, and that this time lag would have to show up in the form of *interference* between the two portions of the light beam when they were brought together into one beam again to strike the target card. They further reasoned that as their apparatus was slowly rotated in its mercury bath to bring the mirrors in exact alignment with the direction of the ether wind or perpendicular to it, a familiar interference pattern of light and dark bands ought to appear on the target card.

This experiment, like Galileo's, was well conceived. By all rights it should have worked—but it failed completely. To Michelson's incredulous disappointment, not the slightest evidence of any interference pattern appeared on the target card when the apparatus was in operation, no matter in what direction they rotated the mirrors. As a colossal flop, the Michelson-Morley experiment rivaled Galileo's failure of almost four centuries before.

Worst of all, Michelson couldn't understand *how* it could have failed. Was the flaw in his apparatus? In his methods of measurement? Like any good scientist he set about diligently to examine his own experiment, hoping to devise more accurate equipment.

Nor were Michelson and Morley the only scientists thrown for a loss by this experiment's failure; the whole world of science was nonplused. No one, at first, suspected the truth: that the very *failure* of this experiment to detect an ether wind actually made it one of the most spectacularly successful and useful experiments in the history of science, for the very failure of the experiment led directly and inevitably to the whole shattering concept of relativity. It remained for an obscure little Swiss-German mathematician, a few years later, to point out the only possible conclusion that made sense: that the reason Michelson and Morley had failed to detect an ether wind was simply because *there was no ether wind.* And with that conclusion as a basic foundation, Albert Einstein then proceeded to propose a strange new theory that was soon completely to revolutionize scientists' way of looking at the universe—and ultimately to revolutionize the everyday world we live in.

The two stories related above have no direct connection with one another. They occurred in different eras of scientific development, and they were concerned with totally different scientific ideas. All the same, these stories together provide an appropriate introduction to a book about the remarkable discoveries and the complex enigmas of the world of modern physics.

For one thing, both experiments were concerned, one way or another, with the nature and propagation of light—still today one of the most puzzling and contradictory phenomena in the known universe. Even more significant, the experiments took place during two different periods in history that were critical to the development of our present-day understanding of the physical universe in which we live. In the centuries before Galileo, scientists and philosophers had made remarkably little progress trying to explain everyday occurrences in nature. It was Galileo more than any other man in history who single-handedly pried open the door to the scientific method of observation, hypothesis, experiment, and conclusion that has led us to virtually everything we know in modern physics today. Slowly the men of science who followed after Galileo began to reject mere philosophical guesswork about what things were and why things happened, and painfully began piecing together a group of rules—so-called laws of physics, or laws of nature that seemed to describe how things worked in the universe. By the time of the Michelson-Morley experiment these laws of nature had been codified, confirmed, modified, tested, retested, built upon, and expanded to the point where they seemed, taken together, to describe accurately all known natural phenomena. Many worthy scientists of Michelson's day actually believed that the end of scientific discovery would soon be at hand, that all the really basic and important laws of nature had already been discovered and defined, so that all that remained for physicists was a sort of mopping-up exercise, a matter of tying up a few loose ends here and there.

The Michelson-Morley experiment delivered a jarring blow to that attitude of self-satisfied complacency. Einstein's relativity theories shattered it completely. Within thirty short years the whole splendid edifice of classical physics, built up over the centuries, was torn to shreds by the work of a few brilliant men who dared to question the validity of the laws of nature as they were then understood, and found them wanting. The names of some of those men have become commonplace family words. But just what they did is not always so clearly understood. Often they have been thought of as "mad scientists"—or at least decidedly odd ones—and their work has been considered too complicated for ordinary non-scientists to comprehend. Often there has been a sort of a vague resentment connected with these men: why was it necessary for them to upset the clean-cut classical laws of nature and to leave things in such a muddle? But those men were not mere protesters and iconoclasts. They had no desire to destroy mankind's hard-won and comfortable picture of an orderly universe functioning according to orderly rules, just for the joy of making a mess. They were simply hard-nosed, sharp-minded, stubborn men who were disturbed about things in the world of physics that they could not explain and who insisted that a law of nature either had to do its job and explain *all* cases that came within its scope, or else it had to be changed until it did.

Contrary to popular opinion, these modern physicists and mathematicians were not prophets, or gods, or magicians. They were human beings with their own special abilities, their own failings, their own tempers and irritabilities and prejudices, just like all other human beings, including you and me. They were living in the same world of sunlight and shadow, war and politics and peace, in which you and I live. But somehow these people were able to look beyond the everyday world of sunlight and shadow, and to probe the puzzles and enigmas of new and different worlds of physics: the incredibly tiny micro-universe of atoms and their nuclei; the incredibly huge mega-universe of far-flung galaxies and cosmic expansion; still another strange mathematical universe of time and space, dimension and light-speed. To the outsider their work in these unfamiliar areas seemed puzzling, contradictory, paradoxical, even flatly incredible. To many it has seemed incomprehensible. But nevertheless, like it or not, comprehend it or not, the work of these men has molded the world that we live in and provoked drastic changes in the very lives we are leading today.

The work of these pioneer physicists and others who followed them is not yet over, not by a long shot. It continues to blossom today in thousands of laboratories in countries all over the world. It continues to shape the world that we live in. It is enormously important work. To understand what, exactly, it is, what it has achieved so far, what it seeks to accomplish today and tomorrow, is to understand more clearly every aspect of our bewildering lives as human beings today. It is to understand better how to control the forces at work today in the physical universe around us. No one in his right mind can seriously deny the overwhelming importance of that work.

But recognizing the *importance* of modern physics is one thing, while *understanding its concepts* is something else altogether. And unfortunately, most people do *not* understand very much about modern physics or the work of the physicist today. Of course, we all recognize vaguely that the solid, material objects of the universe are allegedly composed of multitudes of tiny particles arranged in certain peculiar ways and moving about with some kind of invisible motion. We all know that rockets can go to the moon, and that the spacemen inside them seem to float around with a jolly indifference to gravity. We all know that relativity theory has something to do with space and time and the speed of light (even if we don't know precisely *what* it has to do with these things) and that matter and energy, whatever they may be, are supposed to be the same thing (even if we don't really understand *how* or *why* this might be so). We even know that the conversion of matter into energy is somehow related to hydrogen bombs—whatever they are—and to radioactive fallout—whatever that may be.

But for many of us, our understanding of modern physics is uncomfortably vague. We know that something is afoot but we don't know what. We regard physicists as shadowy figures who are somehow suspect in the mysterious work they are doing. But in all fairness, we cannot blame the physicists for our lack of understanding. Very few of them care for the role of mystery man that has been thrust upon them. Most would like nothing better than to have more people understand more clearly what they were up to. Above all, they would like to be regarded in the same way that most other scientists are regarded: as fairly normal, mundane human beings who happen to be engaged in a fascinating but complicated kind of scientific work. Most physicists feel that they should not be blamed for "creating" hydrogen bombs and radioactive fallout, even if people *don't* completely understand what physics is all about. After all, most people don't understand the stock market either—but they don't hold stockbrokers responsible for economic crashes and depressions. Many physicists have even set out to explain to the general reader, in print, just what they are doing—yet the average layman still finds himself at sea as far as the work of modern physics is concerned.

## THE IMPEDIMENTS TO UNDERSTANDING

It is not just bad luck that this should be so. In fact, there are certain factors that make it almost inevitable that the nonscientist should have trouble understanding what is going on in physics.

One such stumbling block in his path is the matter of language and terminology. Many of the words used by physicists to describe their work lead to confusion because these words mean one thing to the physicist and something quite different to the layman. Precision is the very keystone of research in physics—precision of measurement, precision of calculation, precision of description. When the physicist uses common words such as "speed," "velocity" or "momentum," "heat" or "temperature," he uses these words in a very limited and specific sense. He gives such words *precise definitions*. Most of us, however, understand these words in far more general terms. When we use the terms "heat" and "temperature" almost interchangeably, the physicist accuses us of sloppy, imprecise speech while we accuse the physicist of unreasonable fussiness.

In addition to this—and often because of it—the physicist has been forced to invent completely new terms which mean nothing whatever to the reader who is not acquainted with them through scientific training. When the physicist tries to explain to the layman what these terms mean, he finds himself at a loss to "translate" them accurately and precisely. Too often he gives up the struggle at this point, saying, "Well, *I* know what I mean, but there isn't any way to explain it to *you*!" Anyone who doubts that language and terminology create a real barrier to communication between physicists and nonphysicists needs only to look at a modern journal of physics to be convinced. To the untrained layman, nothing in it will be comprehensible. The gulf of words is enormous.

Another stumbling block, perhaps even more formidable, is the fact that so few nonscientists have any significant acquaintance with one of the physicist's most important working tools: the language of mathematics. Of course, most of us have had *some* training, one way or another, in simple arithmetic, algebra, plane geometry, perhaps trigonometry. Some may even have studied calculus once. But for the physicists, a casual passing acquaintance with math some years back just isn't enough. The physicist uses mathematics as an automatic, constant, indispensable working tool, day after day. Even more important, he uses it as a concise *language*. And many physicists insist that a layman simply cannot understand the work of physics at all without a solid grounding in advanced algebra, calculus, and higher mathematics —that physics by its very nature has to remain a mystery to anyone who is not so equipped.

From the physicist's point of view, this may be very true. It may indeed be impossible for us, as laymen, to understand the minute details of physical theory. It may be quite impossible for us to follow the intricate mathematical reasoning that has been part and parcel of so many of the great discoveries in physics without a great deal of skill and experience in higher mathematics behind us. But such skill and experience is not necessary in order for us to grasp, in broad perspective, the basic concepts and the great general laws of physics as they are understood today. It may take an artist years to learn the correct strokes and the application of great talent and highly trained skill to render a painting on canvas, but his work can be appreciated—at differing degrees of depth—by laymen, fellow artists, learned critics, and so on. Several levels of understanding may be possible. Similarly, great skill and experience in higher mathematics is not necessary in order for us to understand what the great discoveries of physics *mean* in terms of a general description of the universe, or a general prediction of what we can expect to see happen as a result of given circumstances.

To a physicist, of course, simply understanding the laws of physics in general terms is not enough. For him actually to *work* in physics, either in pure research or in developing practical applications for what has already been discovered, a real expertise in higher mathematics is indispensable. And often the physicist who is immersed in his work has difficulty understanding how anything less than a full professional and scientific grasp of all the detail can possibly be of any use to anyone. When he *does* try to "summarize" the work he is doing in simple and nonmathematical terms, he feels forced into scientific inaccuracies and distortions that make his skin crawl. If he is *really* game, he may plow doggedly through to the end and come up with descriptions of his work that he considers at least scientifically tolerable; but they may well be less than crystal clear to the untrained layman. By and large it is far easier and more comfortable for the physicist to say, "It just can't be done without the math" and to let it go at that.

Either way, we have the same result: a failure of communication, and a failure of the average untrained person to understand what the physicist is doing.

Formidable as these stumbling blocks may be, there is still another impediment to a clear understanding of the laws and work of physics, an impediment so imposing that it completely overshadows all the rest. This impediment, simple as it may seem, is nothing more nor less than the limitation of our everyday experience. And when it comes to understanding the laws of nature as they are or may be, our everyday human experience is very limited indeed.

## THE LIMITATIONS OF EXPERIENCE

After all, what do we actually *know* of the world around us? How do we know what we know? From infancy on, we know primarily what we have experienced through our human senses. We know what we can see, hear, smell, taste, and feel. From the evidence of these senses we have developed a "normal" view of the universe around us—a picture, so to speak, that we carry in clear focus in our minds. But this picture is severely limited by the boundaries of our senses.

Over the centuries, by means of various clever devices, we have been able to augment our human senses in a number of ways, up to a point. By using the lenses of a microscope, for example, we are able to magnify—up to a point—objects too tiny for us to see with our unaided eyes. The telescope greatly expands our ability to observe heavenly bodies, up to a point. Various devices can amplify sound waves so that we can hear things we could not ordinarily hear, and certain drugs and chemicals can render us hypersensitive to various sensations of touch, smell, or taste.

Even more cleverly, we have learned to extend the limits of our senses by converting unavailable data into some more available form. For example, sound waves that are beyond audibility can nevertheless be detected by converting them into visual images on the oscilloscope screen. Light which appears monochromatic and homogeneous can be split into its varicolored components by a crown glass or diamond crystal, or a spectrograph. In some cases, artificial amplification of our senses in this way is so much an ordinary everyday part of our experience that we don't even stop to consider that we are not actually experiencing what we seem to be experiencing at all. During a telephone conversation we know that sound waves at one end are really being converted into electrical impulses which are transmitted along conducting wires and then reconverted into sound waves again. But most of us find it easier to imagine that the voice we are hearing is the actual voice of the person talking to us, transported across the miles by some sort of miracle.

Yet for all our cleverness, most of us still picture the universe in the terms that it is revealed to us by our senses, and in no other way. When we try to understand the laws governing the behavior of the universe—the laws of physics—we are really trying to understand those laws *solely in terms of the normal world of our senses*. And this, unfortunately, cannot be done for the simple reason that the universe extends far beyond the limits of human sensory experience.

The physicist knows this and accepts it without qualms. It does not bother him that the natural laws he is trying to define involve objects and forces which he can neither see, hear, feel, taste, nor smell. Some of his work, of course, involves tangible objects in the "real" or "normal" world—but not much of it. Far more of it involves investigations in one of several other "worlds" in which it is *not* possible for him to measure or experience anything by means of his senses. But what may seem perfectly clear-cut, simple, and quite understandable to the physicist seems obscure and incomprehensible to most other people because it does not seem to fit into any frame of reference that they can understand.

What are these "worlds" of physics, these views of the universe, that physicists find so easy to deal with and the rest of us find so obscure? Need they be so obscure? Not if we can find some comfortable and familiar place to start, and then find ways to relate unfamiliar ideas to the things in our experience that we already comprehend. To find such a frame of reference, we must begin by looking for a moment at these different "worlds" in which the physicist works, to see just how they are related to the "normal" universe of our everyday experience.

## THE FOUR WORLDS OF MODERN PHYSICS

Everyone is acquainted with the first and most obvious "world" which physicists have explored: the "normal" universe that we see around us every day.

There is nothing particularly mysterious or frightening about this familiar and comfortable view of the universe. It is the universe of earth and sky, fire and water, in which we see trees growing, animals bearing their young, or the wind moving the grass. This is the universe our senses can explore directly: a world of buildings and oceans and sounds, of tangible objects we can grasp, of light and darkness. Beyond our immediate planet in this universe are the sun, the moon, and the stars that we see on a dark night. This is the world of ordinary sensory experience.

It is important to realize that throughout centuries and millenniums of mankind's existence—until very recently indeed—*this was the only view of the universe that there was.* It was with this world of physics that the classical scientists grappled, trying to search out laws of nature that described the phenomena that were observed by the senses. In this world of physics objects had fixed masses and behaved according to simple laws of mechanics. Objects in motion moved with finite, measurable speeds along paths that could be predicted—paths that were either straight or curved but not both at once. Space was three-dimensional in this universe, described to perfection by the geometry of Euclid that we all learned in high school. Gravity was a downward pull. Light was a phenomenon that could be observed, studied, measured, and manipulated with lenses. Electricity and magnetism were rather mysterious forces, difficult to explain, but observed to behave according to certain consistent, logical, sharply defined rules.

In short, this was a universe in which the classical "laws" of mechanics, heat, light, sound, gravitation, electricity, and magnetism all applied. A long succession of brilliant scientists had labored to discover those classical laws: Galileo, Copernicus, Kepler, Newton, Faraday, and Maxwell, to name but a few. In a sense it was a comfortable and cozy picture of the universe that these men painted over the centuries, and those classical laws *worked,* as far as our human senses could detect. It was not surprising that physicists, toward the end of the nineteenth century, were beginning to think that the work of physics was almost done.

But this comfortable view of the universe was not the whole picture. Not that it was *wrong,* exactly. It just did not cover enough ground. It was nice to have the jigsaw puzzle almost finished, but there were chunks of the picture still unaccountably missing. In order to see even a glimpse of those missing parts, physicists found it necessary to begin regarding the universe in some very different ways than they had ever regarded it before. In fact, they had to explore it as if it were really composed of several quite different worlds all at the same time, with each world superimposed on the others, and each with certain singular rules and regulations of its own.

One of these different views of the universe is the "microcosmic" view, in which all matter in existence is regarded as being composed of incredibly small bits and pieces, elementary particles and wavelets too tiny to imagine and too numerous to mention. In this microcosmic universe, few if any of the classical laws of physics that apply to the "normal" universe as we see it seem to apply. The particles making up this microcosmic universe are far too tiny ever to be observed directly. Some, in fact, are virtually impossible to detect at all. Here the overriding force acting in the "normal" universe (the force of gravity) seems to have little or no power at all; in the microcosmic universe, other forces quite unheard of in the everyday world seem to prevail: the nuclear binding forces that hold atoms together, for example, and various "interactions" discovered to occur between elementary particles. In this microcosmic universe the speeds with which particles or wavelets move seem as incredibly great as the particles themselves are incredibly small. The position and momentum of such particles cannot be accurately measured at all—at least, not at the same time! And the laws of mechanics, those rules that physicists had used for centuries to enable them to predict with (they thought) absolute accuracy what would happen to object A if force B were applied to it, no longer seem even remotely relevant in this microcosmic universe. The laws just don't cover the ground.

The microcosmic view of the universe was originally, in fact, so totally different from our "normal" view that there seemed to be no real relationship between the two at all, at least to a layman. The microcosmic view seemed more of an intellectual abstraction than a real part of our "normal" universe—until the development of hydrogen bombs, nuclear power reactors, transistors, and laser beams made it more and more obvious that the microcosmic universe was indeed "real" enough to change our lives profoundly.

But still another strange and different view of the universe also affects our lives: a "macrocosmic" view, in which our earth and our solar system are themselves mere particles of matter too tiny to mention in a universe that is incomprehensibly large. In this macrocosmic universe, laws of nature have been discovered which cannot be a part of our sensory experience nor even remotely understandable in terms of our "normal" universe. This macrouniverse is so huge that no one yet can really comprehend what relation our minuscule part of it—our earth, our sun, even our galaxy—may have to the whole. Here physicists are concerned not only with the underlying structure of matter but also with the structure of galaxies and clusters of galaxies, with the birth and death of stars, and with the cosmological history of a universe so vast and so crowded with matter, yet apparently so empty, that human minds are at a loss to comprehend or define it. Here are forces at work so subtle we do not even recognize them acting under our very noses, yet powerful enough to form suns and planets out of swirling dust clouds. These forces may be so great that whole galaxies are hurled away from each other as if in a stupendous silent explosion—yet even such forces as these must have had a beginning somewhere at some point in time, and must at some unimaginable future time come to an end.

In this macrocosmic universe—this universe of the inconceivably large—the classical laws of physics again do not seem to apply. Even the Euclidian geometry scientists have used for centuries to describe our "normal" universe seems unable to describe some basic aspects of the macrocosm. The forces at work here are neither the forces of mechanical energy so familiar to us in the "normal" world nor the nuclear binding forces that hold the particles of the universe together. Here is a wholly different universe, unexaminable by ordinary human senses and seemingly quite unrelated to our "normal" universe of experience, yet a very real aspect of the universe as it exists, just the same.

Finally (and much to their own dismay) physicists have come to realize that there is still another view of the universe that must somehow be taken into account, apparently unrelated to the normal, the microcosmic, or the macrocosmic, yet which applies to all three. This is the strange relativistic view of the universe that was inexorably outlined, hypothesized, and then proved valid by Einstein and other giants of twentieth-century physics. In this view the universe is not merely a certain volume of space containing various chunks of matter, but rather, a vast continuum of space and time. It is a universe in which matter and energy must be regarded merely as two different manifestations of the same thing, totally interchangeable from one into the other. It is a universe in which there appears to be one previously unsuspected but inalterable absolute, a single fixed physical limitation that seems to confine the operation of all other forces in the universe: the limitation of the speed of light. Stranger than any other picture of the universe, the relativistic view at first appeared to refute everything that had ever been believed about the classical laws of physics. Yet the evidence that began to accumulate, observed in the other worlds of physics, suggested more and more emphatically that the relativistic view was indeed every bit as valid—and as necessary for understanding how things work—as any of the others.

## AN APPROACH TO COMPREHENSION

Just as scientists realized, some hundred years ago, that the ordinary human senses presented an incomplete picture of the universe, so physicists today recognize that no single one of these differing views of the universe alone is sufficient. All four views—the "normal," the microcosmic, the macrocosmic, and the relativistic—must be taken into account if there is to be any hope of understanding how our universe *really* works. Thanks to their scientific training and experience, physicists can accept this notion, make peace with it, and carry on from there. Those of us who have not had such training and experience tend to balk and stumble. We stumble, for instance, over the whole idea of infinite time and infinite space. True, we cannot see the end of the sky, but our experience with other things tells us that there must *be* an end to it somewhere. We cannot help but think of our "normal" universe as a finite universe, and also a universe with definite fixed limits and boundaries—perhaps very wide ones, but boundaries nonetheless. The idea of infinite extension of *anything* is an awkward and uncomfortable abstraction. So is the idea of a universe without boundaries. In the microcosmic universe, the macrocosmic universe, and the universe of relativity there are no such finite limits and no certain boundaries known, as yet, so we inevitably try, without scientific training and experience to help us, to cram these other views of the universe into the finite limits of our own human experience.

And we find that it can't be done. At first, perhaps, we are confused; we don't quite understand what's wrong. Then we tend to reject these other awkward views of the universe. After all, it is easy to say, "It just can't be understood!" and even easier to say, "What nonsense! These things don't have any real meaning in my world anyway."

Unfortunately, in our world, we are confronted with growing evidence that these views do indeed have real meaning in our world and very real influence in our everyday lives. Furthermore, we discover that understanding something about these strange views of the universe is becoming more and more important to us personally every day. We need to know something about nuclear physics, something about cosmology, and something about space, time, and relativity nowadays just to keep up with what is happening in our familiar "normal" universe!

But how can we avoid trying to cram these other views of the universe into the limits of our own experience? One way would be first to look at each of these four views of the universe separately and distinctly, to see how they developed, where they came from, and where and why they seem to contradict each other. Our modern knowledge of the laws of physics as they are understood today did not appear by revelation overnight. It was accumulated bit by bit over the centuries. Since the days of ancient Greece men have been struggling with a long series of riddles, each seemingly more impenetrable than the last. One by one these riddles have been solved. In the early days physicists and mathematicians were concerned with riddles of the universe as we see it—the normal universe of our everyday experience. Later they began probing into other worlds which they could neither see nor measure. They began thinking in terms of the ultimate atomic structure of matter—the riddles of the infinitely small; or in terms of the ultimate size, shape, and limits (if any) of the universe of stars and galaxies —the riddles of the infinitely large. Finally they began to center in on other phenomena that seemed closely associated with infinite quantities—the riddles of light, of time-and-space dimension, of mass-energy conversion, and of forces of gravity and other stupendous forces operating on a cosmic scale.

To understand where this work has led, we will follow the footsteps of the multitudes of scientists who have tried to answer these riddles. We will try to develop a clear basic understanding of the natural laws governing each of these views of the universe in turn: the normal, the microcosmic, the macrocosmic, and the relativistic. Then, with good fortune, we will try to see how these views meet and join in an orderly, sensible, and *understandable* picture of the universe around us, to understand how this knowledge is already affecting our everyday lives, and to predict where it may be leading us in the future.

# CHAPTER 2

## *The Origins of Physics*

Every day we encounter multitudes of things which ought by rights to give us pause, but which we rarely even notice at all. Some of these things, which we accept as perfectly commonplace, would seem little short of miraculous if we stopped to think about them at all. Others would merely seem remarkable or extraordinary. In truth, however, all of these things are nothing more than simple manifestations of a few basic rules that limit the behavior of objects and forces in the world around us.

Consider a few simple examples. In the morning before we awake on a cold day, a device on the wall goes "click" and somewhere at a distance the furnace turns on, pouring out heat to take the chill off the house before we arise for the day. At about the same time another device on the bedside stand goes "click" and coffee water begins heating in an automatic pot. After a carefully regulated period of perking hot water through coffee grounds, the pot turns itself off, but still maintains a constant temperature so that hot coffee is waiting ready-made at whatever hour we decide that we must face the day. Then, when that hour arrives, still another device sets off an alarm to wake us up so that we can drink the coffee and arise to enjoy a warm house.

Do we marvel at these things? Of course we do not. They are nothing more than common household routine.

But the day's miracles have hardly even begun. Upon arising we walk into the bathroom to shower and shave; the water is already heated for us, and the electric razor doesn't even need to be plugged into a wall socket. But even more remarkable, if we stop to think about it, is the simple and extraordinary fact that we were able to transport ourselves from the bedroom to the bathroom at all, thanks to forces of gravity, inertia, and mechanical principles of leverage, without either having a rope to tow us, falling flat on our faces, or inadvertently leaping out of the bedroom window at the first step.

Later, at breakfast, we listen to the morning news on the radio, broadcast and transmitted from a station perhaps thirty miles away, while a device on the table heats up a network of thin wires to prepare our toast. Both these things are miraculous enough, but no more so than the fact that an hour later we climb into a chrome-plated transportation device weighing almost two tons, and by manipulating a few levers manage to induce it to carry its own weight and ours down the road at a velocity of sixty miles per hour, and then deposit us at the office door without flattening our faces against the windshield. We think nothing of this, but then we seem to be hard to surprise; on the way to work we saw a jet airliner weighing four hundred tons climb gracefully into the air, and we thought nothing of that either. It was an ordinary day.

Leaving the car, we ride on an elevator which lifts us twenty flights up to our office in less than half a minute. Nothing remarkable there. Once at work, we hardly look at the small machine that takes the sound of our voice, converts it into altered molecular patterns on a magnetized tape, and then hours (even years!) later reconverts it on command into the sound of a human voice for transcription. At home again after a day's work we use a somewhat similar magnetized tape to create further magic: By pushing a few buttons we record a television show we especially enjoy, or make an original videotape record of some family event, and then later see and hear this fragment of freshly recorded history any time we desire merely by feeding the tape through the television receiver again. Our eight-year-old son brings in a new toy for us to see, a gyroscopic top which balances at a rakish angle on the sharpened tip of a lead pencil as its flywheel spins madly about. We start to explain the principle of the gyroscope, but the boy cuts us off, blandly remarking that he knows all about it, that it's what keeps our rocket ships from wobbling too badly during takeoff.

We encounter the laws of physics in action constantly, wherever we turn, and think nothing about it at all. This is not really surprising, of course. Throughout our lives we have been both confined and liberated by these natural laws; we are accustomed to their effects even if we don't know what the laws are. We know from experience that certain things always happen in certain ways. Such things we take for granted. We also have learned that nature imposes certain limits on the way things will behave, and that we get in trouble if we try to exceed those limits. Odd as it may seem, one of the earliest of all human insights was man's realization that when he cooperated with nature things generally went well, whereas when he tried to thwart or alter natural patterns, things almost always went badly. Each individual has to learn this fact for himself, to some degree, just as a child learns the limits of his environment. But it is also man's nature that the lesson has to be learned over and over again; throughout history men have tried repeatedly to thwart the laws of nature and get away with it. And we continue to try. The fact that the laws of nature always win in the end has not dampened our enthusiasm a bit!

How have we learned these basic rules and limits of nature which we so easily take for granted in our everyday life? Certainly not from a textbook of physics (although the rules can be found there). Rather, ever since infancy we have been learning them, in a multitude of amazingly simple ways, in the school of practical experience.

## THE SCHOOL OF EXPERIENCE

Not long ago I spent an instructive half-hour watching a four-year-old neighbor boy learn a fundamental law of nature the hard way—and fly into a rage in the process—as he tried in vain to make a wagonload of rocks behave the way he wanted it to.

The child obviously wanted to move the rocks from a nearby gravel bank to the site of some architectural marvel he was working on a hundred yards down the sidewalk. Getting the rocks into the wagon was no problem. He loaded them in one at a time: clunk, clunk, clunk. But getting the loaded wagon rolling was something else again. The child fought and strained and tugged and pulled; then finally, grudgingly, the wagon full of rocks began to move.

Once it started rolling, everything was splendid. The child pulled the wagon along faster and faster. But then he stopped to inspect a bug on the sidewalk, and *whack*! the wagon caught up with him and knocked him sprawling. He got up and looked at the wagon, which once again had stopped moving. Once again he strained and tugged and pulled to get it under way; once again it caught up with him, knocked him sprawling again and, of course, stopped. And once again one furious little boy started tugging at the stalled wagon.

In the course of that single hundred-yard trip, that hapless child got himself knocked down no fewer than six different times as his determination dissolved into wailing frustration. When I stepped outside to ask him what the trouble was, he said it was a bad wagon; when he wanted it to go it wouldn't go, and then once it got going it wouldn't stop. The experience was sad indeed, but no college professor could have found a better way to teach that child one of the most basic of all the laws of nature!

Of course, the child had never heard of Newton's laws of inertia. He might well live to be eighty without ever learning how to express those laws in words, or in terms of mathematical formulas. But by the age of four he had already learned what those laws *meant,* as far as getting along in the world was concerned.

Obviously the behavior of the wagonload of rocks was a frustrating conundrum to that four-year-old, a riddle that seemed to defy understanding. Presently he recognized that he *had to adjust* to the way that load of rocks behaved whether it made any sense or not. That was how it was with a wagonload of rocks. But the riddle still remained. Adjusting to it didn't explain it. Possibly the very existence of this unsolved riddle became a challenge to that boy's ingenuity. We can imagine him later coming back to the riddle time and again, trying to puzzle out why that load of rocks behaved the way it did. We can even imagine him discovering, one day, that the rules that applied to wagonloads of rocks also applied to baseball bats and automobiles and rifle shells and a myriad other things.

There is nothing remarkable about the episode of the boy and the load of rocks. Each of us has encountered the same problem one time or another, and puzzled over the same riddle. In fact, simple as it may seem, this minor episode is a perfect example of the way that men from earliest times have grappled with the mystifying riddles of how things work in our universe and slowly pieced together the basic rules which we know today as "the laws of physics."

## THE EARLIEST PHYSICISTS

Some time long before the dawn of written history, Ug the caveman had a problem.

To us, it might seem a simple problem, hardly worth a second thought. But in the primitive world of Ug the caveman it was a matter of life or death. Early one morning Ug had emerged from his cool, damp cave and set out to hunt for his dinner. While he was away, a downpour loosened a boulder from the hillside above Ug's cave, and it rolled down to block the cave's doorway completely. When Ug returned toward dusk with a fine haunch of hippopotamus that he had rescued from the jackals, he couldn't get back into his cave. Try as he would, he couldn't budge the boulder that stood in his way.

Now brains were not Ug the caveman's long suit, but he knew certain things with terrible clarity. He knew it grew cold at night, so a caveman needed the shelter of a warm secure cave. He also knew that when darkness came, the saber-toothed tiger began to roam in search of a better meal than a mere haunch of hippopotamus. Ug the caveman knew quite clearly that unless he could move that boulder away from the doorway by nightfall, he might never live to see the dawn.

Whether from instinct or from some half-remembered earlier experience, Ug the caveman had a sudden bright idea. He searched for a bough from a nearby tree, broke it off, and wedged one end of it in between the boulder and the doorway. Although he could not budge the boulder before, with the aid of the stick he could move it a little. When he moved closer to the boulder and pushed on the bough, he found it was much harder to move. When he moved farther out to push on the distant end of the bough, the rock moved more easily—but the bough snapped off. Finally, as daylight faded, Ug found a stronger, longer pole, wedged one end in behind the boulder, pushed on the far end—and miraculously, the boulder moved easily away from the cave mouth and bounded down the hillside.

Ug the caveman never thought to wonder *why* he could move that boulder with the aid of a long pole when he couldn't even budge it otherwise.

For Ug, there were more important things to think about: fire and warmth and feast and protection and sleep, for instance. Yet under pressure of urgent necessity this simple caveman had discovered and used, perhaps for the first time in man's history, a simple machine operating according to fixed mechanical principles. It would not be the last time that Ug the caveman would use this new machine—a simple lever and fulcrum—to save his own life, to get things done that he could never otherwise do, and generally to elevate himself from the status of low-class caveman to upper-middle-class caveman.

Of course no one knows who Ug the caveman might have been, nor where he lived, nor when he made his discovery. Like so many other simple machines—the roller, the wheel, the pulley, the inclined plane, the bolus, the bow and arrow, or the siphon, to name a few—the origin of the lever and fulcrum is lost in the mists of antiquity. All these machines were known long before the first human records were kept. Some early physicist discovered each of them. No one knows who. Yet these machines formed the first link between the world of practical experience and the scientific study of physics.

## THE PRACTICAL ENGINEERS

We are so accustomed today to the whole idea of scientific study that we forget that research has not always been a normal part of human activity. It is easy for us to assume without thinking that organized scientific investigation came about as a natural result of man's insatiable curiosity and intellectual vigor. In truth, it is far more probable that the earliest explorations in physics arose either from desperation (as in the case of Ug the caveman) or from man's age-old, insatiable desire to get something for nothing. It was only when practical advantages began to appear that early physicists began searching out the reasons and principles underlying things around them that had always been taken for granted.

Thus, the discovery and development of simple machines almost certainly came about as a result of men trying to get more work done with less effort. Consider the Egyptian pharaoh who wanted to build a huge tomb for himself out on the desert. Naturally, he wanted it as big as possible—but it had to be finished before he died if it was to do him any good. Getting it built raised real problems of engineering and logistics. The stone had to be cut miles away and dragged to the construction site by slaves. Still more slaves had to lift the stone blocks into place. The pharaoh had plenty of slaves, but it still took days and weeks to move a single stone to the tomb and lift it into place, and the harder he worked the slaves the faster they dropped dead on him. To cut smaller stones would take more time and reduce the magnificence of the tomb. But the pharaoh was getting old, and so many slaves could do only so much work and no more.

Or so it seemed. Then some bright young underling came up with an idea. By dragging a rock along on rollers instead of sliding it across the desert, he found that the same number of slaves could transport twice as many stones of twice the size in the same length of time as without the rollers. He found that if the rollers could be moved swiftly from the rear to the front as the rock moved, things went faster than when the load was constantly stopped and started again. Finally, he discovered that by hauling the stones up an inclined plane to the level where they were needed, heavier stones could be lifted higher and faster by fewer slaves than before. The pharaoh had never heard of "mechanical advantage" before in his life, but he could tell a good idea when he saw it; this was practically instant pyramids! So he rewarded his bright young engineer, and rollers and inclined planes became standard operating equipment for pyramid-building.

Later, other pyramid builders discovered other things about the simple machines they were using. They learned, for instance, that a ramp with a long, gentle incline worked better than a short, steep ramp. The stones had to be pulled farther *horizontally* to get them to the required height, but up to a point the ease and speed with which they could be lifted far outweighed the additional distance. When a stone was lifted with lever and fulcrum, they noticed that for some reason the job was easier if the lever arm was long than if it was short. Even the water boys discovered that unless the water bucket was suspended at the exact center of the pole between two carriers, the one closest to the bucket did most of the work.

These were practical observations which led to useful refinements of the earlier simple machines. Inevitably, one day, somebody scratched his head and said, "Now, wait a minute—*how much* easier is the long gentle incline than the short steep one? What makes it easier? And why can't we figure out the *exact* length and slope of the ramp that we need so that the fewest men can raise the heaviest rock to the greatest height the fastest?"

It was when questions of this sort began to arise that the first *scientific* study of the physical universe really began.

We know today that some of these primitive observations resulted in amazingly accurate predictions, and certain primitive techniques proved remarkably useful and durable. As early as 3000 B.C. Egyptian astronomers had learned enough from the cyclic movements of the sun and moon to establish a year as a unit of time 365¼ days long—a far more reliable measure than the annual flooding of the Nile. They even knew that this time measurement was sufficiently inaccurate that a correction of three days had to be made every four hundred years, and the search was on for a device to clock hours and minutes. In matters of calculation, early men gave up counting on their toes in favor of chalk marks on the wall. Presently they invented symbols to stand for various numbers. In ancient Sumeria and Babylon these symbols were used to develop a method of calculation we know today as arithmetic. Still later, in Arabia, it was discovered that when special symbols were used to represent unknown quantities and simple rules of logic were applied, certain kinds of problems could be solved which arithmetic could not handle. Thus the basic techniques of algebra were developed. In those primitive days, too, the abacus was invented as mankind's first mechanical computer—a device so simple yet so efficient that it is still in widespread use in the world today.

Such observations and discoveries did not arise from any philosophic search for the meaning of it all. Those were violent and hazardous days in which to live. A man's average life expectancy was about 26 years the day he was born; he was likely to be too preoccupied with feeding, sheltering, and protecting himself to have much spare time for philosophic ruminations. He needed to understand simple machines in order to build his cities, draw his water, plough his fields, or build monuments to his kings or gods. He needed to know when to plant and when to harvest. He needed arithmetic and algebra in order to hold his own in an era of cutthroat commercial dealings. He did not often ask *why* things around him happened the way they did.

But people in those days did learn one very important thing about the way things worked. They learned that whatever the reason things worked the way they did, *they always seemed to work the same way one time as another*. Whatever laws of nature might be at work, those laws were *orderly*. If a pulley worked one way one time, it would work the same way the next time and the next. When two and two were added up, the result was always four. It remained for a later and more sophisticated civilization to begin questioning just *what* earth, air, and water were really made of, and *why* things worked the way they did. The search for answers continues to this day—but the questions were first asked by the philosophers of ancient Greece.

## THE BIRTH OF MODERN PHYSICS

Most historians of science today concede that the first serious scientific questioning began in the civilization of ancient Greece. In many ways this is strange, because the early Greeks were anything but scientists. They did not regard themselves as "investigators of nature" in the sense of modern scientists, using observation and experiment as their tools. If anything, they considered something as coarse as mundane experimentation to be far beneath their dignity. Rather, the Greeks thought of themselves as natural philosophers, seeking to penetrate the secrets of nature by means of reason and logic. Many notions that could have been proved demonstrably wrong by the simplest of experiments were accepted as true without question, simply because they were philosophically and esthetically satisfying. Debate and logical dialogue were the accepted methods of investigation; great men would argue for months about some point of "natural philosophy" which a modern scholar could have resolved in one minute flat with a good slide rule.

Even so, ancient Greece and her philosophers built an absolutely critical groundwork for the organized body of scientific knowledge about the physical world which was to come later. The Greeks did make certain discoveries about the ways the laws of nature could be investigated. They also showed the world some of the ways those laws could *not* be investigated. It was upon their ideas, discoveries, and errors that the whole structure of modern scientific exploration first arose.

For one thing, the Greeks recognized philosophically that there was order in the universe. Things that happened in nature happened consistently. To them, this indicated that some kind of absolute "natural law" governed the behavior of things. The movements of the stars, the operation of simple machines, the phenomena of heat, light, and sound were not things that occurred capriciously at the will of the gods. There was, the Greeks concluded, a definite cause-and-effect relationship between things that occurred in nature. One thing happened because something else had happened first, and this led to something else with such consistency and regularity that it was actually possible to *predict* what was going to happen next before it occurred, on the basis of what had already happened before.

The Greeks also believed that certain truths about nature could be accepted as obviously true without proof, and then be used as basic *axioms* from which other truths could be deduced by means of logic and reason. These so-called intuitive truths were very fundamental things, so clearly and self-evidently true that they were considered proofs unto themselves—things that "any fool could plainly see." For example, it seemed self-evident that the material from which the earth was made *had* to be composed of certain tiny, indivisible units. The idea of an infinitely divisible chunk of rock simply defied reason. Break it up into smaller and smaller pieces and sooner or later you *must* reach some small basic unit which could not again be divided. Accepting this as an axiom, it followed logically that *all* forms of matter must be built up from an assortment of such individual units. It is hardly surprising that our modern word "atom" was first used by the Greeks to denote a tiny particle of matter which itself could not be further divided.

Again, the Greeks realized that certain shapes and patterns (such as straight lines, triangles, or circles) occurred repeatedly in nature, and that the concept of number or quantity was suggested in nature by collections and sizes of objects. Certain facts about these geometric patterns appeared to be self-evident without proof. Two circles drawn with the same radius *had* to be equal in size. One right angle, by its very nature and definition, *had* to be the same shape as any other right angle.

These conclusions did not arise from careful experiment or measurement. They arose from somewhat casual common-sense observation. Yet on this basis the ancient Greeks began to collect a volume of basic scientific data, and then started to build upon those data by means of logic and deduction. The handful of basic axioms which form the basis for Euclid's system of geometry (a system regarded as the *only possible* system of geometry for almost two thousand years) were never considered subject to proof. They were accepted as self-evidently true. But with these unproven axioms as a foundation, each succeeding proposition in Euclid's system was then subjected to rigorous logical proof. Each new proposition, once proven, then became the basis for still further propositions, until a whole series of rules had been built up which consistently applied to any and all cases within geometric experience.

Such a system of reliable rules was, of course, highly useful. Even more astounding, it was discovered that by using these rules one could actually discover physically meaningful information which could not possibly have been discovered any other way. For example, it was impossible to prove by observation or experiment just what shape of rectangle enclosed by a piece of string of a given length would have the greatest area. One might *guess,* but one could not *prove*. But by means of geometry it was possible to demonstrate beyond any doubt that a perfect square would have the greatest area of any rectangle that could be enclosed by a string of given length. It could be proven geometrically that a perfect pentagon formed by the same string would enclose a greater area than a perfect square, and that the string laid in a perfect circle would enclose the greatest area of all.

Thus the early development of plane geometry resulted in a discovery with staggering implications. By applying logic and reasoning to situations taken from nature it was possible to produce new and hitherto unsuspected knowledge.

Simple as this idea was, it was vital to the growth of physical science. For one thing, it encouraged men to begin observing "taken-for-granted" natural phenomena more closely. The sun, the moon, and the planets moved in the heavens. If one observed closely and then applied reason and logic, surely it should be possible to determine the exact orbits of these heavenly bodies, and thereafter to predict accurately where they would be found at any given moment in the future. Of course, this did not prove to be as simple as it seemed. In the second century A.D. an Alexandrian Greek astronomer named Ptolemy undertook the job in the traditional Greek fashion, and created a misconception that took a thousand years to clear up. Ptolemy assumed as self-evident that the earth itself stood still in the heavens while the planets and the sun pivoted around it. He also assumed that all the heavenly bodies moved in perfect circles, since the circle was obviously the most perfect form of motion for a heavenly body (philosophically speaking). Fitting his observations of the planetary movements into these axioms, he developed a theory to explain the motion of the sun, the moon, and the other planets around the earth.

Unfortunately, later observations of planetary movements never quite fitted into this "Ptolemaic system" he had devised; so Ptolemy and his followers had to refine and modify his theory over and over again through the years. Finally, fifteen centuries later, somebody proved that both of Ptolemy's "self-evident" axioms were wrong, but so great was the stature and authority of those early Greek philosophers that it often took millenniums finally to replace some of the inconvenient theories they propounded.

For all their shortcomings, however, the ancient Greeks' intellectual and logical approach to the study of nature did bear some useful fruit. The Greeks examined an enormous number of natural phenomena and developed logical theories to explain the nature of heat, light, and sound, the operation of the lever and the inclined plane, the factors and forces acting upon fixed or moving bodies, and the nature of work and energy. They catalogued the heavens and created astronomical theories which, incorrect as they were, still provided future astronomers with a solid foundation on which to work. They recognized the three physical states of matter—solid, liquid, and gaseous—even though they completely missed the relationship that existed between those three states.

Above all else, the ancient Greeks proved that man *could* learn how nature behaved, and thus could hope to predict nature's future behavior. As we will see later, physicists today seriously challenge even this idea, and not without reason. But it is pertinent to note that without that concept to guide scientists throughout the centuries in searching out answers to the riddles of the universe, there would be no modern physics today. By proving to their own satisfaction that nature behaved in an orderly manner, and that the truth about nature could be uncovered by human intellect, these ancient explorers opened the door to a two-thousand-year-long assault upon the riddles and conundrums of nature that men had to face in the world about them.

# CHAPTER 3

## *From Philosophy to Science*

By the close of the ancient Greek era of intellectual achievement, the groundwork had been laid for a giant step forward in scientific discovery. But over fifteen hundred years were to pass before that forward step was begun, and another four hundred years before the classical laws of physics governing the "normal" universe of everyday experience were finally outlined.

We should not forget that the physical world those early Greek philosopher-scientists were seeking to explore was the world they saw about them. It was the world they knew from the experience of their own senses, a universe they could see, hear, touch, smell, and feel. Objects in that world had measurable size and weight, and moved at measurable speeds in discernible directions. Everything on the surface of the earth was influenced by a mysterious, undefined force which tended to pull everything in one direction and that direction was down—the force we know today as gravity. Aristotle explained gravity very simply: Every object on earth tended to seek its "natural place," he said, and the "natural place" of all objects was on the ground. Ergo, any object not resting on the ground tended to fall to the surface. It *sounded* good, but as for really explaining anything, even Aristotle must have realized, on occasion, that it was fatuous nonsense. The Greeks knew that when an irresistible force (such as a warrior's battle-ax) struck an immovable object (another warrior's skull, for instance) the battle-ax came suddenly to rest and the skull got crunched. They even knew that the harder the battle-ax hit, the more satisfying the crunch. They were not aware that the total momentum of the ax-skull system remained unchanged by the encounter, nor that the kinetic energy of the ax was largely absorbed by the elasticity of the skull, nor that a certain amount of heat was generated in the process. These things came later.

Again, when the ancient Greeks discussed the atom as the "ultimate indivisible unit" of matter, they were discussing intellectual abstractions that had no real meaning to them in terms of their experience. They knew perfectly well that in the *real* world a bit of sand could be crushed into a fine powder, but that was as close to an "ultimate indivisible unit" of sand as anyone could hope to approach—or needed to, for that matter. They were, perhaps, able to imagine very large or very small objects or distances, but no real concept of *infinity* was possible from their observation of nature around them. They simply had no toe hold for such a concept. Whenever they encountered it inadvertently (as in Zeno's paradox about the runner who could never finish a race because he would first have to run halfway to the goal, and then run half of the remaining distance, and then half of that half, etc., ad infinitum, and thus could never quite get to the goal line) the concept was regarded as precisely what it was called: a *paradox* or conundrum of mutual exclusives which simply did not admit of a solution.

What is more, as the centuries went by, as experimental methods were developed and as devices were found to extend the range of human senses, the study of the natural laws of the universe still remained a study of the "real" world that could be seen, measured, and experienced. Not that there were not clues to the existence of other and unsuspected worlds of physics beyond sensory measurement and experience. The phenomenon of gravity and the phenomenon of light were two such major clues—but they were either ignored completely or examined only in terms of the real world of solid objects and measurable forces. Gravity was studied as a constant force that made things fall to the ground when you let go of them. Light was an unexplained and apparently unexplainable *something* which no one pretended to understand, but which could be manipulated by lenses in a useful manner. When mathematicians began coming up with concepts that had no relationship to the world of experience—the concept of imaginary numbers, for instance—scientists almost invariably tended to distrust the mathematics and the mathematicians, rather than to consider seriously that there might be some aspect of the universe that was completely out of reach of human experience.

Even within such limitations, a long succession of scientists beginning with the Greeks and ending with the nineteenth century physicists actually learned an amazing amount about the nature of the universe and the natural laws that prevailed—or at least, about the universe of human experience. This knowledge was not accumulated suddenly, nor in any steady progression. It pursued no particularly logical course of development; in fact, it developed in a succession of torpid pauses, staggers, and lurches, assisted by a few perfectly incredible coincidences—about the most disorderly history of discovery imaginable.

Along with a solid groundwork of observation, the ancient Greeks had provided such basic tools as a highly developed system of plane geometry and an increased skill in the use of algebra. They also provided a tradition of inquiry. At least they recognized that things were going on in nature that they did not understand, and that these things were worth wondering about. They did not, however, have any workable concept of a *scientific method of investigation,* as we think of it today. They disdained experimentation, and considered their speculations and hypotheses "proven" if they were logically and philosophically pleasing—even when new observations flatly contradicted those hypotheses!

In addition, the Greeks overlooked or ignored many natural phenomena simply because they didn't seem to admit of philosophical explanations. For example, they were perfectly aware of the existence of certain kinds of natural stones to which bits of iron mysteriously seemed to cling, even in opposition to the "downward" force of gravity. The Greeks did not understand *why* an iron swordblade should be "drawn to the lodestone rock" in this manner, nor did they understand why a piece of soft iron rubbed on a lodestone took on some of this curious iron-attracting quality itself. But they never investigated this phenomenon, nor did they ever discover that an iron rod rubbed on a piece of lodestone and then suspended from a string would always assume a north-south orientation with respect to earth.

Similarly, the Greeks were aware that a piece of silk rubbed on a lump of amber tended to repel another piece of silk, and to crackle and spark in the darkness when it was shaken, but this fact seemed to arouse no excitement or curiosity. These people were just not emotionally or intellectually equipped to investigate such phenomena in any kind of orderly fashion. If they had been, the world might be quite a different place from what it is today. What would have happened had the magnetic compass been available to mariners from the time of ancient Greece on? What if Archimedes or Aristotle had begun a systematic investigation of electricity and magnetism? It is useless to speculate; they did not. Nor did they attempt a study of the curious properties of light, although they had certainly observed rainbows in the sky, and most assuredly knew of the brilliant play of colored light in a natural quartz crystal.

## THE RENAISSANCE GIANTS

With such a varied foundation built by the Greeks, we might have expected a steady growth of scientific investigation in the centuries that followed. But in fact, after the decline of Greek civilization progress simply ground to a halt. Practically nothing of scientific significance happened at all for over a thousand years.

Historians have a variety of explanations for this long period of scientific stagnation. Certainly a number of factors contributed. The Roman empire rose to power as Greek civilization declined, and the Romans were neither philosophers nor scientists. Preoccupied as they were with expansion, commerce, politics, and warfare (and, later, with living the good life at the expense of all else) the Romans simply accepted and copied what the Greeks had worked so hard to achieve. They made no effort to investigate or expand Greek ideas about the nature of the universe; they bought them wholesale and passed them on as revealed truth. Later, the Church also played an important role in discouraging new directions of thinking and scientific exploration. Threatened by any unorthodox concepts, the Church would accept only those scientific ideas and hypotheses which seemed consistent with Christian teachings—and for centuries the Church had the power of life or death over anyone within its realm who deviated from these accepted principles.

But perhaps the main reason for the long stagnant period in scientific investigation was the simple fact that the Greeks had painted themselves into a corner with their philosophical and speculative approach to science. The ideas they had developed from their casual observations of nature had already been expanded as far as possible by means of reasoning and logic alone, and they had no alternative approaches to offer. Further investigation along those lines could only lead to further refinement of the same ideas, as the gulf between the "proven" conclusions of those early philosophers and new, more accurate observations of nature grew wider and wider. Then, when Rome fell, the Church became custodian of what scientific knowledge did exist, and most of the thinkers and philosophers of the time were far more concerned with questions of theology than with new or challenging ideas about the nature of the universe. Indeed, for over a thousand years the closest approach to science was the pseudoscience of alchemy, that strange mixture of philosophy, scientific investigation, and mumbo-jumbo whose practitioners sought in vain for the mystical "philosopher's stone" that could turn base metals into gold.

Above all else, no one had found the tools necessary for true scientific investigation, and without the tools there was no place for science to go.

This long, sleepy period did not last forever. In the late 1400s, quite suddenly and for no clear-cut reason, some giants began to appear—men who were to jolt the world of science to its very foundations in the course of less than two hundred years. The names of these men are household words today: Copernicus, Galileo, Tycho Brahe, Kepler, Isaac Newton, Faraday, and Maxwell, to name but a few. Incredibly, after a thousand years' sleep, five of these men were born within the span of 170 years, and four of them lived and worked within the span of a single century.* All were physicists in the broadest sense of the word—investigators and explorers of the nature of the physcal world. Among them, in two centuries, they changed the course of history.

Later we will see in more detail just what discoveries each of these men made and how they made them. Among them, they built up the first orderly and sensible explanation of how things worked in the universe of human experience. Theirs was the world of classical physics. They found answers and then proved them, insofar as they *could* be proved by the senses. At long last they overthrew the ancient Greek tradition of investigation by

* Galileo died in the same year that Isaac Newton was born.

debate and philosophy and established a new tradition of investigation by experiment—a tradition that has persisted to this day.

As we shall see, a great many conclusions of these giants of classical physics have since been found to be incomplete. Many of the "laws of nature" they outlined have been shown to be valid only under special or limited circumstances, not universally valid always, under any circumstance. Some of their conclusions have proved to be flatly wrong. But if their work was incomplete, or limited, or flawed, it provided a solid and *scientific* basis for the work of others. Above all else, these men performed one service of staggering and overriding value to humanity: They forged the missing tool by which men *could* study and hope to understand the nature of this universe, the tool without which modern physicists would have remained as helpless as the ancient Greeks. Today that working tool is known as *the scientific method of investigation.*

Nicolaus Copernicus, a Polish astronomer born in 1473, made the first and probably most revolutionary break with the ancient Greek tradition. Ever since Ptolemy had assumed that the earth was the center of the universe and that everything in the heavens revolved around it in perfect circles, astronomers had been trying to fit the observed movement of moon and planets into the increasingly awkward Procrustean bed Ptolemy had provided for them. When the guest did not fit the bed, they whittled off his legs until he did. They even tampered with the bed, so to speak. Repeated efforts were made to revise the Ptolemaic theory slightly, and each new revision seemed to straighten things out for a while, but always new observations came into conflict with the theory again. The repeatedly modified Ptolemaic system became progressively more clumsy to use as time went on, but no one ever dared question the basic assumptions—that one could use *only* the earth as the center of coordinates in the universe, and that heavenly bodies *had* to move in circles.

Copernicus not only challenged the first of those assumptions, he devised clear-cut scientific proof that it *had* to be wrong. Drawing from a lifetime of his own careful observation he concluded that the sun had to be the center of our solar system, not the earth, and that the earth and all the other planets revolved around the sun. True, our own moon revolved around the earth; but on the other hand it was the earth and not a "celestial sphere" of fixed stars that turned on its axis every twenty-hour hours, producing the apparent motion of the stars, and the movement of the sun across the sky.

It was such a revolutionary concept that Copernicus himself withheld its publication until the very end of his long life. But the Copernican system had one very good thing going for it: It happened to agree splendidly with what had actually been *observed and recorded* of the motions of the various known planets, while the Ptolemaic earth-centered system failed to do so even after centuries of refinement and modification. For a while it even appeared that the system of Copernicus was the final and ultimate answer.

But then, a century later, other astronomers began finding some new discrepancies between theory and observation. The Danish astronomer Tycho Brahe spent decades between 1570 and 1600 in a patient study of planetary motions, using better instruments and more astute observation than Copernicus could command. He accumulated a gold mine of data about movements of the planets, much of which just didn't quite fit the Copernican theory. It remained for a young assistant of Tycho Brahe, a German astronomer named Johannes Kepler, to study Brahe's data in the early 1600s and discover what was wrong.

Copernicus had challenged one of Ptolemy's basic assumptions, that the earth was the center of the solar system. But he had failed to question the other assumption: that the heavenly bodies moved in perfect circles. Kepler realized that the notion that a circle was the perfect path for a planet to follow, while philosphically tidy, did not actually *have* to be true. He began searching for some other path of motion for the planets which might explain the discrepancies between the theory of Copernicus and the things that Brahe had observed. Finally Kepler discovered the truth: that the planets traveled in *elliptical* orbits, with the sun always located at one of the foci of the ellipse. He also found a relationship between the speed with which a planet moved and the distance it lay from the sun. As a planet moved closer to the sun in its elliptical orbit, Kepler found, its velocity increased; when it swung away from the sun its velocity decreased.*

Kepler also noted that planets that lay close to the sun sped around it faster than those far distant, and these differences in the periods of revolution of the various planets could be described in a fixed mathematical ratio to their mean distances from the sun.

It was heady stuff, Kepler's contribution—a real bonanza of new and enormously important information for astronomers. But the work of these men had far deeper significance for the whole world of physics. They had plowed through a roadblock to scientific investigation which had persisted for centuries. The Greek technique of investigation was simple: Apply philosophy, reason, and speculation to casual observations of nature; then arrive at a theory; then somehow cram any newly observed facts into the theory, difficult as that might be. It was not that the Greeks were fools; they simply assigned importance to the wrong things. Copernicus, Tycho Brahe, and Johannes Kepler for the first time demonstrated that *observation and measurement* were the *real* keys to scientific discovery. The Procrustean-bed technique was no good; theory could be accepted only as long as all ob-

* Kepler described this much more accurately by saying that an imaginary line drawn from a planet to the sun would always sweep across the same area of the ellipse in the same unit of time; thus a planet near the sun (i.e., near perihelion in its orbit) would move faster in order to sweep the same area of the ellipse each second as it did when moving more slowly far from the sun (i.e., near aphelion). See Figure 2.

servations substantiated it, and not a moment longer. If careful observations failed to substantiate a theory, then it was the theory that was wrong and not the universe.

## THE RENEGADE OF PADUA

Of all the other giants of those days, it was Galileo Galilei (1564–1642) more than anyone else who established this simple idea once and for all, and forced scientists all over Europe to throw out the accepted conclusions of centuries. Galileo was the father of experiment, *repeatable* experiment which anyone else could duplicate if he wanted to take the time to bother.

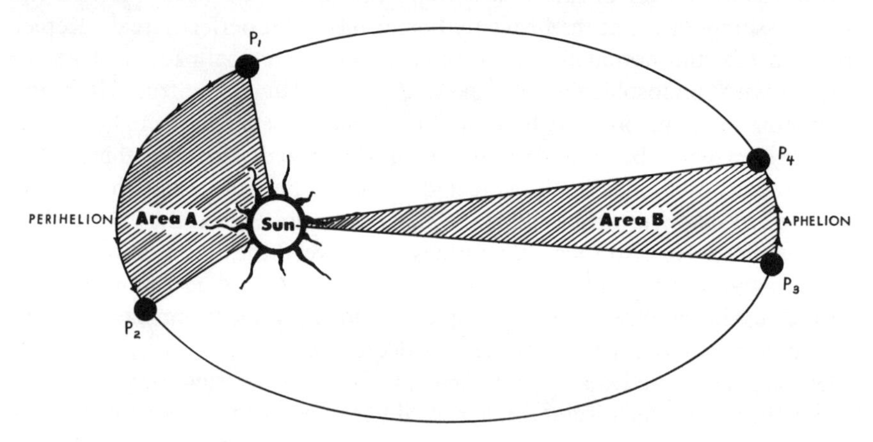


*Fig. 2* Kepler's elliptical orbits. According to Kepler's laws, a planet at perihelion moves faster in its orbit than at aphelion, covering distance from $P_1$ to $P_2$ in the same time required to cover the shorter distance $P_3$ to $P_4$, and Area A of the ellipse equals Area B. The elliptical orbit illustrated is, of course, greatly exaggerated.

He was the father of the orderly statement of principles or conclusions derived from his experiments, setting these conclusions down in simple, flat statements which could then be subjected to further experiment either to prove them or disprove them. Today Galileo's principles are usually encountered as algebraic formulas in textbooks of physics, and thus intimidate all but the brave and determined. This is unfortunate, because these principles, for the most part, are actually nothing more than simple statements of how things change in relation to other things.

For example, we could record a commonplace observation by saying, "Downtown traffic becomes lighter at a constant rate the longer after rush hour that you measure it." We could state the same thing as a simple formula:

$$D = \frac{t}{K}$$

Both statements express a constant ($K$) relationship between two measurables: the volume of downtown traffic ($D$) and the time interval since rush hour ($t$). Expressed either way, the principle could be a useful addition to our knowledge, provided it was confirmed by repeated, independent observations. With it we could *predict* something we might not otherwise have any way to know: that driving through downtown will be easier (and conceivably, safer) the longer we wait after rush hour. To the scientist the principle expressed in a mathematical form is more useful than in words; but to the layman, the words convey the meaning better.

Galileo used both means of expression. Certainly he was a genius at the detailed study of mechanical things that happened in the world around him, and at discovering relationships between one occurrence and another. Above all, he was a master at *generalizing* from one specific case to other similar but different specific cases. He was not content merely to observe that an object dropped from his hand fell with increasing velocity until it struck the earth, nor even just to record accurately the rate of acceleration of that object as it fell. He did not merely conclude that that particular object accelerated at a constant rate throughout the time of its fall. He went a step further; by expressing his observation as a simple law or rule, he then reasoned that he was really describing the acceleration of *any freely falling body anywhere* in relation to the height from which it fell and the length of time of the fall.

Galileo's study of falling objects is probably more familiar to more people than anything else he did. But his work did not stop here. He also studied the behavior of objects rolling down inclined planes, and worked out the mechanical principles of the pendulum—the groundwork for study of all kinds of cyclic or oscillating mechanical motion. He established the basic laws of work and energy, and studied the effects of friction and the various phenomena of inertia—in fact, the whole range of principles of mechanical motion. He investigated such physical phenomena as light and sound, developed and improved the first practical working telescope, and made dozens of other basic contributions to the knowledge of classical physics.

In some areas Galileo failed completely, as in his attempt to measure the velocity of light. He was in almost constant conflict with the Church of his day, and was considered a renegade among scientists—but his two great and indispensable contributions to the future development of physics could not be denied or suppressed. First, he established once and for all that *experiment, measurement, and observation were more valid ways of discovering the truth about nature than intuition and speculation*—indeed, that intuition simply could not be trusted at all. Second, he established firmly the idea that the *workings of nature throughout the universe were uniform.* When something happened in one situation it could be counted upon always to happen the same way in the same situation at another time or anywhere else in the universe. To put it differently, Galileo demonstrated that any valid laws of nature were indeed *laws*; if nature later were observed not to be obeying those laws, this then had to mean that the laws were not quite valid and needed to be redefined, not that nature had turned fickle.

In fact, Galileo had stumbled upon a powerful tool for the investigation of natural occurrences: the tool of scientific method. He had tried it and found that it worked. And close behind him, other men went on to learn just how powerful this tool could really be, properly used. With an orderly scientific method at their command, scientists for the first time could brush aside the cobwebs of ignorance, misinformation, and superstition to begin searching out nature's most closely guarded secrets: the natural laws describing the interactions of matter and energy in the universe.

But what, precisely, *was* this scientific method of investigation? And what, precisely, are these "natural laws" that scientists have been searching out for so long and continue to search out? Much general misunderstanding and confusion about the discoveries of modern physics arises directly from confusion about the scientific method and the nature of "natural law." Before we go further, we would be wise to define these things as simply and specifically as we can, not only to help us understand how things work in the world of our everyday experience, but to better comprehend the other strange and exciting worlds of modern physics as well.

# CHAPTER 4

## *The Methods of Discovery*

To the mind of a child, the universe is narrow and sharply defined. Things which seem enormously complex to us are perfectly simple and obvious to him. Because of this, strange as it may seem, any child knows what "the scientific method" is, although he may not know it by that name. In fact, he uses it constantly in his everyday life. He also understands what "natural law" is, within the confines of his limited universe, with perfectly amazing insight.

For some of us, it may seem unnecessary to discuss the scientific method in simple and specific terms. But simple as the concept really is, it is often widely misunderstood. Because our whole modern concept of natural law' is, within the confines of his limited universe, with perfectly amazing wise to review what it is and how it arose in the most fundamental terms possible. A clear understanding of both the concepts of natural law and scientific method is vital in any discussion of the ideas of modern physics.

How does a child use the scientific method? Consider the case of Johnny, a three-year-old boy with a normal, healthy appetite for chocolate cake frosting. One afternoon Johnny stood watching his mother frost a cake with this brown, sticky stuff that tasted so good. From Johnny's limited view of the universe, it seemed obvious that Mother was spreading that frosting out solely for him, so he reached out and took a large sticky handful off the top of the cake just as Mother finished spreading the last spoonful.

Unfortunately, Johnny's view of the universe was too limited. His mother shrieked, whacked him soundly, and said, "Keep out of the cake frosting." Then to top it off, she washed his hand off before he even got a taste.

To Johnny this was a hard lesson—a new and unnecessary complication in an already overcomplicated world. Obviously, something was wrong with his assumption that the cake frosting was there for him to grab. Yet he remembered distincly that Mother had *given* him cake and frosting to eat at dinner the last time a cake had been baked. What was going on here? *What rules applied?* After mulling it over for a while, Johnny threw out his previous assumption in favor of a new one that seemed to fit the facts better: The frosting was there for him to grab *provided Mother wasn't standing right beside him at the time.*

A few days later when his mother baked another cake, Johnny tested this new hypothesis. This time he waited until Mother had left the kitchen and gone into the dining room to set the table. *Then* he grabbed his fistful of frosting. This time he did get a taste before the reaction came, but Mother, of course, returned too soon. Again she shrieked. Again Johnny got whacked. Again he was admonished, "Keep out of that cake frosting—or you won't get any cake at all!"

Now this was *really* a conundrum. Johnny retired to nurse his wounds and reconsider the data. The frosting was good. Since he liked it, it must be made for him. But whenever Mother saw him reach for it, he got whacked. Maybe the determining factor was whether his Mother saw him take it or not. Well and good; the next time a cake was made he waited until Mother was outside gardening. *Then* he took some frosting from the top, with a little cake thrown in for good measure. Of course, he left a hole in the top of the cake, but if Mother didn't *see* him take it, how could she know? Johnny ate his frosting and then went on about his business, the mystery of the cake frosting finally solved—until Mother came in half an hour later and found the violated cake. Another whacking, another scolding, and no cake for supper that night.

So it was back to the drawing board again for Johnny. Clearly *something* was wrong with his whole approach. The frosting was there for him to eat, *but only at certain times* (i.e., when served at dinner). At other times it was sharply proscribed. To Johnny this made no sense whatever but that seemed to be how things worked. Even more puzzling, when he took the frosting at the forbidden times his punishment seemed *completely unrelated to whether Mother could see him or not.* The business had him baffled—but baffled or not, there was obviously *some* method of detection at work, and that, too, was a part of the way things worked in the world around him.

Johnny might have tossed in the towel, if he had been a little less stubborn and persistent. He might just have accepted things the way they were. But Johnny was a stubborn little boy who *really liked* chocolate frosting. He was not about to accept a law that made no sense to him—"cake frosting is forbidden at some times and permitted at others"—nor to stop trying to outwit the mysterious method of detection that was thwarting him. The situation challenged his young mind, and presently a new approach came to mind. Perhaps he could take the frosting at the forbidden times *if he could do it without being detected.* Hard to achieve? Well, maybe not so hard. First, try taking the frosting from a part of the cake where Mother wouldn't notice. If *that* didn't work, then take it off the top but smear some on the cat's whiskers in order to get Mother confused—

And so the struggle went on.

## THE SCIENTIFIC METHOD

Foolish as this fanciful story may seem to us, we must realize that it was anything but foolish to Johnny. Within his limited view of the universe, Johnny was facing precisely the same sort of problem that adult men have been facing for centuries: finding ways to supply their needs and wants, and to obtain for themselves more comforts and satisfactions in life in return for less effort.

In reaching for these goals men have always found themselves thwarted by a series of inexplicable, baffling, and seemingly senseless rules governing the way things happen in the world, just as Johnny did. Often these rules have seemed to exist for no other reason than to annoy and hamper mankind in the fulfillment of his needs. The man who tried to carry water home from the well in his hands found that he couldn't do it. The stuff ran through his fingers and was lost before he could get ten paces away from the well. Sometimes the rules seemed completely arbitrary and confusing: In the winter a man *could* carry water home in his hands—in solid blocks of ice —while in the summer he could not. There was no sense to it, but it was one of the facts of life he simply had to deal with if he wanted water. Only later was it discovered that it was possible (1) to preserve water in its hard, "carryable" state in the summertime by burying it in deep pits underground in the winter; or (2) to carry the liquid water more efficiently in an earthenware jug than in your hands.

So men, faced with such awkward rules, set out to find ways to get around the obstacles in their way, just as Johnny did. Bit by bit they learned which rules could be by-passed easily and which could not. They learned how to change and control conditions around them. Above all, painstakingly, they learned to *define what the rules actually were,* in hopes of sometime discovering *why* they existed.

In his battle for the cake frosting Johnny was doing exactly what men have been doing for centuries in their fight for survival. Johnny was exploring the "natural laws" governing *his* universe—the narrowly limited universe of his own experience as seen through his eyes. Instinctively, he used a method of exploration which, bit by bit, provided him with useful knowledge and useful results.

Today we call that method of problem-solving *the scientific method.* It involves four critical steps, each taken in turn and each equally important in reaching a satisfactory solution. Those steps, in order, are *observation, hypothesis, experiment,* and finally, *conclusion-drawing.*

How does it work in the hands of a modern scientist? First there must be a question to be answered, some riddle to be solved. The scientist encounters some question which he cannot answer, some phenomenon of nature which he does not understand. He then begins to gather all of the data about this riddle that he can find simply by observing it as closely and carefully as possible. This period of acute *observation* is absolutely vital. Without it the scientist has no basis even to guess what the solution might be.

Second, on the basis of what he observes, the scientist will think of one or two, perhaps several, plausible and possible explanations for the phenomenon in question. These possible explanations are called *hypotheses*. Often one or another hypothesis can be discarded right from the start. Perhaps certain observations obviously don't fit one hypothesis; another may seem highly improbable, even though the scientist can't pinpoint exactly why he thinks so. After eliminating these, he will choose the one remaining hypothesis that seems the most likely of all. Accepting this—at least tentatively—as his *working hypothesis,* the scientist then devises *experiments* to test whether it really does explain the phenomenon or not.

If he is a good scientist, this will not be any half-hearted gesture; he will submit the hypothesis to a real trial by fire. Of course, he will seek out experiments that seem likely to support it—but he will also rack his brains for any possible experiment that *might prove it wrong*. Indeed, the most important step of all in this method of exploration is the cold, deliberate attempt to poke holes in a possible explanation, to knock it to pieces if that can possibly be done. Nor does the scientist trust himself to be sufficiently objective. He knows from long experience how easy it is to fool oneself, to select only favorably loaded experiments, and to try to prove what one wants to believe, whether it is actually true or not. For this reason, the scientist records his experiments for others to criticize. He devises experiments which are *repeatable,* so that other scientists can also do them and compare results. Because, as scientists know, any experiment that cannot be repeated successfully by others with the same results is of no value in proving anything.

Finally, from his experiments, the scientist gathers a large quantity of new data which is then matched up with his working hypothesis. From this, *conclusions* can be drawn. If all his experiments seem to bear out the hypothesis in all respects, and if others come up with the same results, the hypothesis begins to look really promising as the true explanation for the phenomenon. If it survives new tests by others, it takes on new strength. This is not to say that it is *proven,* by any means; at best it may be considered "true until proven otherwise." The scientist is fully aware that new data may be discovered at any time which might raise questions about it or even disprove it altogether. But when sufficient corroborative evidence has been gathered, with no evidence at all to contradict it, the hypothesis attains a state of general conditional acceptance among scientists. It is then considered a *theory*. If the theory stands up to multiple experiments carefully designed to try to disprove it, if it holds true time and again no matter who may test it or in what way, and if no new observations come along to challenge it, it may in time come to be considered a proven *law of nature*.

Notice that the scientist did *not* start with a conclusion and then attempt to bend the facts to fit it. Notice that from the start he accepts that his hypothesis may prove to be wrong even though all of his experiments seem to substantiate it. Given a single bit of contradictory evidence, one single experiment which doesn't fit in even though a dozen others do, and the scientist knows that *something* is wrong. *Somewhere* there is a flaw in the conclusions that were drawn, something that is missing, something that has been misinterpreted. When that happens (and it almost invariably does happen) he must then revise his hypothesis to fit the observed facts, changing it again and again as he goes along until it fully and reliably explains every part of the phenomenon he is trying to explain. Even a hypothesis that has been corroborated to the extent of becoming a theory, and then substantiated to the point that it is considered a proven law of nature is still vulnerable. New discoveries, new observations, new methods of measurement may at any time cast doubt upon it. Even the best-established laws of nature must be revised if newly discovered data demand it. *No* natural law can ever be considered finally and irrevocably proven.

There is no magic in such a method of finding an answer to a problem. Indeed, it is so simple and logical that all of us, scientists or not, use it to some degree or other every day of our lives in solving everyday problems. It is the time-tested method of telling truth from nonsense and proving it. As such, it is the method that has been used in discovering virtually everything we know about our universe and the way in which it works.

But if the scientific method is so simple and logical, why was it such a staggering idea when it first appeared? Probably because it was such a complete reversal of the way ancient scientists and philosophers had done things for centuries. Before the scientific method was devised, these men started with conclusions they had come up with on the basis of meditation, casual observation, and sheer guesswork. Then they wrenched and twisted newly observed facts to fit the conclusions. Up to a point, they got results, too; that ancient method worked splendidly as long as men could manage to ignore the facts that didn't fit in with their conclusions. It took centuries to discover that this was a blind alley, producing more and more wrong answers all the time. It took centuries to recognize that an unbroken chain of cause and effect ran throughout nature governing its happenings. It was not until the scientific method became firmly established that the knowledge of science began to grow and that our understanding of the laws of nature began to expand.

## THE LAWS AND THE LAWYERS

Obviously, as this scientific method came into use, scientists everywhere began to revise their ideas about what, exactly, a "law of nature" was.

Clearly it was not something that could be determined on the basis of intuitive or self-evident "truths," philosophical dialogues, logic, or reasoning. More and more, as the scientific method began to develop, scientists began observing nature more closely to see what was going on that they couldn't understand. Freed from the idea that observed events had to fit into arbitrary molds, they began questioning everyday things that were happening all around them.

A ball, when dropped from the hand, fell to the earth with increasing speed, taking a measurable time to reach the ground. A phenomenon: what was happening here? The "self-evident fact" that a heavy ball fell faster than a lighter ball came under scrutiny, and was found to be neither self-evident nor a fact. A ball rolling down an inclined plane also was seen to roll faster and faster until it reached the ground.

On the other hand, a ball thrown straight up into the air appeared to *slow* to a stop at a certain point in the air, then reverse its direction and begin accelerating downward again. A pendulum swinging freely back and forth seemed to do something strangely similar, accelerating from one extreme of its swing down to the lowest point in its arc, then decelerating to a stop at the other extreme, then reversing direction and accelerating down to the low point again. Another phenomenon, something that always happened in the same way—but could there be a connection between the way the ball tossed in the air behaved and the way a pendulum behaved? If so, what was it? The two things seemed similar but not identical. Could it be possible that both behaved according to the same general principle? If so, then what were the rules? Could the same principle also apply to the behavior of other kinds of moving objects; for example, to the behavior of a ball thrown horizontally in an arc? And what about the size of the ball? After all, the moon was a ball moving through space —but it was not moving freely. Something seemed to bind it to the earth around which it revolved. Was there, conceivably, some similarity in the behavior of that celestial ball and the behavior of a handball thrown in an arc (which also seemed, in a different way, to be irrevocably bound to the earth)? Or was such an idea merely a wild reach? If it *wasn't* a wild reach, if there really *was* some similarity in the behavior of these moving bodies, shouldn't the moon also be included in the ever-broadening general principle that governed the behavior of the falling ball, the thrown ball, and the pendulum?

It was a slow process, this vast exploration of utterly unknown territory. But little by little certain broad general principles were identified—ways of describing things that happened, ways of comparing one phenomenon with another. Whenever such broad principles (or "laws of nature," as we call them today) began to emerge, they were subjected to relentless testing. Generation upon generation of scientists experimented repeatedly to see if these principles really *did* describe widespread or universal phenomena, or whether they applied only in a few special or limited situations, and even then, perhaps, only sometimes.

Under such ruthless scrutiny, many apparently valid "laws" fell by the wayside, disproven by repeated experiment and testing. Others were corroborated again and again—sometimes modified, sometimes broadened, but still holding up under the careful scrutiny of the scientific method.

Some of the strongest, best-proven of these principles are still being challenged today; they are *still* accepted as valid only until proven otherwise. To many nonscientists this persistent effort to disprove things which have been shown to be valid in thousands upon thousands of cases may seem ridiculous. But we know today that many of the laws of nature which seemed to apply to all areas of the universe in the early days of physics have since been shown to apply only to one limited area: the world of everyday experience. They have been proven to be incomplete—or even flatly invalid—in describing events in the microcosmic universe of elementary particles or in the macrocosmic universe of far-flung galaxies. So those "proven" laws of nature have had to be discarded, or severely modified. Indeed, over a period of two thousand years of painstaking observation and experimentation, only the barest handful of basic, universal laws of nature have survived unscathed to this day.

What, exactly, are these few basic laws? What do they say? More important, what do they *mean*? Most of us have only vague memories of these laws as complicated mathematical formulas encountered in high school or college, never very meaningful at best, and now long forgotten. Is it possible for us really to understand those laws now without getting involved in complex calculations and pages of mathematical reasoning? Perhaps so, because the real, incredible beauty of these basic laws of nature is their splendid simplicity. Basically they are nothing more than clear, simple statements of relationships—simple quantitative descriptions of things that have been observed to occur in the universe. They are statements that describe how things work, nothing more.

The few basic laws we are discussing are powerful as well as simple. Evidence collected over centuries supports them. They have weathered innumerable challenges, and face new ones every day. Scientists today, as throughout the last eight hundred years, challenge them by use of the scientific method. To investigate the law of gravitation, for example, a physicist studies the motions of objects of all sorts: planets moving in orbit, balls rolling down inclined planes, feathers set free to fall in a vacuum tube. First the physicist observes and measures. How fast does a given object move? In what direction? Is its speed constant, or changing?

What about its direction? With his observations made, he then tries to think of some rule or principle that can explain all the things he had observed. With such a rule as hypothesis, he tries applying it to other objects in motion under other circumstances. To be useful, the rules must describe how *any* object will move under the influence of gravity. So again and again he tests his rule, measuring it against the actual behavior of a great variety of moving objects.

If he can find such a rule, and if neither he nor other physicists can find even a single experiment that seems to disprove it, eventually the rule or hypothesis becomes a theory, then later is considered a law of nature. Thereupon, the physicist seeks to use this rule to predict how *any* object in *any* gravitational field *anywhere,* will move. He knows the rule may not hold up. Many early "laws of nature" proved to be nothing more than descriptions of isolated, individual events with no application to other similar events. Other such "laws" described a broader range of phenomena but still did not cover *all* phenomena of a similar nature. Such "laws" may well be useful indeed to help make certain kinds of predictions or to solve certain kinds of problems, but they are not really good laws, not because they are untrue (as far as they go) but because they are too limited.

## GOOD LAWS AND POOR

What, then distinguishes a *good* law of nature from a poor or limited one? First, a good law of nature deals with *situations in general,* not with specific cases. It does not, for example, describe the movement of one particular kind, size, or shape of object under certain limited circumstances. This would be little more than a description of a single limited occurrence, perhaps even a single experiment. Rather, a *good* law provides a general description of the movements of objects of any kind, size, or shape under a wide variety of circumstances.

Second, a good law of nature should apply *universally*. If it describes the motion of objects, it should apply not just in any one place or at any one time, but anywhere in the universe at any time, whether in the world of our sensory experience, in the microcosm, in the macrocosm, or wherever; and it should apply to *any and all* objects, no matter how large or small, no matter *where* they are moving or how fast or slow, or in what direction.

Third, a good law of nature should not contain too many exceptions. If a law is hedged with ifs, ands, buts, and whereases, specifying various exceptions in special cases, and applying only when everybody in the world happens to be wearing green neckties, it becomes too complex for any real usefulness. Whenever too much work and effort are required to figure out when a law is supposed to be applying and when it isn't, this is usually a pretty good indication that the law is not really a general, universal description of anything, even though it may seem to have wide application. It becomes just one more of a multitude of relatively insignificant rules of the road, and probably highly vulnerable to challenge and testing anyway, rather than a good, useful law of nature.

Finally, a good law of nature is a *complete and quantitative* statement or description, not just a hazy, indefinite expression of generalities. A law describing the interaction between an object and a force acting upon it must do more than just to state vaguely that something influences something else. It must state in what specific way something influences something else, how strongly, when, and in what direction. It must state these things in exact quantitative terms that do not omit anything significant. This, of course, is why good laws of nature are so often expressed in the form of mathematical equations; that is one of the most reliable ways of being sure that the law describes some relationship completely and quantitatively. The terms of an equation are explicit; a given factor is either included or it is not.

This is also why good laws of nature are often so difficult to express correctly in words: Words can be very slippery indeed when it comes to pinning down *exact* meanings! In this respect, it is inescapable that laws of nature be expressed in mathematical terms if they are to be stated completely and quantitatively—but at worst (in the case of the most basic and fundamental laws of nature) they are expressed in the form of simple equations which require little more than a rudimentary command of algebra or calculus to interpret.*

To understand more clearly exactly what a good law of nature is and how it fulfills the criteria we have discussed above, let us look at one of these basic universal laws more closely. Many of the phenomena that occur continuously all around us are described by one very familiar law of nature—the law of universal gravitation. A baseball arcs toward the earth when we throw it. A coconut drops from a palm tree. Water runs downhill when left to its own devices. The moon and planets move in their orbits according to a reliable, repetitive pattern.

According to legend, the law of gravitation was discovered by Sir Isaac Newton when an apple fell from a tree and hit him on the head. Whether any such thing actually occurred is a moot question, but the idea that Newton "discovered" this law in any blinding flash of revelation is pure nonsense. Strictly speaking, Newton did not "discover" the law of gravitation at all; it had been obvious for centuries that some orderly principle

* For all of this, the *meaning* of these basic laws can usually be conveyed without recourse even to this simple level of mathematics. In this book we will venture onto the high seas of simple algebra on occasion—but the author is convinced that many readers tend to reject and pass over even simple mathematical equations they can understand perfectly well, so we will seek to avoid them wherever possible.

lay behind the behavior of moving objects near the surface of the earth. Galileo had long before made accurate measurements of how fast objects fell to the ground when released, and how their speed kept increasing steadily as they fell. Copernicus had already observed that the moon revolved around the earth at a certain velocity without flying off into space. What Newton *did* achieve was to demonstrate that *the same natural law that described the movement of falling bodies on earth also described the movement of the moon around the earth, or of the planets around the sun.* Newton showed that these objects all moved as they did because of a simple, universal relationship between every object in the universe and every other object: that *every object in the universe attracts every other object in the universe;* that the force of attraction between any two given objects is always dependent upon the masses of the objects and the distance between them; and that this attractive force between any two objects in the universe could be calculated according to a precise mathematical equation.

In other words, the things that happen "according to the law of gravitation" actually do nothing of the sort. They are natural occurrences, which happen. The law itself is nothing more than a *generalized description* of all these events and phenomena put together—the words (or mathematical formulations) that we use to describe with extreme precision the attraction that demonstrably exists between any two objects. The description also states precisely how strong the force of attraction is, in what direction it acts, and how it may change from place to place.

It is important that we realize clearly that this "law of gravitation" that we are talking about is a simple description of events, so to speak. It does *not* say *why* every object in the universe attracts every other object. It does *not* say from whence the attraction arose, how long it has been there, nor—for that matter—whether or not it will still be there tomorrow. All it says is *this is the way it is, this is how things work as far as anyone has been able to observe and measure so far.*

The same thing exactly can be said about all the other great universal natural laws. The "law of inertia," for example, simply describes certain observable characteristics of objects in motion and objects at rest. In brief, this law says: *Any object in motion tends to remain in motion in a straight line at a constant velocity, and any object at rest tends to remain at rest, unless acted upon by an outside force.*

Such a "law of nature" is a very useful thing to have around. It allows us to know what behavior we can expect of objects at rest and of objects in motion. It allows us to make accurate predictions about *how* objects, whether at rest or in motion, will behave (1) if an outside force is applied; or (2) if *no* outside force is applied. The law of inertia is indeed useful—but it does not even attempt to explain *why* an object at rest remains at rest, nor *why* an object in motion remains in motion.

Well, then, *why do they?*

Nobody knows why.

Nobody knows why every object in the universe attracts every other object, either. Nobody knows *why* the force of gravity exists, nor even *what it is,* nor what inertial force is, nor why it exists. What is more, it took the scientists of the world *centuries* of banging their heads on the wall before they began to realize that *it simply didn't matter why*—or, rather, that to ask *why* is to ask a fruitless question.

And here we discover a final reason that the ordinary man in the street finds it so hard to understand what the modern physicist is doing. Most people who have thought about it at all have assumed without any question that physicists were trying to find out *why* matter is made up the way it is and *why* forces act the way they do, and it just isn't true. Scientists gave up asking *why* centuries ago. Indeed, they regard such questions as stumbling blocks and blind alleys, rather than the proper and legitimate business of the scientist.

Admittedly, this is a rather startling and unsettling observation when we first encounter it. Isn't that what any study of science, and above all the study of physics, is supposed to be all about? Isn't its whole purpose really to explain *why things happen the way they do in the universe?* Certainly the ancient Greeks thought so. They devoted an incredible amount of time and energy trying to dream up plausible reasons why, and then *trying to prove them.* And as an intellectual exercise, this was great. It stimulated the mind no end. It developed the rules of logic and argument to a razor edge. But as far as finding out *what was happening* was concerned, the search for reasons why drove the ancient Greeks into a blind alley. For a thousand years after the Greeks, scientists kept falling into the very same trap. In their search for reasons *why* things happened they got nowhere. They didn't even learn too much about *what* was happening. Only when they decided to shelve the question of *why* things were happening did they begin to forge ahead in accurate, useful observation and description of *what was going on in the universe in what way,* whatever the reason why might be.

This is not to say that the question of *why* is somehow forbidden and improper. A great many fine scientists today are still very much concerned about the *why* of things as well as the *how*. Perhaps this *is* indeed the ultimate goal of science, to find out *why*. If so, then scientists still have a long way to go because the one cannot come before the other. There is little hope of ever finding out the *why* until *what is actually happening* has been fully and accurately described—until we know *how* matter is constructed, *what* energy is, *how* matter and energy interact not just in our universe of everyday experience but in all the microcosm, in all the macrocosm, in all space, throughout all time, always, under all circumstances conceivable or even inconceivable. And even if physicists were to begin right now with all the knowledge that has been already been accumulated, this would even today be a whale of a large oider.

Of course over the centuries great progress has been made. We certainly know more of what is going on (or of what probably is going on, or at the very least, of what *seems* to be going on) today than the ancient Greeks did. We have better tools to work with—more accurate instruments for observation and measurement, more fruitful methods of investigation, newer and more useful mathematical techniques. And thousands of scientists all over the world today are working to refine these tools even further.

Furthermore, what we do understand today of how the universe works has paid almost unbelievable practical dividends. This knowledge has allowed us to take giant strides in fulfilling our everyday human needs better. It has allowed human beings everywhere to live more comfortable, varied, useful, and interesting lives. It has also shown us what men *can* potentially do with matter and energy in order to improve living conditions and control environment. The *application* of our knowledge of how the universe works has closely followed the discovery of *how* things work, revolutionizing the world we live in. And the promise held forth by modern physics for even more useful and revolutionary applications of new knowledge in the future is staggering.

Unfortunately, all parts of the picture are not so bright. If our knowledge of physics has paid great dividends, it has also created frightening liabilities. Man has found the means to destroy himself and all his works in one great bloody ball of fire. Nature is not fooling around; raw energy can be found or produced in nature in perfectly staggering supply for Man to use for survival or for self-destruction, whichever he elects. If men are clever enough to snoop out her secrets and discover how to use them, nature seems to say, they had better learn quickly how to harness and control the energies they are capable of releasing, if they hope to survive. But nature doesn't *care*. There is a great deal of universe available, and what men do or fail to do in this particular little corner of it will hardly cause any cosmic upheaval.

So we search out information about *how* things work, not *why*. All of the great, basic laws of nature which have survived all challenges over the centuries are nothing more than simple, general, universal, quantitative descriptions of how things happen in the universe—descriptions of the built-in characteristics of matter and energy, and of the way objects and forces interact with each other. Many laws once thought to be "basic" have been shown to apply perfectly only within certain limits; such laws have had to be altered, or discarded, or modified to take into account new observations. *There is no law of nature known today, no matter how "good" or "strong," that can be considered to be finally and absolutely proven.* At best it can be considered "unlikely to be proven wrong."

Of course, it is always a shock when a long-established, apparently basic, and immutable law of nature is overthrown. That was the reason Albert Einstein's work was so profoundly disturbing—and challenging—to the world of science. He challenged laws of nature long believed to be basic and universal, established beyond possible doubt, and *he made his challenges stick*. As a result, like it or not, physicists were forced to revise their outlook about what, exactly, those laws were really describing and how things actually did work. In later chapters we will see just which of these laws Einstein challenged and just why his challenges could not be brushed off. Certainly that one man, more than any other in history, forced scientists to make an agonizing reappraisal of how things really worked in the universe—a reappraisal which is still going on today.

Einstein also combined certain other categories of "basic," "universal" laws into one, thus showing that they were far more "basic" and "universal" than anyone had ever imagined. For example, he combined space and time into one entity, and he demonstrated that the two of the most basic and universal of all known laws of nature—the law of conservation of matter and the law of conservation of energy—were in fact two aspects of the same law, an insight into the true nature of matter and energy that opened up profound and frightening vistas. Ironically, even Einstein himself was chagrined by some of the logical implications of his own work, and refused to the end of his life to accept them, even though he could not find a way to disprove them—a heartening indication that one of the greatest genius minds to appear in all history was still a *human* mind, sharing the same human emotions and human frailties that apply to you and to me.

## A CHECKLIST OF NATURAL LAWS

When we come down to final cases, we discover that the basic, universal laws of nature that have held up over the centuries in the face of all challenges (even if modified or altered somewhat) are relatively few in number. There is nothing mysterious about them. They can be listed and stated very simply, and understood without resort to great technical knowledge or the skills of higher mathematics. In every case, these laws are "rules of the road" and nothing more—merely *descriptions of what happens, descriptions of characteristics of matter and energy*. To provide a clear focus for later chapters, we can list these laws in a simple chart, as follows:

I. *The laws of motion*

a. *The law of universal gravitation* is a quantitative description of the force of attraction that exists between any two objects anywhere in the universe, and of how that force affects the objects. It states that *every object in the universe attracts every other object,* and describes exactly *how* this attractive force can be calculated in the case of any two objects, whether on earth or in the far-flung reaches of space. It also provides a precise statement of *how powerful* that force will be in any given case.

b. *The law of inertia* describes certain built-in characteristics of all material objects in the universe. It states that *any object at rest tends to remain at rest, and any object in motion tends to remain in motion in a straight line at a constant velocity, unless acted upon by some outside force.* This law originally served to explain a variety of characteristics of objects at rest and objects in motion that seemed to have nothing to do with the force of gravity. But today, as we shall see later, there is at least good reason to suspect that these two "basic" laws—the law of universal gravitation and the law of inertia—may really be no more than two different ways of describing the same phenomena. Unquestionably there are certain ways in which they are similar and intimately related; and there are certain circumstances in which it is completely impossible to be certain which of the two is actually the proper law to use to describe a given phenomenon.

II. *The laws of conservation* describe other characteristics of matter and energy, and apply to certain other phenomena associated with the motion of objects.

a. *The law of conservation of matter-energy* states *that matter or energy may be changed from one physical state or form to another, and may be converted reversibly from one to the other, but that the combined total of matter and energy existing in the universe can neither be increased nor decreased.* Here, again, two original conservation laws (the law of conservation of matter and the law of conservation of energy) have been combined into a single, all-encompassing principle, since it was found both theoretically and experimentally that matter and energy are two totally interconvertible forms of the same thing.

b. *The law of conservation of momentum* states that *the total momentum of any interacting system is always conserved.* This law, together with its "twin sister" corollary, the *law of conservation of angular momentum,* is probably the single most firmly established and basic of all known laws of nature, yet it is quite unfamiliar to most nonscientists, and is not even always clearly understood by scientists. To make sense of it, we must understand precisely what such terms as "momentum," "angular momentum," and "interacting system" mean to a physicist. Thus our discussion of this law will be deferred to a later chapter.

c. Other conservation laws, including laws of *conservation of electrical charge, conservation of nucleon charge, mirror symmetry, and electrical charge symmetry,* as well as others, are all essentially statements of properties or characteristics of matter or energy which *ultimately remain unchanged,* always, under all circumstances, no matter what happens—as far as physicists know today.

III. *The laws of thermodynamics* are intimately related to the macroscopic or statistical behavior of molecules of matter in motion, and describe how *heat* or *thermal energy* is transferred from object to object or particle to particle. These are essentially nothing more than the *law of probabilities* as applied to the behavior of molecules en masse. In brief summary, the laws of thermodynamics state:

a. *That the total entropy of any closed system must always either remain unchanged or increase.* From this we can deduce that:

1. *The natural direction of heat flow is from a hot area to a cold area, and this direction of flow cannot be reversed without the aid of some outside source of energy*; and that

2. *The natural direction of energy transformation is from mechanical energy into heat energy.* Again, our discussion of the puzzling concept of *entropy,* related both to heat content and molecular disorder, must be deferred to a later chapter.

In addition to these laws of classical physics, we must add to our list the concepts of two great theories of modern physics, considered "laws of nature" by many physicists, and four great forces which are known to act upon objects at a distance and to create "fields of force" through which they work:

IV. *The theories (laws) of relativity,* worked out by Albert Einstein and others, are based on the notion that there is no such thing as *absolute* motion in the universe, that all natural phenomena observed in the universe can be described only relative to the observer and may be described differently by another observer in a different location. The distinctions between the *special theory of relativity*, and the *general theory of relativity,* and the overwhelming significance of these revolutionary theories in the world of physics, will be discussed in detail in Part III of this book.

V. *The theories (laws) of quantum mechanics* deal with the behavior of subatomic particles making up atoms of matter, and are based on the concept that energy always occurs in tiny but finite packets or "quanta" which represent the smallest quantities of energy that can be interchanged between particles. Part V of this book will deal with quantum theory in considerable detail.

VI. *The laws of forces and fields* are concerned with four universal forces capable of acting upon objects or particles at a distance (i.e., without physical contact) through the medium of energy fields:

a. *Gravitational forces,* acting most strongly between very massive objects in the universe;

b. *Electromagnetic forces,* encompassing the attractive or repulsive forces of magnets and of electrically charged particles;

c. *Weak nuclear binding forces,* arising from certain "weak" or "relatively improbable" interactions between subatomic particles in very close proximity to each other;

d. *Strong nuclear binding forces,* arising from certain "strong" or "highly probable" interactions between subatomic particles in very close proximity to each other.

Such a list of natural laws, even so encapsulated, may seem dreadfully complex at first; yet when we consider it, we see that the list is really not all that long. Approached logically, against a comprehensible background or frame of reference, it will appear less and less complex, and more and more comprehensible as we go along. But at this point we must recognize clearly that *all* these laws are descriptions of *what* and *how*, not discussions of *why*. When we ask why, even today, we are in trouble. And the physicist more than anyone else grows weary of the age-old questions which he still cannot even begin to answer:

*Why can't matter either be created or destroyed?* We don't know—but nobody has ever been able to do it yet.

*Why can't energy either be created or destroyed?* We don't know—but nobody has ever been able to do it yet.

*But who says it can't be done?* Nobody says so—except the thousands upon thousands of scientists who have tried and failed with monotonous regularity over the centuries.

*Well, what is matter?* We don't know. We're still trying to find out.

*What is energy?* We don't know. We're still trying to find out.

More than any others, those last two questions are the real challenge to modern physicists. It is because these two questions have not yet fully been answered that physicists remain at work today, after thousands of years of study and in spite of their inability to explain why. And it seems likely that they will remain in business for some time to come.

What *is* matter? What *is* energy? There is no better way for us to start considering the riddles of the "normal" universe of human experience than with these questions in mind. In succeeding chapters we will see what answers the classical physicists came up with as they tried to describe the universe they saw and the way things happened in it. In the process we will see how some of these great laws we have listed above were first defined, and see more clearly exactly what they mean.

# Part II

## *The Universe of Classical Physics*

# CHAPTER 5

## *Assumptions, Observations, and Measurements*

By the last quarter of the nineteenth century the first three groups of the great natural laws outlined in the last chapter had been defined, proved, entrenched, and fully accepted throughout the world of science. These are the laws we think of today as the "fundamental laws of classical physics" —the law of universal gravitation, the conservation laws, the laws of inertia, and the laws of entropy and thermodynamics. As little as seventy years ago it was firmly believed that these laws alone described virtually every phenomenon and occurrence that existed in all the universe. Arising as they did from a thousand years of labor, they were regarded by the physicists of the 1880s and 1890s with great satisfaction; through these great "unshakable" laws, it appeared that all the workings of the universe had really been pretty well defined and described.

Today, of course, we know that those who held this pleasantly self-satisfied attitude were in for a rude awakening. They were sitting on a powder keg, with Michelson and Morley all set to touch off the fuse. But from one point of view, physicists of that day were perfectly right. They could hardly be blamed for assuming that the universe *they* knew—the universe of everyday human experience—was all the universe there was. And so far as that "normal" universe was concerned, these classical laws of physics were indeed perfectly valid. Except for a few loose ends, such as the question of the "ether wind," and the puzzle of the nature of light, these laws did indeed describe virtually everything that had ever been observed in nature.

What is more, those same great laws are just as valid today in describing and predicting things in the "normal" universe as they were then. They still work. They still apply. We still need to understand these laws in order to understand the things we see happening around us. We still use them as powerful tools in solving our everyday problems of mechanics, optics, acoustics, or electrical engineering.

More to the point here, we need to understand these laws in order to understand *what was found wrong with them and why*. It was a painful reappraisal of these classical laws, beginning at the end of the nineteenth century, which led directly to the explosive revolution in physics which took place later. Our everyday world in turn has been revolutionized by that great scientific explosion. It will be worthwhile reviewing those classical laws to see better how and why that revolution in physics came about.

## NOTHING STARTS FROM NOTHING

As we have seen, these classical laws of physics evolved from centuries of observation and experiment. They did not, however, arise from scientific investigation alone. Earlier we saw that Euclid could never have developed his elegant and complex system of plane geometry out of thin air. He started with a few basic assumptions which he considered self-evident, so obviously true that they required no proof to support them. Then, on the basis of these assumptions (or *axioms*, as we call them) he created his system of geometry. Beyond these axioms, *every single proposition* required rigorous logical proof before it could be accepted as a valid basis for further propositions.

Euclid's axioms were simple. They included such statements as "The shortest distance between any two points is a straight line"; or: "Any two lines which, extended indefinitely, never meet at a single point are parallel." Those axioms have stood the test of time splendidly well. Within the limits of plane geometry, they have never been disproved. They are taught today in precisely the same terms as Euclid taught them. True, we know now that there are other useful systems of geometry in which the shortest distance between two points is *not* a straight line, and in which parallel lines may indeed intersect one another, but Euclid's axioms still apply perfectly well *within certain practical limits*. Within those limits they are still enormously useful in solving everyday geometric problems, as any engineer, architect, or surveyor can testify.

By the same token, the great classical laws of physics outlined in the last chapter were also based upon certain assumptions that were accepted without question as self-evident. The first of these assumptions, so obvious as to seem ridiculous to mention, was the *assumption of reality*. In order to explore anything about the universe, the scientist first had to assume that the universe really did exist—that *he* existed, that other people existed apart from him, and that the earth, the planets, the stars, and all the rest of the universe also existed in fact.

Second, the scientist had to assume that *he could learn something* about the universe by means of his senses—by seeing it, feeling it, smelling it, and measuring it—and that this was the *only* way that he could learn anything about it. This assumption, of course, conflicted with the ancient Greek notion that something could be learned about the universe by means of logic and reason alone. Modern physicists disagree on another basis; they argue, for example, that pure theoretical mathematics has, in fact, suggested or predicted any number of previously unsuspected phenomena which simply could never have been suspected or detected any other way. Modern physicists also realize that observation by means of the senses has some built-in problems: Often the very act of measuring something alters the thing that is being measured. But such considerations were quite unknown to the early physicists. Both Galileo and Newton assumed without question that their sensory observations were the only avenue to learning anything, and that their measurements yielded valid information. And just as well, too; without those "self-evident" assumptions, the scientific study of nature could never have gotten off the ground!

Third, early scientists assumed that *there was regularity in the universe* or—to put it another way—*that the universe was orderly.* This meant, simply, that everything that happened in the universe happened in accordance with certain natural laws, and that there were no exceptions to those laws ever, anywhere. The implications of this assumption are a bit more profound than first meet the eye. Essentially, this was an assumption that *cause and effect* ordered the universe. It implied that if scientists could discover *all* the natural laws there were, understand them perfectly, and then apply them to things that were currently happening in the universe, they should then be able to predict precisely *what would be happening any time in the future, anywhere in all creation.*

To take a simple example, suppose a cannonball is fired at a given instant. Suppose a good scientist then takes *precise* measurements of the mass of the ball, its muzzle velocity as it leaves the cannon, the air resistance it encounters in flight, the direction and angle the cannon is tilted, the windage affecting the ball's trajectory through the air, and the force of gravity acting to pull it toward the ground. If cause and effect prevail, the scientist should then be able to predict, *in advance,* precisely where and when the ball will strike the earth, and with exactly what force.

Or suppose a scientist has a box a foot square containing twenty Ping-pong balls, and gives the box a vigorous shake so that the balls go bouncing from the walls and colliding with each other in an apparently chaotic scramble. But is it really chaotic? Not so, if cause and effect prevail. By applying the natural laws describing the movements of material bodies and the forces acting upon them, the scientist should be able to predict in advance precisely where any given Ping-pong ball would be found ten seconds after the box was shaken, provided that its location before the shake was known and all other factors (including the force and direction of the shake) were measured exactly. It is granted that accurate measurements in this case might be fantastically difficult, and that the calculations involved would be a tall order even for a high-speed computer. But, at least in theory, *a correct and accurate prediction could be made.*

The assumption that all things occurred as a result of cause-and-effect relationships in an orderly, regular universe was very comforting to the early physicists. It was also very useful. Without this basic assumption it is unlikely that any of the classical laws of physics could have been derived at all. There was, of course, no way for those early scientists to guess that a young man named Werner Heisenberg, living centuries later, might successfully challenge the assumption's validity. For the world of physics those early workers were investigating, it was a perfectly valid assumption—and it remains just as valid today, *within certain practical limits,* in our understanding of things happening in the "normal" universe of our everyday experience.

A fourth basic assumption was that *all the laws of nature apply with equal validity throughout all regions of the universe, throughout all time.* In other words, it was assumed that the universe was *uniform* from one part to another, and from one time to another. Thus, an object was assumed to have the same mass in some far corner of a distant galaxy as it had here on earth. It would respond to a given force in exactly the same way, no matter where in the universe the force might act upon it. Once again, this was an utterly necessary assumption if any kind of "rules of the road" were to be outlined at all. And again, this assumption proved perfectly valid within certain limits, but was challenged at the beginning of the twentieth century. We know today that things which occur in our "normal" universe of experience do *not* necessarily occur under all conditions in all parts of the universe. Objects or particles accelerated to velocities approaching the velocity of light, for example, behave quite differently than the same objects or particles traveling at low velocities.

Finally, early physicists assumed that *time was uniform, flowing steadily from past to present to future.* It was assumed that natural phenomena always occurred in chronological sequence. Natural laws would be the same today as they were yesterday, and still the same tomorrow, or next year, or next century. Just as things were assumed to happen through cause and effect, the sequence of cause and effect was always assumed to be forward in time. Of course, there was no logical *proof* of this assumption. It was taken for granted on the basis of experience. It was a very necessary assumption, if scientists were to have any hope of finding any orderly patterns in the apparent chaos of natural occurrences. Even today this assumption is considered valid for almost all practical purposes. Of course, we might argue that we have not really been observing nature very long, and have no way of being *certain* that natural laws do not gradually change in some way with the passage of great stretches of time. What is more, certain very peculiar phenomena observed recently in the microcosmic world of elementary particles actually do appear to be taking place backward in time! Thus modern physicists may ultimately find that even this assumption is only valid within certain limits.

But however vigorously or successfully these basic assumptions may have been challenged in modern times, they were unchallenged and unchallengeable in the days when the classical laws of physics were being outlined. Together they formed a solid foundation on which scientists could begin, with the use of the scientific method, to observe and describe things they found happening in the universe around them, and to begin to formulate, however imperfectly, the first orderly system of natural law.

## OBSERVATIONS AND MEASUREMENTS

Armed with these assumptions, the first task of the early physicists was the observation, description, and measurement of natural phenomena. But for this work not to be wasted, the things being measured first had to be defined, useful methods of description found, and units for measurement agreed upon.

If we look closely at the classical laws of physics we see that all of them were concerned with *matter* (the material substance of the universe) and with various kinds of *motion* of material objects resulting from the action of certain forces. Indeed, the whole work of classical physics has been summed up as *the study of matter and motion.* But how can we describe a material object to distinguish it from some other similar but different object? How can motion of an object be described, or a force acting upon it be defined?

First of all, a material object to qualify as such must occupy a certain amount of space; we can observe and measure the space that it occupies. It has *linear dimensions*—length, width, and height—which we can measure and record if we can agree on what units to use. It has other characteristics of *shape* which can be described—roundness, for example, or flatness, or angularity. It has *consistency*—hardness, softness, solidity, fluidity, firmness, doughiness, etc.

Other physical characteristics can be noted: If we bend it it may change shape or not; if it does change, it may spring back to its original shape when we stop bending it, or it may remain bent. If we drop it, it may shatter into pieces, it may bounce, it may splash, or it may just go *whonk* and sit there. Even an "object" which lacks any or all of these physical characteristics (a certain volume of gas, for example, may have no set limits of linear dimension, nor any of the other characteristics we have mentioned) will still have weight and occupy space; we might hesitate to call it an "object" but it certainly qualifies as "matter" and can be described in *some* unique and distinguishable fashion.

Nor need we stop here. Other measurable characteristics can be used to describe a given object. Its temperature can be measured and recorded. So can its position with reference to other things around it—the particular point in space that it occupies can be identified. Even two objects that are virtually identical in every way can be distinguished one from the other by pinning down their respective locations in space through linear coordinates;

two objects cannot occupy *exactly* the same space at exactly the same time. (This is true even of a mixture of two gases in a closed container. We may have trouble telling one from the other and even more trouble separating them, but no single molecule of gas occupies the same space as any other molecule.) We can further describe some objects in terms of color, others in terms of roughness or smoothness, scratchiness or slipperiness.

Finally, we can describe and identify a given object by stating *what, if anything, it is doing at a given time*. It will either be at rest, or in some kind of motion. If in motion, it may be moving in a straight line in a given direction, or in a circle around a fixed pivot point, or some combination of the two. It may be moving at a constant velocity, covering the same distance during each unit of time, or with *positive acceleration* (moving faster and faster with each unit of time) or with *negative acceleration* (moving more and more slowly—decelerating—with each unit of time). If it is accelerating positively or negatively it is either doing so *regularly* (with a constant increase in acceleration or deceleration for each unit of time) or irregularly (as if the force causing the motion were varying in strength from moment to moment).

Certain of these things were soon found to be easy to describe and measure while others proved suprisingly difficult. Physicists soon discovered, for example, that while it was relatively easy to describe regularly accelerating motion, describing irregular or varying acceleration merely caused confusion and did not contribute any more knowledge. It was easier to describe an object moving with a constant velocity than with a velocity which kept changing all the time. Angular motion in a circle was more convenient to study than in a stretched-out ellipse. Furthermore, the simpler forms of motion seemed to be the more basic ones that occurred in nature anyway, while more complex motion could almost always be broken down into combinations of two or more simpler forms. For example, the most commonplace form of accelerated motion occurring in nature was the downward acceleration of an object under the influence of gravity. This was soon found to be a perfectly regular acceleration if the object was allowed to fall freely. Thus scientists sought out "idealized conditions" in which to study the behavior of objects at rest or in motion, and would always try to set up their experiments so as to rule out all useless and irritating irregularities right from the very beginning.

The early physicists also soon realized that certain characteristics of matter and motion were very slippery to describe and thus could cause all sorts of confusion, while others were far less complicated to use. The *weight* of an object, for example, could vary a great deal depending on how far from the center of the earth it was when its weight was measured. A steel ball would weight a tiny bit less atop a 10,000-foot mountain than at sea level, and would weigh less by far if weighed on the surface of the moon, since its weight would be directly related to the gravitational force acting on it. To use its weight to describe it accurately, you therefore always had to specify where it was weighed—which was a bore. Far better to measure a characteristic which did *not* change from place to place—for example, the *amount or quantity of matter* in an object, the characteristic which came to be called the object's *mass*. While an object's weight might vary from place to place (thanks to varying gravitational forces acting on it) the quantity of matter that it contained—its mass—would be the same wherever and whenever it was measured.* Thus, early scientists preferred to describe an object according to its mass rather than its weight, just because it was simpler and less confusing.

Once we have settled on some reasonably reliable characteristics to use in describing an object and how it is behaving, we must next agree on some units of measurement so that someone else measuring the same characteristics independently will be able to understand what exactly we have measured and how. Suppose scientist A measured the mass of an iron ball in kilograms, while scientist B chose to measure it in cocos (one coco being equivalent to the mass of an averaged-sized coconut). One says the ball has a mass of 10 kilograms, while the other insists that its mass is 19.6 cocos. Obviously the two are going to have trouble communicating their findings—or even comparing them. If they both knew that they were measuring the mass of the same iron ball, they might well deduce that 1 kilogram is equivalent to 1.96 cocos—but the two scientists, being equally lazy, might wrangle forever about who was going to have to do the work of converting the other's units. Don't laugh—the world of science has had precisely this sort of problem on its hands since time immemorial, and is still struggling with it today. In the United States weights are measured in pounds; in England the unit is the stone; in France, the kilogram. In measuring linear distances you can take your choice among inches, feet, miles, rods, furlongs, centimeters, angstroms, astronomical units, or light-years. That is just in the United States alone; other countries offer further possibilities.

Fortunately, most scientists got tired of having to convert continuously from one unit to another, and settled among themselves upon the *metric system* commonly used in Europe and much of the rest of the world for all kinds of measurements. In this system linear distances are measured in meters, centimeters, millimeters, or kilometers; weights and masses are measured in grams and kilograms; areas are measured in square centi-

* Or so it seemed. Today, of course, physicists know that the mass of an object can also vary quite noticeably, if the object is moving at a high enough velocity. But nobody knew that in the days when the basic laws of motion were being studied, and even today no tangible object in our "normal" universe ever moves fast enough for us to worry about measurable change in its mass. For all intents and purposes, in the universe of human experience, we are still quite safe in saying that the mass of an object is constant wherever it may be.

TABLE 1

| Unit | Abbreviation or Symbol | Metric Equivalent | Common Equivalent |
|---|---|---|---|
| *Length* | | | |
| 1 kilometer | km | 1,000 meters | 0.62 miles |
| 1 meter | m | 100 centimeters | 39.37 inches |
| 1 centimeter | cm | 1/100 meter | 0.39 inches |
| 1 millimeter | mm | 1/1,000 meter | 0.04 inches |
| 1 micron | u | 1/1,000 millimeter | |
| 1 angstrom unit | A | 1/10,000,000 millimeter | |
| *Area* | | | |
| 1 square meter | $m^2$ | 10,000 square centimeters | 10.76 square feet |
| 1 square centimeter | $cm^2$ | 1/10,000 square meter | 0.155 square inches |
| *Fluid volume or capacity* | | | |
| 1 liter (fluid) | l | 1,000 cubic centimeters | 61.02 cubic inches *or* 1.057 liquid quarts |
| 1 milliliter (fluid) *or* 1 cubic centimeter | ml / cc | 1/1,000 liter *or* 1/1,000,000 cubic meter | 0.061 cubic inches |
| *Mass and Weight* | | | |
| 1 kilogram | kg | 1,000 grams | 2.2 pounds |
| 1 gram | g or gm | 1/1,000 kilogram | 0.035 ounces |
| 1 milligram | mg | 1/1,000 gram | 0.015 grains |
| *Other* | | | |
| 1 parsec | (distance having a heliocentric parallax of one second) | | 3.26 light years *or* 206,265 times the radius of earth's orbit *or* approx. 19.2 trillion miles |
| 1 light-year | (Distance light travels in one year in a vacuum) | | 5,878,000,000,000 miles (approx. 5.9 trillion miles) |
| 1 astronomical unit A.U. | (Mean distance of sun from earth) | | 93,000,000 miles |

meters, square meters, etc.; and fluid volumes in cubic centimeters, etc. Table 1 sets forth the units, metric and otherwise, which are most commonly used in the world of science today, together with their equivalents. Throughout this book we will proceed to use these units even though some of them may be quite unfamiliar (see Table 1).

In this chapter we have discussed a number of factors which entered into the first scientific exploration of the laws of physics, both from the viewpoint of the early scientists and from the retrospective view of present-day knowledge. Of course those early physicists did not have retrospect to help them. But in spite of this, and in spite of all the flaws in their knowledge and impediments to their work that we recognize today, some eight hundred years ago a real and fundamental exploration of nature began. It proved to be incredibly fruitful. Armed with a few "self-evident" assumptions, a few ideas of how objects and forces might be described, and a few basic units for measurement, a handful of scientific pioneers with the first glimmerings of a scientific method of study to work with began to observe and measure certain physical characteristics of matter, and to study the manner in which certain forces seemed to affect and alter the ordinary motion of material objects.

Their "self-evident" assumptions, as it happened, were mostly wrong. They were not entirely sure how to describe what they were observing, nor even exactly what they were trying to describe. Their measurements were sorely limited by the built-in boundaries of five human senses. Yet in a few hundred years they built up an amazing groundwork of valid observation, and developed a truly awesome structure of natural law based on that observation. And in addition, all unknowing, they laid the foundation for a vast scientific revolution which is continuing on to this very day.

# CHAPTER 6

## *The Riddle of Falling Objects*

Matter and motion were the first order of the day. Galileo sought to explain the behavior of material objects in motion. So did Copernicus. So did Newton and a dozen others. To describe the enduring and predictable characteristics of matter in the universe, and to describe the various forces that might act on material objects and what happened when they did—that was the goal. The earliest scientific explorations of physics were aimed at finding universally valid descriptions of what would always happen to any kind of material object, anywhere in the universe, when it was acted upon by one or another kind of force.

We know today that there are a variety of forces that can act upon material objects. Sometimes the result of such action is a change in the physical state of the object in question: A cube of solid ice under pressure turns into liquid water. Sometimes a change in the shape of the object can occur: A rubber band stretches; an impacting bullet flattens. Sometimes the object itself may be unchanged, but a change occurs in the direction or manner in which it is moving (or not moving, as the case may be).

Many of the forces we are speaking of are part of our everyday observation. The action of human muscles to move an object from one place to another, perhaps with the aid of a lever or a pulley, is one such force. Friction is another force that influences the behavior of objects. Centrifugal and centripetal forces affect objects moving with angular speed, that is, rotating on an axis or following a curving path of motion. Of all the forces affecting all material objects in the universe, perhaps the most commonly recognized—and still the most mysterious—is the force of gravity. We could pick no better place than this to begin our exploration of the nature of matter and motion.

Gravity must certainly have been among the first natural forces that mankind became aware of at the dawn of history. Surely it was the first that man tried to explain, define, and put to practical use. Throughout the centuries gravity has remained the most constant and omnipresent of all forces men have had to grapple with in their everyday world. Gradually its characteristics were explored and defined and it was indeed put to practical use; yet even today physicists are as much at a loss as ever to explain *what exactly it is* or *why it exists in the first place.*

We can readily see why gravitational force would have been the most persistently observed and studied of all natural forces in human experience. Even Ug the caveman (Chapter 2) knew that if he let go of an object held in his hand, it would fall to the ground, and that it would hit his toe a good solid whack unless he got his toe out of the way. He soon learned that it didn't matter what the object was or how much it weighed (if he disregarded leaves sailing upward in the breeze); *any* object that he let go of would immediately begin moving, and always in the same direction: down. Indeed, Ug the caveman came to learn that any object that was not already on the ground and was not supported off the ground in some way would fall down in the same general fashion. He might even have been clever enough to notice that it was the wind—a gust of moving air which he could feel—that pushed the leaf upward. On a still day even the leaf would fall down.

Thousands of years probably passed before anyone seriously tried to measure *how fast* a released object fell under the influence of this mysterious force that pulled unsupported objects downward. But even Ug the caveman must have been puzzled by this force and wondered about it, if he was at all perceptive. He knew that *he* could force objects to move in one direction or another by pushing them, striking them, pulling them, or throwing them. He could usually even predict roughly how far a given object would move in what direction when he did the pushing. But this strange, invisible force that acted on objects that nobody had pushed was something else again. It was as if all the objects on earth were somehow attached to the ground by invisible cords constantly straining to pull them down. Ug found that he could "defy" this mysterious force, at least temporarily, by hurling a rock, for example, straight up into the air. But he also found that that rock would then inevitably rise more and more slowly, come to a stop in the air, and then turn around and come back down again with just as much force as he had thrown it upward. In the long run, gravity always won!

We can also see how early men might have made some perfectly common-sense assumptions about this invisible force—"obvious" assumptions that were considered self-evident, things which "everybody knew" were true. Efforts to explain *why* objects fell to the ground didn't get very far, but even to Ug it was obvious that objects must naturally *belong* on the ground; otherwise why would they keep falling down there? Millenniums later even the great Aristotle couldn't improve much on this "self-evident fact" that things fell to the ground because that was their "natural place."

Similarly, common sense said that heavy objects fell to the ground faster than lighter ones. After all, they were clearly heavier; therefore they *had* to fall faster. What was more, since no one had ever produced a vacuum in those days, casual observation actually "proved" that light objects fell more slowly than heavy ones: A leaf floated gently down, while a rock fell *thud!* And it stood to reason that the harder a rock was thrown in a horizontal direction, the longer time it would stay in the air before falling to the ground.

Nobody really knows if Galileo ever personally stood atop the tower of Pisa and dropped a ball of iron and a ball of wood simultaneously, as the legend goes, to see which struck the ground first. There are spoil-sports in modern physics who claim that he probably never did. But it seems certain that various early ordnance experts and artillery men had already begun to

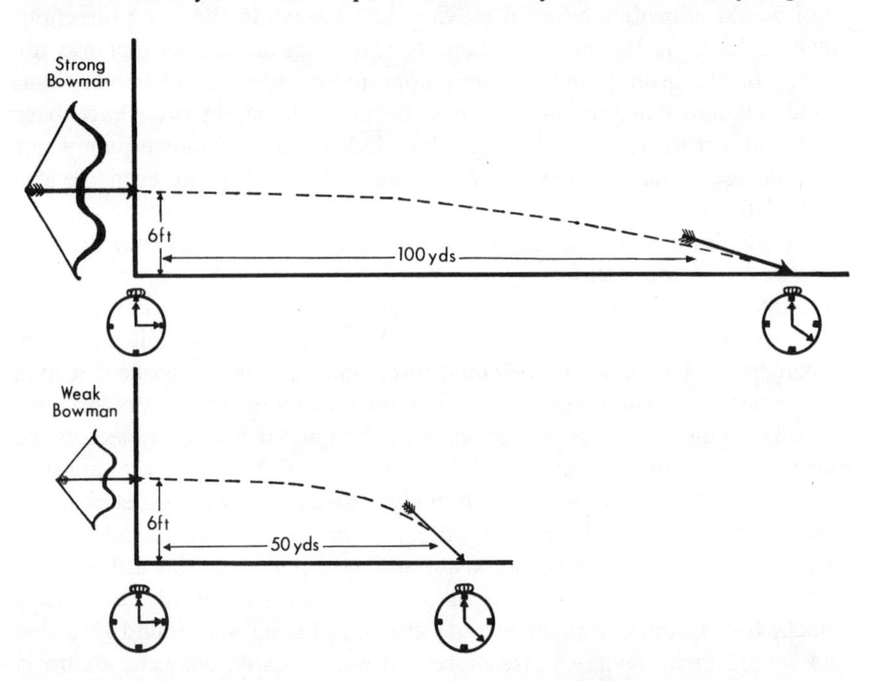


*Fig. 3* Fast arrow travels *twice as far* as slow arrow in the same time, but both arrows, shot simultaneously, will strike ground at the same instant.

poke holes in some of the "self-evident truths" about the force of gravity and its action on moving objects, long before Galileo ever appeared on the scene. Surely some early bowmen or catapultists must have realized that an arrow shot from a bow, or a stone from a catapult, began to be pulled earthward the very instant it started on its path through the air, and that two such objects thrown horizontally at the same instant would ultimately come to earth at *precisely* the same time, no matter how hard one might have been thrown compared to the other. The object thrown harder would *travel farther horizontally* before it hit ground than the one thrown more gently, but it would *remain in the air* no longer (see Figure 3).

Those early artillerymen must also have discovered that the only way to extend the ultimate range of a projectile, such as an arrow fired from a bow, was to tilt the arrow up *away* from the earth before it was released. Even then, there was a point of maximum gain: the range of the arrow could be extended by tilting its trajectory up to an angle of about 45 degrees; tilt it any higher and it began to *lose* range rather than gain it. Essentially, however, they must have come to realize that *any* object thrown, hurled, or tossed at or near the surface of the earth ultimately became a *falling* object, sooner or later, thanks to the inexorable pull of gravity.

Whether Galileo did or didn't stand atop the tower of Pisa is unimportant. What *was* important was that Galileo was the first to demonstrate two key facts about the behavior of falling objects: First, that except for the influence of air resistance, the *mass* of an object had nothing to do with the rate of speed at which it fell; in other words, that the downward acceleration of gravity was precisely the same for heavy bodies as it was for light bodies; and second, that the gravitational acceleration of falling bodies was always uniform and constant at any given place that it was measured.

As I. Bernard Cohen once pointed out: "In studying the science of the past, students very easily make the mistake of thinking that people who lived in earlier times were rather more stupid than they are now." It is easy, for instance, for us to consider Aristotle "stupid" for declaring that objects fell down because their "natural place" was on the ground, or to regard his contemporaries as "stupid" for accepting such an "explanation" of the force of gravity. Likewise, we may think Aristotle rather dull for teaching that objects of different weights would fall at different speeds without making any effort to prove such a contention. But in all fairness, we must remember that Aristotle inherited a philosophical point of view of the universe and a philosophical attitude toward exploring it, not a scientific attitude. It was not because of stupidity that Aristotle scorned experimentation; it was simply that no one had yet begun to realize how crucial experimentation might be to the support and proof of physical theories. Indeed, we might well have expected him to champion human reason and logic as far superior to the brute labor of experiment as a method of investigating nature. We might also have expected him, as a philosopher, to be far more concerned with *why* things happened than with trying to observe exactly *what* happened or *how*.

Doubtless there were many people before Galileo who had private reservations about certain of Aristotle's pronouncements. But Galileo in the seventeenth century was Aristotle's challenger. Galileo *was* an experimenter. From the results of his own and others' experiments, he set out to describe the *what* and *how* of gravity. But in trying to observe how fast objects fell and with what sort of acceleration, Galileo faced two problems that made any experimentation difficult. First, there was the problem of air resistance, which he recognized clearly as a factor that had confused the earlier Greeks. The second problem merely compounded the first: Objects falling freely in air fell so fast that it was all but impossible to make accurate measurements of *how* they fell over short distances; and when the distance of the fall was increased so as to provide longer time periods for measurement, the effect of air resistance became more and more exaggerated.

What about air resistance? We can see for ourselves how confusing this must have been to early observers of the forces of gravity simply by holding a glass marble and a Ping-pong ball side by side at the same height and then releasing them at the same time. Although the fall to the floor may be only four feet, there is no question but that *the marble strikes the floor sooner than the Ping-pong ball.* We can repeat this experiment innumerable times and the same thing will happen every time. Furthermore, if we stand on a table and increase the distance of the fall from four to eight feet, we will find the difference in the times the two objects reach the floor to be even more exaggerated.

From this simple experiment we could draw either of two different conclusions, one correct, the other incorrect. We might conclude that Aristotle was right: The heavier of the two objects (the marble) fell faster and therefore struck the floor sooner. We could argue that we had experimented and measured carefully and observed this behavior with our own eyes. On the other hand, we might conclude that both objects *would* have fallen at the same speed except that something or other impeded the fall of one object and not the other.

On the face of it, the first conclusion seems to make more sense—even though it is the wrong one. Indeed, it might seem to be the *inevitable* conclusion, if we did not know that the space between the point of release and the floor was not empty, but was actually filled with a very real substance—air—which could conceivably impede the downward fall of light objects. But suppose we didn't realize this, and accepted the first conclusion. Where would it lead us? It would lead us straight into trouble the moment we looked farther than the special circumstances of our original experiment.

Suppose, for example, that we repeated the same experiment with a strong draft blowing horizontally through the room. Would we observe the same results? Of course not; we would then see the Ping-pong ball fall to the floor *at an angle* instead of straight down. It might reach the floor halfway across the room if the breeze was strong enough! Again, suppose we used identical-sized marbles, but with one made of glass and one of lead. Now, oddly enough, we might not detect *any* difference in time of fall, either from four feet or from eight, even though lead is just as much heavier than glass as the glass marble was heavier than the Ping-pong ball. Well, could it be the fact that the Ping-pong ball was hollow that made the difference? Perhaps we substitute a tennis ball, which is clearly heavier than a marble, but hollow like a Ping-pong ball, and try again. And again find that the marble and the tennis ball strike the ground at virtually the same time.

Thus, if we had accepted the first (and seemingly inevitable) conclusion —that heavier objects fall faster than lighter ones—further experiments would soon force us to conclude that gravity made no consistent sense whatever. Those experiments would seem to show that sometimes heavier objects fall to the ground faster than lighter objects, but that sometimes they don't—in other words, that gravity is a completely quixotic force that behaves one way one time and another way another. We would even have to conclude that it sometimes pulled objects straight down and sometimes pulled them sideways at an angle. If that first conclusion were correct, it would have to mean that gravity was a totally whimsical and inconsistent force, subject to physical rules that kept changing with the conditions.

One of Galileo's greatest contributions to science was that he could not and would not believe that the laws of nature were whimsical, inconsistent, and ever-changing. Nature, he insisted, was consistent if nothing else. The same laws that governed the behavior of *one* free-falling object had to govern the behavior of *any* free-falling object anywhere in the universe. The laws of nature, he was convinced, were simple, logical, and reasonable; when something was observed to occur that made a natural law *seem* whimsical or inconsistent, it was only as a result of some unsuspected factor which the observer was missing, and thus was failing to take into account.

Galileo, for example, realized that *something* other than a consistent, unchanging, downward gravitational force was influencing the behavior of various falling objects—but what was the missing factor? Obviously, objects falling through the air met with a certain amount of resistance just from the air itself; they literally had to *push aside* the air through which they were falling. Heavy, compact objects managed this with little difficulty, and thus showed little effect from air resistance; but light, feathery objects had more trouble shouldering their way through the air and were noticeably retarded in falling.

So air resistance had to be ruled out somehow if the effect of gravity, free of any other influence, was to be tested and measured. But how could this be done? Galileo realized that the *only* way to *really test and measure* the behavior of falling objects would be to measure their fall in some way in which air resistance could have no effect whatever, because they were falling *not through air, but through a perfect vacuum*. Unfortunately, though Galileo could *imagine* a perfect vacuum (i.e., a container in which all air had been pumped out, so that it contained *no air at all*), he had no way actually to *create* such unheard-of conditions.

So what could Galileo, or *any* scientist, have done? It would have been easy to quit and go home. Instead, Galileo found an ingenious way to dodge the problem and come up with the correct answer in spite of it. The device he used, now a time-honored method in the study of all sorts of physical phenomena, was the device we know today as the "thought experiment."

Galileo couldn't create a vacuum in which objects of various weights could be observed falling, but he could *imagine* such conditions. By doing so, and then by imagining how objects would fall under such "ideal" (though unachievable) conditions, Galileo came up with several shrewd guesses:

First, he guessed that all falling bodies would fall in the same way, and if released together, would fall together and reach the ground together, regardless of their respective weights.

Second, he guessed that these objects would fall with *constant acceleration*—that is, that they would increase in downward speed steadily, with the same increase in speed each successive second until they struck the ground.

Now admittedly these were guesses about what would happen under certain ideal circumstances. But armed with these guesses, Galileo set about diligently testing them in real experiments in which he made allowances for the complications that would have to arise because of air resistance. He found that the results of these experiments, once the necessary allowances were made, *coincided exactly* with the results he had imagined in his thought-experiment.

In other experiments he tried in other ways to minimize the effect of air resistance—by using large, dense objects, for example. He tried to improve the accuracy of his time measurements by using counterweights to slow down the velocity with which objects fell. Again and again, his observations coincided with the results he expected from his thought experiment.

Now this all sounds very neat. But was this real experimental proof that Galileo's guesses were right? Of course not. Galileo didn't have actual experimental proof. Suggestive supporting evidence, yes. Proof, no. And precisely because he *didn't* have actual proof at a time of growing skepticism toward taking things for granted, other scientists set about to *obtain* actual experimental proof. Galileo had no vacuum with which to test out the "ideal case" of his thought experiment. He very probably doubted that such a vacuum could ever be achieved. But some years later another Italian scientist named Torricelli discovered that a vacuum *could* be created. Torricelli took a long glass tube sealed at one end and filled with mercury, and then inverted it open end down into a pan of mercury. He discovered that the mercury in the glass tube fell to a certain level and then stabilized, leaving an empty space at the sealed top end of the tube. Torricelli reasoned that since that "empty" space up there in the end of the tube had originally been filled with mercury, and since no air could have leaked in when the tube was inverted, that space *really was* empty—it had to be a vacuum!

Today, of course, we know that Torricelli's vacuum was not 100 per cent perfect. That "empty" space actually did contain a few atoms of vaporized mercury, perhaps a few molecules of oxygen, nitrogen, or carbon dioxide from air that had been entrapped in the mercury and was drawn out by the pressure of the vacuum. But Torricelli's vacuum was so very close to the perfect vacuum needed for Galileo's "ideal case" that the imperfection was totally negligible—and Galileo's guesses were substantiated. Later, air pumps provided more convenient vacuums in tall glass pipes. The crucial experimental *proof* of Galileo's conclusions finally came a century later when Sir Isaac Newton released a bit of goose down and a gold coin simultaneously at the top of an evacuated glass pipe and found that even this pair of objects fell side by side, with constant downward acceleration, and struck the bottom of the tube at the same time.

We have gone into this seemingly simple and insignificant matter at such length for two very important reasons. First, it is an excellent example of how exceedingly difficult it was for those early physicists, groping in the dark, really to pin down anything so that they could say, *"This is true and we can prove it."* Galileo saw the problem, guessed the answer, paved the way—yet brilliant men searched another *hundred years* to find actual experimental proof.

But more important, we can see in this example how answers had to be torn away from nature by hook or crook, guile, cleverness, and ingenuity. Many nonscientists find themselves very uneasy with such scientific procedures as Galileo's "thought experiments" and "experimental proofs" based upon "ideal conditions" that don't actually exist. How can anyone really *prove* anything this way? It is all very nice for a physicist to guess that in an ideal case two objects will fall together in a vacuum, regardless of their weight, and with constant acceleration. It's even nicer to be able to prove this by experimental observation later, once you can create a vacuum in a tube to test it out. But where does this really get us? The fact is that our world is *not* a vacuum. It is a world covered with a blanket of air, and in this *real* world, objects of different weights (or of different sizes or shapes) do *not* fall together. Leaves flutter down in the breeze, while apples fall *thud!*

From the nonscientist's point of view, this objection is perfectly valid. We do not live in a world of ideal conditions, and it is in our *real* world that we have to cope with the behavior of things. What, in fact, we come up with is not one but two sets of rules: the rules that govern the *ideal* case, and the rules that govern the *real* case. In the ideal case, two objects do fall together. They *do* increase in speed with constant acceleration so that the longer they fall the faster they are moving and the harder they hit bottom. In the real case, however, not only do objects of different weights, sizes, or shapes fall at different speeds because of air resistance, but *any object* that falls freely meets with air resistance. The more it accelerates and the faster it falls, the greater the air resistance it meets and the greater the resultant "friction lag."

Indeed, if it were possible to drop a marble out of an airplane at 30,000 feet and then to measure its pattern of acceleration, we would find that it accelerated downward at a constant rate per second only for the first part of the fall, up where the air is thin. As it fell into denser layers of atmosphere, its acceleration would increase more and more slowly until at a certain point it wouldn't increase its acceleration any more at all; it would just continue to fall at precisely the same velocity, without further acceleration, until it struck the ground. Furthermore, in this *real* world, something else would happen that Galileo never dreamed of in his "ideal" case: *The falling marble would heat up* as it shouldered its way down against air resistance. Much-faster-moving objects, such as meteorites entering the earth's atmosphere, actually heat to incandescence and vaporize before striking the ground. We can calculate that the marble falling from 30,000 feet would not become *that* hot before it reached ground, but we might have quite a job actually documenting experimentally what *did* happen to it under these "real-world" conditions!

Perhaps we could study these problems more conveniently by dropping the marble down through a medium that offers even more resistance than air: measuring its fall down through a long tube of water. This, at least, would be a manageable experiment. So suppose we release an iron marble at the top of a thirty-foot tube of water and measure its acceleration as it falls. What would we see? We would see precisely the same thing happen as we had *guessed* would happen to the marble in our thought-experiment. The marble's downward acceleration would *not* be constant; it would decrease second by second until the marble was falling down the water column with a uniform speed, and it would then maintain that uniform speed until it reached bottom. Furthermore, if we had delicate enough instruments for measuring temperature, we would discover that the temperature of the whole system of marble-falling-through-water would have *increased* a measurable amount. In this case, however, it would not be the ball alone that increased in temperature, but ball, water, tube, and all!

We know today that each of these curious observations can be explained completely and accurately in terms of our knowledge of modern physics. We know that air resistance (or water resistance, as the case may be) are in fact forms of *friction*—one of the natural forces we have yet to discuss. We know that any falling object at the beginning of its fall possesses a certain form of energy—"potential energy"—which is steadily converted into another form of energy—"kinetic energy"—throughout the period of its fall. We also know that as the falling object encounters the opposing resistance of friction during its fall, part of its accumulating kinetic energy is converted by that opposing frictional force into still another form of energy—*heat.* We further know that what happens to the heat energy depends on the heat-conducting or heat-insulating qualities of all materials concerned: iron ball, water, tube, etc.

If it seems that we are getting farther and farther away from our discussion of the force of gravity and its characteristics as we examine this "real" case, indeed we are. We are getting into a whole complex snarl of considerations that are only peripherally related to gravity and its effects. And it doesn't matter that all these other physical considerations are beside the point; *we can't escape them* in the "real" world. The only possible way we can extricate gravitational force alone from this mess and examine it by itself is by imagining "ideal" situations which we *know* are flatly impossible, in hopes of *isolating* the thing we are trying to study from a wildly confusing array of other things we really don't want to have to deal with right then.

Of course, such isolation of phenomena in "ideal" cases or "thought experiments" may not always be practical in terms of the "real" world, but it is a perfectly legitimate procedure if we know what we are doing, and above all, *it allows us to learn things we couldn't learn otherwise.* What is more, sometimes those "ideal" and "imaginary" conditions are not so impractical as they seem. Maybe Galileo couldn't imagine a vacuum existing; but today we are sending real human beings into a part of the universe where relative vacuum does indeed exist: the space beyond earth's atmosphere that we have been so busily exploring lately! Try to tell an astronaut that there are no practical applications of the laws of gravitation under "ideal" conditions of free fall, and see what he says as he frantically tries to retrieve a pencil that has slipped from his fingers and is merrily floating around the inside of his space capsule. Try to tell the first man who breaks his anchor cable during a "space walk" outside his ship that the laws of gravity, the laws of motion, and the laws of inertia have no "real" meaning! You might find him eagerly offering to trade places with you—if he and his teammates weren't quite so busy trying to apply those very laws to grapple him back aboard ship before he embarks on a long, cold, and endless space walk to nowhere.

In fact, we will see throughout this book more and more areas in which the "ideal case" of the scientist and the "real world" of the man in the street impinge, and we will find more and more reasons why each one of us, scientist or not, needs a clear understanding of basic laws of physics, however "impractical" or "useless" they may seem. The time when these laws can be ignored with impunity is passing fast. Tomorrow this will be a body of knowledge each one of us will need—perhaps even for survival.

Galileo and the men who followed after him clearly recognized the part that air resistance played in obscuring what the real characteristics of gravitational force were. They got around this first by setting up imaginary thought experiments involving ideal conditions, then by actually creating those ideal conditions as best they could, and then making the necessary corrections in their observations and calculations; and they found that *it worked*. They ended up with answers they couldn't have obtained any other way. Gradually the use of thought experiments involving ideal conditions became one of the most widely used and most fruitful of all techniques of scientific investigation.

We will use this technique ourselves again and again throughout this book to make various points. We need not feel the least nervous about it. If such thought experiments are used with care and integrity they do not in any way invalidate actual experimental data that may be obtained, nor the conclusions drawn from them. On the contrary, they may well *show the direction* that actual experiment must take to produce any usable results whatever. Virtually all natural physical phenomena are not simple occurrences, but are complex combinations of many forces acting at the same time to produce a given result. The scientist is constantly laboring to try to separate one such force from another so that he can study each force separately instead of trying (usually in vain) to analyze everything that is happening all at once.

Of course, once various laws of nature are known, then it may be possible to examine two or more forces acting in different ways at the same time, and successfully predict the net result on the basis of what is already known about each of these forces. This is particularly true today when modern computers can handle the mathematical leg-work so simply and quickly. But in searching out a natural law that is *not* yet known or understood the scientist always seeks to study just one factor or force at a time. Thus even though Galileo recognized that air resistance was a force acting to *modify* the behavior of objects falling under the force of gravity, he had to sidestep this "extraneous" force temporarily in order to make sense of the phenomenon he was trying to examine.

The other major problem in studying gravitational force was the simple fact that objects fell so fast that it was all but impossible to measure time intervals accurately, or actually to observe or measure the state of things at various points in the course of an object's fall. Galileo and later physicists got around this problem by dreaming up a number of ingenious devices for *slowing down* the movement of objects acting under the force of gravity—by "diluting" gravitational force, so to speak. One way of doing this was to study the movement of objects sliding or rolling down a ramp, rather than trying to measure their behavior in free fall. Galileo had already guessed that Aristotle was wrong, and that any two objects released simultaneously would fall together (under ideal conditions) at the same velocity and with the same acceleration, regardless of their comparative weights. He further guessed that the acceleration of falling objects (that is, the increase in their downward speed per unit of time) would remain constant second after second, or minute after minute for as long as they continued falling. Galileo found times and speeds hard to measure when objects were allowed to fall freely; but by rolling polished metal balls down a parchment-coated trough held at an angle of 20 degrees from the horizontal, he found that the *time required* in the downward movement of the ball from the top of the trough to the ground was greatly lengthened (and thus more easily measured) because the ball also had to roll horizontally down the length of the trough. Indeed, if the trough were held at a *very* slight incline to the ground, and if it were long enough, the steel ball might take as much as 15 or 20 seconds to move the vertical distance of a single inch from the top of the trough to the ground (see Fig. 4).

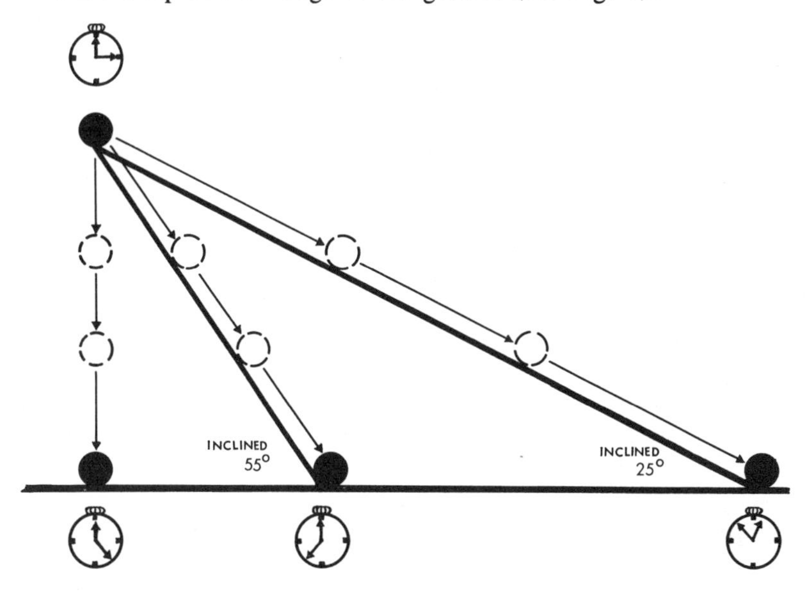


*Fig. 4* Use of an inclined trough to "dilute gravity" by slowing down the motion of a ball falling under the force of gravity. The shallower the incline, the more the ball's rate of motion is slowed down.

But were the natural laws governing the movement of a ball rolling down an inclined plane the same as those ruling its behavior falling straight down? Galileo didn't know; but he *guessed*—correctly—that exactly the same rules would apply. He then proceeded with experiments, rolling many different-sized balls of widely differing weights down identical troughs parallel to each other, and found (1) that they always rolled side by side, regardless of their weights; and (2) that they always accelerated at a constant rate and reached the ground at the same time, with the same final velocity. By raising the ends of the trough higher and higher, he found that the ratio of time required per unit of distance traveled always remained the same: The same rules seemed to apply no matter to what angle the troughs were tilted. Tilt the troughs almost vertical and the balls would roll faster and reach the ground sooner—but *they would still roll together* and their final velocities would still be in the same proportion to the vertical distance they had traveled.

Obviously, it didn't matter a whit whether the motion being studied was that of an object falling freely, or rolling down an inclined plane: The rules were the same, and the results were the same. In fact, the tilted troughs were nothing more than a device to "dilute" or *slow down* the effect of gravity on a falling object, much as a motion picture can be slowed down by varying the speed the machine is running. Nothing basic is changed; the man *seems* to leap over the barrel much more slowly than before, but the distance and pattern of his leap, from beginning to end, remain in exact proportion to the time that the leap seems to take!

Other means were eventually devised to dilute the effect of gravity. The movement of a pendulum, for example, was one way of studying the behavior of a falling body in "slower motion" than if it were falling freely to the ground. Again, an ingenious device called "the Atwood machine" was invented—essentially a cord going over a pulley with weights on either end, so that the weight on one end counterbalanced the weight on the other. When a weight was released from a platform at the top of a marked scale on the Atwood machine and allowed to fall, its fall was not "free," but was impeded by the counterforce of a slightly smaller weight that had to be drawn up the other side of the scale. Even so, the same gravitational rules seemed to apply (in slowed-down proportion) to weights on the Atwood machine as to objects in free fall.

## GRAVITATIONAL PULL VS. THE "OBJECTNESS" OF THINGS

Thus, by finding ways of diluting, or *modifying the effect* of, gravitational force on freely falling objects, early observers began to learn a great deal about the effects of gravity. As a fringe benefit, they also discovered that the effects of gravity on objects were essentially the same whether the object was suddenly released to fall freely, straight down to the ground, or whether the fall was impeded by counterweights, inclined planes, or whatnot.

But these early observations of gravity revealed something else of very great importance: that there was a fundamental difference between the measurable downward gravitational pull on a given object—what we call its *weight;* and the basic, irreducible amount of matter that the object contained—its *mass.* We raised this question of the weight versus the mass of an object briefly back in Chapter 5 when we spoke of the trouble that arose in trying to use the weight of an object as a reliable identifying characteristic. Now we can see more clearly what the problem was when we consider some actual observations recorded by those early scientists.

They discovered, as we have seen, that the gravitational pull on a given object could vary, depending upon where measurements were being taken. For example, an object falls to the ground a tiny bit more slowly at the top of a high mountain than if its fall is measured at sea level. The same object even falls more slowly at sea level somewhere on the equator than it would at sea level in northerly latitudes. Now, this observation was quite unexpected and perplexing. Scientists had always assumed that a given object would have the same weight (i.e., be pulled downward by gravity by the same amount) no matter *where* the object was located—but now simple observation proved this wrong! An object suspended, say, from the same spring balance in two different places might register two different weights.

Simple observation also showed that another characteristic of the object, similar to its weight but not quite the same, remained *unchanged* wherever the object might be. If the object were placed on a rough horizontal surface, *exactly the same* amount of force would be required to drag it horizontally across the board for a given distance, *no matter where the experiment might be done.* In short, early physicists discovered that an object's weight, always assumed to be a constant, was in fact variable, while another quality of the object related to its weight (but not the same) *did* remain constant and did *not* change from place to place.

Obviously this second quality would be more useful in describing the object than its weight—but what *was* it? Early physicists were hard put to say. An object's weight was nothing more than a measurement of the amount of downward tug exerted upon it by the invisible but apparently omnipresent force of gravity. Why this amount of downward tug should vary from place to place for the same object was a mystery, but vary it did. The quality of the object which did *not* seem to change, even when its weight did, was its resistance to being dragged horizontally and *that* quality could be counted upon to remain constant anywhere—certainly a more useful quality to measure than one that kept shifting around from place to place!

That constant quality of a material object, its resistance to being dragged horizontally which was so similar to its weight, yet not quite the same, had no name. Even today it is a difficult concept for most of us (including fledgling physicists!) to grasp. Actually, we might call that quality anything we choose. We could speak of it, for instance, as an object's "basic objectness"—in other words, the *basic amount* of the object that was there. We could demonstrate that the "objectness" of a lump of iron—the actual amount of iron that was there in the lump—would remain the same anywhere we might measure it, be it at sea level, on a mountain top, or on the surface of the moon, even though the *weight* of that lump of iron might be 10 per cent less on a mountain top than at sea level, and 85 per cent less on the moon! As long as none of the iron was allowed to rust away, or be dissolved by acids, or something of that sort, the "objectness" of the lump would always remain constant any place.

Early physicists grappling with this concept chose a different term than "objectness" to describe it. They spoke of this constant, unchanging quality of an object as its "mass" and defined mass as "the amount of matter present in an object." That amount of matter might *weigh* more or less in different places; where gravitational force was strong, it would weigh more, and where gravitational force was weak it would weigh less—but the amount of matter would remain the same. Later they even speculated that the mass of one object could be compared to the mass of another object made of the same material (at least theoretically) by counting the number of atoms in the two objects and comparing the results. Thus, they speculated, two lumps of pure iron with *exactly* the same number of iron atoms in them ought to have *exactly* equivalent mass.

Today the modern physicist would agree that this speculation would probably be absolutely true and universally applicable anywhere in the universe, so long as both lumps of iron happened to be at a temperature of absolute zero, and were completely at rest. We now know that the mass of an object *does* vary—to an infinitesimal degree—with changes in temperature, although we would have quite a job measuring the change. We also know that an object's mass increases, if only infinitesimally, the moment it is set in motion. Thus if one of the lumps of iron is moving, even very slowly, while the other remains at rest, the mass of the moving lump will be *slightly* greater, even though the two lumps contain exactly the same number of iron atoms. And if one of the objects is moving very fast indeed—close to the speed of light—its mass will increase enormously, although it *still* has the same number of atoms as the object at rest.

As we will see later, these become exceedingly important considerations under certain circumstances, but from our standpoint here, we can ignore them as quibbles. In our everyday world, the universe the early physicists were exploring, we can regard the "objectness" of an object—its *mass* as opposed to its *weight*—as a constant unchanging measurement of the amount of matter in that object wherever it may be on this planet or any other. But we recognize that the tug of gravity upon the mass of an object (i.e., the object's weight) may be quite different at different places on earth, and still more different on other planets or in orbit in space, where gravitational forces may be stronger or weaker than on earth. Modern space science has fortunately helped make this concept clear to most nonscientists. We can readily grasp that there is obviously just as *much* of an astronaut present when he is walking about on the surface of the moon as there was when he was entering his space capsule at Cape Kennedy. Yet the moon's gravitational pull upon him is only one-sixth that of the earth's, so that his 200-pound weight on earth becomes a mere 34 pounds.

Thus, Galileo and his contemporaries made many observations of the force of gravity acting upon falling objects, and evolved the very useful concept of an object's mass as distinct from its weight. From his observations, Galileo came up with a number of general statements describing how gravity affected objects, whether they were falling freely, swinging like a pendulum from a pivot point, or rolling down an inclined plane. These "general laws of falling objects" were a far cry from the great law of universal gravitation that Isaac Newton would later derive—but at least they were a respectable start. What these laws of Galileo really were, in effect, were the laws describing *local gravitational effects* on objects allowed to fall under special circumstances: objects, for example, suspended close to the earth's surface and then allowed to fall.

We will see later how Newton took these "local" laws and extended them to the behavior of all objects anywhere in the universe. But first, we must consider certain other forces affecting the motion of objects that drew the attention of Galileo and the other early physicists. Here even more perplexing riddles were encountered—in particular, the riddles of *friction* and of *inertia.*

# CHAPTER 7

## *The Riddles of Friction and Inertia*

It was a great step forward when early physicists finally pinned down a few general rules that reliably described the behavior of material objects in certain familiar patterns of motion. These rules could be used, for example, to predict how a solid object such as a rock would move if it were allowed to fall freely from the hand or to roll down an inclined plane (assuming, of course, that these things were occurring somewhere near the surface of the earth). With some minor modifications these same rules helped to describe other, related forms of motion—the movement of a swinging pendulum, for instance, or the behavior of a rock tossed into the air. When they were applied at various geographical locations on the earth's surface (on a mountain top, at sea level, etc.) these rules of motion furthermore demonstrated why a distinction had to be made between an object's *weight* (a measure of the gravitational force acting on it at a given moment, which could vary greatly depending on where it was weighed) and the object's *mass* (a measure of the quantity of matter it contained, always the same under ordinary circumstances of human experience).

We must remember that these rules made no attempt to explain *what* gravitational force was, nor *why* it existed. They merely *described how gravitational force had been observed to act.* They showed gravity's effect on the motion of material objects at or near the earth's surface, and some findings were a little surprising. For example, it was found that the earth's gravity exerts a continuous force upon objects *whether they are in motion or not.* An iron ball held in your outstretched hand will not fall to the ground as long as your hand continues to support it. It remains "at rest"—yet gravity is constantly acting upon it just the same, tugging it groundward, so to speak. You can even feel the downward pull. But what, then, keeps the ball from falling? Obviously, your hand must actually be *pushing the ball upward,* against gravity, with a force exactly equal to the gravitational force pulling it downward. Similarly, when a man stands "at rest" on the ground, the surface of the earth must be pushing upward against his feet with a force equal to the gravitational force seeking to pull him down toward the center of the earth.* Of course, in this context the net result in either case is a state of equilibrium—neither ball nor man is set into motion by either force acting on it—and this whole discussion may seem rather pointless. But as we will very soon see, this notion of force and counterforce always existing together was later recognized as an extremely important concept in physics, and we will be encountering it repeatedly as we go along.

Gravity, however, was not the only force observed by early physicists to act upon objects and influence their behavior, nor was the falling motion of an object as a result of gravity the only form of motion that was studied. Indeed, a number of forces quite unrelated to gravity were observed (or at least postulated) and investigated, even as early as the Greeks. Among these were two very puzzling kinds of force which seemed to exist side by side quite universally, apparently closely related to each other, but acting upon objects in directly opposite ways. One of these forces was a *resistant force* that was observed to make moving objects slow down—or to obstruct their movement. The other seemed to be a *propelling force* that kept moving objects moving in their paths in direct opposition to the resistant force. Between the two, the behavior of these opposing forces seemed extremely confusing and contradictory to early investigators.

Today we know that only one of these two puzzling "forces" actually exists; the resistant force, which we know as "friction." But we can easily see why early physicists were trapped into assuming that a "propelling force" also had to exist to keep a moving object moving. Consider the problem as they saw it: A light object, such as a feather or a light piece of wood, falls to the ground much more slowly when released than a dense object such as a lead ball. Why? Even the Greeks recognized that some resistant force in the air, some "air resistance," impeded the fall of lighter objects, while more dense objects were affected by it less, if at all. They could also see that another force acted to pull the feather or piece of wood down toward the ground in spite of the air resistance: the force we know as gravity. As long as the gravitational force pulling the object downward

* By the physicist's logic, the weight of an object (as distinguished from its mass) is actually *defined* as the precise "equal-but-opposite" upward force necessary perfectly to counterbalance the downward gravitational force acting on the object, thus preventing it from accelerating downward. To confuse things further, the physicist will argue that neither the ball nor the man under discussion is really "at rest" at all, but in each case is "in motion" as a result of the two external forces acting upon them—the downward gravitational force and the upward force acting in the opposite direction. However, since these forces are exactly equal in magnitude and exactly opposite in direction, the physicist would speak of the resultant "motion" of the ball or the man as "static motion." Unfortunately, such technical distinctions in terminology (i.e., "motionless motion") sound like pure double talk to most laymen; in this book we will settle for more comprehensible (if less sophisticated) terms.

was greater than the air resistance impeding its fall, the object would eventually reach the ground.

But suppose the same light piece of wood were thrown horizontally through the air. Here again air resistance would impede its flight—but now no force of gravity is acting to pull it or push it along in a horizontal direction. What then keeps it moving long after it has left the hand? What keeps an arrow moving against air resistance long after it leaves the bow? To the Greeks it seemed self-evident that *some* sort of propelling force had to be pushing or pulling an object through the air as long as it continued moving. They even devised some colorful (if spurious) explanations of how this "propelling force" might be generated. They speculated, for example, that the air that was pushed out of the way by the tip of a flying arrow flowed back along the shaft and filled a vacuum that the arrow left behind it, thus shoving the arrow onward from behind. We know now that such a thing would be tantamount to lifting ourselves by our own bootstraps—quite impossible—but it seemed a plausible idea to the Greeks. And it seemed even more plausible if one compared the "air resistance" impeding an arrow in its flight with the "ground resistance" that was encountered when an object was pushed along the ground. If a sculptor wanted to move a block of marble from one side of a hall to the other he would have to push it or drag it forcibly every inch of the way; the moment he stopped pushing, the block would stop moving. So why shouldn't an arrow flying through the air require a continuing force to keep it moving against "air resistance" just as much as the marble required a continuing force to keep it moving against "ground resistance"?

The Greeks made many casual observations of the way such resistant forces acted to slow down moving objects, whether they were moving in air, on land, or in the water. They also speculated about the nature of the "propelling forces" they thought necessary to keep such objects moving. It remained for Galileo and later physicists of the seventeenth and eighteenth centuries to define just what these "resistant forces" were and how they behaved—and to show that the mysterious "propelling force" of the Greeks simply didn't exist at all.

## THE FORCES OF RESISTANCE

Today we know that the "resistant force" that slows a feather's fall or makes a chunk of marble difficult to push across the floor is a familiar and universal force that arises any time that an object moves in physical contact with another object. We call this resistance *frictional force,* or more simply, just *friction.* The "air resistance" which impedes the fall of a feather is nothing more than the frictional force created when the air molecules surrounding the feather literally rub against it as it falls—the force of gravity has to *pull the feather down* through the air. A much greater frictional force has to be overcome when the sculptor (by sheer muscle power!) pushes his heavy block of marble across the rough stone floor. In fact, an obstructing force is actually created the moment he even *attempts* to slide the block, even if he can't make it budge; that force reaches its maximum at the point that the block begins to move. But the block would not even *start* to move unless the sculptor could apply enough force to *exceed* the maximum frictional force obstructing its movement, at least by a little bit!

This obstructing force, called friction, tends to impede the motion of any object rubbing against another. But the amount of friction arising in a given case depends upon many different factors. The rougher the surfaces of the two objects that are in contact, for example, the greater the friction when one of the objects tries to slide past the other. Similarly, the more tightly the two surfaces are pressed together, the greater the frictional forces created when one moves relative to the other—whether the surfaces are pressed together by gravitational force (as when a brick is "pressed" against the sidewalk by gravity) or by some other force (such as a clamp pressing two pieces of wood together, for instance). Oddly enough, frictional force does *not* depend upon the surface areas in contact; just as much force would be required to overcome frictional resistance and slide a brick along the sidewalk if the brick were lying on its narrow edge as if it were lying on its broad side. On the other hand, if the narrow edge of the brick were ground smooth from sliding along the sidewalk, then less frictional resistance would arise when it was moved than if it were lying on a rougher edge.

Equally important, the amount of friction present in a given case will vary greatly depending on whether the objects in contact are in the solid, the liquid, or the gaseous state. In a later chapter we will discuss a number of commonly occurring "states" or physical forms of matter in more detail, and examine more closely just how one state differs from another. For now it is sufficient to point out that in solid matter the component atoms and molecules are bound together in a more or less rigid crystalline structure, so that when two solid surfaces grate against each other each holds its shape stubbornly, and the friction created is comparatively great. In a liquid the molecules are less rigidly bound and can usualiy "give" a little. Thus, a boat can move through water with far less friction than if it were dragged across the ground. As for a gas, each molecule is quite free of all the others and can easily be "pushed aside," so that the frictional force arising when a solid object "slides past" matter in a gaseous state is generally very small indeed. Thus an arrow flying through the air is "impeded" very little by the friction of the air molecules rushing past it—but *some* friction still arises. Meteorites plunging into our atmosphere from outer space, for instance, encounter enough frictional resistance to heat them to incandescence, and we see them at night as brilliant "shooting stars."

Friction, in short, is a resisting force, surprisingly ubiquitous, which tends to obstruct or impede the motion of objects in relation to one another, with the *amount* of frictional force present in any given case depending heavily on certain specific physical characteristics of the objects involved. But what about the "propelling force" necessary to keep objects moving in spite of frictional resistance? In order to see more clearly exactly what effect friction has upon the motion of objects, and to separate fact from fallacy in regard to the mysterious "propelling force" of the Greeks, a simple imaginary experiment will be helpful.

Suppose we had a device, much like the spring-operated shooting mechanism on a pinball machine, with which we could give a carefully calibrated push to any moving object we chose (so long as it was of manageable size). Suppose we satisfied ourselves, after repeated tests, that this device would always move our test object in a certain specified direction with a certain fixed force for a uniform period of time, say for one-tenth of a second. In other words, our "shooter" would deliver an *impulse* to the test object that would be exactly the same every time it was applied.

Then imagine that we use this "shooter" to deliver an impulse to a small block of wood, shoving it horizontally across a rough table top. What would be likely to happen? If the impulse were forceful enough, we would expect the block to continue moving across the table for a short distance after the impulse was over before it came to a stop.

Next, suppose we used the same shooter to move the same block of wood across a highly polished counter, and then across the slippery surface of a frozen pond. We would find the block sliding farther on the polished counter than on the rough table top, and farther still on the frozen pond before it came to rest. But why? In each case the shooter delivers precisely the same impulse to the same block of wood; why then does the block move perhaps ten inches in the first case, ten feet in the second, and ten *yards* in the third?

Obviously what is happening has nothing to do with the wooden block alone, nor with the shooting mechanism, nor with the impulse it delivers. The behavior of the block in each case depends on the interaction between the block and the particular surface upon which it is sliding at the time. In each case the moving block is met by a resisting force—friction—which tends to slow it down. Indeed, with such an imaginary setup as this, we have attempted to eliminate all factors that might affect the movement of the block *except* frictional resistance. We see that when the block is sliding and scraping across a rough wooden surface, friction is great and the block slows down quickly. Much less frictional force impedes the block's movement when it slides across the polished counter top, even though solid is still rubbing on solid. As for the icy surface of the pond, if we had keen microscopic vision and looked very closely we would see that the block was not really sliding on ice at all, but on a thin film of water melted from the ice under the block. In such a case frictional resistance is far less than in a case in which a solid was sliding on a solid. (We can also see why a lubricant such as oil or grease can often reduce friction between solid surfaces so effectively. Such a lubricant not only fills in the cracks and crevices on both surfaces and thus smooths out the roughness, but also forms a thin slippery film of liquid between the surfaces in contact.)

This kind of experiment tells us a good deal about the nature of friction —the resistant force that slows down the movement of a moving object. We would expect friction to arise any time we moved an object relative to another (or even attempted to move it). We might even be able to guess quite accurately *how much* friction would arise in a given case. If we had actually measured the velocity of our wooden block from moment to moment while it was still moving in each of the three test shots of our experiment, we would have found a constant decrease in its velocity every second after the impulse was over, but the *rate* of decrease would vary according to the amount of frictional resistance. Thus the wooden block sliding on rough wood would slow down and come to a stop very abruptly. When sliding on the polished counter top, the velocity of the block would decrease at a more leisurely constant rate. When the block was sliding on ice its velocity would decrease comparatively little with each passing second.

But what is the "propelling force" that keeps pulling the moving block along against the resistance of friction? We have seen that when friction is reduced (as in the case of the block sliding on ice) the block continues moving for a remarkably long distance before it stops. What, then, keeps it moving? Or to ask a more pointed question: What would the block do if there were *no frictional resistance impeding it at all*? Common sense tells us that if any kind of "propelling force" continued to push the moving block and no resistant force of any sort acted to impede it, the block would *continue to accelerate* indefinitely after the impulse had ended. And yet no one has ever seen any such thing happen.

## THE FORCE THAT WASN'T THERE

Of course, in Galileo's time physicists could only imagine conditions in which a moving object encountered no friction at all. The best they could do was to set up experiments under *relatively* friction-free conditions, and then attempt to extend their findings to conditions which to them were totally imaginary and unobtainable: conditions of no friction whatever. We approached friction-free conditions in our own imaginary experiment when we had our shooter push a block of wood across the glassy surface of a frozen pond, and found that it continued to move for a comparatively long distance—perhaps ten yards—after the impulse before frictional resistance had brought it to a stop. Suppose in the same situation that we substituted a block of dry ice for the wooden block, taking care that the two test-objects had exactly the same weight. What would happen if we pushed the block of dry ice across the frozen surface of the pond with the same impulse we had applied to the wooden block? We might well not believe our eyes, for the block of dry ice would continue to move and move and move, fifty yards, a hundred yards, perhaps as far as half a mile, before it finally came to rest! Again, if we had very keen microscopic vision, we would discover the reason very easily. Dry ice is nothing other than frozen and solidified carbon dioxide gas, and where the dry ice touched the surface of the pond, a thin layer of frozen carbon dioxide melted from the block into its gaseous state, so that the block was really sliding across the ice on a thin layer of carbon dioxide gas.

Here we have reduced friction to about as bare a minimum as we can attain here on earth. The frictional resistance between the carbon dioxide block and the icy surface of the pond is very slight indeed, but still enough to decrease the velocity of the moving block of dry ice slowly, bit by bit, at a constant rate. Indeed, if we were to compare the effect of friction on the block of dry ice on the pond with its effect on the block of wood on the pond, we would find that each of these objects slowed down at a rate exactly proportional to the amount of friction present. The less friction, the less rapidly an object would slow down. And in the case of the dry ice, it appeared at first that the block might just go on moving at the same velocity and in the same direction forever, the frictional resistance was so very slight.

But one thing the block of dry ice did not do: Even under these practically friction-free conditions, *it did not accelerate in the slightest*. And indeed, by this time we suspect very strongly that there is no propelling or accelerating force pushing a moving object against friction. But to be certain, suppose we find a place where there is virtually no frictional resistance of any sort acting to slow down our test object. Suppose, for example, that we could transport our shooting device and the wooden block into outer space far from any sun or planet, and there use the shooter to launch the block still farther into space, free from contact with any other surface. Under such circumstances, with no frictional resistance possible, we would certainly expect any "propelling force" acting on the block to make it *accelerate* (that is, to move continually faster and faster) after it left the shooter. But if we actually measured the velocity of the block under such circumstances after launching it with the shooter, we would find no such thing happening. True, the block would not slow down in the absence of frictional resistance or any other impeding force. *But neither would it speed up*. Once the impulse was applied, the block would continue to move in the same direction at exactly the velocity it had attained by the end of the impulse, and would keep on moving at this same constant velocity literally forever—neither accelerating nor decelerating—unless or until it encountered *some* outside force acting upon it either to speed it up, slow it down, or to change its direction in some way.

Of course, in Galileo's time physicists could only *imagine* conditions in which a moving object encountered no friction at all. The best they could do was to set up experiments in which friction was reduced to the barest possible minimum, and then attempt to extend their findings to ideal conditions of no friction whatever which, to them, were totally imaginary and unattainable. But even so, Galileo very soon shrewdly guessed that the mysterious "propelling force" the Greeks had imagined had to exist to keep moving bodies moving *simply did not exist at all.* He saw clearly that a moving object, once it had been started moving by some force or another, tended to continue moving without any need for any additional propelling force, just as long as no outside force acted upon it in any way. He also saw that the opposite was equally true: An object sitting undisturbed at rest would not suddenly start moving of its own accord, but would remain at rest until some outside force—some impulse—acted upon it to start it moving. That impulse might be the powerful force of a bowstring against the end of an arrow, the short, sharp blow of a hammer on a peg, or the steady, continuing pull of gravity on an object released from the fingers—but *some* outside force had to act to get an object moving.

Galileo performed many experiments testing the behavior of objects at rest and objects in motion, and found that invariably his results seemed to support the truth of his shrewd guesses. Other scientists also experimented along the same lines; Leonardo da Vinci, for example, described frictional resistance simply but accurately in his notebooks. Finally, a hundred years after Galileo, Sir Isaac Newton summarized all these observations of the behavior of bodies in motion and at rest in the form of two simple but sweeping *laws of motion* which he believed applied to all objects, whether they were moving or at rest, anywhere in the universe:

1. *Any object in motion will continue in motion in the same direction and at a constant velocity unless acted upon by some (resultant) outside force.*
2. *Any object at rest will remain at rest unless acted upon by some (resultant) outside force.*

Taken together, these laws have come to be known as the "*laws of inertia.*" First investigated by Galileo and finally formalized by Sir Isaac Newton, they were among the first of a very few sweeping laws of nature which became the foundation for all the modern work in physics which came later. The great importance of these early natural laws lay in the very fact that they were believed to be accurate, universal descriptions of relationships that occurred in nature. They were believed to be *general* (i.e., to apply to *any* object, no matter how large or small, with no exceptions) and *universal* (i.e., applying to objects at any location anywhere in the universe no matter how near or far away).

Furthermore, these laws were believed to mean *exactly what they said.* When they spoke of an object "at rest" they meant an object at *complete* rest or inaction; when they spoke of an object "in motion in the same direction and at a constant velocity" they meant an object in perfectly uniform motion. The laws recognized that any number of outside forces might be acting upon an object at any time, but that a change in that object's state of rest or motion would only come about if there were a *net resultant force* at work. Two exactly equal forces acting on the object in precisely opposite directions, for example, would in effect cancel each other out and have no effect whatever on the object's motion; they would exert no *resultant* force. Finally, these laws stated explicitly that only forces *external* to an object could affect its motion one way or the other, a very significant qualification, as we will soon see.

In short, the laws of inertia simply stated that all objects would always continue indefinitely in the same state that they were already in, whether that be a state of rest or a state of uniform motion, until *some* external force came along to change things—and that no change in an object's velocity or direction could possibly come about unless it was caused or brought about by some external force.

We know, of course, that a variety of forces can and do act upon objects to bring about changes in their velocity or direction of motion. We have seen that frictional forces can slow a moving object down to a stop, or even act upon an object at rest to keep it at rest if another force attempting to make it move is not great enough to overcome the frictional resistance. We know that gravitational force here on earth will act on a flying arrow to pull it down to the ground eventually, no matter how swiftly it leaves the bow, and we can imagine that any gravitational force anywhere in the universe would act as an "outside force" capable of changing or modifying the velocity or direction of any object within its reach. Still another force capable of acting on an object is the kind of force that the shooter in our imaginary experiment applied to the wooden block: essentially a *collision* of one object (the shooter) with another object (the block) producing what we spoke of as an "impulse"—a force acting on an object in a given direction for a given time.

But why are the laws of inertia so extremely important and basic to an understanding of how things work? If we look to see how these laws affect the behavior of moving objects in the face of action of each of these three kinds of "external force" in a variety of situations we will discover some rather surprising things—things which have a direct application to our everyday life here on earth.

We have already seen the effect that friction has on a moving object;

acting as an outside force, frictional resistance will slow a moving object down, and may alter the direction of its motion as well. In either case, the object's *velocity is changed.*

But how does a moving object behave when the outside force acting upon it is a gravitational force alone? Here on earth if we set our wooden block moving horizontally through the air, gravitational force would pull it down to the ground so that its path in flight or *trajectory* would be a curving line. In this case, gravity as the "outside force" acts on the moving block to change its direction of motion, pulling it out of the straight-line path it would follow if no such force were acting on it.

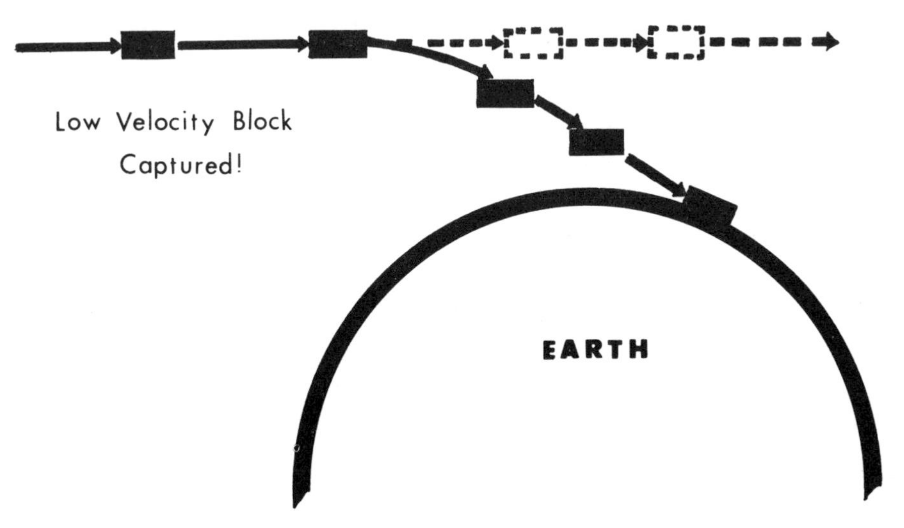


*Fig. 5a* Trajectory of an object moving through space and "captured" by earth's gravitational field.

Suppose, however, that we knew that the earth's gravitational field extended far out into space beyond its atmosphere (as, of course, it does) and imagine that our wooden block were moving through outer space in a straight line with a constant velocity and were to approach within the earth's gravitational field. Here again, gravity as an external force would act upon the moving block to change its velocity and direction, and again (if the block's original velocity were not very great) it would follow a curving path down to the surface of the earth, "captured" in effect by the earth's gravitational force (see Fig. 5a).

This is clear enough; but we know that the earth has a moon. Suppose that the moon also has a gravitational field, however great or small it may be in comparison to that of the earth, and suppose that our wandering block approaches and passes between the earth and the moon at just exactly such a point and at just such an angle that at all times the gravitational pull of

Earth on one side would be exactly equal and exactly opposite to the gravitational pull of the moon on the other side (see Fig. 5b). How would the block behave under these circumstances? We can see that it would not change in direction to curve either way, with both forces equally balanced —but what about its velocity? Surely, we might think, these opposing forces acting on the block would at least tend to slow it down some in its flight.

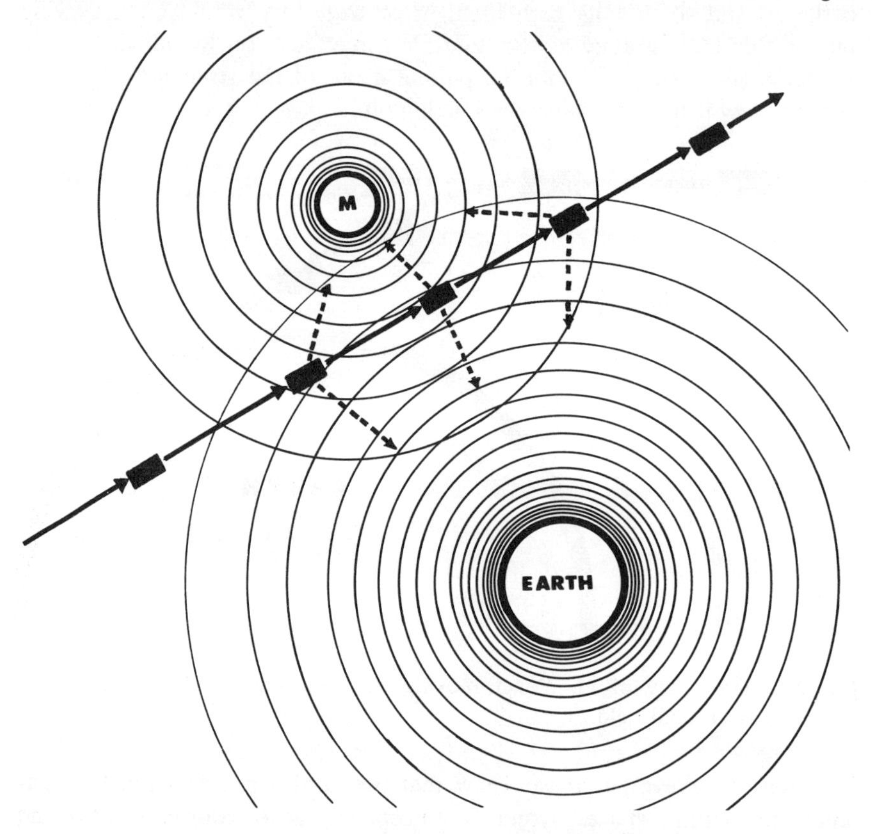


*Fig. 5b* Object passing between earth and moon. The gravitational tug of the earth on the object always equals that of moon, so net effect of gravity is zero.

But nothing of the sort would happen. Assuming no other forces were acting upon the block, the *resultant* outside force of the two opposing gravitational fields would be zero. According to Newton's laws of inertia, therefore, there could be no *resultant* force acting on the block at all and it would continue placidly on its way without any change whatever in either its direction or velocity!

So far we have been talking about objects moving at a uniform, relatively low velocity. What would happen if higher velocities were involved? We have seen that our wooden block approaching and entering earth's gravitational field from outer space and traveling at a low velocity was forced to curve sharply from its straight-line path and nose-dive down to the earth's surface. What about the other extreme? Suppose the block approached from outer space and entered earth's gravitational field at an extremely *high* velocity. Earth's gravity would still be an outside force acting upon the block, so that its path would have to curve toward the earth. At the same time, the block's velocity might be high enough to carry it on beyond the pull of the earth's gravity and ultimately free again. In such a case the block's direction and velocity would be *modified* by its encounter with this outside force but it would "escape capture" and continue hurtling into space beyond the earth, traveling in a new direction and with a new velocity (see Fig. 5c).

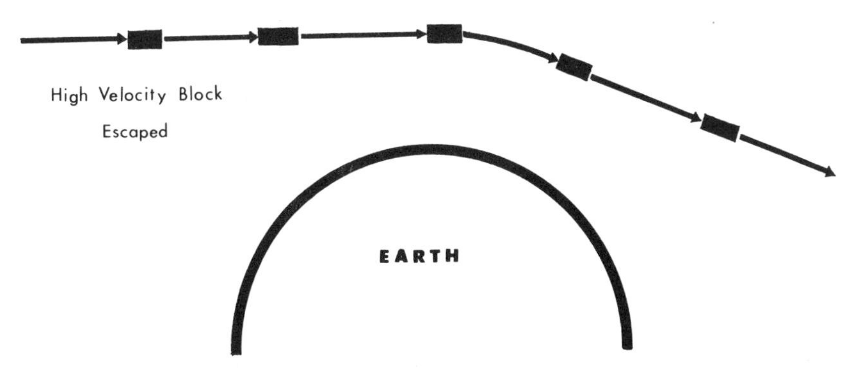


*Fig. 5c* Object moving at high velocity changed in velocity and direction by earth's gravity, but not captured.

But suppose the block were traveling at a velocity somewhere between these two extremes. Suppose its velocity as it approached the earth's gravitational field was just enough to balance the gravitational pull of earth upon it. Then the block's path would be curved toward the earth each instant, but its velocity would still remain high enough that at each instant it sought to continue in a straight line or tangent away from the earth on its own path. If the gravitational pull of the earth at each instant was exactly balanced by the velocity of the block seeking to travel in its own inertial straight line, the block could neither quite fly off into space again free of earth's gravitational pull, nor could the earth quite pull it down to the surface. Rather, the block would continue to curve toward the earth, and curve toward the earth, and curve toward the earth, and continue indefinitely revolving around and around the earth at the same velocity and the same distance away from the surface—in effect, a new satellite in a permanent orbit (see Fig. 5d).

At first glance this might seem like a highly fanciful idea, but if we think about it we see that there is nothing fanciful about it. After all, earth has a somewhat larger satellite, the moon, which travels in a stable orbit around the earth at the same average distance and with the same average velocity that it did thousands of years ago. We can begin to see why Galileo, and later Johannes Kepler, and ultimately Isaac Newton, after studying the behavior of gravitational forces on earth, began looking to the sky and wondering if the moon's wandering path around the earth was not a result of a cosmic tug-of-war between earth's gravitational pull upon the moon and the moon's continuing struggle to pull away from earth and hurtle out to space in a straight line according to the dictates of the laws of inertia. We will return to this line of thought again presently, because it will lead us to a better understanding of how Newton ultimately arrived at the laws of *universal gravitation* (as opposed to the laws of local gravitational effects

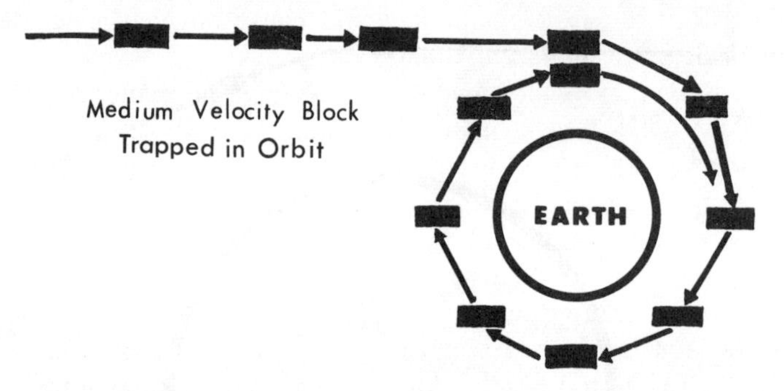


*Fig. 5d* The capture of an object in orbit, a balance between earth's gravitational attraction and the object's own inertial (centrifugal) force due to its original motion.

which Galileo worked out). However, before we abandon our block of wood traveling through space, obeying the laws of inertia, we should turn our attention for a moment to the question of what constitutes an "outside force" capable of acting upon an object at rest or an object in motion and what is not, in fact, an "outside force." Again, the answers may not be quite what we would at first imagine.

Suppose, for example, that instead of launching our wooden block into space we transport the block of dry ice instead, and send it traveling in a straight line at a constant velocity in a place where no gravitational force could reach and no frictional force existed. Suppose, however, that the block of dry ice (merely a chunk of frozen carbon dioxide gas) should happen to move into an area of space which happened to be unusually warm so that the dry ice evaporated completely into an expanding cloud of carbon dioxide gas. The original block, indeed, would cease to exist altogether, transformed into something quite different! Would this not be a violation of the laws of inertia? Surely such a transformation from within the "object" which was *not an outside force* would nevertheless act to alter its direction or velocity, would it not?

The answer is that it would not. True, the original block of dry ice is transformed into a cloud of gas, and the molecules of gas in that cloud, stimulated by the increased temperature, might diffuse and move every which way, creating a thinner and thinner cloud of carbon dioxide gas of greater and greater diameter. But in the absence of any outside resultant force acting upon it, that cloud of gas *as a whole* would continue to move in exactly the same direction and at exactly the same velocity as if it were still an untransformed block of dry ice. In other words, the encounter with this area of heated space might bring about an internal change or transformation *within the system,* but it would cause no change whatever in the behavior of the system as a whole.

In fact *nothing* that might occur within a system, no matter how forceful, could alter the over-all behavior of the system as a whole, no matter how much that system's shape or appearance might be changed. When Newton spoke of "some outside resultant force," he meant some force arising strictly from *outside* the moving system, not some force that arose in any way from within the moving body itself.

We can see this more clearly, perhaps, if we imagine our moving object not as a block of dry ice traveling through space but as an old-fashioned Bolshevik bomb with the fuse lit. After the bomb travels along for a period of time at a constant velocity in a perfectly straight line, the fuse burns down and the bomb explodes, throwing fragments off in all directions. Certainly *this* would appear to be a case in which an "inside force" had a grave effect upon the direction and velocity of the bomb!

But once again, appearances deceive us. Of course, the fragments of the bomb would go flying in all directions, some backward in the direction the bomb was coming from, some forward, some off to the sides, up or down. But the total net result of the explosion would add up to nothing more than a change in the over-all shape of the bomb. Just as we had a solid block of dry ice suddenly transformed into an expanding cloud of gas, here we would have a bomb that suddenly "expanded" violently in all directions, with pieces thrown out at a hundred different velocities. But if we were to add up all the velocities, directions, and masses of every single one of the resulting fragments we would come up with a net result of zero: no change in the over-all velocity of the bomb fragments *taken as a whole.*

Of course, each of those bomb fragments individually would continue to move in its own direction at its own velocity forever unless individually acted upon by some outside force, but the *system* (which previously was a nice solid Bolshevik bomb but now has been transformed into an expanding cloud of shrapnel) would continue to move *as a whole* in the same direction and with the same velocity as the original bomb was moving.

We could look at the explosion another way. It would be as though the bomb, before its explosion, had been enclosed in an infinitely elastic, stretchable plastic envelope. After the bomb exploded, the envelope containing all the fragments would be stretched out into a completely different shape than the original bomb, but in spite of this, *the entire envelope and its contents* would continue moving at its original velocity and in its original direction until an *outside* force affected it.

Now suppose that if instead of a round Bolshevik bomb, our floating object had been a bomb in the form of a tube open at both ends with one end pointing in the direction of its motion and the other end pointing in the opposite direction. Suppose that instead of breaking the bomb up into shrapnel, under these circumstances the explosion simply blew a cloud of hot gas out of both ends. There would still be no change in the velocity or direction of the *system as a whole* even though an expanding cloud of gas was blossoming from either end of the tube. Then, finally, suppose that the front end of this tube-shaped bomb were plugged up so that the only escape for the explosive gases would be from the rear end. In this case, when the explosion occurred, the shell of the bomb (which now would essentially be a rocket) would certainly show a change in velocity: It would suddenly leap forward as a result of the gust of hot gases escaping out the open rear end. The shell of the bomb would end up with a much higher velocity going one way while the escaped gases would move with an equally high velocity in the opposite direction. But even in this extreme case the velocity and direction of the *system as a whole* (bomb shell plus released gases) would remain completely unchanged; if the shell and the gases were all enclosed in an infinitely elastic stretchable bag, we would see a greater and greater distention of the bag in the fore-and-aft direction, but except for this distortion in shape the entire system would continue moving just exactly as it was moving before the explosion.

If we think about this for a moment, we see that here is the reason that we cannot lift ourselves up by our own bootstraps no matter how hard we may try. We may exert all sorts of effort, straining our muscles and tugging for all we are worth, but all of the force that we are applying in an attempt to pull our own feet off the ground comes from within the system. There is no force from outside the system (i.e., from outside our bodies and our boots) which is acting to lift us, and consequently we can bring about no net change in our position. Of course, if a block and tackle were attached to the ceiling and we ran a rope through our bootstraps and then through the pulleys we could hoist ourselves up off the ground with no trouble, but the force that is lifting us is coming from outside of the body-and-bootstrap system; in fact, we are utilizing the upward pull of the ceiling on the block and tackle (the ceiling is obviously outside the system) in order to provide ourselves with "some place to stand" or, more accurately, "something to hang on to." (This is assuming that we consider the "system" in this case to include only ourselves and our bootstraps; if we included the building and its ceiling in the "system," then hoisting ourselves up with a block and tackle would constitute nothing more than moving ourselves around within the system, and we would not bring about any net change in the position of this new and larger "system" unless we could find something for the building to hang on to.)

As we have seen there are a number of clearly distinguishable "outside forces" that can affect the behavior of objects in motion and objects at rest. Gravity is one such force. Friction is another. Both are forces that can alter the direction of a moving object or alter its velocity once it has been set in motion. Other outside forces can have similar effects: A magnetic field, for example, can exert a powerful outside force upon certain kinds of moving objects—objects composed of iron, for example, or objects which happen to possess a certain peculiar characteristic known as "electrical charge" (a characteristic we will discuss in detail later). But perhaps the most common of all outside forces that we observe in our everyday world is a force that is easy to identify, easy to understand, and highly instructive to consider when we are attempting to figure out the peculiar behavior of objects in motion and at rest. This is the force of one object striking another —the simple phenomenon of a collision. We know from long experience that the velocity and direction of a moving object can be altered very sharply and radically if it happens to run into some other object. What is more, in our everyday world Newton's laws of motion should allow us to predict in advance precisely what the end result would be any time one object collided with another, providing only that we know the masses of the objects, their velocities, and the directions in which they are moving.

# CHAPTER 8

## *Push and Push Back: The Riddle of Collisions*

*What happens when an irresistible force meets an immovable object?* Surely every one of us must have puzzled over this moldy old grade school conundrum at one time or another. The unwary ones may even have accepted the far-from-satisfactory answer that "heat is produced" and dismissed the matter without further thought. On closer inspection, however, we can see that the question itself is absurd: Its very terms are self-contradictory. A truly "irresistible" force would have to be a force so completely overwhelming that it would be capable of moving any object it acted upon, even if that "object" contained all the existing matter in the universe. A truly "immovable" object, on the other hand, would have to be an object so incredibly massive that no imaginable force could possibly make it budge, no matter how enormous that force might be. It is hard enough for us to imagine either such a force or such an object existing in the absence of the other; obviously the two could not possibly exist simultaneously in the same universe!

But absurd or not, this conundrum is still intriguing. For one thing, it forces us to pause and reconsider exactly what we mean when we speak of a "force" acting upon something. For another, it implies certain things about the behavior of objects in motion that such great early experimenters as Galileo and Sir Isaac Newton came to believe were universally true, after years of thoughtful experiment and observation of things in the world around them.

First, this conundrum implies that *any* change in an object's state of rest or motion can occur only if some outside force acts upon the object. In the last chapter we saw how the great classical physicists were led inexorably to this conclusion. Indeed, they went a step further: They began to realize that the only way a "force" could be identified at all was by observing and measuring the effect it had on the motion of some object. A cannonball flying through the air exerts no force upon anything (except a few air molecules) *until it strikes a target somewhere.* Neither does a rough warehouse floor exert any frictional force upon a packing case stored upon it until someone comes along and tries to push the packing case. Odd as it may seem, a force cannot really be said to exist at all until it somehow acts upon some object, and then its magnitude can only be deduced—indirectly, as it were—from the *change that we observe in the object's motion as a result of the force acting upon it.* And if we are tempted to challenge this slightly sneaky idea, we need only consult our own everyday experience for proof. We see leaves fluttering and treetops swaying and deduce from this that a wind is blowing. We discover that a gravitational force is present by its action upon a vase that we drop on the floor. We detect frictional force only when we attempt to slide one object across another. In each case, both the force and its magnitude are identified solely by the change in an object's motion resulting from the action of the force.

But our conundrum has an even more subtle implication. It suggests to us that the *amount* of change in an object's state of rest or motion must be directly related not only to the *magnitude and direction* of the force acting upon it, but also to the *mass* of the object—that is, to the object's *inertia,* its built-in resistance to any change in its motion. A truly "irresistible" force acting upon any object would cause the maximum possible change in that object's motion, regardless of the object's mass. Such a force would be able to accelerate the entire mass of the universe up to the speed of light in the direction the force was acting. It would just as readily be able to slow a beam of light (which has *some* mass, as we will see later) from light-speed down to a standstill, if the direction of the force happened to oppose the direction of the light beam. At the other extreme, a truly "immovable" object would offer so much resistance to *any* change in its state of rest or motion that no force could budge it even an inch, no matter how great that force might be.

Happily, we are unlikely ever to witness either of these outside extremes, but we can see that any interaction between an object and a force acting upon it *must* fall somewhere between these extremes of total immovability and total irresistibility. Time and again the early experimenters found that any change occurring in the motion of any given object was always directly proportional to the magnitude of the force acting upon it—the greater the force, the greater the change in the object's motion. On the other hand, they also found that the more massive the object, the more it resisted change in its state of rest or motion, and the less change any given force would be able to bring about when acting upon it.

Today this proportional relationship between an object's mass, the force acting upon it, and the change that occurs in its state of rest or motion seems self-evident. Common sense tells us on one hand that a given force will influence the motion of a comparatively light object more than it will a more massive one, but that on the other hand a given object's motion will be influenced more by a powerful force than by a weaker one. If we tried to play Ping-pong with a golf ball, we would have trouble getting the massive ball across the net with a standard lightweight paddle. If instead we used a paddle made of lead, we might be able to play the game all right using a golf ball, but we would be likely to knock an ordinary standard Ping-pong ball into the next county on the first serve.

Finally, our irresistible force–immovable object conundrum suggests that whenever any force is brought to bear upon an object to push it around, the object must counter with a resisting force of equal magnitude acting in the opposite direction. An object being pushed literally *pushes back*. In the extreme case, our imaginary "immovable" object would be just as unwilling to be moved at all as our "irresistible" force would be unwilling to be resisted, so at best we would end up with a Mexican standoff. To imagine a more familiar example, we might substitute a twenty-pound sledge hammer for our "irresistible" force and a brick wall for our "immovable" object. If we then swung the hammer against the wall with all our strength, we might expect the hammer's force at the moment of impact to "move" the wall, at least to some degree: A few bricks would be chipped or cracked, perhaps even crushed. But the sledge hammer would not continue to plow its way through the wall, unaffected by the encounter; we would see it *bounce back* from the impact, perhaps with so much force that it is torn out of our hands!

But in such a case, what force could possibly be acting upon the hammer to bring it to a halt against the wall and then thrust it violently back in the opposite direction? This could only happen if the wall were to exert some counterforce against the hammer at the same time the hammer hits the wall. If we could somehow measure the forces acting at the *moment of interaction* between wall and hammer, we would find the wall's impact on the hammer to be exactly equal to the hammer's impact on the wall, but acting in the opposite direction. Furthermore, if we were to experiment with other such interactions or collisions, we would soon find that an equal but opposite counterforce is *always* present any time any force acts upon any object. In somewhat simpler terms, we can say that *for every action* (of a force upon an object, for example) *there is an equal but opposite reaction* (of the object resisting the action of the force, for example).

If this idea seems confusing and obscure, take heart: Even physics majors find it difficult to grasp. Part of our trouble is that we tend to overlook or ignore the "reaction" part of the equation that is always present when forces and objects interact in the world around us. We just don't ordinarily *think* in terms of the wall hitting the hammer back, even though it obviously does so, any more than we think of the ground pushing upward against our feet, or of the rough floor pushing back against the packing case we are trying to move. Yet when we hold a steel ball at rest in an outstretched hand for a while, our tired arm muscles soon tell us that we have been pushing the ball upward at the same time and with the same force as gravity has been tugging it downward. Even if we fail in our efforts to move the packing case so much as an inch across the floor, we nevertheless know that we have been pushing with might and main against *some* real force that is opposing our efforts and preventing the box from moving. When we fire a rifle, we expect the bullet to be driven forward by the force of the explosion and imbed itself in a tree—but we also feel the rifle butt slam against our shoulder in recoil. The only reason that the rifle does not fly backward as far and as fast as the bullet flies forward is simply that the rifle itself has much more mass to be moved backward than the bullet has to be driven forward, and even at that we are likely to end up our target practice with a black-and-blue shoulder if we aren't careful.

The fact is that all of these characteristics we have been discussing of the interactions of forces and objects can easily be observed and confirmed in the world around us every day, *providing we know what to look for*. For the most part, people rarely look. They simply take these things for granted without even trying to describe what is actually happening. And if we find these characteristics of moving objects hard to pin down when we try to describe them, we can take comfort that the early physicists found them just as hard to comprehend, if not harder. Galileo spent decades trying to figure out how to describe the behavior of moving objects with accuracy. From his experiments he came to recognize *all* the characteristics that we have been discussing—but he never did find a way to express them as concise rules which could then be applied to the motion of *all* objects, large or small, anywhere in the universe. It remained for Isaac Newton, starting where Galileo and others had left off, to work out three simple, general statements which he believed accurately described all possible forms of motion throughout the universe. Today these statements are known as Newton's laws of motion, and can be briefly summarized here:

*Law I: Any object in a state of rest (or of uniform motion in a straight line) will remain at rest (or in uniform motion in a straight line) unless acted upon by some external resultant force.* (We have already seen some of the implications of this statement, and we will soon see more.)

*Law II: When an external force acts upon an object, the change in the object's motion is proportional to the force and occurs in the direction that the force is acting.*

*Law III: When any force is brought to bear upon an object, an equal force is brought to bear acting in the opposite direction;*

or:

*For every action there is an equal but opposite reaction.*

(We will consider the implications of laws II and III more closely later in this chapter.)

At this point we must remind ourselves once again that Isaac Newton did not come up with these three principles by means of any divine revelation, nor did he regard them as irrevocable "laws of nature" at the time he formulated them. He merely considered them as useful working rules, tentative conclusions based upon hundreds of years of observation and experiment. In effect, he was saying: "These rules seem to describe what happens any time a force acts upon an object, or any time one object interacts with another. As far as we know now, these rules *always* apply, with no exceptions. Let's consider them to be true until some new evidence shows up to prove them false."

This was surely a reasonable stand to take, and scientists of the day accepted it. But as time passed, no such new evidence showed up; repeated crafty attempts to find exceptions to Newton's "working rules" invariably failed. By the beginning of this century, most physicists had come to accept these rules as broad, universal laws of nature, essential to any understanding of how things work in the universe.

This is not to say that Newton's laws of motion became any easier to comprehend as time went by; to this day physicists themselves cannot fully agree upon just how the laws of motion should be interpreted. As casual bystanders we might be tempted to ignore them as obscure and meaningless, if it were not for the fact that these rules are profoundly important to us in the conduct of our daily lives. Any time we sip a cup of coffee, throw a baseball, walk to the grocery, or slam the garage door, we are in fact utilizing the laws of motion to fulfill our needs, whether consciously or not. Those laws guide and limit virtually every move we make. What is more, they enable us to make useful and accurate predictions about things that have not yet happened. Every time we drive a car around the block we embark upon a multitude of half-conscious computations, judgments, and predictions of what is going to happen next, all based upon the laws of motion. And when we see a sand-lot baseball flying through the air toward our biggest plate-glass window, we do not need Isaac Newton to tell us that that window is going to be smashed to shards unless we can stop the ball before it reaches its target.

To understand more clearly just what the laws of motion actually mean, and to see how they enable us to predict how things are going to work in the world around us, it will be helpful to examine more closely one of the most familiar and commonplace of all the interactions we observe every day: the collision of one moving object with another.

## THE COSMIC POOL TABLE

What actually happens when one object crashes into another? In our everyday experience it often is difficult to say, precisely, because of the red herrings that lead us astray. In some collisions the colliding objects are shattered. In others one object or another may be bent out of shape, altered beyond recognition, heated to incandescence, even vaporized! We have already seen how such extraneous factors as air resistance or friction interfered with early observations of the influence of gravity on falling objects, leading to many confusing or downright misleading experimental results. If we wish to concentrate solely on the effects of collision forces upon moving objects, we must rule out all other forces and effects, as far as possible. In short, we must try to imagine a "perfect collision" of two moving objects occurring under ideal conditions. In actuality we could never hope to find the "ideal test objects" necessary to fulfill such rigorous specifications, but we can find a very close approximation from our everyday experience: the collision of two ordinary billiard balls on the smooth green felt of a large pool table.

There are several reasons that billiard balls lend themselves so splendidly to our purpose. For one thing, most of us are already familiar with their collision behavior from personal experience. We know that when two billiard balls collide, they inevitably bounce away from each other, with some alteration in the velocity and direction of motion of each ball. We even recognize from experience that what we see happening after a collision of two billiard balls depends a great deal upon the velocity and direction of motion of each ball *before* the collision and upon the angle at which the collision takes place. In short, we recognize a cause-and-effect relationship between the conditions before the collision and the new conditions after it has occurred, and we already have *some* idea of what to expect when billiard balls collide. All we really need to do is fill in the details of what we actually observe under a variety of collision circumstances.

Furthermore, the physical properties of billiard balls lend themselves well to our needs. For one thing, billiard balls are substantially massive objects, unlikely to be affected much by such minor forces as air resistance, cross-wind drafts across the pool table, or whatnot (whereas Ping-pong balls, in contrast, would be). For another thing, we can safely assume that any two billiard balls we might choose would be very nearly identical in mass—certainly nearly enough that we could ignore the effect of any minor differences.

Again, because of their respective masses (and their attendant qualities of inertia—i.e., resistance to any change in their state of rest or motion) we can expect billiard balls to behave very much the same as objects in free fall, relatively unaffected by frictional or gravitational forces. We expect a billiard ball at rest on the table to remain at rest unless some outside force starts it moving. Once it is set in motion, however, we expect it to continue rolling (at least for a while) at a relatively constant velocity in the direction the force acted upon it, unless it is again acted upon by another external force.

Of course, we acknowledge from the beginning that these things are only *approximately* true. Billiard balls *are* influenced by earth's gravity, as we would soon learn if we dropped one on our toes. But if our pool table is perfectly level, gravity would influence any one ball exactly as much as any other, so the effects of gravity would be canceled out, as far as our experiment was concerned. As for friction, we know that it will indeed slow down a rolling billiard ball a bit at a time, but not enough to cloud our experiment. As long as the balls are moving at a fair velocity and we observe their behavior over relatively short distances, we can imagine their movement to be virtually frictionless.

Finally, billiard balls provide an ideal sort of collision to observe. When they strike each other, only a tiny surface area of one actually comes in contact with the other and then only briefly: The impact is nearly instantaneous, and the balls bounce freely away from each other almost immediately after colliding. What is more, billiard balls will not be significantly deformed at the instant of collision, as two soft-rubber balls would be, nor do they have any tendency to stick together when they strike, as two balls of well-chewed bubble gum might. A physicist would speak of a collision between billiard balls as comparatively *elastic*—that is, a collision in which virtually all of the force of the collision is transmitted directly to the colliding objects almost instantaneously upon impact, so that very little energy is dissipated into heat or exhausted in the physical deformation of one or both of the objects.

In short, billiard balls offer a reasonable approximation of the "ideal" collision conditions we are seeking. So what happens when one billiard ball collides with another? First, let's imagine that we have two shooting devices, such as the one we used in the last chapter, installed at opposite ends of a pool table and use them to start two billiard balls rolling toward each other in a straight line, each ball moving with exactly the same constant velocity as the other but in opposite directions. When the two balls reach the exact center of the table we are not surprised to see them collide with each other—*smack*!—and then bounce smartly apart again. Fine, but *what actually happens during the collision*? How do the velocities and directions of the balls after collision compare with their velocities and directions *before* the collision?

We could make some shrewd guesses without even measuring, just on the basis of common sense and experience. First, we would expect the balls to collide head-on, since they were approaching each other dead-ahead on the same line. Further, since each ball has the same mass as the other and is moving toward the other with the same velocity we can imagine that each will strike the other with identical force at the moment of impact, so that whatever happens to one ball as a result of the collision will also happen to the other in mirror-image fashion.

Similarly, we might guess that certain other things might be observed and measured:

1. Since the balls are moving toward each other head-on, each will obstruct the movement of the other at the moment of impact, so that for

a split second during collision the balls will be standing motionless side by side at the center of the table. Each ball will have brought the other to a complete halt.

2. Since neither ball can pass through the other, and since each exerts an equal force on the other, the two balls will be pushed away from each other in opposite directions as a result of the collision. Thus the direction of motion of each ball will be exactly reversed.

3. Since the force that each ball can exert on the other depends upon the mass of the ball and its velocity at the moment of impact, and since the balls approaching collision have identical masses and equal velocities in opposite directions, we would expect the balls to bounce away from each other after collision at *exactly equal velocities* in *exactly opposite directions*—and to continue moving away from each other with equal constant velocities until some other outside force (such as the end of the pool table) forces another change.

4. Finally, since each ball approaching the collision is carrying a certain amount of energy with it (let's call it "energy of motion" for the moment) and since practically none of that energy of motion is lost in heat or expended in deforming the balls during the collision, we would expect each ball to throw virtually all of its energy of motion into pushing the other ball away at moment of impact. As a result, the velocity imparted to each ball as a result of the collision will be exactly equal to the velocity of the other ball before the collision, but in reversed direction; and since the velocity of each ball before collision is equal to that of the other, each will bounce away after collision with exactly the same velocity it carried into the collision.

In short, the net result of our imaginary "ideal" collision should be a complete reversal of the direction of each ball, in mirror-image fashion, with the after-collision velocities of the balls precisely equal to their before-collision velocities but directed in opposite directions. Indeed, we might see the same result (at least as an illusion) if we rolled one billiard ball into a head-on collision with its own image in a mirror!

Of course we must bear in mind that this "ideal" collision would never actually come about on a real pool table. Because of the frictional resistance of the table top, the balls would not be moving toward each other at *perfectly* constant velocities, but rather with constantly decreasing velocity, however slight the decrease. Their collision would not be *perfectly* elastic, because some of the energy carried by each ball would be consumed in correcting a slight distortion of the surface of each ball caused by their impacts. Each ball would, in fact, flatten slightly at the point of impact and then spring back to normal shape immediately after. Some energy (not much, but some) would be converted into heat at the same time—each ball would become slightly warmer at collision point. Thus, in reality, we know the balls will move away from each other with a *slightly* lower velocity than they had the instant before collision and then (thanks again to friction) will continue to slow down as they move away from each other in opposite directions. We could eliminate friction completely only if we could stage the collision somewhere in outer space; and even there we would have to conjure up completely undentable billiard balls and arrange a collision in which no energy whatever is converted to heat before we could achieve a *perfectly elastic* collision in which the balls really would retire from each other at precisely the same velocity they had had before the collision.

Indeed, the more closely we compare a real-life pool table collision with our imaginary one, the more red herrings we find. But the truly amazing thing is not how far away from ideal results we would come in an actual billiard ball collision, but rather, *how close we would come!* If we had actually measured velocities and energies throughout our considerably-less-than-ideal pool table collision, we would have found that what *actually happened* there approached our ideal predictions surprisingly closely—so closely, in fact, as to suggest strongly that *if ideal conditions had been possible,* our experimental results would have been *precisely* as we had predicted. Thus such imaginary experiments, although impossible to actually perform, can still be a perfectly valid way—sometimes the *only* way—to learn what is true and what is not.

But in imagining impossibly "ideal" conditions for our billiard ball collision, we have made one quite unwarranted assumption. We have assumed, because the balls bounced away from each other in mirror-image fashion, that each ball must have transmitted all of its energy of motion to the other, and vice versa, during the moment of impact. But how can we be so sure there was an *exchange* of energy? Suppose we had interposed a thick plate of steel in the middle of the table, so that each of the balls struck the steel plate instead of the other ball (see Fig. 6). In such a case both balls would have rebounded exactly as if they had collided with each other, even though nothing could possibly have been transferred from one ball to the other. How do we explain that? And if there *was* an exchange of some sort between the balls when they actually collided, *what exactly was exchanged?*

Obviously, as long as both balls are of identical mass, approaching collision with equal velocities but moving in precisely opposite directions, we will have trouble answering these questions and determining exactly what *does* happen. We can clarify the question by repeating the experiment under slightly different conditions. Suppose this time we insert a lead weight into the center of one of the balls so that ball A has twice the mass of ball B. Everything else we keep the same. We beef up the shooter pushing the heavy ball A so that both balls are again set moving toward each other just as before, with equal constant velocities in precisely opposite directions, on a head-on collision course. What effect will the doubled mass of ball A have on the results of the collision?

Once again we will see the two balls smack together at the center of the table. Once again they will bounce away from each other in opposite directions—but this time we would find that the more massive ball A would bounce away at *only half its original velocity,* while the less massive ball B would spring away at *twice its original velocity!* And we would observe the same thing every time we repeated the experiment.

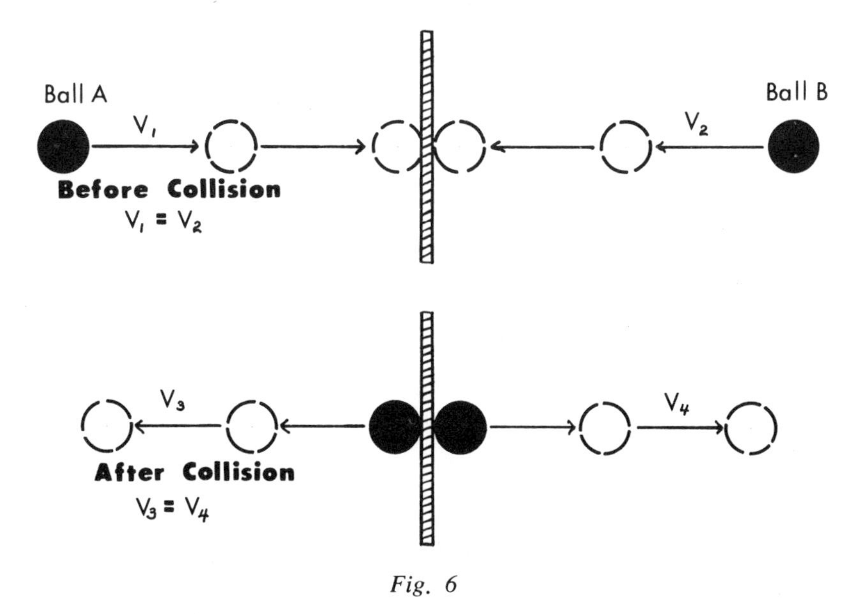


*Fig. 6*

Well, what is happening here? At the instant of collision, massive ball A exerts a force—call it force A—on less massive ball B, while ball B simultaneously is exerting a force, force B, on ball A (see Fig. 7). As a result of these forces acting, each ball is stopped and its direction of motion reversed—the after-collision course of each ball is in the direction of the opposing force of the other ball acting on it. All this is the same as before, except that in this collision the *after-collision velocity of each ball has been changed.* The after-collision velocity of the more massive ball A has been reduced by half, while that of the less massive ball B has been doubled, as though ball A gave up some of its velocity to ball B at the instant of collision.

What could account for this change? Clearly it must be the difference in mass of the two balls, since nothing else had been altered in the second experiment. And indeed, accurate measurements would reveal that the *change in velocity* of each ball was inversely proportional to the mass of the ball: Massive ball A ended up with half its former velocity after collision, lighter ball B with twice its former velocity. If we did the same experiment with ball A bearing four times the mass of ball B, we would find the four-times-as-massive ball A bouncing away from the collision with only one-fourth of its former velocity, while ball B would have *four times* its former velocity after collision, and so on for any difference in mass of the balls that we might arrange.

From this, it might seem that "might makes right" in the case of collisions: The big guy pushes the little guy harder and farther than the little guy can push the big guy when the two run into each other. But is the mass of the billiard balls the *only* factor that can alter the results of their collision? Another variation in our experiment will help us find an answer.

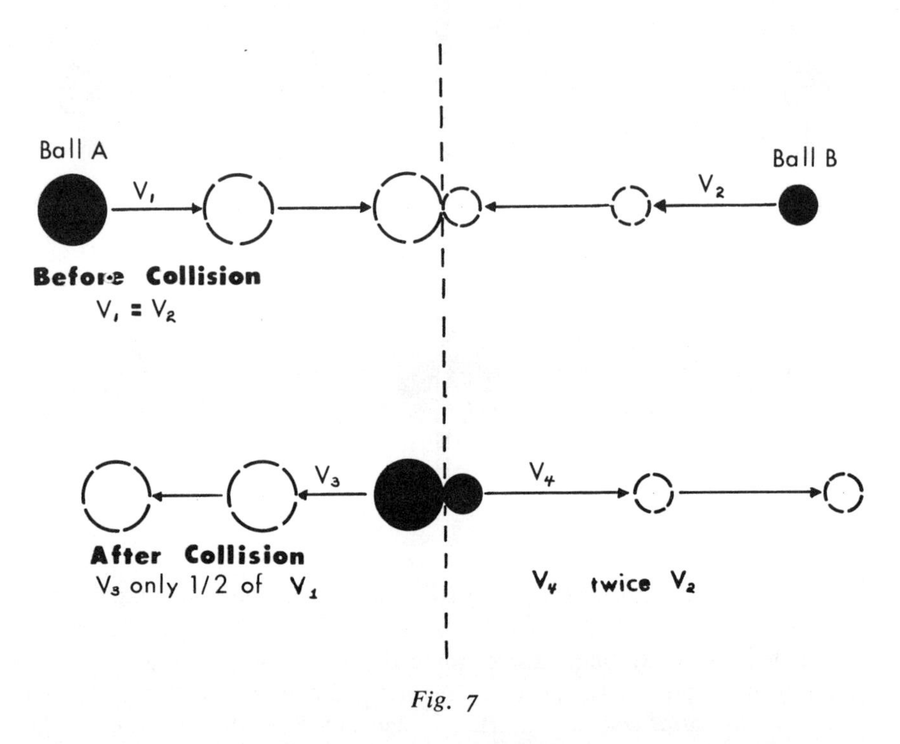


*Fig. 7*

To find out, let us make a different kind of change in the ground rules. This time imagine that ball A and ball B are identical in mass again but that this time we set ball B rolling into the collision with *twice the velocity* of ball A (see Fig. 8). What will happen in this case? Once again the balls will collide and bounce away in opposite directions, but this time the slower-moving ball A will bounce away with twice the velocity than it had coming into the collision, while the faster ball B will bounce away with only half its former velocity.

Here again it would appear that the faster ball B has *given up* or *transferred* some of its higher velocity to the slower ball A at the instant of collision, but this time the exchange could not be blamed on any difference in the mass of the balls. The only difference that could possibly account for this exchange is the difference in the before-collision velocities of the balls. If we ran a multitude of tests in which ball B entered the collision at higher and higher velocities compared to ball A, we would find that the after-collision velocity of ball A would increase in direct proportion to the velocity of ball B before collision.

Thus we begin to see that what happens in an ideal instantaneous head-on collision between two billiard balls (or any other freely moving objects) must depend not only on the *masses* of the colliding objects, but also upon their respective velocities at the instant of collision. Furthermore, we see

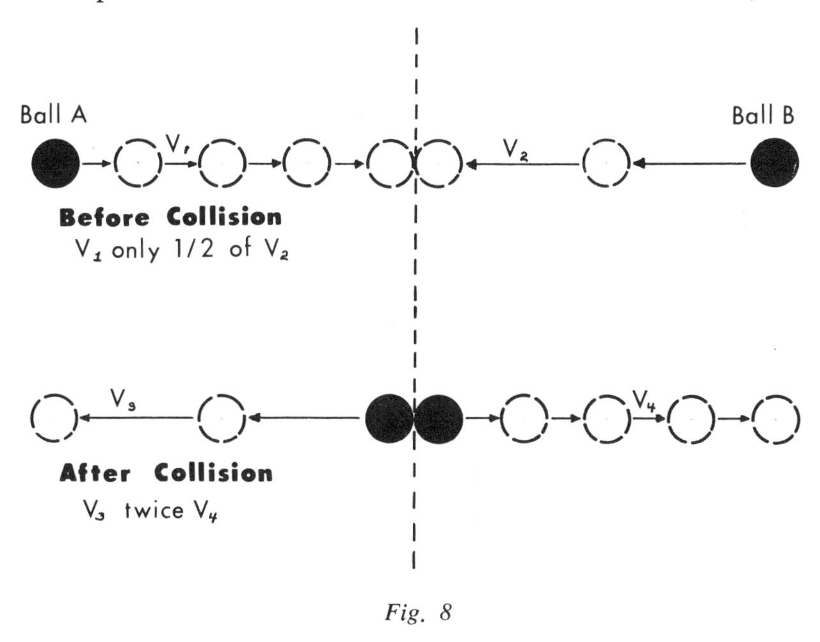


*Fig. 8*

that in every collision, each ball applies a force to the other which is proportional to both its mass and its velocity, and that each ball then reacts in accordance with the force applied to it by the other.

So far we have limited ourselves to a very special kind of collision in which the billiard balls approach each other from exactly opposite directions, with no spin or "English," and collide squarely with each other head-on. Thus the only change in direction of the balls that we have seen after collision has been 180-degree reversal of their directions along the same straight line. But what would happen if we set the balls rolling toward each other at an angle? Would the direction each ball was moving at the moment of impact have any influence on the results? Any billiard enthusiast can tell us the answer: The angle of impact of two billiard balls has a very important effect upon what the balls will do after collision. But how much effect, or what sort?

Suppose we take ball A and ball B, identical in mass, and start them rolling toward each other from two adjacent corners of the pool table, so that they will collide at a 90 degree angle, taking care that both balls approach collision at precisely equivalent velocities (see Fig. 9a). What will happen when they collide? Once again, each ball will exert a force upon the other at the instant of collision: Ball A smacks ball B and ball B smacks ball A. Once again, these collision forces result in a change of direction for each ball. Ball A, approaching from the left, bounces away from the collision to the left as though it had turned a square corner of 90 degrees. Simultaneously, ball B approaching from the right would bounce away to the right in mirror-image fashion. But the velocity of each ball after the collision would be exactly the same as it was before. The only result of the collision would be a *change in direction of motion of each ball*—exactly what we saw in our first experiment when balls of equal mass collided head-on while approaching each other with equal velocities.

With further experiment, we would obtain similar results no matter at what angle the balls approached collision. If the angle is very narrow (as in Fig. 9b) the balls will diverge after collision at an equivalently narrow angle; if the angle is very wide (Fig. 9c) they will diverge at an equivalently wide angle after collision.

Obviously, *direction of motion* of the colliding balls is important to what happens in such "glancing blow" collisions. Does the mass or the before-collision velocity of the respective balls also play a part? To find out, we can vary the circumstances of the collision as we did before, first making ball A twice as massive as ball B, then making ball B approach collision at twice the velocity of ball A. In such instances we would discover—perhaps to our surprise—that alteration of the masses of the balls, or of their initial velocities, or both, has *no effect whatever* on the angle they bounce away from each other. This seems to depend solely on the angle at which they approached each other. But we would see, once again, the same apparent "exchange" of velocities of the balls in relation to their masses and before-collision velocities: the higher-velocity (or more massive) ball would appear to transmit some of its velocity to the lower-velocity (or less massive) ball.

Finally, consider one other situation. Suppose ball A, with mass identical to ball B, is not moving at all. We simply place it at rest in the center of the table and then start ball B rolling toward it along a straight line at a constant velocity. Here again a head-on collision would occur, but we would see a curious thing happen. At the instant of collision, the moving ball B would stop dead, while ball A, formerly at rest, would bounce away from the collision with precisely the same velocity as ball B had had before the collision.

What has happened here? At first it might seem that only the stationary ball, ball A, had any force acting upon it. It was just sitting there minding

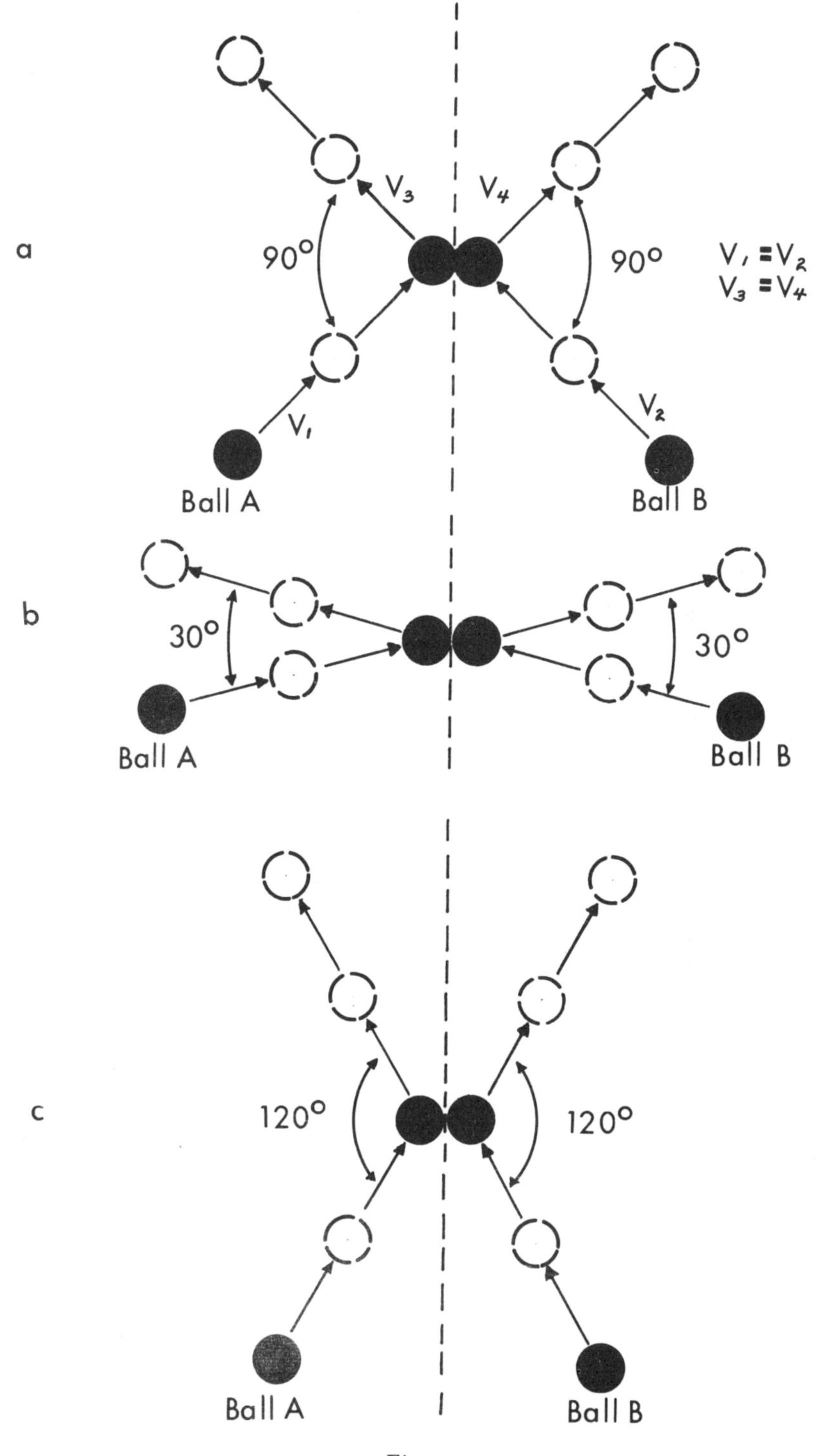
a
$V_3$
$V_4$
90°
90°
$V_1 = V_2$
$V_3 = V_4$
$V_1$
$V_2$
Ball A
Ball B
b
30°
30°
Ball A
Ball B
c
120°
120°
Ball A
Ball B

*Fig. 9*

its own business when ball B ran into it, so to speak, and started moving only when the collision force of ball B was applied to it. But if we think carefully, we see that ball B must also have had a force applied to it. After all, before the collision *it was moving,* whereas after collision it was not. *Some* force must have acted upon it in a direction opposite the direction it was moving to bring it to a halt—and that force could only have been applied by the stationary ball A at the instant of collision.

Thus once again we see that a force and an equal but opposite counterforce must both have been present at the moment of collision. Ball B pushed stationary ball A with sufficient force to set ball A moving, while ball A pushed moving ball B in the opposite direction with sufficient counterforce to bring it to a stop. What is more, if the velocity that ball A "acquired" in the collision was the same as the velocity that ball B "lost," then the force and counterforce must have been equal in magnitude but acting in opposite directions.

What would happen if the stationary ball A were twice as massive as the moving ball B? We would expect the lighter ball B to have much less effect on the twice-as-massive ball A than it would have if the balls had the same masses, as long as ball B approached at the same velocity as in the previous test. And our expectation would prove correct. As before, moving ball B would come to a halt at the instant of collision, while stationary ball A would bounce away, but this time ball A would have only half the velocity after collision that ball B had before it. Does this mean that ball B exerted only half as much force on ball A this time? Not at all; it exerted exactly the same force as before, but that force had to act upon twice as much mass in ball A as before, so that the change in ball A's velocity could only be half as great.

In all of these imaginary situations, we have simply been watching Newton's law of motion at work. Ruling out such confusing influences as frictional and gravitational forces and rotation, or spin, which might have affected the motion of our billiard balls, we have seen that the changes in motion and direction of the balls occurred in a predictable and uniform manner according to some sort of specific rules or principles. Indeed, in each case the collision results seemed to depend directly upon the relationship between at least three easily distinguished factors: the respective masses of the billiard balls, their respective velocities as they approached collision point, and the respective directions in which the balls were moving.

Galileo and other early physicists came to realize that these three factors—mass, velocity, and direction of motion—each played a vital role in determining the behavior of all material objects known to them, whether at rest or in motion. They realized that some outside force had to act on an object if its state of rest or of motion was to be altered. They also realized that the *period of time* during which a force acted upon an object also played a part in determining *how much* change was to come about in the object's state of rest or motion, whether the time was a split second (as in our billiard-ball collisions) or prolonged over many minutes (as when we exert a force to push a piano across the room an inch at a time). These early scientists groped for ways to express the relationship between these factors in the form of broad universal principles which might serve as good rules of thumb to help them understand all forms of motion they might encounter—in other words, as tentative laws of motion.

Newton succeeded better than any before him in piecing together the observations that had been made, making shrewd guesses as to what these working rules had to be, and then experimenting further to test the rules as he worked them out. The laws of motion that he finally set forth seemed to describe the motion behavior of *all* known objects, whether at rest or in motion, and whether moving in isolation or interacting with other objects.

It was a breathtaking scientific breakthrough to have discovered such simple yet apparently universal natural laws as these. Newton's laws of motion must have seemed like a haven in the storm for scientists of the day—a patch of solid, reliable ground appearing at last in the vast quagmire of confusing observations, superstitious dogmas, and conflicting theories that beset the early physicists. But in working out his laws of motion, Newton made yet another discovery that was perhaps even more staggering: He discovered that there were certain properties of matter or energy that seemed *never to change, ever, anywhere in the universe,* no matter what forces were brought to bear and no matter how objects might move about or interact with each other. For reasons unknown, certain baseline characteristics of the universe seemed always and invariably to be *conserved,* unaltered since creation and perpetually inalterable.

## THE VITAL CONCEPT OF CONSERVATION

It seemed that *matter,* for example, could neither be created nor destroyed by any force or interaction in the universe. Matter could be moved about, altered in shape, even forced into chemical reaction with other matter, but the total quantity of matter in the universe always remained totally unchanged. Similarly, *energy* seemed always to be conserved: It could be changed from one form to another, even transmitted from one part of the universe to another, but none could ever be destroyed and none freshly created. Even the discovery, later, that matter and energy were really two manifestations of the same thing did not alter this rule of conservation; the two separate conservation rules were simply combined into one that was more all-inclusive.

The idea that the universe has certain unchanging and inalterable properties may seem commonplace to us today, but in Newton's time it was by no means self-evident. The concept evolved bit by bit over the centuries as suggestive experimental evidence began to accumulate. As it was, the most familiar conservation laws were not the first to be affirmed. Long before physicists were convinced that neither matter nor energy could ever be created or destroyed, Newton's study of objects in motion began turning up evidence that yet another very important universal property was always conserved—the property of moving objects which physicists call *momentum.*

We ordinarily use the word "momentum" very loosely as a vague reference to the apparent ability of something to continue moving of its own accord against some kind of resistance. We say, "The halfback plowed through to the goal line on his own momentum," or "The train picked up momentum as it raced down the grade," or even "Let's get the job done before we lose our momentum."

To the physicist, however, "momentum" has a much more precise and specific meaning. To understand what the physicist means when he says, *"The momentum of any closed system is always conserved,"* and to see why such an odd-sounding and unfamiliar natural law should be so important to our understanding of how things work in the universe, we must return to our imaginary billiard ball experiments once again and consider some things which we deliberately ignored before.

First, suppose we place two billiard balls with identical masses at one end of a long pool table, and start them rolling simultaneously toward the far end of the table by means of two shooters. Instead of using identical impulses to push the balls, however, let us imagine that ball A is pushed twice as hard as ball B, and thus rolls toward the end of the table with twice the velocity of ball B. Which ball then has the greater force as they roll toward the end of the table?

Instinctively we might feel that ball A must obviously have twice the force of ball B—but our instinct would be wrong. In point of fact, neither ball would have *any force at all* as it rolls along (assuming "ideal" conditions of no friction and no gravity), *until it collides with something.* Then and only then would either ball exert a force upon whatever it happened to collide with.

Even so, we see ball A hustling down the table, increasing its lead over ball B with every passing second. Surely ball A must have twice as much of *something*! But what? If we could review our experiment in slow motion, we would discover the answer. To begin with, each ball was at rest at one end of the table. The velocity of each was zero. Then a shooter exerted a force on each ball, and in each case that force was applied to each ball for a brief but measurable period of time. In each case, this force-applied-for-a-period-of-time had a similar effect: The instant the force was applied to either ball, the ball's state of motion began to change in the direction the force was applied. The shooter's force caused each ball to begin to accelerate from a state of rest to a state of progressively swifter motion, and the acceleration of each ball continued *until the shooter stopped pushing it.* The instant the accelerating force of the shooter stopped acting on either ball, that ball stopped accelerating and thereafter continued to roll down the table at precisely the velocity it had attained when the accelerating force stopped acting on it. In the case of ball A, that velocity was exactly twice the velocity ball B had attained, so that during each subsequent second ball A continued to roll twice as far down the table as ball B rolled during the same second.

At this point we must be wary or we will fall into a trap. Because ball A attained twice the velocity of ball B, thanks to the action of its shooter, we might be tempted to conclude that ball A's shooter exerted twice the force that ball B's shooter exerted. But this does not necessarily follow. As we have seen, the velocity ball A attained depended not only upon the size or magnitude of the force its shooter exerted upon it, but also upon the *length of time* that force continued to act. A gentle force acting on ball A for two seconds would impart *precisely the same* velocity to the ball as a force twice that magnitude acting for only one second. Thus we see that the outstanding performance of ball A in our experiment compared to ball B depended not on the force of either shooter alone, nor on the time either shooter was acting alone, but on the force of the shooter multiplied by the time it was acting. We have already spoken of this combination of force times time as the *impulse* of the shooter on the ball, and we can clearly see that the impulse that started ball A moving had to be twice the magnitude of the impulse that was applied to ball B.

We can begin to glimpse what a mare's nest the early physicists found themselves in when they tried to define what a "force" was, or specify what effect a "force" might have on an object. They could see that a "force" always had to be associated with a specific direction—it would obviously be impossible for a single force acting on an object to move it in opposite directions at the same time, for example, or to move it in all directions at once. They could also see that one "force" might be of greater size or magnitude than another, but how could this be demonstrated? How could they ever pin down what a "force" was when the effect of its action on a given object invariably got all scrambled up with the *length of time* that it acted?

Newton was one of the first to recognize that a "force" was really an intangible and abstract concept somehow related to observed changes in the velocity or direction of material objects, rather than a tangible "thing" that one could isolate, measure, and describe. He also recognized that the only possible way a given force could be measured—or even be identified as existing at all—was in terms of some observable and measurable change in the *state of rest or motion of some object.* Thus, the only way we could describe or measure the force of the shooter that moved ball A (over a period of time) from a state of rest to a state of motion (i.e., rolling at a constant velocity down the pool table) would be to describe it in terms of *what* it managed to move—the mass of ball A—and the final velocity that mass had achieved by the time the force stopped acting on it.

In other words, like the pessimistic fisherman who assumes that there are no trout in the stream until one tugs on his line, we must assume that no force is present at all until we observe some change occurring in some object's state of rest or motion. When we *do* see such a change occurring, we know some force of some magnitude is at work bringing about the change. We can measure the *length of time* that force is acting on the object by starting a stopwatch the instant we see some change in its motion begin to occur (whether the change be acceleration, deceleration, or merely a change in the direction of its motion) and then stopping the watch the instant we see the change in the object's motion cease. Then since we know that a force was acting and for how long, and since we know the object's mass and can compare its velocity before and after the force acted upon it, we can calculate the magnitude of the force that brought about the change. We can also calculate the total resultant change in the object's motion, as well as determining *how fast* the force caused the change to take place—the rate of change in the object's motion.

We have already seen these ideas in action on our imaginary pool table. The impulse (i.e., the force times the length of time it acted) starting ball A down the table was clearly twice as great as the impulse brought to bear on ball B. Due to the double-strength impulse propelling it, ball A ended up with twice the resultant velocity that ball B acquired. In our earlier experiments with colliding billiard balls we ignored the time factor involved when two balls smacked together and bounced apart because the time involved was so short that the collisions seemed to take but an instant. Now we realize that the time involved, however short, is nevertheless an important part of the picture if we are talking about measuring the forces acting upon the balls during their collisions, or if we are trying to measure comparative changes in the velocities of those billiard balls.

You will also recall from those billiard ball collisions that we recognized that something was *transmitted* or *exchanged* from one ball to the other and vice versa when they collided and bounced away, but we couldn't quite pin down what, exactly, that something was. The reason for our confusion was simply that we were trying to juggle too many interrelated factors at once. Early physicists had the same problem, and tried to simplify their concepts of what happened when forces acted upon objects to bring about changes in their motion by combining certain closely related factors and using special blanket terms to describe these combinations. We have already seen that the concept of a force acting upon an object cannot be separated from the length of time during which it acts, so the term *impulse* was used to indicate "a given force multiplied by the length of time it acts on an object" or more simple "force $\times$ time."

Similarly, physicists realized that an object with a given mass moving at a constant velocity possessed a certain formidable capacity as it moved along its way. Although it had no "force" connected with it, it possessed the capacity to bring a force to bear upon any other object it happened to encounter or collide with. Just *how much* of such threatening capacity was possessed by any given object moving at any constant velocity depended on *both* the mass of the object and what its velocity happened to be. An object of little mass moving at high constant velocity might have the same "capacity to exert a force on something else" as an object with much greater mass moving at a much lower constant velocity.

In either case, this "capacity to exert a force" would be altered the instant the object in question actually collided with another object and began, so to speak, to "put its capacity to work" by exerting a force on the other object. It would, in fact, lose its "capacity to exert a force" by actually exerting that force, while the other object that was struck would *gain* the same amount of "capacity to exert a force" in the course of the collision, and carry it off with it as it was pushed away by the collision force, bouncing away in a changed direction with a changed constant velocity. But, of course, that other object also *had its own* "capacity to exert a force" which it *lost* to the first object by exerting a counterforce during the collision at the same time it *picked up* the first object's "capacity to exert a force." In short, in the course of the "push and push back" of a collision between these two objects, each gave up its "capacity to exert a force" to the other, and each moved away from collision with a new direction and velocity dictated by the "capacity to exert a force" the other ball had had prior to collision, and had brought to bear during the collision.

If we consider this closely, we notice four interesting things: First, in the course of the collision, there was obviously an *exchange* of capacity to exert a force between the colliding objects; what one object lost, the other gained, and vice versa.

Second, the final or resultant velocity and direction of each object *after* the collision was determined by the capacity to exert a force possessed by the other object *before* the collision, and vice versa. If the capacity to exert a force of the first object—that is, its combined mass times velocity—was greater than that of the second object, the force actually exerted by the first object upon the second in the collision would cause a greater change in the second object's velocity and direction than the force exerted by the second object could bring about in the resultant velocity and direction of the first object. But if, as it appears, the little guy gets shoved around by the big guy in this encounter, we must recognize that the little guy takes on the big guy's "capacity" to shove the next guy around in the course of some similar encounter later. If "might makes right," then there are some equalizing compensations involved!

Third, in such a collision the end results depend not on the comparative masses of the colliding objects alone, nor upon the velocities or directions of the colliding objects alone, but upon the *combined* mass times velocity of each object. In any interaction it is this combined property of mass times velocity of each interacting object—in other words, the capacity to exert a force of each object—that makes the difference and determines the results.

Finally, and more significant than anything else, *the total combined capacity to exert a force of both objects is precisely exchanged* from one object to the other and vice versa in the collision, but *none of it is lost, nor is any new capacity acquired.* If we think of the two colliding objects as a "closed system"—if we imagine, for example, that they are the only two objects existing in the whole universe, with nothing else whatever capable of influencing their motion behavior in any way—then we could say that *the total capacity to exert a force of any closed system is always conserved*—or, if we preferred, *in any closed system the capacity to exert a force always remains unchanged: It can neither be diminished nor increased* (*i.e., created or destroyed*).

If we find something strangely familiar about these statements, it is no wonder. What we have reasoned through and expressed here is nothing more nor less than a conservation law—in this case, the law of conservation of capacity to exert a force. We defined capacity to exert a force as a property of any object, big or small, moving at a constant velocity anywhere in the universe, and saw that that property was equal to the mass of the object multiplied by its velocity. We also saw that that property was not a tangible "thing" but rather a quality or capacity possessed by any moving object which could only be "used" or "spent" if and when the object came into interaction or collision with another object somewhere, sometime. All the same, that shadowy capacity had very real meaning in terms of what might happen in the universe. The baseball flying toward the picture window has a very real and threatening quality about it, even though it is perfectly harmless flying through the air: It has the capacity to exert a force sufficiently great to smash the window if some other object (such as our hand) is not interposed before the ball and window reach collision point.

For simplicity physicists use a different term to describe the threatening mass-times-velocity capacity of a moving object. What we have called "capacity to exert a force" physicists call *momentum*; and they define the momentum of any moving object, just as we have, as *the mass of the object multiplied by its velocity.* Thus, obviously, any object sitting at rest (and thus possessing no velocity) has no momentum—or more accurately, its momentum equals its mass multiplied by zero velocity, which like anything else multiplied by zero equals zero. The moment a force (such as the collision force of another billiard ball striking the first) begins to act on the object, its "state of motion" (in this case its state of rest) is changed. The object begins moving, its velocity increasing steadily from a state of rest (zero velocity) through a period of constantly increasing motion (acceleration) as long as the collision force continues to act on it. And as the object's velocity increases from velocity zero to velocity one to velocity two to velocity three and so forth, instant by instant it acquires momentum from the object colliding with it, and that acquired momentum (the object's mass times its velocity at any given instant) steadily increases from initial momentum zero to momentum one, to momentum two to momentum three and so forth, instant by instant, as the collision force continues to act upon it. This acquired momentum, of course, does not appear like manna from heaven; it is *acquired from* the colliding object, and that object's momentum is steadily given up throughout the collision in precise balance with the *momentum acquired* by the other object. And again, if the colliding objects together were considered a "closed system" isolated from any outside force, there would be an *exchange* of momentum between them, but *no net gain or loss of momentum* in the course of the interaction. The total momentum of the system would be conserved.

From the above, we can see that the period of time the force acts is important in a collision of objects, because it is during that time that the momentum of each object is changing. The impulse of a collision (force times the time the force acts) is the factor that determines the *change in momentum* of either object in a collision, so that the *magnitude* of the force acting on an object can at last be pinned down and identified as equal to the *change in momentum* of the object per unit of time the force acts. In short, the force acting on an object in any collision is equal to the *rate of change* in the object's momentum.

Granted that this seems a long and tricky way around to identify the magnitude of a given force acting on something—but it is the *only way* a force in action can be separated from the length of time it acts. Similarly, the concept of momentum, its exchange between interacting objects and its conservation within any closed system, is the *only way* we can really describe what happens when one object pushes another. We can see now that some of the ordinary, everyday uses of the word "momentum" are more accurate than we realize. The line-rushing halfback does indeed have "momentum" (mass times velocity) as he plunges through to the goal, and the momentum he loses in the plunge is gained by opponents he scatters in his wake. A train does indeed "pick up momentum" as it rolls down the grade: With the same mass, its velocity increases, so momentum is being acquired, in this case from the gravitational field which is pulling the train down the grade. Later the train will lose momentum back to the earth's gravitational field as it rolls up the grade on the other side of the valley. But in the "closed system" composed of the train and earth's gravitational field, momentum is merely exchanged between train and gravitational field; it is never increased or decreased within the system.

Finally, we see how the concept of momentum explains our earlier billiard ball experiments. Before, we spoke of the collisions as "almost instantaneous." But in each case the collision forces lasted long enough for momentum to be exchanged between the colliding balls. Before, we recognized that *something* seemed to be exchanged between the colliding billiard balls, at least in some cases (as, for example, when one ball had twice the mass of the other, or twice the velocity of the other) but we could not say quite what it was that was exchanged. Now we see that it was momentum that was exchanged between the colliding billiard balls, and that exchange took place in *every* collision, not just certain kinds. In fact, the outcome of the collision was determined in every case by the exchange of momentum between the balls during the collision. But we only saw outward evidence of an exchange in those cases where one ball had greater momentum (due either to greater mass or greater velocity) than the other; when momentum was equal on either side of the collision each ball acquired exactly the same momentum it lost and no outward evidence of the exchange seemed apparent—except, of course, that the balls changed direction.

It would be foolish to pretend that physicists created anything new when they began using the word "momentum" to mean "mass times velocity of a given object." They merely gave a new name to a combination of two already known measurable quantities. But new or not, the concept of momentum proved extremely useful in trying to sort out how objects in motion behave and what really happens when they collide or interact. Since so much of early physics had to do with studying the characteristics of matter and motion, the concept of momentum was a powerful tool. But the discovery that momentum was always conserved in any "closed system" interaction between objects—that is, in any interaction unaffected by any outside force—was an enormously important gain for science. And of all the natural laws worked out by Newton and other classical physicists, the law of conservation of momentum remains one of the strongest even today.

Most of us have heard of various "laws of conservation" off and on since early grade school, but what exactly are "conservation laws"? Why are they considered so important and so powerful? Actually, there is no magic connected with them; the conservation laws are simple statements that there are a few distinctive properties of matter and energy that *never change,* no matter what happens. It is this simple fact and this alone that makes them so important to anyone trying to unravel and explain all the myriad peculiar things happening all around us in the universe. If nothing else was clear to early scientists, it was clear that we lived in a universe of constant, bewildering movement and change. The sun, moon, and stars moved; earth's surface changed from moment to moment; forces acted upon objects. Physicists attempting to describe in some sort of orderly, sensible fashion just how things worked in this constantly changing universe desperately needed something solid to hang onto, some firm ground that *never* moved. They needed a few things that were invariably stable and unchanging to use as a baseline against which to measure other things. The conservation laws provided such a baseline.

Understand again that these laws were not blindly accepted as gospel. They were constantly being tested and challenged, and the more challenges they survived, the more important they became. The law of conservation of momentum very early became the most powerful and important of them all, and remains so even today. For one reason, the law covers all sizes, sorts, and varieties of moving objects known to Man, and all sorts and varieties of forces acting on them, no matter what the force might be. The law says that when two objects interact, whether they be galaxies or subatomic elementary particles, their total combined momentum before interaction will still be present after interaction. No matter what forces, changes, upheavals, or holocausts the interaction itself may involve, nothing occurring within the interacting system can *ever* alter the total momentum present. Thus the law is especially powerful because it applies equally to all the different worlds of physics; it crosses the boundaries between the cosmic universe, our everyday world of experience, and the microworld of nuclear physics, as valid in one world as another.

Of course physicists have tried to challenge the law of conservation of momentum repeatedly and tirelessly throughout the years. They are still trying today as new knowledge is gained, as new and puzzling phenomena are observed and recorded. So far the law has survived every challenge; *no one has ever, even once, found a single exception to it.* That is not to say that no one ever will, but such a flawless record makes us wonder. For now, at least, it remains one of the strongest, most fundamental, and universal laws of nature ever discovered. In the next chapter we will see more clearly just how important it really is, and why.

# CHAPTER 9

## *Motion, Momentum, and Universal Gravitation*

From the very beginning in this book it has been our contention that the great laws of physics, however difficult they may be for nonscientists to comprehend, nevertheless have very real and practical significance in our everyday lives. If knowledge of these natural laws is not of practical use to us, if they do not help us understand things that are happening all around us in the world of our personal experience, then we would do well to leave them to the scientists to understand. The fact that a revolutionary concept proposed by Isaac Newton three centuries ago opened up sweeping vistas to the scientists of the day is not enough; as laymen and nonscientists trying to grasp such a concept, it is perfectly reasonable and proper that we should ask: So what? How does this affect *my* life today? *What use is it* for me to understand this obscure and confusing idea?

We should not feel embarrassed, therefore, to ask just such questions about the whole perplexing concept of conservation of momentum that we have been discussing. So physicists recognize this natural law even today as powerful and well established, unshaken so far by any challenge—so what? How does this law touch our lives? What use can we make of it? What does it allow us to do that we couldn't do without it?

That the concept of momentum and its conservation is hard to grasp we cannot argue. It is one of the most difficult ideas we will encounter anywhere in this book—so difficult, in fact, that many elementary textbooks of physics side-step it entirely. Yet the fact remains that the law of conservation of momentum (or certain of its consequences) touches our lives continuously. Whether we are aware of it or not, we are using this natural law constantly in our everyday encounters with the universe around us. Specifically, we use it to *help us predict what is going to happen next,* on the basis of what is happening now, and our predictions are so accurate that we rarely indeed come up with the wrong answers. When we do, for the most part, it is only because conditions are somehow unfamiliar or unusual.

We live in a world of motion in which forces of all kinds are continually acting upon material objects. We pick up a pen, kick a football, watch leaves fluttering in the wind, see the moon rise and set; very few things in our lives remain motionless and unchanged for very long. The concept of momentum tells us how to predict the future behavior of objects, or groups of objects, at rest or in motion as a result of forces acting upon them. The billiard player wins or loses according to his skill in predicting precisely what will happen next to all the balls on the table if he applies a certain force on a certain ball in a certain manner in a certain direction. An automobile driver may live or die according to his ability to predict with split-second accuracy just what will happen on a fast freeway if he accelerates his car a certain amount in a certain direction at a certain time. Momentum concerns us all.

What is more, the concept of momentum simplifies our everyday calculations immensely in a variety of ways. To take a single crucial example, it shows us that we can accurately predict the *over-all behavior* of a whole group of dissimilar objects as a result of the action of some outside force upon that group, even if that group or "system" of objects is very large and spread out over huge areas of space, with individual objects within the group all moving in different directions at different speeds in the most complicated fashion imaginable.

We might never be able to calculate what would happen individually to each object in such a group as a result of the action of the outside force, and we wouldn't need to. The concept of momentum shows us that we can find the correct answer for the group as a whole by imagining that the combined masses of all the objects in the group are concentrated in a single imaginary point in space—the "center of mass" of the group—and then calculating what would happen to that single imaginary massive point as a result of the outside force acting on it. But how would we know our answer was correct if we couldn't actually add up the varying behaviors of the individual parts? We would know simply because it is a long-established law of nature that the total momentum of any isolated group or "system" —that is, the combined mass times velocity of all individual "members" of the system—is always conserved, so that the system as a whole will always behave *as if* all its mass were concentrated in a single point moving at a single velocity and direction, and thus possesses a single unchanging resultant "group momentum."

## CENTER OF MASS AND "AVERAGE BEHAVIOR"

The idea that any object (or any group or "system" of objects we choose to name, no matter how large and diverse) might have all its mass concentrated in some single imaginary point in space which then behaves *as if* it were the whole object (or the whole group of objects) is important enough to bear closer inspection. There is some faint aura of double talk about this notion, some slippery quality that makes us draw back and say, "Now, wait a minute. Is this really true?"

A couple of examples may convince us that it is.

First, suppose we have some awkward, lopsided object like a hickory baseball bat to experiment with and want to determine its mass so that we can measure what happens to it when it is thrown or dropped. Since we know that an object's mass will equal its weight if the measuring is done at sea level, we can easily discover the baseball bat's mass—the exact quantity of matter it contains—simply by weighing it on a good spring balance.

Now consider that it doesn't matter just *how* we go about weighing the bat. Its mass would be precisely the same whether we laid it horizontally across the pan of the balance, suspended it from the balance by a thread, or somehow balanced it on end in order to weigh it. If we wanted to do things the hard way we might balance it horizontally or vertically on the point of a thumb tack, or even on the ultrafine point of a needle—same value for its mass in any case. Balanced horizontally on a needle point it would look "off center" because one end of the bat is thicker and heavier than the other, but we would still come up with a constant value for its mass.

With the bat balanced that way on a needle point, however, we have a singular situation. The only contact between the delicately balanced ball bat and the weighing device is at a single tiny spot—that point of the needle; yet the spring balance registers the same weight as if the bat were lying flat on the pan. Obviously the force of gravity is pulling the bat down on the needle point *precisely as if* all the mass of the bat were concentrated at that single point of contact and all the rest of the bat had no mass at all. At the same time, the needle is pushing upward against the bat as if all the bat's mass were concentrated on that single contact point. Clearly there is some mysterious, dimensionless "point in space" somewhere within that ball bat which acts as if it contains the bat's entire mass!

To the physicist, this imaginary point somewhere within the baseball bat is known as its *center of mass*. Laymen are more familiar with the term "center of gravity" since we are accustomed to measuring objects here on earth within earth's gravitational field where the mass of an object is roughly equal to its weight. But we can see that an object floating somewhere out in space might be "weightless" in the absence of gravitational forces but would still have the same mass there as anywhere else in the universe. In other words, our baseball bat would always have the same *center of mass* wherever it might be, whereas it might well have no center of gravity. Furthermore, in any experiment involving the baseball bat, any place in the universe, we could always treat the bat as if all of its mass were contained within a single point.

Now suppose we are interested not in one object alone but in a *closed system* of two or more objects taken together. For our purposes a "closed system" might consist of any wild combination of two or more related objects that we care to pick, as long as we agree to think of those objects taken together as behaving as an independent and isolated group, cut off in some way from the rest of the universe so that all other objects or forces are by definition "outside" the system. We can imagine a "closed system" as something akin to an ordinary artistic mobile hanging from the ceiling by a thread; its sundry parts may twist and turn individually, but anything that happens to it as a whole must be a result of some force acting from outside it. Just as we cannot lift ourselves by our own shoelaces, a mobile cannot jump loose from its mooring because of any chance concerted action of the individual parts making it up. Yet the very fact that a mobile, however large and complex it may be, can be suspended from a thread attached at a single point suggests that such a closed system of objects has a center of mass just as the baseball bat had.

The fact is that virtually *any* group of objects, no matter whether in motion or at rest, no matter whether close together or scattered all over the universe, can be considered a "closed" or "isolated" system so long as we are willing to exclude everything else in the universe other than the group's constituents as being "outside." And for any such closed system of objects a center of mass exists for the system—an imaginary point in space which, if we could find it, would behave exactly as if it contained all the mass of the system moving with a velocity equal to the resultant combined velocities of each of its constituent parts.

Consider our solar system, for example. Here we have a confusing collection of planets moving around the sun in their various orbits at various distances and with varying speeds, many of them equipped with their own satellites whirling around them in even more perplexing fashion. At first glance this unruly collection of celestial rubble would seem to show little evidence of cohesion as a closed system of objects. Yet sure enough, the entire solar system has a center of mass located somewhere near the surface of the sun, and no internal movement of the sundry component planets, individually or in concert, can cause that center of mass to move an inch. The fact that the solar system's center of mass is moving through space, just as the solar system *as a whole* is moving through space, is a result of forces from outside the solar system acting on the system as a whole (or on its center of mass) to change its velocity and direction.

The same can be said for any other closed system, whether it be an exploding bomb sending fragments out in all directions, a pair of billiard balls colliding and bouncing away from each other, a collection of oxygen and nitrogen molecules moving about at random inside a closed container, or any other group of objects you care to mention. In any such case, since we know that the momentum of the system remains unchanged by anything going on within it, we can predict how the system *as a whole* will behave by regarding the entire system as a single pinpoint-sized object located at the center of mass of the system which behaves in accordance with the *average behavior* of all of the system's component parts.

One important implication of the law of conservation of momentum is that if no outside force is acting upon a closed system of objects, then there can be no change in the velocity of the center of mass of that system. If no outside force of any kind were acting upon our solar system, for example, then all the motion of all the planets and their satellites, each with its own individual momentum, could have no effect whatever on the velocity of the solar system as a whole. Its center of gravity would remain at rest if it were at rest, or would move at whatever constant velocity it always had had. The net effect of all that planetary motion would be zero —no change. If the sun had four thousand massive planets moving about it instead of nine or so, the net effect of all that motion would still be zero. Nothing that happened *within* the solar system could budge its center of mass in the slightest. But if an *outside* force, however small, began to act on the system, things would be different. The velocity of the system as a whole (as represented by its center of mass) would immediately be affected, changing its motion in proportion to the magnitude of the force and in the direction the force was acting.

Of course we know that this is precisely what *is* happening to our solar system. It is not in a state of rest, but is constantly moving *as a whole* in an orbit around the center of the galaxy. And as we will soon see, the outside force that moves our solar system is the resultant of two conflicting forces: a gravitational force tending to drag the solar system in toward the center of the galaxy, and an opposing centrifugal force tending to drive the solar system away from the galactic center on a straight-line trajectory.

If we look at this idea more closely, we will see that this implication of the law of conservation of momentum is nothing more than a restatement, in slightly different terms, of Newton's first law of motion: Any object (or the center of mass of any closed system of objects) which is at rest will remain at rest unless acted upon by some resultant outside force; and any object (or the center of mass of any closed system of objects) which is in motion will remain in motion in a straight line at a constant (i.e., unchanging) velocity unless acted upon by some resultant outside force. The fact that we substitute the center of mass of a closed system of objects for a single object, or substitute the average behavior of the constituents of a closed system for the constant velocity of a single object doesn't alter the law in the least. If anything, it strengthens and reaffirms the law, showing us that it extends to the behavior (as a whole) of closed systems of objects as well as to the behavior of individual objects at rest or in motion.

Thus we can see that the law of conservation of momentum is very closely related to the laws of motion, and together with them helps us understand what to expect when forces act upon objects or groups of objects around us. It provides a powerful tool for predicting whether or not a given object will remain at rest under given circumstances, or in what manner its velocity and direction will change in response to a given force. It buries forever the common-sense idea that a continuing force is needed to keep a moving object moving, yet at the same time it forces us to recognize that any *change* in an object's velocity or direction must be the result of the action of some outside force, whether we are aware of the force or not.

It was such a simple realization as this, so hard come by after years of experiment and observation by the classical physicists, that enabled Newton to recognize at last that the force of gravity which pulled objects to the ground when they were released from his hand might be the same force that kept the planets moving in their orbits, and to extend Galileo's limited concept of gravity as a purely local phenomenon into the great law of universal gravitation which Newton finally defined.

To understand how the one group of concepts led to the other in those wonderful days in history, however, we must first fill in a few important bits of background that are still missing. Up to now we have gotten away with using certain key terms without defining them too fastidiously. Now we must consider certain fine shades of meaning which we previously ignored. In particular we need to understand precisely what we are saying when we use such terms as *speed, velocity, direction,* and *acceleration.*

## THE SIGNIFICANCE OF DIRECTION

We have already seen how a word like "momentum" can have a broad, general meaning in common usage, yet mean something far more precise and specific to the physicist. Oddly enough, there are also certain pairs of closely related terms which the layman may use interchangeably, as if they were synonyms, but which have very distinct individual meanings to the physicist, and still other terms which do not necessarily mean what we assume they mean at all.

Take "speed" and "velocity," for example. Ordinarily we use these words interchangeably to describe how fast something is moving. We assume that "a speed of sixty miles per hour" means precisely the same as "a velocity of sixty miles per hour"—and meanwhile, the physicist cringes. To him the words have distinctly different meanings.

But when should we use "speed" and when "velocity"? The difference is all a matter of *direction.*

"Speed" is properly used as an *abstract* description of a state of motion, whether swift or dawdling, high or low. Speed is measured in terms of distance traveled per unit of time: A car may move at a speed of sixty miles per hour, a snail at a speed of two millimeters per second. A star's light rushes away from its source at a speed of 186,000 miles per second, which we must agree is a pretty high speed. But even though it is concerned with the description of an object's motion, the term speed tells us nothing whatever about *the direction in which the motion is occurring.*

The term "velocity" does, and therein lies the difference. Velocity is properly used as a *specific* and *complete* description of an object's motion, telling us not only its speed but also the direction it is moving. In fact, the term velocity is defined as *speed in a specified direction.* When the physicist speaks of an object's speed he is describing the size or magnitude of its rate of motion in the abstract, without reference to anything else in the universe. When he speaks of the object's velocity, he is coming down to earth, so to speak: he is describing the magnitude of its rate of motion with specific directional reference to *something else,* whether it be to the ground, to the azimuth, to himself, to another observer, or whatnot. And this directional reference to something else always implies speed in a stated direction.

Thus velocity is measured in terms of distance traveled per unit of time in a given (or understood) direction with reference to something else. A car moves with a velocity of sixty miles per hour north (with reference to the ground); part of the sun's light moves with a velocity of 186,000 miles per second toward the earth (with reference to the sun); a Saturn V rocket must achieve a velocity of seven miles per second away from earth's center in order to "escape" from earth's gravitational field and carry astronauts to the moon.

This seemingly quibbling distinction between certain quantities which include a directional element and others which do not is actually so important in exact, descriptive sciences such as physics and mathematics that scientists use special terms to distinguish them. Quantities which have *no* directional element are called "scalar quantities" because they can be fully described by a number indicating magnitude alone, or represented by a point on a scale. Speed is one such nondirectional scalar quantity; so is time, which has no direction (at least not in terms of our three familiar linear dimensions). The mass of an object is likewise a scalar quantity, for its magnitude does not depend on the direction the object is moving; in fact, an object's mass remains the same even if the object isn't moving.

On the other hand, quantities which *do* include an inseparable directional component are called "vector quantities," and are always described both by a number indicating magnitude *and by a specific direction,* like a signpost saying "10,000 miles to Nowhere Much," with an arrow attached. Velocity is one such vector quantity; so is momentum since it is defined as the mass of an object times its velocity. So also is acceleration, defined as the rate of change in an object's velocity (i.e., the total change in an object's velocity divided by the interval of time in which the change occurred). After all, it would be impossible to describe an object's acceleration without specifying in which direction the acceleration took place. For clarity we might emphasize the distinction between scalar quantities and vector quantities in a simple table (Table 2):

TABLE 2

| *Scalar Quantities* | *Vector Quantities* |
|---|---|
| *Speed:* | *Velocity:* |
| 60 miles per hour (no direction) | 60 miles per hour, *thataway* |
| *Mass:* | *Momentum:* |
| 3,000 kilograms (no direction) | 3,000 kilograms x 60 miles per hour, *thataway* |
| *Time:* | *Acceleration:* |
| 10 seconds (no direction) | 6 miles per hour, per second, *thataway* |

Clearly there is a distinction, then, between scalar quantities and vector quantities. But why all the fuss about it? To the nonscientist the difference may well seem pointless; after all, it is perfectly true that the need to distinguish between speed and velocity rarely occurs in our daily life. Normally we can see the direction most objects are moving with reference to ourselves or to the ground and see little need to specify. Even though we almost always *mean* "velocity" when we speak of "speed," nobody gets confused. But the distinction becomes very important in physics when we recall that the most fundamental natural laws describing the motion of objects—the laws of motion—all include very careful reference to the direction an object is moving. They tell us that any object in motion will remain in motion at a constant velocity *in a straight line* (that is, in the direction it is already going) unless acted upon by some outside force. When an outside force does act on an object, its velocity will change *in the direction the force is acting*. And for every action of a force on an object there is always an equal reaction *in the opposite direction.*

We have already seen examples of these principles in action. We have also seen that any time a force acts on an object steadily over a period of time and thus produces a steady or uniform change in the object's velocity, that change in velocity per unit of time—that is, the rate of change in velocity—is spoken of as "acceleration."

Here we have a case of a word which means something more than we may think. Ordinarily we think of acceleration only in terms of an *increase* in an object's velocity per unit time. To the physicist, however, acceleration means *any* kind of change in an object's velocity per unit of time. If a force acts on a billiard ball to increase the "speed" aspect of the ball's velocity, the resulting acceleration is called "positive acceleration." But if some force acts to slow down the "speed" part of the ball's velocity, the rate of change is still called acceleration—in this case, "negative acceleration" or more colloquially, "deceleration."

The notion that a force might cause an object to "accelerate to a stop" seems a little ridiculous at first, but this is only because we normally ignore the full meaning of the term "accelerate." The fact is that objects are "accelerating to a stop" all the time. A billiard ball does this when it collides with another and then bounces away in the opposite direction. An automobile does the same thing when we step on the brake; so also does a rock which we toss into the air. But if the idea of negative acceleration seems a bit odd, there is still another form of acceleration that is even more peculiar.

Remember that an object's velocity is its *speed in a given direction.* Remember also that when an object's velocity is changed by the action of a force, its acceleration is a measure of its *change in velocity per unit of time* that the force is acting. But since velocity is a vector quantity with a "speed" part inseparable from a "direction" part, a force can cause a change in an object's velocity merely by *altering its direction of motion slightly*—by pushing the object off course, so to speak—without either increasing or decreasing its *speed* in the slightest. A strong crosswind, for example, acting on a sailboat could change the boat's velocity from 10 knots due north to 10 knots northeast. The "speed" part of the boat's velocity would remain the same; only the "direction" part is changed—but this action of the crosswind would still result in a true change in the boat's velocity! And by the same token, the amount of change in the boat's velocity per unit of time that the crosswind is acting is a true acceleration of the boat even though the speed of the boat is unchanged.

But what can we call this kind of acceleration? Obviously it cannot be either positive or negative acceleration since the boat is neither speeding up nor slowing down. A new word is needed to describe such "sideways acceleration" or acceleration arising solely from change in a moving object's direction. The term generally used in physics is "angular acceleration": the acceleration of an object along a curving path as a result of a uniform and continuing change in its direction, describable and measurable in terms of an angle of a circle.

Once again we see that *direction of motion* plays a critical role any time we attempt to describe the behavior of moving objects as a result of forces acting upon them. And the concept of angular acceleration immediately draws our attention to a form of motion we have barely considered so far. The laws of motion and the law of conservation of momentum very nicely enable us to describe and predict the behavior of objects as long as they are moving in straight lines, but what about the multitudes of objects in the universe which normally move in curves, parabolas, ellipses, or circles? Must we find a whole new set of natural laws to describe such motion? Fortunately not, for we shall see that the old laws apply perfectly well with certain minor but significant modifications.

## ANGULAR VELOCITY AND CENTRIFUGAL FORCE

Earlier we employed "ideal" billiard balls on an imaginary "ideal" pool table to see how objects moving in straight lines behave in accordance with the laws of motion when acted upon by various forces. We also saw that when our ideal billiard balls collided, momentum might be exchanged between the colliding pair but the total momentum of the "closed system" of two colliding balls was always conserved, the momentum lost by one was gained by the other, and vice versa. Indeed, the laws of motion seemed to suggest that the motion of such objects *had* to be in a straight line. But suppose now that we alter our billiard ball experiment a bit. Suppose once again that we have one target ball resting motionless in the center of the table—call it ball A—and then use our shooting device to start ball B, equal in mass to ball A, rolling toward it on a collision course. But suppose that this time we have anchored ball A to the table by attaching a six-inch bit of thread to it and tacking the other end of the thread firmly to the table top. Now what happens when ball B smacks into ball A?

Certainly not the same thing that happened the first time we tried this experiment! Earlier, when neither ball was attached to the table in any way, you will recall that moving ball B came to a halt when it collided with stationary ball A, while ball A was sent rolling away from ball B by the force of the collision. In other words, the two balls exchanged momentum in the collision; the originally stationary and momentumless ball A picked up all of ball B's momentum while ball B became stationary and momentumless. But this time something rather different happens. Once again moving ball B collides with stationary and momentumless ball A. Once again momentum seems to be exchanged in the course of the collision, for Ball B comes to a halt, while ball A starts moving—but this time, *not in a straight line*. This time, like a dog on a leash, ball A moves away from collision *along a circular path* with the tacked-down end of the thread as a pivot point and the six-inch length of the thread as the radius of the circle (see Fig. 10).

But what has happened to Newton's first law of motion? It claims that an object in motion will remain in motion *in a straight line* unless acted upon by some outside force, doesn't it? Indeed it does. But here, ball A, once set in motion, is continuing in motion along a *circular* path. How can this be? The collision seemed identical with the one before. The billiard balls are the same ones we used before. Why are the results of the collision different this time?

The answer, of course, is right before our eyes in Newton's first law. Ball A set in motion by the collision would remain in motion in a straight line *unless acted upon by some outside force*. The fact that we see ball A moving in a circle rather than a straight line indicates that there must be some outside force making it do so, whether we recognize the force when we see it or not. But what could the force be? There is only one way such a force could be acting: through the thread attaching the ball to a fixed point on the table six inches away!

If we look at this collision through our slow-motion lens and examine the forces acting on ball A instant by instant from the time of collision on, we see clearly what is happening. The instant before collision ball B is approaching collision point with momentum equal to the ball's mass times its velocity. Ball A at that instant has *no* force acting upon it; it is at rest, with a velocity of zero and hence a momentum equal to zero.

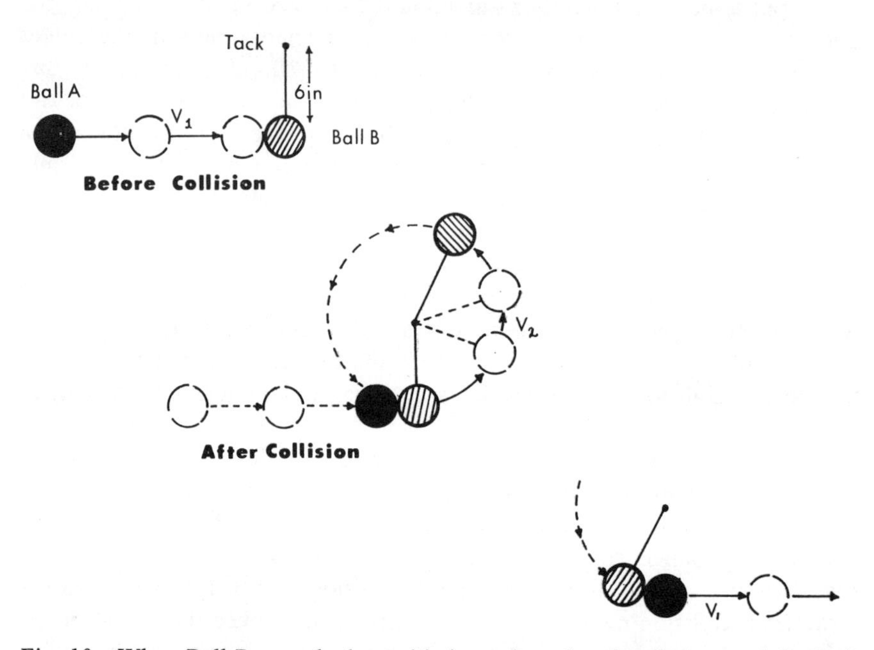


*Fig. 10* When Ball B, attached to table by a thread and tack, is struck by Ball A, straight-line velocity V1 is transformed into *angular* velocity V2. When Ball B completes its circular "orbit" and strikes Ball A, angular velocity V2 is re-transformed into straight-line velocity V1.

Then in the course of the collision ball B exerts a force on ball A setting it in motion, while ball A exerts a counterforce on ball B bringing it to rest. Ball B has lost its velocity (and hence its momentum) while ball A has gained that lost velocity in a straight line in the direction ball B had been moving; it has picked up ball B's lost momentum. So far everything is the same as in the previous experiment. But this time the moment ball A starts to move off in a straight line with its newly acquired momentum, a new force begins acting through the thread to pull it off course. That new force in fact begins *tugging* ball A in toward the pivot point at the same time that ball A's inertia keeps tugging it away from the pivot point along the straight-line tangent that the ball would follow if the in-pulling force were not acting upon it.

The result? A compromise. Ball A at each instant moves in the only direction permitted by the two opposing forces acting upon it. This means that at each instant it is pulled off course exactly as much as it is tending to move back on course. Neither the off-course force nor the on-course force can overcome the other, so the ball follows a resultant circular path, constantly pulled in by the thread and pulled out by its own momentum (see Fig. 11a). Assuming, as we did before, conditions of no friction and no

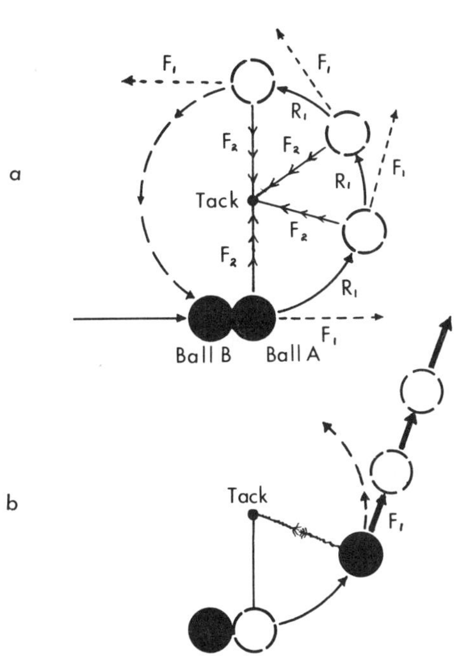


*Fig. 11* F1 represents the inertial force acting on Ball A; F2 is the centripetal (in-pulling) force exerted through the thread; R1 is the resultant path of motion of the ball. If thread breaks, centripetal force can no longer act, so Ball A then follows course determined by inertial force F1.

gravity, we could say that the circling ball is "caught in orbit" around the pivot point, and would continue circling the pivot point forever unless some other outside force acted upon it, or unless one or the other of the two opposing forces acting upon it suddenly quits acting.

Of course the latter case is not impossible. Suppose that the thread tugging the ball in toward the pivot point were a slightly frayed thread, and suddenly broke under the continuing strain of tugging at this stubborn ball. The instant that the thread breaks, the in-pulling force can no longer act on the ball. So what will it do then? Obviously, it will then have only one force acting upon it—the collision force which gave it straight-line velocity and momentum. Thus, if the thread snaps, the ball will continue moving in a straight line out at a tangent to the pivot point from the instant that the in-pulling force ceases (see Fig. 11b).

But barring such an "accidental" occurrence, we have ball A trapped in a circular path of motion by the effect of two forces acting upon it from the time of collision onward. What can we say about its velocity (which must include an element of direction), or about its momentum gained from ball B at time of collision? What happens to our principle of conservation of momentum, which insists that the momentum lost by ball B must equal the momentum gained by ball A? Earlier we saw both velocity and momentum as straight-line qualities, always associated with direction of an object's motion in a straight line. What can we say about these things in the present case in which ball A's motion follows a circular path?

First, if we could measure carefully, we would find that the speed part of ball A's velocity after collision is *constant*, just as the former speed part of ball B's velocity was constant before collision. Ball A's speed in its circular orbit is neither increasing nor decreasing. The direction part of its velocity, however, is *changing constantly,* instant by instant, as it is forced to move off course. Thus we must say that ball A is *accelerating* constantly around a pivot point at a distance of six inches even though its speed neither increases nor decreases. It undergoes a constant *angular acceleration* as it moves. But what can we call the odd kind of velocity ball A has acquired—a velocity in which the direction keeps uniformly changing from instant to instant? Because this is a *special kind* of velocity, physicists give it a special name and call it "angular velocity."

We can see that the angular velocity of ball A is different from straight-line velocity because the direction part of its velocity is uniformly changing. But what controls the speed part of its angular velocity? The collision force of ball B originally set ball A in motion in the first place, and contributed a straight-line element to its resulting angular velocity "in orbit," but the inward tugging of the thread must also contribute something. And with further experiment, we would find that it does. If the thread holding ball A in orbit were shortened to three inches—half its original length—we would find the speed part of the ball's angular velocity to be twice what it was before; but if we increased the thread to twelve inches—twice its original length—the speed part of ball A's angular velocity would be only one-half what it was originally.

In other words, wherever *angular* velocity is concerned, the speed element is inversely proportional to the distance between the moving object and the pivot point, i.e., to the radius of the circle. When the radius is made smaller, the object speeds up proportionally, and when it is made large, the object slows down proportionally. But distance from object to pivot point is not the only factor influencing the speed part of the object's angular velocity; the object's mass also plays a part. If we substituted a ball with twice the mass for ball A, its speed in orbit would be only half that of ball A, whereas if we used a ball only half as massive as ball A, it would travel in orbit at double the speed achieved by ball A.

If this all seems to sound vaguely familiar we ought not to be surprised, for we are talking about a relationship between an object's mass and its velocity (in this case, its *angular* velocity) which sounds a great deal like our old friend momentum. In discussing straight-line motion, we saw that the momentum of a moving object was equivalent to its mass times its (straight-line) velocity. We also saw that in our present billiard ball experiment, moving ball B had straight-line momentum at the instant before colliding with the motionless but "captive" ball A, and gave up that momentum during the collision. Ball A at the same time took on something very much akin to the momentum lost by ball B but not quite the same because the newly acquired velocity of the captive ball A was angular velocity. Thus ball A acquired a "different kind" of momentum equivalent to its mass times its angular velocity which was greater or smaller according to the distance of the ball from the pivot point.

Just as the "captive" ball A acquired a "different kind" of velocity from ball B in the collision, a velocity described as "angular velocity," it also acquired ball B's momentum transformed into a different kind. This different kind of momentum is also given a special name, "angular momentum," since its velocity factor is angular velocity. In either case the difference arises from the motion of the object in a circular path rather than along a straight line due to the continuing action of an inward-pulling force upon the object. Thus, if we define an object's angular momentum as its mass times its angular velocity, we would then find that all of Newton's laws of motion apply to an object moving in a circular path around a pivot point exactly as if the object were moving freely in a straight line.

Furthermore, we would find that angular momentum is conserved in any closed system of objects moving with angular velocity, just as straight-line momentum is conserved—and that straight-line momentum can be *translated* into angular momentum and vice versa any time an object moving in a straight line interacts with an object capable only of angular or rotational motion. In fact, if we followed out the results of our imaginary collision between straight-moving ball B and fettered ball A on our friction-free and gravity-free pool table, we would see an amazing thing: Ball B colliding with ball A would set it into angular motion in a circle around its pivot point while the impact would bring ball B to a halt. But with all ball B's momentum translated into angular momentum of ball A, ball A would swing completely around its circle to smack in turn into the back side of ball B. In this impact ball A's angular momentum would be completely given up to ball B, translated back into straight-line momentum, so that ball A would halt and ball B would continue on its interrupted straight-line course in the same direction it had been traveling before its first collision, and *at the same velocity it had originally!*

Clearly, momentum and angular momentum are completely equivalent except that the latter includes an additional "rotational movement" factor related to the distance between object and pivot point—the radius of the circle. In a case in which an object has angular velocity, and thus angular momentum, the "mass" part of the momentum constitutes an inertial force urging the object out on a straight-line course; and indeed, the physicist speaks of the mass of an object in rotational motion as its "moment of inertia." The force applied to the rotationally moving object through the thread that holds it to the pivot point is an off-center force acting on the object, not changing its speed but pulling it constantly off its straight-line course; such a force is called a "torque" (i.e., a twisting force pushing the object off balance and thus into rotational motion). And the laws of motion and conservation of momentum apply equally to straight-line motion and rotational motion, as long as all the factors in rotational motion include the radius-of-a-circle factor and an angle factor in place of a distance factor.

## ANGULAR MOTION AND NATURAL LAW

The whole concept of rotational movement of objects, and the notion that angular momentum had to be analogous, in some way, to straight-line momentum, had kept the early physicists puzzled and confused. To Newton, however, the problem was not "How are the two different?" but rather "How are they the same? What are the common denominators?" Above all, Newton was convinced that the universe was *orderly* and that such natural laws as existed were few, simple, and broadly applicable. Motion, he must have reasoned, was motion, and he could not have believed for a moment that one set of rules would apply to one kind of motion and another different set of rules to another kind. When he saw the close analogy that seemed to exist between rotational or angular motion on the one hand and straight-line motion on the other, and when he saw that rotational motion seemed to be no more than a *variation* of straight-line motion which occurred when a second force in addition to straight-line inertial force was acting on a body, he recognized that his laws of motion must apply equally well to either form of motion so long as a "rotational motion factor," so to speak, was taken into consideration in the case of rotational motion.

Furthermore, this idea seemed to be confirmed from observations in nature, for it was obvious to all physicists in those days that angular motion was in fact present in nature at least as commonly as straight-line motion. Indeed, it had become clear, in the centuries before Newton, that angular motion was the rule rather than the exception, at least so far as the relative motion of the sun, the moon, and the planets was concerned. For almost a thousand years men had believed firmly that the sun, the moon, and all the planets moved in some kind of complex circular fashion with the earth at the center of them all. In 1543, after a lifetime of study, Nicolaus Copernicus had finally overthrown this deeply entrenched and hallowed idea: He had demonstrated that the observed facts simply did not fit the theory. He contended, rather, that although the moon did indeed revolve in circles around the earth, the earth and the other known planets turned in even more huge circular orbits with the sun as their "pivot point."

Copernicus was so shaken by the enormity of this idea, and so aware that the full prestige of Aristotle and the full earthly power of the Church supported the old earth-centered theories of Ptolemy that he waited until the year of his death to publish his own contradictory findings. Even then they were by no means universally accepted by contemporary scientists. But the real strength of his observations lay in the simple fact that anyone with a good telescope could *repeat them* and come up with the same answers he had come up with, and his theories set such great astronomers as Tycho Brahe in Denmark to work studying the heavens again.

After Brahe's death in 1601, his student Johannes Kepler pieced together the huge volume of observations Brahe had made, and confirmed once and for all that the planets followed curving paths of motion around the sun. Their rotation was not in a perfect circle, however, as was the case with our "captive" billiard ball; instead, Kepler showed that the earth and the other planets moved in an oval or elliptical path around the sun, with the sun not at the center but occupying the position of one of two foci of each such ellipse.

These observations had all become known during Newton's lifetime or even earlier; he was perfectly aware of them when he was working out his universal laws of motion, first published in 1687. He realized that such massive objects as the earth or the other planets *ought* to be moving in straight lines and indeed would *have* to be unless some hitherto unrecognized force were acting to "pull them off course." The observed fact that they were moving not in straight lines but in great curving orbits was not a matter of divine whim, but the *result of a force* that was continuously tugging planets away from the straight-line courses they would otherwise be following. We saw the same thing, on a small scale, with our captive billiard ball. It did not follow a circular path because nature had suddenly changed her mind about straight-line motion, but because straight-line motion was rendered impossible for that billiard ball by a force acting on it through the thread to pull it in toward the thumbtack pivot point. Similarly, Newton saw that by all rights earth's moon *ought* to be flying out into space away from the earth along a straight line, and the fact that it obviously was not doing so could mean only one thing: that *some* force was at work continuously tugging the moon in toward the earth, against its own better judgment, so to speak.

At the same time, Newton thought about other examples of angular motion which were observable closer to home. It was known that a projectile such as a rock hurled in a horizontal direction began falling toward the earth the moment it was released from the hand. Its straight-line momentum kept it moving horizontally, all right, but some other force simultaneously pulled it off course toward the ground. Thus the rock's trajectory was never a straight-line path, but rather a curving line representing the resultant instant-by-instant compromise between two opposing forces. Obviously, the "other force" that opposed the rock's straight-line inertial force by tugging it down toward the ground was nothing more than Galileo's old, familiar force of gravity. If the rock were hurled a second time with greater force than before (and thus with greater straight-line momentum and greater straight-line velocity) it would be able to travel considerably *farther* before gravity finally brought it to earth, but it would still begin to fall off course toward the ground the instant it was released. Its trajectory would be a longer, flatter curve than before, but gravity again would win.

But now suppose we took that rock to a high mountain top, and launched it horizontally with such an exceedingly high velocity that the curving surface of the earth fell away from it just as fast as gravitational force pulled it downward. Imagine at the same time that there was no atmosphere, so that the rock was unopposed by any air resistance. What would happen then? When would the rock finally reach the ground?

The answer was clear to Newton: *It wouldn't*. In such a case, that rock would continue to travel around and around the earth, always seeking to follow the straight-line path away from earth's surface dictated by its inertial force, yet always tugged earthward by a second force, gravity. Because of its very high velocity, that rock would *never* fall downward far enough nor fast enough to strike the earth's surface, even though it would continually be falling toward the center of the earth just as any tossed rock falls toward the center of the earth. Thus the rock's *direction of motion* would continually be changing so that it would continually be *accelerating downward* without any change of speed—perpetually accelerating toward the center of the earth.

Newton realized that the continued acceleration of a rock under these imaginary circumstances could be nothing other than *acceleration due to gravity*. If the rock were slower moving, with less straight-line momentum, gravitational force would presently tug it down to earth. If it were faster moving, with greater momentum, it would pull farther and farther away from earth's center and presently "escape" on a straight-line course out into space. But if the rock's outward inertial force *exactly counterbalanced* gravity's tugging, that rock would continue to circle the earth in orbit, over and over again, indefinitely.

Very much, indeed, the way the moon circles the earth in orbit, over and over again, indefinitely . . . !

It is easy now to see where Newton's reasoning was taking him. It was obvious that *some* unrecognized force was acting across empty space to hold the moon on a leash, so to speak, forcing it to travel in perpetual orbit around the earth instead of flying off on a straight-line tangent into space.

Could it be possible that that "unrecognized force" was the *force of gravity* pulling the moon continually earthward even as it struggled continually to move out and away from earth? Could it be possible that earth's gravitational force was not just a local phenomenom occurring on the surface of the earth, as Galileo had imagined, but a force that could act upon objects across such enormous expanses of space as the distance from earth to the moon?

Newton saw that it *had* to be possible, no matter how incredible it seemed. The moon's motion in orbit followed the laws of rotational motion perfectly, so long as a factor for earth's gravitational pull was always taken into account. But Newton also realized that far more was afoot than just this, if his own laws of motion were really valid. Gravitational force could not be an exclusive property of the earth alone, for if the earth were exerting a gravitational force on the moon, influencing its path of motion, the moon must simultaneously be exerting an equal but opposite gravitational force on the earth!

The idea must have seemed preposterous, but Newton was too immaculate a scientist to accept a law of nature when it proved convenient and disregard it when it became clumsy. He may even have seen proof that the moon had gravity tugging away at the earth in the waxing and waning of the tides: whenever an ocean was facing the moon, the tides were low, as though the whole mass of the ocean's water were drawn up into a bulge by some powerful moonward force; a few hours later, when the ocean was turned away from the moon that force was relaxed, allowing the tides to rise at the ocean's edges. But if the moon also had gravitational forces at work, why didn't the earth revolve around the moon instead of vice versa? Obviously because the mass of the earth was so much greater than that of the moon that the moon's gravity could not overcome earth's momentum as readily as earth's gravity could overcome the moon's.

The idea was awkward, but it fitted in with other observations. The earth perhaps held the moon on a controlling gravitational leash—but in the case of earth's motion around the sun, it was *earth* that seemed to be on a leash, moving around the sun in a perpetual elliptical orbit so huge that it took the earth a full year to make one circuit. This could only mean that the sun also possessed a gravitational force—a force so powerful it could reach out to affect the motion of planets millions of miles away from it. But like our captive billiard ball moving in a faster circle when the thread was short and more slowly when the thread was long, the sun's effect on the planets seemed related to the distance they were away. The inner planets Mercury and Venus circle the sun at higher angular velocity than the earth does, while Mars's angular velocity is lower, and far Jupiter's velocity in orbit is ponderously slow.

Bit by bit, Newton fitted the pieces together. Gravity was not an exclusive property of earth, nor was it a one-way street, but a mutual force of attraction existing between any two celestial bodies, a force precisely the same in nature as the familiar "local" gravitational force existing between the proverbial apple over Newton's head and the earth below him. But as a final stroke of genius, Newton went one step further: He realized that gravitational attraction existed not just between the sun, the moon, and the planets, but that it was a universal property of *all* objects—a force of attraction which exists, quite literally, between *any two objects anywhere in the universe,* no matter how large or small they may be, no matter how close together or far apart, no matter where they are located nor how fast or slow they are moving.

But the *magnitude* of the gravitational force between any two objects, Newton realized, is not necessarily the same as between two other objects; it varies from one situation to the next. By applying his laws of motion to what was known of the orbital behavior of the sun, the moon, and the planets, he searched for some consistent principle or relationship by means of which anyone could calculate the gravitational force between any two objects anywhere. He found, for one thing, that the magnitude of the force depends upon the respective masses of the objects: Massive objects attract each other more strongly than less massive objects. Again, the distance between the objects makes a difference, regardless of how massive they are: The farther apart they are, the less they attract each other. In any case, the force of gravity is always exerted in a straight line between the objects in question—or between their centers of mass—and in any case a fixed, unchanging number, a "universal gravitational constant" is needed in order to *compare* the magnitude of the force between one pair of objects with the magnitude of the force between any other pair of objects, no matter how dissimilar they may be. It was this gravitational constant that Newton calculated which permitted him to set the masses of any two objects, the distance between them, and the gravitational force between them in the proper proportion—in effect, to provide a "yardstick of perspective" for comparing the gravitational force between two grains of sand located 200,000 miles apart, for example, and two earth-sized planets located 200,000 miles apart. Without such a constant the magnitudes of these two gravitational forces could not be compared in terms of actual values. It might be possible to say that the force present in one case would be larger or smaller than in the other case, but not *how much* larger or smaller.

These calculations were laborious, but as in so many cases it was the basic idea, not the mathematical details, that threw open the door. The idea that every object in the universe attracted every other object in the universe was breathtaking in its sweeping implications; the task of determining *how much* attractive force existed between any two given objects on the basis of their respective masses and the distance between them was more of a mopping-up exercise. Newton finally worked out a famous formula relating the masses of any two given objects, the distance between their centers of mass and the magnitude of gravitational force that existed between them in simple mathematical terms that could be applied universally to any two objects one might choose. In its simplest form, this formula states that the gravitational force $F$ that exists between any two objects anywhere (call them Mass 1 and Mass 2) always *increases* in direct proportion to the product of their two masses (Mass 1 times Mass 2) and always *decreases* in direct proportion to the distance between their two centers multiplied by itself. This unvarying proportional relationship can be expressed for any two objects, using the symbol $\propto$ to indicate proportionality, as follows:

$$F \propto \frac{\text{Mass } 1 \times \text{Mass } 2}{R \times R} \quad \text{or} \quad F \propto \frac{(M1 \times M2)}{R^2}$$

but to obtain specific values for the gravitational force $F$ existing between any two specific objects, it is necessary to use Newton's universal gravitational constant $G$ as a proportionality constant in an algebraic equation:

$$F = G \times \frac{(M1 \times M2)}{R^2}$$

Most of us have encountered this familiar "inverse square" law of universal gravitational attraction between objects at one time or another. Once the gravitational constant $G$, always the same in all cases, had been measured, this equation made it possible to calculate the amount or magnitude of gravitational force that existed between any two objects in the universe. Essentially, the law of universal gravitation as expressed above means simply this:

1. That an attractive force exists between any two objects in the universe;
2. That in any given case the magnitude of the force depends directly upon the product of the two objects' masses according to a specific rule, so that the larger the product of the two masses, the greater the gravitational attraction between them;
3. That in any given case the magnitude of the gravitational force between two objects also depends directly upon the distance between the two masses according to a specific rule, so that the force decreases by a factor of the square of the distance. In short, the force between two objects is greatest when they are the closest together, and falls off in geometric progression as the objects move apart;
4. That in any given case the force exerted upon one object by the other is equal in magnitude to the force exerted by the other object upon the first, and this mutual force of attraction is always exerted along a straight line between the centers of mass of the two objects.

This law of universal gravitation was a revolutionary concept when Newton first expressed it, not merely because it enabled physicists to *measure* gravitational forces existing between the planets and thus predict the effect those forces would have on planetary motion, but even more because it demonstrated that the force of gravity, once thought to be a strictly local phenomenon affecting only the motion of objects on or near the earth's surface, was in fact a universal force that existed between any two objects in the universe to one degree or another. What was more, Newton's law of gravitation explained clearly why gravitational forces seemed to be active in some cases and absent in others. If the gravitational force of the sun were great enough and could extend far enough into space to pull the earth away from its normal straight-line course and force it to move in a curving "captive" orbit around the sun, why weren't all sorts of smaller objects on the earth's surface pulled together by their mutual gravitational attraction? The answer was that the masses of such small objects were so small that the gravitational attraction between them was unmeasurable, far too small to produce enough force to overcome the inertia of the objects and make them move toward each other.

But was it possible, then, to prove that any force really existed between two objects of comparatively small mass? Indeed it was; in his laboratory in England in the year 1797-98, almost a century after Newton published his theory of universal gravitation, Henry Cavendish set up an experiment to demonstrate whether gravitational attraction between two objects could actually be measured in the laboratory or not. For his test objects he used two heavy iron dumbbells each balanced and suspended from a thin wire in a glass container which excluded any air currents which might cause the dumbbells to turn. After the dumbbells had become perfectly still, Cavendish gently moved them closer together, and found that at a certain distance apart the metal bells could be seen to attract each other, and the amount of attraction could actually be measured by the movement of a needle-pointer attached to each wire. Not a force of great magnitude, this attraction, but a force just the same—a measurable force that could be nothing but the attraction of each dumbbell for the other. And to top it off, Cavendish found that the magnitude of the attractive force he measured was almost precisely the magnitude predicted for dumbbells of this mass held at this distance apart, according to Newton's equation!

Gravitational attraction between objects was real, beyond doubt, and Newton's equation for calculating the magnitude of the attractive force in any given case was valid, but the force of attraction between small objects is so very tiny that it has no effect on the motion of the objects. If one of the objects is very massive—as massive as the earth, for example—and one very small—only as massive as an apple, say—each will attract the other with equal gravitational force, but the *effect* of the earth's gravitational force upon the apple's motion is far greater than the effect of the apple's gravitational force on the earth's motion, so that the apple is pulled toward the earth far more readily than the earth is pulled toward the apple.

Similarly, the motion of the earth is affected more readily by the gravitational attraction of the far more massive sun than the sun's motion is affected by earth's gravitational pull. But here distance too plays its part. Earth's gravitational effect on the motion of its nearby moon is greater than the more distant sun's effect on the moon's motion, so that the moon is held in orbit around the earth. The distance between earth and moon is so much less than the distance between moon and sun that the earth-moon gravitational force is much greater than the moon-sun gravitational force and thus the moon orbits the earth and not the sun. If by some disaster, however, the earth were suddenly to vanish completely from space and thus suddenly cease to exert its gravitational attraction upon the moon, the moon would then shift its allegiance to the sun, responding to the sun's gravitational pull. Indeed, the orbits of all the planets and their various satellites are as they are solely as a result of the interplay of conflicting gravitational forces acting upon them. Not only does the sun exert a powerful gravitational attraction on the earth; each of the planets, large or small, exerts its own gravitational attraction on each of the other planets (and vice versa) in a continuing cosmic tug-of-war.

In fact, sometimes the gravitational effect of one planet upon another can actually be observed to shift or alter a planet's "normal" orbit around the sun, at least temporarily. We know, for example, that the planet Neptune is "perturbed" in its usual orbit around the sun from time to time because of the conflicting gravitational attraction of its neighboring planet Uranus at such times as both planets happen to be on the same side of the sun at the same time and approach comparatively near to each other in their orbits. We also know that the orbit of Uranus is similarly "perturbed" by Jupiter's gravitational attraction under similar circumstances.

As a matter of fact, it was the observation of a mysterious "perturbation" or temporary alteration in the normal orbit of Uranus around the sun that first convinced astronomers of the 1700s that another still-undiscovered planet must be present somewhere in space out beyond Uranus. From time to time the nice, orderly orbit of Uranus seemed to go temporarily out of whack for no apparent reason; astronomers reasoned that only a strong gravitational force between Uranus and some other massive planetary body passing comparatively nearby could possibly throw a planet the size of Uranus even temporarily off course. By measuring the amount that Uranus was bounced around by this shadowy stranger in space, astronomers were able to predict not only that the planet Neptune had to be out there, but could even make surprisingly accurate predictions of the shape of its orbit, how far beyond Uranus it lay, how long it took to make one turn around the sun, and how massive it was. And when the planet was at least spotted in the telescope and studied, virtually all these earlier predictions proved to be correct!

The final cataloguing of Newton's laws of motion, the twin concepts of conservation of momentum and conservation of angular momentum, and the discovery that gravity was a universal force that kept the planets in their orbits and existed between any two objects anywhere in the universe were among the greatest achievements of a Golden Age of classical physics. In the foregoing chapters we have discussed at some length how these discoveries were made and how these concepts grew, because taken together they formed a coherent picture of how things work in the world of our everyday experience, the only world of physics those early scientists could examine. It may have seemed that Sir Isaac Newton received the lion's share of the credit for expressing these great, basic concepts, but Newton himself once wrote, "If I have seen further than other men, it is by standing upon the shoulders of giants." And indeed, these concepts arose not from the work of any one human mind but from a great intellectual ferment that had been evolving for centuries, beginning with Galileo, embracing the lifework of such giants as Copernicus, Tycho Brahe, Paracelsus, and Johannes Kepler, and was reaching full flower in the time of Newton.

Out of this ferment came a few concepts, a few simple but sweeping working rules that seemed to describe a great deal that was known to be happening in the universe, and that seemed always to be verified whenever they were put to test. These laws of classical physics strongly confirmed the ancient assumption that everything happened in the universe in an orderly fashion. They predicted the end results of the motion and interaction of objects and the effects of forces acting upon objects, no matter whether here on earth or in the far reaches of space, no matter whether the objects or the forces were large or small. Because they seemed so universal in application, they were exceptionally useful laws, for they explained what was happening and allowed men to calculate what would happen next.

There was nothing about these classical laws of nature to suggest *why* they existed. In working out the law of universal gravitation, Newton made no attempt to explain *why* massive objects attracted each other, nor what caused the force of gravity to exist in the first place. The law merely *described what happened*. It stated under what circumstances gravitational forces existed, and what effects these forces had upon the motion of objects whether on earth or in the farthest reaches of the heavens. And when it came to helping physicists understand the world they saw around them, these laws seemed powerful indeed. Indeed, they were so apparently universal, and seemed to describe so many natural phenomena that physicists of a hundred years ago began to feel that they were approaching the end of the road to discovery. The ancient dream of discovering a few simple, interrelated principles which together would describe *everything* that happened in the universe seemed at least to be coming within reach. Only a few knots still remained to be unraveled.

Or so it seemed. As we will see, it was not to be quite that simple; those complacent physicists were heading for a rude awakening, their dream soon to be shattered. But before we consider the strange discoveries that shattered that dream, we must take time to look more closely at the nature of the physical matter that composes the universe as we see it, and discuss a strange and puzzling manifestation of matter which has only been clearly recognized as such in the last hundred years or so—a form of matter that we know more familiarly as "energy."

# CHAPTER 10

## *The Forms and Shapes of Matter*

In the previous few chapters we have been discussing in some detail quite a wide variety of abstract ideas and esoteric concepts. We have been concerning ourselves with frictional forces, air resistance, and other forces difficult to pin down precisely. We have been considering the behavior of imaginary objects existing under imaginary conditions in the course of imaginary experiments. We have also been dealing with that most mysterious and peculiar force of them all, the force of gravitation, capable of reaching its shadowy fingers across millions of miles of space in order to juggle planets around in their orbits, yet so weak that its presence between two small objects in an earthly laboratory can be measured only with difficulty, if at all. We have encountered a curious quantity known as "momentum" which can, it seems, be transferred from one object to another under certain circumstances, and even be translated from one form to another, but can never be created or destroyed. We have explored the historical development of certain deceptively simple natural laws, and have seen some surprising implications of these laws as they were understood and interpreted by physicists in the seventeenth, eighteenth, and nineteenth centuries.

Many of these concepts, admittedly, have been difficult to grasp, and even more difficult for us to relate to the "real" world of earth, sky, and sea in which we live—the world of our own experience. And in pursuing these concepts we have, again admittedly, by-passed some very practical and fundamental aspects of the universe that would help us greatly in making sense out of the more esoteric concepts. It may be with a sense of relief, therefore, that we now come down to earth for a while and consider one of the most basic and commonplace of all facts that we must cope with every day: the fact that the physical universe we live in is made of material objects composed of a very solid and tangible stuff known as *matter*, and that it is material objects of this sort which we constantly see acted upon, altered, or influenced by the wide variety of forces we have been discussing.

Earlier we had a great deal to say about various kinds of objects at rest or objects in motion. For the most part we selected objects that were familiar to us all in one form or another. But we neglected to pause and consider the "matter" of which those objects were composed. It is important now that we do so before we move on, for the matter making up the physical universe exists in a bewildering variety of shapes and forms, and possesses some surprising and extraordinary characteristics which we need to know about.

Of course in our everyday lives we rarely give much thought to the nature of material objects all around us. Most of us feel relatively confident dealing with chairs, books, drinking water, or sledge hammers. We are natively aware that matter exists in various forms and have learned to use the varying characteristics of different forms of matter to our own best advantage. When we sit down on a chair, we expect it to support us, whether comfortably or uncomfortably, and when we spill a cup of coffee on the rug we expect it to make quite a mess and be difficult to recover intact in its original form. Practically speaking, we know pretty well how far we can go with any given object on the basis of the form of matter that composes it: we use steel girders rather than wooden beams to support our bridges, because we know that steel has a degree of strength that wood lacks; we would not normally try to wash with a dry bar of soap any more than we would try to dry our hands with sandpaper; and we can predict with some accuracy that the man who tries breathing water instead of air is likely to find himself in trouble.

All such "common-sense" considerations are a natural and normal part of our everyday working knowledge of the universe around us. We don't have to be told these things. But in addition, we are also aware that much that happens in the universe around us depends very heavily upon certain particular characteristics of matter in one form or another, and that the effect that a given force may have upon an object may vary widely according to what form the matter making up that object happens to be in. The thirsty man on the desert will draw small comfort from a canteen full of water vapor, although the month before he was bemoaning the flash flood that swept away his house. The wings of a small airplane are specially built to provide the lift necessary to get the plane off the ground in regions where earth's atmosphere is the most dense; the wings on the new supersonic transports must be quite remarkably different if those planes are to fly at great speeds and high altitudes, while wings of any sort on the lunar excursion modules designed for soft landings on the moon would be totally useless. All three of these craft are designed to move men from one place in the universe to another in response to various kinds of driving forces, but it is the form of matter *through which* they must move that determines the details of their structure.

But if we feel reasonably confident of the predictable nature of the mundane material objects around us, we must bear in mind that not everyone has been as confident as we are either in past ages or even in modern times.

The ancient Greek philosophers quite seriously questioned whether matter existed at all, except as the creation of their own imaginations. With the evolution of experimental science (and particularly of experimental physics) this particular notion seemed pointless to pursue and was generally forgotten. Scientists came to assume as a basic working axiom that matter was indeed real and did indeed exist quite independently of the influence of the human mind. Yet, strangely enough, in modern times physicists have come full circle and once again are questioning seriously whether matter as such has any real physical existence that can ever be specifically defined, and if so, they are wondering precisely what in the universe actually is "matter" as such and what isn't. For practical purposes, of course, we must go along with the assumption of the classical scientist that matter is real and does exist, but we will be able to understand the uncertainty of the modern physicist, and the truly remarkable discoveries he is making because of such concerns, far better if we pause briefly to review the shapes and forms of matter as it was regarded by the physicists of Newton's time and later.

## THE ANCIENT ELEMENTS

Since time immemorial men must have been worrying about the shape or form in which substances in the natural world around them were to be found, and about the elementary substances they felt must in one combination or another go to make up all those substances. Even the prehuman ancestors of man were apparently aware that some substances were more practically useful than others. When those ancient creatures first took up hand weapons to help them in the hunt, they seem to have found the long bone from the foreleg of certain antelope to be a more satisfactory club than objects made either of wood or of stone. And even the later cavemen must have been aware on a very practical level that various natural substances were more useful in one form than in another, and that some substances, under certain circumstances. could be altered or changed into quite different substances while others could not. A fine dry log thrown into a fire would flare up brightly, then shrink, crack, snap, turn glowing red, and ultimately change first to a charred black lump and then to a powdery ash; yet the stones on which the fire was built remained unchanged. They learned that water poured on a fire would vanish into a cloud of hot white vapor, but that in the midst of this process the fire would go out. Then there was the fire itself, a mysterious and unpredictable thing, sometimes flaring up in long orange flames, sometimes glowing red, sometimes merely giving off heat. Certainly fire was a strange and fearsome thing to those ancient men, capable of warming them and frightening off predators on cold nights, yet equally capable of crippling or killing them if they got too close to it—and who was to say what fire was?

Of course no one knows how much those early men actually thought about such things, except that various primitive societies worshiped fire as a deity. Later societies in Egypt and Babylon merely used it as a controllable tool and were clearly aware of the practical utility of a wide variety of substances which could be shaped and changed in multitudes of ways to make them more useful. Metals such as copper, tin, lead, or gold could be smelted out of various kinds of rocks. Weapons and utensils could be fashioned, first out of rock or clay and then out of metal. Wood could be carved and shaped to man's will, and animal hides could be converted into durable leather.

By the time of the ascendancy of the Greeks in the ancient world, men had come a long way from the caveman's vague musings about the nature of matter. Nor were all of the ancient Greeks hung up in sterile debate about whether matter was real or not. By Aristotle's time three basic ideas had become well established about the nature of matter. All three were to guide men's thoughts for millenniums, and all three retain a certain degree of validity even to this day. First, there was the conviction that all the material substance in the universe was composed of various quantities and combinations of a relatively few basic "elementary substances." Second, there was the notion that these elementary substances were discontinuous by their very nature; that is, that although they might be divided and subdivided into extremely tiny particles, eventually one would reach a final "indivisible unit" of each basic substance which could no longer be subdivided. And finally, there was the fundamental idea that these elementary substances existed in a certain fixed quantity in the universe which could neither be increased by any means nor destroyed in any way.

Of course the elementary substances as described by the ancient Greeks were far different from the "elements" as we understand them today. The Greeks believed that all material objects in existence could be accounted for in terms of combinations of four basic "elements": earth, air, water, and fire. They regarded mud or clay as a combination of earth and water, for example, while the substance issuing forth from a sulphurous hot spring was composed mainly of water and fire with a little earth thrown in. Furthermore, they argued that various special circumstances could either separate the component "elements" of a substance fairly readily or else change one "element" into another.

For example, the Greeks believed that when a log was burned it gave up its "fire" component (you could see the flames rising up) as well as its "air" component (you could see the smoke) until nothing remained but the log's "earth" component (the pile of charred ashes) which could no longer be changed by further burning.

Later, as metallurgy developed and as alchemists began heating, dissolving, or mixing substances in hopes of transforming them into precious metals, it became harder and harder to explain observed changes in matter purely on the basis of the four basic elements of the Greeks. Even so, the idea persisted for centuries, as did so many other philosophical ideas of the Greeks. The idea of the four basic elements became related and entangled with the ancient medical notion that health or illness was determined by a balance or imbalance of four "humors" within the body, and it was not until well into the Renaissance that these ideas were finally reluctantly discarded. Yet even today we see a curious and prophetic symbolism in the Greek's idea of four basic elements. Certainly three of the four could be seen to represent the three forms of matter most familiar to us today—solid, liquid, and gas—and in modern times we might even regard the fourth Greek element, fire, as symbolic of energy, now considered by physicists as merely another manifestation of matter.

The second idea of the Greeks, that matter could not be endlessly subdivided into infinitely smaller and smaller portions, but was basically composed of tiny but finite indivisible particles, proved even more significant. This was a purely intuitive idea, unsupported by a single scrap of experimental evidence; yet it appeared repeatedly throughout the centuries as scientists again and again tackled the problem of defining the exact nature of these tiniest of elemental particles. On repeated occasions in history physicists and chemists thought they had at last found the answer, but as we shall see, the mystery of the exact nature of the elementary particles composing matter is even more perplexing to modern physicists than it was to the alchemists of a thousand years ago.

Finally, the Greeks became convinced that while matter might be changed from one *form* to another, there was no way that any part of it could ever be destroyed, nor could even the tiniest fragment of matter be created which did not already exist. Once again this was an intuitive idea; the Greeks simply assumed that all the matter the universe ever had contained or ever would contain was already there in one form or another, and that the total quantity could never be altered. Of all these ancient ideas, this one above all seemed vulnerable to experimental challenge. It was by no means a self-evident truth; indeed, it often seemed to be contradicted by casual observation of things appearing in nature. Over the centuries, as the scientific method of investigation began to evolve, scientists made repeated attempts to prove that matter could be created from nothing, or that it could be destroyed under certain circumstances—but these attempts invariably came to grief. Even today continuing attempts in modern laboratories continue to fail, although modern-day understanding of the inseparable interrelationship between matter and energy has forced certain modifications or extensions of the ancient concept.

In the days of Isaac Newton, however, and even up to the middle of the last century, no one had even begun to suspect any intimate relationship between matter as such and energy as such. Up to that time physicists had largely been concerned with the universe as they could observe and study it directly—the universe of sensory experience—and nothing in that world had yet appeared to suggest that any such relationship between matter and energy existed. As far as classical physicists were concerned, all the matter in the universe was believed to exist either in a solid or a fluid state, with the fluid state further subdivided either into liquid form or gaseous form. Each of these forms or states of matter was familiar and observable. Each had certain specific characteristics all its own. In terms of our everyday world the individual characteristics of these three major states of matter are as critically important to us today as they were to the early scientist. It will be worthwhile to see what common characteristics these states or forms of matter share, and to see in what ways they are distinctly different from each other and why.

## FORM AND SUBSTANCE: THE SOLID STATE

We need only a glance around us to confirm that the most substantial part of the world around us is composed of matter in the solid state. Solid objects are so commonplace to our experience that we might even be tempted to assume that most of the universe is made up of matter in the solid state. In this assumption we would be wrong, but there can be no doubt that from the standpoint of functional utility matter in the solid state is by far the most useful to man and provides him with the most durable and permanent of the artifacts he needs for his everyday life.

This is a direct result of certain common characteristics of matter in the solid state. Solids, for example, tend to be rigid and hold their shape under considerable stress and strain. Whatever the components of a solid bit of matter may be, those components seem to be locked into a durable formation and resist being shoved out of that formation. Although certain kinds of solids which we know as metals can be hammered or bent or pulled from one shape to another, solids in general cannot be compressed very much nor stretched very much. They tend to retain their shape without external support and resist (to a greater or lesser degree) the action of forces to break them into fragments or to twist them out of shape.

Furthermore, solids characteristically possess a quality known as *elasticity,* a tendency, when acted upon by some distorting force, to spring back to their original shape as soon as the force is removed. We ordinarily think of elasticity in terms of some object such as a rubber band, which quite obviously can be greatly distorted or compressed, and which has an equally obvious powerful tendency to pull itself back into shape when the force is removed. But the behavior of a rubber band is actually an extreme and dramatic example of the quality of elasticity possessed to a much lesser degree by innumerable other solid objects which we would not ordinarily think possessed it at all. A solid ivory billiard ball, for example, would hardly impress us with its elasticity; yet it demonstrates surprising elastic qualities. When it is struck by another ivory billiard ball, both balls momentarily "give a little" or flatten at their point of contact, then immediately spring back to their original shape as they bounce apart. Even such a brittle solid object as a sheet of plate glass will bend and buckle slightly in a gust of high wind and then spring back to its original shape as soon as the wind declines.

But just as solids possess varying degrees of elasticity, they also possess varying limits to their elasticity. A brittle object like a plate-glass window has a very narrow limit of elasticity; let a gust of wind distort it just a whisker beyond its limit and it will shatter. Even a highly elastic object such as a rubber band will break if stretched too far, and once broken cannot then be restored merely by pushing the broken ends together.

We know today that the relative brittleness, malleability, rigidity, or elasticity of a solid object is directly related to the manner in which its component molecules and atoms are locked together. In many solids these components are bound tightly into a rigid crystalline structure which permits very little distortion. The internal atomic structure of a quartz crystal, for example, is such that if it is struck with a hammer it will shatter into smaller and smaller fragments of quartz, even down to a very fine powder; yet each fragment will retain the same basic crystalline structure that it possessed as a part of the original chunk. Metals too possess characteristic crystalline structures, but in many metals the interatomic bonds holding the crystals together are comparatively weak, so that metals such as gold or silver can be hammered and shaped quite readily without shattering. In the case of a highly elastic substance such as rubber, the component molecules are made up of extremely long serpentine chains of atoms only loosely bonded one to another so that the atoms and molecules can be stretched apart, yet bound firmly enough that they tend to pull each other back together once the distorting force relaxes. Still other solid substances such as phosphorus, certain forms of sulphur, caramelized sugar, beeswax and glass have no distinguishing internal crystalline structure at all and are called *amorphous* substances from the Latin term meaning quite literally "without shape." Some such substances, in fact, come so close to an indistinct no-man's land between matter in the solid state and matter in the liquid state that we would be hard put to specify whether they should be classified as very soft and formless solids or very dense and viscous liquids.

Solids as a group, however, do generally exhibit some degree of crystalline rigidity; their component atoms and molecules tend to remain locked together in fixed formations relative to each other, capable of some very limited localized motion or vibration but incapable of moving freely out of the vicinity of their immediate neighbors. This characteristic is a result of powerful binding forces between the atoms which hold them relatively close one to another with insufficient atomic or molecular motion to overcome these bonds. As a consequence, matter in the solid state is generally more dense than matter in the liquid state (with a few notable exceptions such as mercury).

But in addition to these more or less familiar characteristics, solids have certain other characteristics which we might not ordinarily consider. Although they retain their shape, they also tend to expand in volume when heated, and to contract in volume when cooled. This expansion or contraction of solids as a result of temperature variations is ordinarily so slight that we fail to notice it, but it is a very real occurrence just the same. It would be perfectly possible, for example, to use a narrow iron rod as the temperature indicator in a thermometer if the instrument were equipped with a microscopic temperature scale to register changes in length of the rod with changes of temperature, and if we had a microscope with which to take readings. And, in fact, the property of metals to expand and contract with temperature variations is actually utilized in the temperature-regulating mechanism of many delicate thermostats in which an expanding strip of metal bends sufficiently to break contact and shut off a furnace when room temperature reaches a given level, and then bends back and reestablishes contact to turn the furnace on again when temperature of the strip has dropped to a certain point. For the same reason, railroad tracks are laid with end-to-end gaps between strips of rail at regular intervals so that the tracks will not buckle in the heat of the summer sun.

Furthermore, solids characteristically melt into liquid form if heated sufficiently, providing that some other chemical or physical change does not take place first at a temperature lower than the melting point of the solid. As we know, it is easy to melt ice or paraffin in a saucepan on the stove with the use of comparatively little heat. Glass or even iron will melt at considerably higher temperatures, but if we try to melt a stick of wood, for example, the cellulose of which it is composed will break down and separate into water vapor, carbon dioxide gas, and carbon before it will melt, and the carbon itself will combine with oxygen from the air and be dissipated before its melting point can be reached.

Among solids which can be melted into liquid form by heating, we notice a curious thing: With the singular exception of one solid—frozen water or ice—the solid form of a substance tends to be more compact, more *dense,* than the same substance in its liquid form. The reason again is related to the internal molecular structure of the substance; when a solid has been heated sufficiently for its molecules to break free of the intermolecular binding forces that held them rigid, the distance between the individual molecules is increased so that fewer molecules are present in a given volume of the liquid than in the same volume of the solid.

The very close relationship between the internal atomic structure of various solid materials and such physical properties as hardness or softness, elasticity or brittleness, is especially well illustrated in certain solid substances which can exist in two or more solid states with quite different physical characteristics. There would seem to be little relationship between the crystalline hardness of diamonds, the powderiness of charcoal, and the slippery "greasiness" of graphite; yet all three are not only examples of matter in the solid state, but are actually different forms of the same pure element. In diamonds, the carbon atoms are locked together in an exceptionally rigid crystalline formation; in charcoal powder the pattern of the carbon atoms is far less dense, while in graphite the carbon atoms are held together loosely in flat sheets which slide readily over each other. The element phosphorus may occur as a soft yellowish-white material with a waxy consistency which can burst into flame spontaneously when exposed to air, or in the form of a dry, dull-red powder which is stable in air and ignites only at a much higher temperature. Metallic sodium, one of the most violently chemically active of all the elements, is as soft as butter and also burns spontaneously when exposed to air; yet in chemical combination with other substances forms exceptionally stable compounds such as sodium chloride, sodium carbonate, or sodium sulphate, most of which exhibit very distinctive crystalline characteristics.

Finally, solids in general are found to *lack* certain identifying characteristics of liquids or gases. In general, solids do not tend to evaporate as liquids do, nor to diffuse from one area of space to another, as gases do. Furthermore, solids in general are not good solvents. It is possible to mix a quantity of one solid such as flour with a quantity of another such as sand, but we do not end up with a solution of flour in sand or a solution of sand in flour. Two solids mixed together in such a way may interreact chemically with each other to form a new substance which is different from either ingredient, but unless a chemical interreaction occurs the particles of two solids mixed together remain intact, with particles of one lying side by side and intermixed with particles of the other. In the case of the flour-and-sand mixture, it would be perfectly possible physically to separate the mixture again with the aid of a good enough microscope and delicate enough tools, picking out the sand grains and putting them in one pile and the particles of flour in another, to end up with precisely the same quantity of flour and sand unmixed which had originally gone into the mixture.

It may seem that we have been hedging a bit in this discussion of the characteristics of solid materials by using such terms as "in general," "as a group," and "usually." And, in fact, we *are* hedging a bit, but for a good reason. For one thing, the myriad solid substances with which we are familiar exhibit such an enormous variety of different identifying characteristics that it is difficult if not impossible to find any single identifying characteristic common to all solids. The best we can do is to enumerate a few general characteristics commonly shared by most solids, while freely admitting that some exceptions exist in the case of each particular characteristic. Second, as we have seen, there are many substances which seem to fall in a gray area which lies somewhere in between matter in a solid state and matter in a liquid state. Graphite, for example, has certain liquidlike characteristics of slipperiness or greasiness, while taffy appears for all the world like a solid substance until you start chewing it.

Finally, and perhaps most baffling of all, there are a number of solid substances, with more and more appearing in recent years, which are not *quite* solids, or at least do not behave quite the way other solids behave. Among these substances are those which, while "solid" enough, possess a characteristic known as *plasticity*—a capacity to be molded, tugged, or pulled into a permanently altered shape by the action of comparatively gentle forces acting steadily over a period of time. A stick of sealing wax which would shatter into fragments under the sudden force of a hammer blow will bend into a 90-degree angle if one clamps one end in a vise and leaves a weight hanging on the other end overnight. A rod of glass can be molded with gentle pressure when heated red-hot in a flame, without ever actually melting. A modern example of this class of solid is the vinyl plastic phonograph record which will bend and curve out of shape if left out in the direct sunlight or stored too close to a hot-air register.

Perhaps the strangest of all examples of a solid which is not exactly solid is the silicone putty recently marketed as a children's toy. A ball of this stuff is solid but pliable, easily molded with the fingers. A ball of it left on the table will gradually flatten out over a period of time into a sort of a puddle under the effect of gravity. If we drop it on the floor it will bounce like a rubber ball, but if we whack the same piece with a hammer it will shatter into a dozen pieces. If we try to tug it into two pieces gently it will stretch and stretch almost without limit, yet if we wrench it with a sudden shearing force it will snap. If we press it gently with a thumb it will flatten out to reveal a perfect fingerprint impression, and if two portions are snapped apart they can be rejoined again simply by placing the broken ends in contact.

Substances such as these certainly have some "solid" characteristics, but in some respects behave more like extremely thick and viscous fluids. The chemists and physicists of Newton's time who were busy classifying substances into three main categories of solid, liquid, and gas were not acquainted with many such peculiar substances and largely tended to ignore those few they did know. But in our modern age of shirts that melt on the ironing board and shoes made of impermeable plastic that still allows air to circulate through it, we encounter a great many "peculiar" solids and other in-betweens. To complain that these substances are impossible to classify is to blame the shoe for the faults of the foot.

## THE CONVENIENT QUALITIES OF LIQUIDS

If much of the commonplace usefulness of matter in the solid state derives from its strength, rigidity, and durability of structure and from its built-in resistance to the effects of stresses and strains, the equally convenient qualities of matter in the liquid state derive in large part from the very absence of these same peculiar characteristics. A liquid may be so thin as to have practically no substance at all (as with ethyl chloride, an aromatic fluid which evaporates so fast that a finger dipped into it will be dry by the time it is pulled out) or thick, viscous, and sticky (as with honey, for example, or axle grease); but whether thick or thin, the one common characteristic of all liquids is the *absence* of any rigid internal crystalline structure. The molecules of a liquid are free to move around, over, above, or below their neighboring molecules with far more freedom than the molecules in a solid.

Consequently, liquids have no fixed and rigid shape of their own but can be poured freely from one container to another, spread out in a puddle on a flat surface, or be physically separated into portions which subsequently can freely rejoin and commingle, but ultimately take the shape of whatever container they happen to be deposited in. It would be impossible to describe the "shape" of a glass of water except by describing the shape of the vessel formed by the containing walls and the bottom of the glass. You may pour sufficient water into the glass to have a column of water several inches high and the width of the glass—but if you then smash the glass you end up not with a column of water but with a puddle on the table.

Liquids and solids have certain distinctly similar characteristics, and others which are sharply differing. Like solids, liquids cannot readily be compressed by squeezing or crushing forces, nor stretched when subjected to pulling forces. Indeed, they have no quality of elasticity whatever. They do tend to expand in volume when heated, or to contract in volume when cooled, much as solids do, but usually to a greater degree. Like solids, liquids generally have clearly distinguishable borders and to remain within coherent boundaries of volume unless physically separated into smaller portions by outside forces; but unlike solids, many liquids tend to evaporate quite readily, converting bit by bit into a gaseous state even at temperatures far below their boiling points.

But perhaps the single most unique characteristic of liquids—not to say one of the most fortunate, from the point of view of earth's living creatures—is their ability not only to mix freely with other substances, but actually to *dissolve* other substances so that the substance being dissolved (the "solute") and the liquid which has dissolved it (the "solvent") can no longer be separated without a great deal of bother. This ability to dissolve other substances is something quite apart from the ability to form chemical combinations with other substances, although many liquids are capable of that also. In a true solution, both the dissolving liquid and the substance dissolved in it remain distinct and autonomous chemical entities, each essentially unchanged except for being so thoroughly mixed and interspersed that physical separation has become difficult if not impossible. And indeed, the dissolving of one or another substance in a liquid brings about such a thorough intermixture that the dissolved substance or solute is broken down into molecule-sized particles, and often even those molecules are further separated into electrically charged fragments of molecules called *ions*. And because such a thorough intermixing of substances in solution takes place, the solution of some substance in a liquid often brings about a change in the physical characteristics of either the dissolved substance, the dissolving liquid, or both, even when no chemical alteration in these substances occurs.

Thus, for example, when water (a colorless, tasteless liquid) and salt (a crystalline solid) are intermixed in solution, the salt loses its "solid matter" property to become part of the liquid solution, while the water loses its tasteless property and assumes a salty flavor. Furthermore, the dissolving of salt into water produces a solution that freezes at a temperature several degrees below the freezing point of pure water. But even in spite of these changes, if the proper steps are taken at least one or the other of the intermixed substances *can* be separated and recovered in its original form. If a solution of salt and water is allowed to sit in a flat dish, exposed to air, the water will gradually evaporate leaving a residue of solid crystalline salt on the dish, and careful measurement would show that the residue contained *all* of the salt that had been mixed in the water, and that it was chemically indistinguishable from the salt before it was placed in solution. Recovering the water free of the salt is quite a different matter, but even this can be accomplished in a clumsy fashion by boiling the solution dry, collecting the water vapor in a separate container, and then distilling it back into liquid form by passing it through the coils of a condenser.

Just because the ability to dissolve other substances is a commonplace characteristic of liquids does not necessarily mean that any given liquid will dissolve any other substance we mix into it in any quantity that we desire. Liquids in general are highly individual and selective with regard to which substances they will dissolve and which they will not, and in regard to the quantity of a given substance they will admit to solution. A given liquid may dissolve one substance quite readily, yet not dissolve another at all. Even when a substance can be dissolved by a given liquid, the liquid may dissolve only a tiny amount of it before becoming "saturated," whereas it will dissolve enormous quantities of some other substance without becoming saturated. Thus the capacity of a given liquid to dissolve other substances is comparable to a country's immigration quotas: A country may admit a large number of people of one nationality, only a few people of another nationality, and none at all of a third, according to fixed quota laws. If a given quota of immigrants is exceeded, the country simply refuses to admit any more and sends them back home—just as a liquid that is saturated by a substance it has been dissolving simply refuses to dissolve any more. You can dissolve surprising quantities of salt in a tumbler of water but, if you continue adding salt, a point will be reached at which any additional salt crystals will simply sit surrounded by the solution in the bottom of the glass and will not dissolve no matter how much you stir nor how long you wait.

Usually when we think of liquids dissolving substances, we think of a solid substance being dissolved by a liquid. But in truth, a liquid may dissolve surprising quantities of various gases, or may even dissolve another liquid. Fish and other water-breathing fauna depend for their lives on the oxygen gas dissolved in water, either from the contact of the water with the oxygen and air, or from oxygen released by water plants as a by-product of photosynthesis. In the case of two liquids in solution—water and ethyl alcohol, for example, or gasoline and motor oil—it is difficult to say which liquid dissolves which, so we speak of such liquids as "mutually soluble." Even in such cases, however, each liquid will have an upper limit, a saturation point, above which it will no longer dissolve more of the other liquid.

In many mutually soluble liquids (again, such as ethyl alcohol and water) the saturation points may be infinitely high, so that either will dissolve an infinite quantity of the other; but in the case of two liquids which are only very slightly soluble in each other—water and ethyl ether, for example—only a very tiny amount of one liquid (the ether) will dissolve in the water, while virtually no water dissolves in the ether. In such a case, when excess ether is added to a container of water, an "interface" between the ether and water will form, with the lighter liquid (in this case the ether) rising to the top and the water sinking to the bottom.

There are, of course, instances in which one liquid will not dissolve even a tiny amount of another, or vice versa; in such a case we speak of the two liquids as "mutually insoluble." Any two such liquids placed in the same container will simply separate one from the other with the more dense liquid beneath and the less dense liquid forming a layer on the top with a clear interface at the line of demarcation. Water and mineral oil, for example, are two such mutually insoluble liquids and literally cannot be forced to dissolve one another unless an emulsifier is added.

So far we have been talking about a true solution of one liquid in another (or lack of it). But there are other ways by which two liquids can be intermixed that do not involve a true solution of one in another at all. If we mix olive oil and vinegar together, one liquid may *seem* to dissolve in the other at first, but if the container holding this "solution" is allowed to stand, the two liquids will gradually separate, with the vinegar gradually sinking to the bottom and the oil rising to the top. Obviously such a combination is not a true solution at all. In such cases we speak of the two liquids as being "miscible"—capable of being mixed together quite thoroughly by stirring or shaking, but never forming a solution. Finally, further to confuse things, there is yet another kind of "phony solution" in which certain substances which are insoluble in a given liquid nevertheless divide into such tiny particles when they are mixed into the liquid that the particles remain permanently suspended in the liquid and evenly distributed throughout its volume, never settling out or separating even though they are not really dissolved. Such a mixture is called a "colloidal suspension." A familiar example is homogenized milk, in which the butterfat has been mechanically broken down into such tiny particles in the watery liquid of the milk that it remains suspended. Even skim milk is essentially a permanent suspension of butterfat droplets in water; although the butterfat is not dissolved in the water, the only way to separate it out from the water is by addition of some such agent as lemon juice which causes the butterfat particles to cling together in a sticky curd which can be filtered or centrifuged out of the water or from which the water can be decanted.

Finally, liquids are quite as capable of dissolving gases, sometimes in great quantity, as they are capable of dissolving solids or other liquids. Cola drinks and other carbonated beverages depend upon the ability of water to dissolve quantities of carbon dioxide gas under pressure and reduced temperatures for their distinctive tanginess. Here we see a splendid demonstration of the fact that the amount of gas dissolved in a given volume of liquid depends upon increased pressure and decreased temperature. Try drinking a bottle of Coke sometime after it has been left standing open in the hot sun for a few hours, and see whether things go better or not.

The capacity to dissolve other substances is a distinctive identifying characteristic of any matter in liquid form; it is also a highly convenient characteristic which permits liquids to lend themselves to all sorts of practical uses. Literally hundreds of products that we use each day, from the coffee we drink in the morning to the soap we wash the dishes in to the fuel we burn to the perfumes and colognes we use, all are solutions of one substance dissolved in another. And of all known liquids, water is perhaps the most ubiquitous solvent, willing and eager to dissolve an endless variety of other substances. We depend on this vast solution-forming capability of water for our very lives; not only are all the cells in our body constantly bathed in nutrients dissolved in water in the body and giving up waste products to be dissolved and carried away, but it was the presence on the face of the earth millions of years ago of a warm sea of water containing salt and other dissolved substances that provided a medium in which life on our planet first came about or was even possible.

But if water and other solution-forming liquids are important to man (and perhaps innumerable other forms of life throughout the universe) there is a third commonplace state of matter, more quixotic than either solid or liquid, which is equally deserving of our attention.

## THE EFFERVESCENT GASES

Just as solids and liquids have certain characteristics in common and other characteristics sharply in contrast, liquids and gases are similar in some ways and dissimilar in others. In fact liquids and gases are so similar in so many ways that scientific classification frequently lumps them together as "fluids," but their differences are such that we rarely have difficulty distinguishing matter in one state from matter in the other.

Like a liquid, matter in the form of gas has no fixed shape of its own, but tends to fill and take the shape of any closed container into which it is placed. Also like a liquid, a gas can flow freely from one place to another, and demonstrates the same sort of swirls and eddies as a liquid when it is flowing from one place to another. Similarly, quantities of two different gases enclosed together in a container will diffuse and intermix uniformly one with another, much like two mutually soluble liquids, but here we encounter a major dissimilarity. Unlike liquids, gases do not dissolve other substances, nor are gases fussy about which other gases they will mix with. While mutually insoluble liquids simply will not mix at all, any gas will intermix with any other gas placed in the same container, each behaving precisely as if it were the only gas around (unless, of course, the two gases enter into chemical reaction and form quite different substances which may not be gaseous in nature at all. Both hydrogen and oxygen are gases, for example, but when mixed together in the same container may combine explosively to form a liquid, water).

The most singular characteristic of matter in the form of a gas is, in fact, that it has no physical coherence whatever. A quantity of a liquid released from a closed container will flow and change its shape just as a gas will, but it will also maintain a coherent delimited physical form of some kind even if that form happens to be one or more puddles on the floor, each of which has physical coherence. A gas, on the other hand, will show no physical coherence at all if released from a closed container; it will diffuse freely out of the container without limit, and freely intermix with any other gas that happens to be around. In the absence of any confining forces, it will continue to diffuse and expand indefinitely. If it were not for the confining force of gravity acting to limit the diffusion and expansion of the gases in earth's atmosphere, those gases would long since have been dissipated, just as our atmosphere's hydrogen and helium were, by diffusion through a vast expanse of empty space in the universe around us, and would still be diffusing. This is precisely why the planet Mars is believed to have as sparse and tenuous an atmosphere as it has, completely devoid of very light gases such as hydrogen and containing only a tiny amount of oxygen: Mars's gravitational force, far weaker than earth's, has been able to "contain" only the heavier gases in its atmosphere while the lighter gases have leaked away a bit at a time over the ages.

Indeed, the only time that a gas could be said to have a coherent physical shape upon release from an enclosing container would be in the event that it was released in an environment in which it is surrounded by a liquid in which it is not soluble. In such a case, the gas would rise to the surface of the liquid (having less density than the liquid) and in the course of rising would be contained in spherical bubbles. But even this is not a case of a volume of gas assuming a "natural" physical shape; when released into the container of liquid, it is effectively "enclosed" by the pressure of the liquid on all sides of it boxing it in—exchanging one container for another, so to speak. Conceivably the same things might hold true in the case of a very light gas such as hydrogen that is released from a closed container into into an environment of a very heavy, dense gas under pressure—an environment such as might be found in the heavily compressed and bitterly cold atmosphere of methane and ammonia gas near the surface of the planet Uranus. Under such conditions the hydrogen might conceivably be confined in the form of "bubbles" within the dense gaseous atmosphere of such a heavy planet; but more likely even under such extreme circumstances there would be plenty of room between the molecules of the "confining" heavy gas to permit the hydrogen plenty of room simply to diffuse and mix into the surrounding atmosphere.

Ordinarily we think of a gas as an effervescent stuff without form or substance, but like any other form of matter any gas has mass and occupies space. The weight of our earthly atmosphere pressing against the ground at sea level is equal to almost 15 pounds for every square inch of the earth's surface; and if we have sneaky doubts that the gases in our atmosphere indeed occupy space, we need only watch a sky diver float to earth under his parachute. Obviously *something* that occupies space is making that parachute balloon out, while the parachute equally obviously is moving down *through* something which has to be pushed aside in order for it to descend.

Furthermore, matter in gaseous form tends to expand in volume when heated or contract in volume when cooled, just as solids or liquids do, except that the expansion or contraction of a gas under these influences is far more marked and dramatic. The air in a blown-up air mattress, for example, can expand enough to split open the seams if the mattress is left out unprotected in the hot sun; but if the outside air cools enough at night, the air inside the mattress may contract so much in volume that more air must be added for comfortable sleeping. Finally, gases possess one unique characteristic that both solids and liquids lack: gases are "compressible." Just as a gas can and will diffuse and expand without limit unless contained or confined by some outside force, so a gas can be squeezed or compressed by an outside force, its volume diminishing in direct proportion to the pressure exerted by the force. But there is a bottom limit to a gas's compressibility. From Newton's third law we know that any gas that is being compressed by an outside force is at the same time pressing back against the force that confines it with an equal force. Thus if a gas's pressure outward against the confining walls of a container into which it is being pressed exceeds the containing strength of those confining walls, the containing vessel will burst. Alternatively, if the confining walls of the container are strong enough and the compressing outside force great enough, molecules of the gas will be forced so close together that interatomic forces of attraction can take over and the gas may "condense" into a liquid which will then no longer be significantly compressible.

The physicists of Newton's day were convinced that all the matter in the universe existed in one of these three major states—either in the solid state, as a liquid, or as a gas. Today, of course, we know that matter can exist in certain other more exotic states under special circumstances. In a number of physics laboratories, for example, modern physicists work with hydrogen atoms in an extremely rarefied gaseous state in which the nuclei of the atoms are stripped of their electron components. These particles, nothing more than naked hydrogen nuclei or "protons," are confined within powerful magnetic fields and are artificially accelerated to great speeds. Matter in such a state as this, which we might think of as a superrarefied, superheated gas is known as a "plasma" and can be considered as a quite separate and unique state of matter characterized by its own peculiar properties. Of course, here on earth it may require a $35 million collection of machinery in order to convert a tiny amount of ordinary gaseous hydrogen into a plasma state in which it is maintained for only 1/1,000 second, but even here matter *can* be converted into such a state for long enough at least to demonstrate that it exists. Elsewhere, most of the visible matter in our universe (stars, etc.) exists in the plasma state, and it is entirely possible that the universe contains more matter in the plasma state in the unimaginable reaches of space between the stars and galaxies than exists in all the other states put together in more familiar corners of the universe.

Similarly, modern astronomers are convinced that the universe also contains uncounted multitudes of dark, cold aggregates of burnt-out star ash, formed from the densely packed nuclei of atoms that once fueled stars that are now long dead, all their available energy expended. Such compacted nuclei with their electrons stripped away would have to form incredibly dense matter unlike any solid ever encountered in our earthly experience. It has been calculated that a cubic inch of such hypercondensed star ash would have a mass of tons, and such material would certainly have to be considered a separate and distinct state of matter. So would the strange substances both liquid and solid that have recently been studied in modern low-temperature physics laboratories. As we will see later, when the temperature of certain substances is reduced to within a few degrees of absolute zero—the point at which all molecular motion is believed to stop—these substances suddenly take on physical and electromagnetic properties totally unlike any other substances known. Here again we might consider such exotic substances as existing in a separate and distinct state of matter.

For the sake of completeness we need to acknowledge that matter can exist in such bizarre and exotic states under certain special conditions or extreme environments. For practical purposes, however, there is no need right now for us to concern ourselves with matter in these peculiar states; for the moment we will concentrate on the conclusions reached by the classical physicists about the nature of matter based on their knowledge of the three major and commonplace states of matter they knew, dealing with the others at a more appropriate place.

## CONVERSION AND CONSERVATION

Today we know that the differing characteristics of matter in the solid, liquid, or gaseous state are directly related to the internal atomic structure of the substance. The elementary units of a substance in the solid state are indeed "locked together" in more or less rigid geometrical patterns, and even when the geometrical structure of a solid is temporarily distorted by one kind of stress or another those elementary units can pull back into their original pattern when the stress is relieved. The atoms of a liquid are not locked together in quite this unyielding fashion; yet they are still tightly enough assembled, with sufficient binding force between them to give the liquid a coherent volume even if the liquid flows freely and takes on the shape of the container holding it. We will have much more to say about these interatomic binding forces in a later chapter, for the precise nature of these forces remains one of the major problems modern physicists are still wrestling with.

In a gas, however, the elementary units making it up are far more widely separated than in either a liquid or a solid and are free of the effect of internal binding forces which are present only when elementary particles are in comparatively close proximity to each other. Thus the atoms or molecules of a gas are free to move quite independently of one another and can move at random within the confines of any container holding them, capable of being pushed closer together by external pressures but also capable of diffusing without limit as long as no containing force acts upon them.

Of course the physicists and chemists of the seventeenth and eighteenth centuries did not know anything to speak of about the submicroscopic structure of matter in any state, although the basic idea that all matter was composed of small indivisible particles had been kicking around for over a thousand years and was soon to be revised and expanded by such men as John Dalton. But if they did not know precisely why matter could occur in three quite different states, those early scientists gathered together a remarkable amount of information about how matter behaved in each of the three major states, and about the conditions necessary to change or convert matter in one state into another state.

Most familiar substances seemed to exist in nature in one or another state by preference under normal conditions, but a great many substances could be converted from one state into another more or less readily and consistently under certain specified conditions. Gold normally was found in the solid state in nature, but if heated to a certain unnaturally high temperature it would change into a liquid. Another heavy metal, mercury, was already a liquid in its natural state, but if heated to a sufficiently high temperature could be converted into a noxious violet-colored gas, while sulphur dioxide which occurred in nature in the vicinity of sulphur deposits as an evil-smelling, pungent gas could easily be condensed into a colorless oily liquid when it was cooled and compressed. A few substances, such as the resinous sap of pine trees or the tallow used in making candles, might be found in either solid or liquid state in nature, depending upon the prevailing temperature. But of all known substances, there was one—and only one—that could normally be found in nature in any of the three major states depending upon the particular circumstances that prevailed.

This substance, of course, was water. It could be found in its solid state in the Arctic or Antarctic ice packs, or in rigid sheets that formed on rivers and canals even in the Temperate Zone during the winter. As a liquid, it fell as rain, emerged as underground springs or flowed down mountainsides in cascading torrents, while in its gaseous state it could be observed as water vapor in hot springs and geysers, or saturating the air in regions with moist climates.

Even the earliest scientific observers recognized that the particular state of matter of any given substance seemed to be a function of its temperature. A solid, heated sufficiently, would melt and become a liquid, providing that it did not undergo some chemical change or breakdown in the process, and the resulting liquid when heated still further would presently begin to boil and become a gas. It was also observed that each substance that could be converted from one state to another had its own characteristic temperatures at which the change would take place, and while these "change-of-state" temperatures might vary widely from one substance to another, a given substance could be relied upon to change from solid to liquid, or liquid to gas, or vice versa, quite consistently and reliably when heated or cooled to the appropriate temperature. Thus it became customary in describing various substances to list as *physical properties* of a given substance its freezing point and its boiling point, in the case of a liquid, its melting point and vaporization point in the case of a solid, or its condensation point and freezing point in the case of a gas whenever those temperatures could be measured under standardized pressure conditions (i.e., at normal atmospheric pressure at sea level). Of course, certain of the transition-point temperatures could not be measured at all for certain substances because they underwent various kinds of chemical alterations before melting point or boiling point could be reached; coal would burn before it would melt, for example. With other substances transitions from one state to another required such extremely high or extremely low temperatures that it was difficult to achieve them. Many metals simply could not be heated to a high enough temperature to measure their boiling points except in a vacuum, while gases such as hydrogen or oxygen had such extremely low condensation points that twentieth-century technology was required to cool them down sufficiently to liquefy them, and helium gas would not condense into a liquid until its temperature was reduced to within four degrees of absolute zero, to – 265 degrees Centigrade!

Among the considerable variety of substances that could be studied in two or more different states, certain other interesting general characteristics were observed. For one thing, most substances would change in an orderly manner from solid state to liquid state, and then from liquid state to a gas, and vice versa, providing necessary temperatures could be reached before chemical changes in the substances occurred. Some few substances, however, seemed to ignore this orderly rule: Solid crystals of iodine, for example, would "sublimate" directly into a gas when heated without ever passing through a liquid state, and iodine vapor when cooled sufficiently would sublimate directly back into solid crystals.

Similarly, a chunk of dry ice (frozen carbon dioxide) would evaporate directly into carbon dioxide gas without passing through a liquid stage, at least under ordinary conditions. Yet under quite extraordinary laboratory conditions, and with a great deal of effort, carbon dioxide gas can be cooled down under sufficient pressure to force it into a liquid state before it freezes. Just why these particular substances happen to deviate from the general rule nobody knows—nor cares, for that matter, since we really have no practical use either for elemental iodine in the liquid state or for liquefied carbon dioxide.

But another even more curious and quite unique exception to the general rule has far greater importance—the very existence of life as we know it on earth depends upon it. Most substances in the solid state are more dense than the same substances in the liquid state, and even less dense in the gaseous state than in the liquid state. A striking exception to this rule is water, which through some fortuitous happenstance is significantly *less* dense a degree or two below its freezing point than it is in liquid form. The extreme good fortune of this curious variation from the general rule is easy to see: If ice were more dense than water at the freezing point, and were thus heavier than water, all of the lakes, rivers, and even oceans on the face of the earth would have frozen solid from the bottom up during the cold seasons, and once frozen would never again have melted completely in most regions, so that any form of life which required warm salty seas in which to develop could never have survived long enough to propagate.

Over centuries of observation of matter in its various states scientists came to recognize many curious variations from what seemed to be the normal rule, but at the same time they began to recognize one characteristic of any kind of matter regardless of its state to which no exceptions of any sort were ever found. A given quantity of matter could be changed from solid state to liquid state or from liquid state to gaseous state, or vice versa; it could be ground up or evaporated or compressed by any number of physical forces, and might undergo any number of chemical combinations or break down into a variety of chemical compounds; but regardless of what was done to it, no matter what kind of interaction in which it might be involved, the total *quantity* of the matter in question always remained the same. None was ever destroyed and no new matter was ever created.

Often in the course of chemical reactions between substances, totally new and different substances would be formed with strikingly different appearances and properties. Sometimes new and insoluble substances would be formed and precipitate out of solutions as a result of chemical reactions, and not infrequently gaseous by-products of chemical reactions might be inadvertently released to diffuse into the atmosphere and be lost to the four winds. But whenever truly meticulous measurements were made and great care was exercised to collect all the end products of some physical change or chemical interaction of matter, the total quantity of matter that resulted, in whatever state it might be found, was invariably found to be precisely equal to the total quantity of matter that existed before the physical change or chemical reaction.

This idea that the total quantity of mass of matter was always conserved in any kind of interaction was by no means self-evident to the casual observer of nature. It was, in fact, vigorously disputed by a great many very excellent physicists and chemists as late as the eighteenth and nineteenth centuries. In many cases it was extremely difficult to take accurate enough measurements to tell whether the principle was valid or not, and even as more and more evidence of its validity accumulated this principle, which was known first as the theory and later the law of the conservation of matter, became one of the most widely challenged and tested of all the classical laws of physics precisely *because* there were so many kinds of interactions in which it *seemed* that a certain amount of matter had been destroyed or had appeared out of nowhere. But by the beginning of the twentieth century the law had been so thoroughly and repeatedly tested and proved that scientists regarded it as a very rock of stability, fully as reliable as Newton's laws of motion or the law of conservation of momentum.

Later, as we will see, the law of conservation of matter had to be modified—or, more accurately, expanded—to include certain manifestations of matter that had never even been dreamed of previously; but with those necessary expansions the law today remains as valid and unshakable as ever. New challenges to its validity even now continue to arise with tiresome regularity, but by now the law has withstood so many such challenges that it seems unlikely ever to be shaken unless some totally unsuspected and unpredictable item of knowledge is suddenly unearthed. Even the most vigorous of the recent challengers, a group of astronomers and cosmologists led by Dr. Fred Hoyle of England, have recently begun to hedge their bets and question the validity of their own challenges. Conceivably one day some key item of new knowledge *will* appear and a successful challenge *will* be mounted, a challenge the law cannot answer; but so far as is known today, it remains as one of the very few physical invariables in the universe. And as we shall see as we learn more of the changing and bewildering world of modern physics, any invariable truth at all is a pearl of great price.

Without question, the history of scientific investigation in the last four centuries has been in great measure the history of multitudes of observers gathering a huge quantity of knowledge of the many different characteristics of matter in the solid, liquid, or gaseous state. But so far we have ignored one characteristic of matter of which scientists in the last four centuries have become increasingly aware. Many substances could be converted from one state of matter to another; substances could be dissolved in one another; substances could be encouraged to form chemical combinations with one another, mixed with one another, forced to expand and diffuse freely or to be massively compressed.

But any time that any such physical or chemical change was brought about, it was first necessary that *something be done,* that certain requirements be fulfilled, before such changes or interactions would take place. In order for *any* change or interaction to occur, *energy* had to be applied in one form or another.

Indeed, even in the case of objects interacting with each other in keeping with Newton's laws of motion and the law of conservation of momentum, energy inevitably seemed to be involved one way or another in any change whatsoever. So it was not surprising that the same scientists who were observing and studying the various states of matter and confirming the law of conservation of matter again and again found themselves simultaneously observing and studying the characteristics of another far less tangible entity in the physical universe—an entity known as energy—

defining it, discovering the various forms in which it manifested itself, unearthing the relationship existing between energy in one form and energy in another form, and ultimately discovering that just as the universe seemed to contain a fixed and inalterable quantity of matter which could be changed in form but which could neither be created nor destroyed, so also the universe seemed to contain a fixed and inalterable quantity of energy which could be converted from one form to another but could neither be created nor destroyed.

But whereas matter could be pinned down, pinched, squeezed, measured, and manipulated, the study of the nature of energy proved to be a far more elusive and frustrating game. The search still continues today in the laboratory of the modern physicist, but he could not even have begun without the groundwork that was laid by generations of classical physicists before him.

# CHAPTER 11

## *The Manifestations of Energy*

Of all of the concepts that have evolved from the experiments and observations of physicists since the time of Newton, perhaps one of the most crucially important, yet most confusing and obscure to the average nonscientist, is the concept of energy in its various manifestations. And once again we find that the major barrier to understanding is semantic. We are in trouble from the first with our use of language.

We have already encountered more than once the gulf that exists between the common usage of terms and the more precise scientific use of the same terms. Remember, for example, the trouble we had when we tried to define precisely what a "force" was. We found it a rather vague entity variously described as a "push," a "pull," and "impulse" (that is, a force acting over a period of time), or even existing in a fuzzy and indefinable "field," as in the case of gravitational force. When it came right down to fundamentals, we found that the closest we could approach defining what a force might be was in terms of the effect it had (whatever "it" was) when it acted upon some object to cause some change in its motion. To many of us this seemed suspiciously similar to defining the "haves" and the "have-nots" as "those people who have" and "those people who don't have," respectively. Unsatisfactory as this may be, it is fairly typical of what happens any time we attempt to discuss an abstract concept in concrete terms: We can do only as well as our language permits us to do.

Now we encounter the same difficulty when we attempt to describe precisely what energy is. To the nonscientist the term inevitably brings a variety of vague and nonspecific images to mind. The dictionary defines the word "energy" as "vitality of expression"; "the capacity of acting"; "power forcefully exerted"; or "the capacity for doing work." But then, what exactly does "vitality of expression" mean? What is a "capacity" for doing anything, or even an "ability"? Ordinarily we tend to equate energy in our minds with somehow stirring around and getting things done; but we also hear of "suppressed energy"; we speak of "mental energy" or read of a modern painting "radiating energy," and so on into the night.

Of course all of these various uses of the word have one thing in common: they all suggest some sort of capacity for doing *something*. But what is the "something" that energy implies the capacity for doing?

Rather than get ourselves thoroughly snarled up in words, it might be better for us to recognize here and now that energy as a physicist uses the term has a more specific and well-defined meaning than any of the commonplace connotations assigned to the word. To the physicist, energy is a natural phenomenon, not a thing but a *concept*. He relates the concept of energy specifically to a *capacity for doing work*: a capability for changing the motion of an object, for forcing a substance to change from one state of matter into another, or for bringing about an interaction between substance A and substance B. Within this limited meaning of the word, the physicist further regards energy as a capacity that exists in a number of different forms, and a capacity which can, in any given form, be nailed down precisely and measured in some kind of comprehensible unit.

## THE CASE OF THE BROKEN TOE

One of the most common and familiar forms of energy that we encounter in our everyday lives is simple *physical or mechanical energy*—the form of energy that is constantly being acquired or released by physical objects in mechanical motion. But how can we *define* mechanical energy? Rather than try to define it, first let us see an example of how it can be acquired, how it can be released, and how it can be measured and described in meaningful units.

Consider the following commonplace situation: A man finds a good-sized rock lying in his driveway. Picking it up with one hand, he raises it four feet into the air, intending to toss it aside. Unfortunately, before he can throw it he loses his grip, dropping it on his foot and breaking a toe.

Now what has happened here? No matter what words we use to describe this particular sequence of events, certain basic things are clear. First, it is obvious that at the beginning the rock represented no direct immediate threat to the man (although it might possibly have damaged his car if he had driven over it). It was merely sitting there in the driveway minding its own business with no capability to roll off the driveway, leap up and fly, or anything else. We could say that at the beginning the rock *possessed no mechanical energy at all,* and would not have acquired any if the man had simply let it alone.

But when he picked the rock up and lifted it to four feet above the ground, the whole picture changed as a result of this action. In doing this the man instilled in the rock a capacity that it did not have while it was resting on the driveway: the capacity to strike his toe with a certain measurable force. Of course, in the instant before it slipped from his fingers while it was suspended four feet above the ground, the rock had not yet done any damage; it had merely acquired a "capacity" or "potential" for doing something. We could then say that the rock had acquired energy. We could even measure the energy it had acquired in any arbitrary units we wished to use, perhaps choosing units that were clearly related to some one of many possible things the rock had acquired the capacity or potential to do. We could say, for example, that at the instant before it slipped from the man's fingers four feet above the ground the rock had been "charged" (like a storage battery) with one broken toe's worth of energy. In saying this we would imply that the rock possessed enough energy—in the event that it was dropped—to break one of the man's toes, but not enough to break two or four or six, nor so little that it could not break at least one.

Now granted, physicists would not ordinarily find "one broken-toe's-worth" a particularly useful or versatile unit for measuring or describing a quantity of energy, although it is a perfectly valid unit for us to use under these circumstances. Instead, they have found certain generally useful words to describe the form or forms of energy involved in this sequence of events, and have selected units for measuring it that are somewhat more universally relevant and practical. The rock sitting on the driveway clearly had no capacity to do anything on its own; it had no energy. In lifting it four feet above the ground against the pull of gravity the man, by virtue of the effort expended by his muscles, instilled in the rock a "potential capacity" to break a toe.

This as-yet-unexercised capacity might be spoken of as "potential energy." As long as the man held the rock suspended and motionless, that potential capacity remained entirely potential; the rock was quite as incapable of doing anything there under those circumstances, suspended by the man's grip on it, as it was when it was resting on the driveway—*provided the man didn't let go of it.* The instant that he *did* let go of it, things changed abruptly. The rock's "potential energy" was immediately transformed into a different, more active mechanical energy which did indeed have the capability of breaking a toe when it struck it. Physicists would speak of this "energy-in-action" or "released potential energy" as "kinetic energy." The inert, immobile, and utterly harmless "one broken toe's worth" of potential energy possessed by the rock when it was held suspended four feet above the ground was very rapidly transformed into "one broken-toe's-worth" of kinetic energy by the time it reached the man's toe—and it is quite obviously the rock's kinetic energy, not its potential energy, that the man had to thank for the broken toe he received.

Was this transformation from potential energy to kinetic energy something which occurred instantaneously when the rock was dropped? Not quite, as we can easily demonstrate. Suppose, for example, that the man happened to have his foot on a two-foot-high apple box at the time the rock slipped from his fingers. In such a case the rock would have fallen only two feet instead of four when it struck his toe, and would not strike it with enough force actually to cause a fracture. It might sting a little, but at the collision point only half the potential energy the rock had acquired in being lifted had been transformed into kinetic energy. If the rock were stopped at that point two feet above the ground and again held suspended it would still possess some of its potential energy—an amount we might describe as "one broken-toe's-worth minus 24 inches" of potential energy which could yet be transformed into kinetic energy in the event that the rock were allowed to fall the remaining two feet to the ground.

Indeed, if we think this through carefully, we see that the potential energy the rock acquired when it was lifted four feet above the ground would not be transformed instantaneously into kinetic energy the moment it was dropped, but that the conversion of the potential energy into kinetic energy would take place gradually and steadily throughout the length and time of the rock's fall, so that if it were stopped at any given place between its release point and the end point of its fall it would there have a ratio of potential energy to kinetic energy directly proportional to the distance it had fallen at that point, and to the time it had taken to fall.

In virtually all examples of mechanical motion of objects or interaction of moving objects we see precisely the same interchangeability of potential energy (unreleased capability to do something) and kinetic energy (energy of action in which the capability to do something is released) in operation. We can see even more clearly the relationship between kinetic "energy in action" and potential or "stored" energy if we return to our imaginary "ideal" billiard table on which frictional forces, gravitational forces, and other red herring forces have been conveniently eliminated for our benefit. Imagine, then, rolling a billiard ball on our table in a straight head-on collision course with the perfectly elastic springy cushion at the far end of the table. As the ball is moving toward the cushion it has a certain amount of kinetic energy or energy in action. It also has a measurable momentum equal to its mass multiplied by its velocity. After it strikes the cushion, we see it rebound in the opposite direction with precisely the same momentum it had before (except that the direction of its velocity has been reversed). Furthermore, if we could measure its kinetic energy on the rebound, we would find that it possessed precisely the same kinetic energy moving in the opposite direction as it had before striking the cushion.

But what happened during the collision? Obviously, in order for its direction to have been changed, the billiard ball striking the cushion must have been slowed down and ultimately brought to a complete stop, then speeded up and thrust away again in the opposite direction. But what happened to the ball's kinetic energy during this process of slowing down, stopping, and speeding up again? At the moment the ball was at a dead stop a split-second photograph would have revealed that the cushion touching the ball had been compressed and distorted out of shape. Sequential split-second photographs taken subsequently would show the cushion expanding again to resume its normal shape and thus pushing the ball away in the opposite direction. But where did the cushion get the energy to push the ball away? We said that the rock sitting dead still on the driveway possessed no energy at all; would it not also be true that at the split second when the ball has come to rest against the cushion, at a dead stop, and the compressed cushion is in completely motionless contact with it, that this whole ball-and-cushion system is at that split second of time completely without any energy at all?

The answer, of course, is no. What actually happens is that as the ball strikes the cushion it begins to lose its kinetic energy steadily, and has lost it completely by the time it has come to an absolute stop. But that kinetic energy has not been *destroyed*. Rather, it has been transferred to something else, namely the cushion, and *converted* into a form which we might call "energy of compression." But since the cushion is elastic and seeks to return to its normal shape, it must have acquired a potential capacity to push the billiard ball away while it is in the process of recovering its normal shape. Thus the energy of compression that the cushion has acquired is just another name for potential energy which could be converted again into kinetic energy when the compressed cushion begins to push the ball away.

We can see this transfer of energy from one object to another and from one form to another very clearly if we regard the ball-cushion collision in detail as a sequence of events much the same as the case of the man picking up the rock. The instant *before* the ball strikes the cushion, the cushion possesses no energy at all, and 100 per cent of the energy in the closed system of ball-and-cushion is in the form of the ball's kinetic energy. The instant that the ball encounters the cushion and begins pressing it in and deforming it, some part of the ball's kinetic energy is being transferred to the cushion and stored there in the form of energy of compression. The more the ball presses in the cushion, the more its kinetic energy is so transferred to the cushion and converted into energy of compression or potential energy until the moment that the ball has finally come to rest.

At that point the ball has *no* energy of any kind left, while the cushion has 100 per cent of the ball's previous kinetic energy invested in it, so to speak, but totally converted from the form of energy in action (kinetic energy) into the form of energy of compression (potential energy).

But at this instant the ball-and-cushion system is clearly not stable. The ball may be at a dead stop and thus possess no energy at all, but the cushion it is compressing has been deformed by the ball's pressure. Because of the cushion's elasticity, it seeks to snap back to its normal configuration again. Although it possesses all of the ball's former kinetic energy in the form of potential energy, there is nothing to prevent this potential energy from immediately being triggered and allowed to begin changing back into kinetic energy again. Thus the compressed cushion with its potential energy is in precisely the same unstable condition as the rock was in during the instant *after* the man lost his grip on it: The potential energy is there and nothing restrains it from being released.

So how is this unstable state of affairs resolved? In the ball-and-cushion system, a complete reversal of the conversion and transference of energy takes place. The cushion begins to convert its potential energy into kinetic energy while simultaneously transferring it back to the billiard ball once again. And the fact that careful measurement would show that the billiard ball rebounding from the cushion has the same kinetic energy (under these ideal, frictionless conditions) as it had before its collision with the cushion must mean that at the moment it had come to rest *all* its kinetic energy had been totally converted and stored in the cushion as potential energy, and then *all* recovered once again so that there was no energy lost or destroyed, and none created, at any time during the interaction.

From this ball-and-cushion example we can see a very interesting characteristic about energy—and a universal characteristic. We saw that the energy of compression was nothing other than potential energy, "energy stored and available for use," so to speak, as opposed to kinetic energy or "energy already in action." We can see that in this interacting system of ball-and-cushion (or in any other interacting system in the universe) energy within the system can be converted from one form to another and back again, and can be transferred from one object or part of the interacting system to another and back again—*but the total amount of energy in the system remains constant despite these conversions and transferences.* This means, in effect, that kinetic energy and potential energy in an interacting system are completely interchangeable and completely equivalent in quantity or magnitude. They could be interchanged in part or *in toto,* but whatever part is interchanged in either direction must always be exactly equivalent in quantity on one side of the interchange as it was on the other side, regardless of what form it is in.

## ENERGY, FORCE, AND THE SEMANTIC BARRIER

Physicists of the 1700s and 1800s clearly recognized the differences between kinetic energy and potential energy, and were fully aware that one could be converted into the other or transferred from one object to another. But they were not by any means certain that some energy was not lost or gained in such transfers, nor for that matter were they convinced that *all* the energy in the universe was a constant quantity that could never be changed. If anything, a great many of their laboratory experiments seemed to demonstrate that there *was* in fact some destruction of either potential or kinetic energy within such interacting systems of objects in motion as billiard balls striking cushions, bullets striking wooden barriers, or railroad trains colliding in the night.

One reason for confusion in this area was that scientists in those days had not yet learned how important it was to be fastidious in their use of words and definitions. They commonly got the everyday meanings of various terms confused with their *technical* meanings. In everyday conversation we often use such words as "energy," "force," "work," and "power" more or less interchangeably. For the most part, of course, this works out splendidly; we succeed in getting our ideas across and save having to bother with absolute precision. When we are struggling trying to wrench a nail out of a plank with a hammer, it makes little difference whether we say, "I'm trying to work this nail out," or "I'm trying to force this nail out." No one cares whether we say, "I haven't the power to budge it" or "I haven't the energy to budge it." But when it came down to describing physical phenomena in precise terms, science was in trouble with such vagueness and had to agree to use one particular, specific word to describe a particular, specific thing or concept. Thus in present-day physics the word "force" is an abstract term referring to something that causes an object to change its velocity in some way. Even physicists fight their own intuitive inclination to think of a "force" as a "push," a "pull," or a "tug" and instead make themselves think of a "force" as something acting upon an object for a certain period of time, measurable only in terms of the resulting change in the object's momentum.

Similarly, we ordinarily think of "work" as practically anything that causes us to exert ourselves and make ourselves tired, but to the physicist "work" is done only when a force moves an object through a certain distance, with the amount of "work" always calculated by multiplying the distance the object has been moved by the force that moved it. In some ways such a definition seems contrary to common sense: A man might completely exhaust himself trying unsuccessfully to push the Washington Monument one foot north, but the physicist would say that he had done no "work" at all unless he actually succeeded in moving it. "Power" is an even fuzzier word in common usage, implying some potential for doing something but really carrying a wide variety of different common meanings. To the physicist, "power" has only one meaning: the rate at which work is done. The man who succeeds in pushing the Washington Monument one foot north in half the time it takes another man would be said to have twice the power of the other man. In physics power is always represented as some form of work divided by the time it takes to accomplish the work.

Finally, whenever work is done (that is, when some force has moved a body through a certain distance) we usually find that energy has been changed from one form to another in the process. Thus, when the man let the rock slip out of his hands, the force of gravity performed work on the falling object and its potential energy was transformed into kinetic energy. Similarly, when the billiard ball struck the cushion, the force of the rolling ball performed a certain amount of work in compressing the cushion (as its kinetic energy was transformed into potential energy), and then in reverse fashion the cushion performed an equal amount of work upon the billiard ball in shoving it away as its potential energy was converted back to kinetic energy again.

Vagueness about the use of such terms made it difficult for early physicists to assess really clearly what was happening in their experiments. The idea that the entire universe might contain a certain finite total amount of energy in some form or another which could never be destroyed nor increased had an ancient and respectable history; it seemed that men had wanted rather badly to believe in the law of conservation of energy for centuries! Yet repeated attempts to prove that energy was always conserved frequently led to failure, whether because of crude or inaccurate measurements, through confusion about just how energy should be defined, or through overlooking extraneous forces that were at work in the experiments. As a consequence, the principle of conservation of energy remained a theoretical dream rather than a well-established law of nature for a long, long time.

Yet one type of evidence, of a rather negative kind, seemed consistently to substantiate that energy *was* always conserved. Continuing attempts to *create* energy—to get energy for nothing, so to speak—invariably failed. Throughout the history of science we can find a long, sad succession of men with all sorts of ingenious ideas for building perpetual motion machines—machines which, once started running, would keep on running and continue doing work without any additional input of energy. Some of these hopeful invèntors were no doubt charlatans, but many others were perfectly earnest in their conviction that there really *was* some way to get mechanical energy for nothing, if only they could just figure out how. But the fact that such attempts always and invariably failed must presently have convinced more and more scientists that in fact there was just *no* way that this could possibly be done, and that some immutable law of nature would be violated if there were.

## THE OTHER FORMS OF ENERGY

In spite of the multitudes of skeptics, Newton and many other physicists of his day had soon become convinced that mechanical energy—potential energy and kinetic energy—was always conserved, and had the courage to base their lives' work on this conviction. But sure as they were that mechanical energy was conserved, even these men were not so clearly certain that *all possible kinds* of energy were likewise conserved.

Of course, in those days there was no knowledge of the internal structure of atoms, of the kinds of energy that served to bind the particles in atomic nuclei together, nor of the forces of attraction and repulsion that exist within atoms and between atoms. But there *were* certain kinds of energy recognized other than kinetic and potential energy. *Heat* was one such form of energy. *Electromagnetic energy* came to be recognized, although it seemed to have no connection whatever with potential or kinetic energy. There was also clear evidence that *chemical energy* also was present in the interaction of various substances to form new chemical compounds. Just what these forms of energy were was not understood, but they were recognized to exist. Other forces such as friction, cohesion, and surface tension were observed and studied, and recognized to involve energy in one way or another, but again no one seemed able to build the vital idea-bridge between the potential and kinetic energy present in interacting systems of objects and these other forms of energy.

In a way this is surprising, for some of these other forms of energy were regularly observed to appear as red herrings in orderly scientific study of conservation of mechanical energy under laboratory conditions. Very often, for instance, the total energy of an interacting system seemed to be measurably less after the interaction than before, apparently indicating that a portion of the mechanical energy in the system had been destroyed; yet heat would simultaneously appear during the interaction. A wheel spinning on an axle, for example, could be given a carefully measured amount of rotational (kinetic) energy by applying a carefully measured force to it, but no matter how well the axle might be lubricated, the kinetic energy of the spinning wheel would gradually seem to be dissipated and the wheel would slowly stop spinning, while at the same time the hub and the axle would invariably become hotter and hotter.

To us today the connection seems obvious, but it was not in those early days. Nevertheless, by the end of the eighteenth century a good deal of interesting data about these other forms of energy began to turn up. In the year 1800 Alessandro Volta invented the battery and proved beyond doubt that certain chemical reactions could result in the production of an electric current—in other words, that chemical energy could be converted into electrical energy. But the electric current produced in this way could in turn produce heat if it was passed through a high-resistance wire; it could even produce light, if the wire became hot enough to glow. Furthermore, if a wire carrying an electric current were wound in a coil around a steel rod, the flow of electrical energy through the wire induced magnetism in the steel rod, which could then produce mechanical motion of iron filings or nails. Conversely, mechanical energy through friction in a generator could produce electricity, and a full circle of conversion of energy from one form to another could be demonstrated: electricity to magnetism to mechanical motion to friction to electricity!

At roughly the same time other observations were made. In 1882 a European physicist named Thomas Seebeck demonstrated that heat could produce an electric current directly when applied to the junction between two different metals. Just at the turn of the century the American Benjamin

Thompson (who became notorious in Europe as "Count Rumford") observed that when a cannon barrel was being bored with a drill, so much heat was produced that the process had to be cooled repeatedly in order to avoid melting either the drill bit or the cannon barrel or both. Finally, during the 1840's, James Prescott Joule of England actually *measured* the amount of heat that could be obtained from a given amount of mechanical energy and showed that the conversion of mechanical energy into heat could take place without any loss.

Thus by about 1850 it appeared that the true implications of the law of conservation of energy were finally becoming clear. It was not just the potential energy and kinetic energy in a mechanical interacting system that were interchangeable and always conserved; *all* forms of energy were interchangeable, and the total energy in any form in a given system was always conserved. The reason that an accurate balance of mechanical energy before and after a collision could not be achieved was simply that part of the kinetic energy of the colliding objects was converted into heat and dissipated into the air, thus warming up the room but becoming difficult to measure, rather than being converted 100 per cent into potential energy. Friction was recognized as a force that "stole" useful mechanical energy from a mechanical system or a machine by converting it into useless heat energy. In fact, it was finally demonstrated that energy in any form could by the employment of the proper means be converted or transformed into energy of any other kind, and that this could be done without any loss of energy whatever. And so another conservation law became a solid cornerstone of modern physics, one of the very few invariables that physicists came to count upon.

But the last chapter in the development of the law of conservation of energy was not yet quite closed for, as we have seen, any *law* of nature can stand only as long as it answers all challenges, and in the early 1900s this natural law came up against a challenge that seemed to shake it to its very roots. In a way, this seemed a catastrophic blow, for by then the idea of conservation of energy was well entrenched as one of the most elegant, satisfying, and useful of all laws of nature and perhaps least vulnerable of any. But at about this time a number of chemical substances were discovered which we now know contained "radioactive" elements, and which seemed to have some strange characteristics that would have been quite impossible if the law of conservation of energy were indeed true. For one thing, some of these radioactive substances seemed constantly to give off heat, even though they were not apparently interacting with any other substances in any way. Even worse, some of these substances were found to hurl chunks of themselves away quite spontaneously, thus reducing the mass of the substance that was left. This clearly did not jibe with classical ideas of how matter ought to behave; physicists were well enough acquainted with such things as dynamic explosions, but this was something else again. The chunks that were thrown off by the radioactive substances were in many cases hurled away with incredibly high energy, with no indication of just where all that energy came from.

Then, further to confound the experts, when techniques were devised to make careful measurements it was found that the total mass of such a substance before a chunk of itself was hurled away was significantly *greater* than the mass of the substance remaining after an emission plus the mass of the particle it had emitted. This seemed to physicists very much like subtracting two from five and ending up with only 2½ as a result. Here was a case in which the total mass of a system seemed to *decrease* by a tiny amount in the course of an interaction, while at the same time a rather enormous amount of energy seemed to appear unbidden out of nowhere! This kind of natural occurrence simply did not add up in terms of conservation of mass *or* of conservation of energy, yet the fact that it indeed occurred forced physicists much against their will to begin to wonder if these conservation laws were really as valid as they had been thought to be after all.

Most of us know the end of the story. Albert Einstein came to the rescue with an idea that is now familiar to us all: that energy itself has mass, and conversely that solid matter is in reality nothing more than another form of energy. Granted that energy does not have *very much* mass, and granted that a very tiny quantity of matter is equivalent to (or represents) a perfectly staggering amount of energy. But even so, the concept explained how mass could be "lost" and energy "created" in cases when radioactive substances spontaneously hurled pieces of themselves into space. The "missing" mass after such an occurrence was later found experimentally to be exactly the amount of mass necessary to produce or be converted into the amount of energy the emitted particle flew off with.

Later we will discuss in more detail the significance of Einstein's familiar equation for the *equivalence* of mass and energy $E = mc^2$. For now it is enough to point out that modern physicists regard mass and energy as essentially two manifestations of the same thing, "convertible" from one to the other and back again under various special circumstances; and to note that the amount of mass "represented" by a given amount of energy is very tiny, whereas the amount of energy "represented" by a given mass of matter is extremely large.

In modern times physicists do not speak of the law of conservation of matter or the law of conservation of energy separately, simply because they can't. For all practical purposes the two laws have been combined into a single all-inclusive law which can be stated as follows:

*In any closed system the matter and/or energy that it contains can neither be created nor destroyed; all the matter and/or energy of such a system invariably remain constant.*

And we must remember, whenever we think of that law of nature, that the entire universe may someday be found to be a closed system.

## MAXWELL'S DEMON AND THE PUZZLE OF ENTROPY

Today we know that the law of conservation of mass and/or energy still remains one of the strongest of all natural laws. We also know of the wide variety of forms in which energy can be found, and that it is possible to convert energy in one form into energy in another form without loss. A list of the major forms in which energy is encountered would include the following:

- Mechanical energy
  - Potential energy
  - Kinetic energy
- Chemical energy
- Electrical energy or, more accurately, electromagnetic energy
- Heat energy (including heat produced by friction)
- Atomic and nuclear energy (including the energy produced when matter is converted into its equivalent amount of energy)

We have seen that it is possible to convert energy from any one of these forms into energy of any other form without ever either creating any new energy (thus increasing the total amount present in the universe) or destroying any energy (thus decreasing the total amount of energy in the universe). In many cases, little or no difficulty is encountered in converting energy from one form to another, and at least theoretically it should be possible to convert 100 per cent of a given amount of energy in one form into any other form desired if the proper technique for accomplishing this is known.

Yet when we get down to the practicalities of actually doing this, we soon discover somewhat surprisingly that here is a case in which nature seems to play favorites. While energy in one form *can* be converted into energy in any other form, it is much more difficult to convert it into certain forms than into certain other forms, while energy in any form seems to be converted so very readily into certain other forms that we have difficulty avoiding those forms and converting it instead into the specific form that we want.

It seems very much as though the various major forms of energy occupy comparatively higher or lower rungs on a ladder, so that it is relatively easy to convert higher forms of energy into lower forms, but exceedingly difficult to convert lower forms of energy into higher forms. And of all forms of energy known, the form that occupies the lowest rung on the ladder—the form of energy that all the other forms can be most readily converted into and the form that is most difficult to convert into any other form—is in many ways the least useful form of energy that we know: *heat energy*.

In fact, energy in the form of heat persistently keeps turning up at times and in places where it is simply not wanted at all. We discussed at some length the conversion of kinetic energy to potential energy and back again, and found that under imaginary ideal conditions 100 per cent of the kinetic energy of a system could be converted to potential energy and then 100 per cent of the potential energy could be converted back to kinetic energy. But any time such a conversion is attempted under *real* conditions, some of the energy involved in each conversion is invariably converted into heat and dissipated into the atmosphere, whether we like it or not. Thus when we have a wheel spinning around an axle, its kinetic energy will gradually be converted to heat and the axle and hub will warm up. Then eventually the axle will cool and the heat energy will be dissipated into the air.

This does not mean, of course, that the energy is lost. It simply means that the air near the wheel becomes a little warmer than it was before—that the molecules of atmospheric gas in the vicinity of the wheel begin moving faster. Those fast molecules will collide with other molecules to set *them* moving faster, or move away from the vicinity of the wheel, so that presently the warmed-up molecules will be mixed with other molecules and disseminated more and more widely. Ultimately the kinetic energy of the wheel "lost" to heat will have been disseminated evenly throughout all of the gaseous atmosphere surrounding the earth and will have served simply to warm that atmosphere up a tiny bit. Similarly, a mechanical device operating under water will lose a certain amount of its mechanical energy to heat which will go to warm up the water in which it is operating. But again, presently, the tiny amount of increase in the water's temperature will warm the atmosphere by a tiny amount.

Other conversions of energy seem to produce the same phenomenon. A great many of the chemical reactions that take place, either in laboratories, in industrial processes, or in the cells of living organisms, result in transformation of a great deal of potentially useful chemical energy into essentially useless heat energy. Nor is the loss always too tiny to measure; recently scientists studying the pollution of rivers and lakes in heavily industrialized areas have discovered that heat-producing industrial processes using river or lake water as a means of getting rid of unwanted heat have in some cases actually raised the temperature of the water in those streams and lakes to such a degree that the normal native plant and animal life can no longer survive. Again, any time electrical energy in the form of electric current passes through one of a variety of substances that offer "resistance" to its passage, part of the electrical energy is converted into heat which will then serve to heat up the air around the substance offering the resistance.

But doesn't this conversion of other forms of energy into heat work just as well the other way? If any form of energy can be converted into any other form, then it must be possible to convert heat into potential energy or kinetic energy or electrical energy or chemical energy, mustn't it? The answer is a guarded yes, except that it doesn't quite work that way. Heat *can* be converted into other forms of energy, given the right circumstances. We might, for example, use heat to raise a quantity of water to the boiling point, and then use the resultant steam under pressure to produce mechanical energy. This is how a steam engine works, and steam engines certainly have proved to be useful devices in mankind's history. *But they have also proved to be somewhat inefficient machines.* Only about 60 to 70 per cent of the heat energy used to heat the water in the steam engine's boiler ever actually gets converted to mechanical energy. But in the opposite direction, if sufficient friction were present, it would be quite possible and indeed very easy to convert virtually *all* of a given quantity of mechanical energy into heat. In 1852 William Thompson, Lord Kelvin, expressed it very succinctly: "It is impossible to invent an engine which simply takes in heat and converts it entirely into mechanical work." On the other hand, it would be very easy to invent an engine which would take mechanical energy and convert it entirely into heat.

As physicists came to recognize this odd peculiarity about heat energy as compared with other kinds of energy, it became increasingly clear that this was not merely an insignificant curiosity. There seemed to be a definite "natural" direction in which energy sought to flow when it was converted from one form to another, just as there is a "natural" direction in which water seeks to flow. Of course it is possible to force water to run uphill, either by pumping it or by arranging one or another kind of device to thrust or carry it uphill, but it flows uphill only "against its will," so to speak, and in opposition to its "natural" direction of flow. On the other hand, given the slightest opportunity, water will flow downhill without requiring any effort on our part at all.

Similarly, heat energy can be forced "against its will" to be converted into other forms of energy, but only at the cost of a great deal of effort, while other forms of energy can be converted into heat given the slightest opportunity. Ordinarily, heat will always tend to flow from an area of high heat content into an area of lower heat content—from a hot area to a cold area—any time two such areas with different temperatures are side by side. A hot area *can* be heated up further while an adjacent cold area becomes colder, but again only at the cost of a great deal of energy *added* to the system. For example, we could open the door of a household freezer in the middle of a warm room, and then set the controls so that the freezing unit inside the freezer becomes even colder while the door is still open, while the room becames even warmer as a result of heat being pumped out of the freezing compartment and dissipated into the air. But this would require a very considerable input of additional energy (in this case electrical energy) to operate the freezer. It is certainly not the "natural" way that heat would tend to move under the circumstances unless it were *forced* to move that way.

Indeed, the "natural" tendency of heat to move from a warm area to an adjacent colder area was soon recognized as a means by which heat energy could be made to do work. In a system in which a quantity of gas with a high heat content (that is, with molecules moving with high kinetic energy) is adjacent to a collection of gas with a low heat content (in which the molecules have a much lower average kinetic energy) physicists thought of the energy in the system as "ordered" to a certain extent because the gas molecules could be made to do work just by leaving them to themselves to intermingle until a state of equilibrium between the "hot" molecules and the "cold" molecules had occurred. In such a state of equilibrium, the system would have become "less ordered" since there was no natural tendency for the hot molecules and the cold molecules to separate themselves out on two sides of the fence again—in other words, since there was no longer any way that the gas molecules could do work when left to themselves.

Physicists use a rather peculiar word to describe the *degree of disorder* of the energy in such a system of hot and cold gases which had been allowed to intermix toward equilibrium: They speak of *entropy* as a measure of the degree of disorder of the energy in such a system. Thus, if we could start with all the hot gas molecules in a container on one side of a partition with a hole in the center, and all the cold molecules on the other side of the partition, the energy of the molecules would be sharply separated or highly "ordered" and the system would have very little entropy. After a while, however, hot gas molecules and cold gas molecules would intermingle through the hole in the partition. The energy in the system would become more and more disordered, which would mean that *the entropy of the system would increase.*

Indeed, in *any* localized or small-scale closed system in which the conversion of energy from one form to another involves loss of some of the energy to heat, the entropy of the system is increased to some degree. It is as though of all the energy present in such a system in all its various forms, the energy occurring in the form of heat forms a sort of pool or reservoir of essentially useless energy, in a form relatively incapable of doing work; and as quantities of energy in other forms are purposely or inadvertently converted into heat energy they are "lost" into this ever-increasing reservoir of relatively useless heat energy, so there is increasingly less and less energy left in the system in forms capable of doing work, and no effective "natural" way for this gathering quantity of heat energy in the system to be converted back into useful forms again. And as the heat energy (relatively disordered and useless) piles up, the entropy of the system steadily increases.

Well, we might ask, so what? There's plenty of energy around. Who cares if the energy in one or another small closed system becomes increasingly disordered and useless? Is this something to worry about?

Many physicists of the mid-1800s thought it was, for they *extended* what they had observed in small closed systems out to what seemed to them a logical extreme: They began arguing that the same steady and inexorable increase in entropy occurring in such small closed systems must also be occurring *in the entire universe*. Indeed, they speculated that all the energy in the universe was gradually, bit by bit, irrevocably being converted into heat energy, that the entropy of the universe was thus steadily increasing, and that ultimately, in the far distant future, the entire universe would come to an end in a dismal sort of "heat death" with all the stars, planets, and everything else in it fried to a crisp and converted into heat energy!

To understand more clearly why such an apparently outlandish idea was seriously debated by physicists right up to the last decade or so, let's see how the increase in entropy in a closed system works, with its inevitable concomitant decrease in the amount of useful, work-capable energy, by means of a simple analogy. Imagine a very rich man with a vast quantity of money in his pocket—hundred-dollar bills, fifty-dollar bills, ten- and five-dollar bills, and one-dollar bills, as well as half dollars, quarters, dimes, nickels, and a few pennies. Then imagine this man going about his daily life trading money for goods, and then retrading the goods for money. Suppose also that the prices he pays for goods and the prices he receives in exchange for goods are frequently odd amounts—a bicycle costs $35.78, a bottle of perfume $10.18, a package of razor blades $0.99. Under these circumstances all sorts of conversions of money would occur. One-hundred-dollar bills would be broken into tens, twenties, and fifties. Half dollars would be broken into quarters and dimes, and—inevitably, because of the irregularity of prices—dimes and nickels would be converted into pennies.

In fact, in virtually every exchange a larger-denomination coin or bill would be broken down into a smaller one, and in virtually every exchange the man would end up with a few more pennies than he had before. We might say that over a period of time, he would find himself losing more and more of the larger-denomination bills and coins and accumulating heaps and piles of pennies. Now, suppose in this strange country that it was perfectly possible for one to gather up pennies and trade them for nickels, dimes, quarters, or dollars, but that everybody else had too many pennies as it was and really didn't want any more. Thus it was always far more difficult, if not impossible, to trade pennies for dollars than it was to buy goods with the dollars and end up with a few more pennies.

Finally, suppose that in the entire country there was only one bank that was required by law to trade dollars for pennies, and that bank was only required to take ten dollars' worth of pennies from any one person on any one given day, and anyway the bank was located two thousand miles away from where our rich man lived. The end result of such a system is obvious: Bit by bit the man would accumulate more and more of his money in the form of pennies which he couldn't use very well, and have less and less of it in the form of useful hundred-dollar bills, ten-dollar bills, half dollars, and quarters. We could even imagine that if something didn't happen presently to revise this badly misbalanced monetary system, not only our man but all the other merchants in the nation would eventually, bit by bit, end up with no useful form of currency whatever and all commerce would grind to a halt, quietly smothered in pennies.

Of course, in this analogy, the rich man, his country, and the country's monetary system were all a closed system—what happened there would not necessarily happen in another country on another planet. But physicists of the mid-1800s saw no reason why this principle of entropy increase—which seemed more and more to be a natural law—should not be *universalized* or extended to the entirety of the universe. Thus they wrestled with this puzzling notion that all the energy in the universe was gradually, bit by bit, being converted into comparatively useless heat. Surely there must be some way that the universe's "monetary system" could be revised! One man who came up with an answer was James Clerk Maxwell, the brilliant student of Michael Faraday, but the answer he found was in the form of a clever paradox which simply seemed to indicate that no such "revision" could ever be hoped for. In a study of the dynamics of heat energy published in 1871, Maxwell postulated a cylindrical vessel much like a large tomato-juice can divided into two sections, A and B, by means of a rigid dividing plate with a hole in it. Then Maxwell imagined that a demon was standing guard by the hole in the separator, a demon with eyes so sharp that he could follow the motion of every molecule in each portion of the container and determine whether it was moving fast or slow. Maxwell's demon was then instructed to open or close the hole between section A and section B of the container in such a way that *only the slower molecules could pass from section A into section B* and *only the faster-moving molecules could cross from B to A* (see Fig. 12).

Presently the sorting activity of the demon would begin to show measurable results: More and more of the swifter molecules would become concentrated in section A of the container, while more and more of the slower ones would become concentrated in section B. Since the "temperature" of a gas is merely a measure of the average speed of its moving molecules, the temperature in section A would begin rising while the temperature in section B would begin falling. What Maxwell's demon would actually be doing, in opening and closing the hole between the two sections of the container in this discriminatory fashion (and without expending any work whatever, since he was a demon) would be to *reverse the natural flow of heat* into adjacent areas and make hot molecules gather in one section and cold molecules gather in the other. Thus he would be creating a system in which the "order" of the heat energy present becomes greater and greater, and in which the entropy of the system steadily decreases.

What was wrong with Maxwell's demon? Why was it not possible for the entropy of such a system to decrease rather than increase—for the heat energy to seek a higher level of order rather than a lower level? The simplest answer, of course, would be to ask Maxwell to produce his demon—but unfortunately no such entity has ever been known to science, either then or now. Even aside from that, the bargain was just too good to be true. The sorting activity of Maxwell's demon would, in fact, have constituted a perpetual motion machine, because the flow of heat from the warmer region A of the vessel to the colder region B could be made to do mechanical work. This, after all, is what makes a steam engine do work.

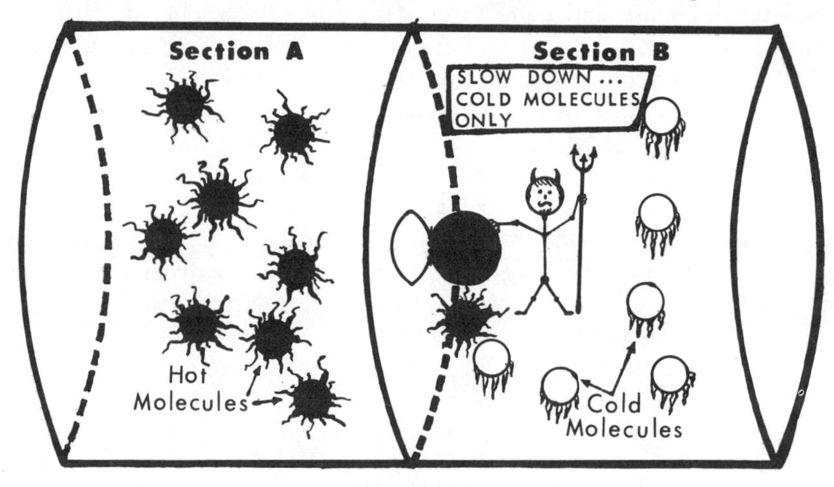


*Fig. 12* Maxwell's demon, guarding the porthole to permit only fast, hot molecules to cross into Section A and only slow, cold molecules to enter Section B. Doubtless he became a demon for violating the Laws of Thermodynamics!

But the real reason that Maxwell's demon could never work, we know today, is not because such a device could not be constructed, but because even if it *could* be constructed with the aid of miniaturized integrated circuits, the demon would *still have to expend energy to find out the velocity of each molecule in order to do the sorting.* And the energy needed to do this and proceed with the sorting would turn out to be *exactly* the amount of energy needed to operate a heat engine to pump energy from one side of the vessel to the other! (We will see just *why* energy is needed to find out the velocity of each particle in a later chapter.)

Thus Maxwell's demon proved to be nothing but a fanciful fallacy. But even without a demon to guard the hole between section A and section B of the vessel, is it not possible that more and more of the hot molecules might, *of their own accord,* speed through the hole into section A while more and more cold molecules, *of their own accord,* found their way through into section B, so that section A would still get hotter and hotter while section B got colder and colder?

Possible? Yes—there would be nothing to prevent any given molecule from crossing through the hole in either direction. But now we are talking not just about *possibility,* but about *probability* as well. Given such a container divided into two sections by a partition with a hole in it, it would be perfectly *possible* for all of the millions of faster gas molecules to move one by one through the hole into one portion, and for all the millions of slower, colder molecules to find their way one by one into the other part of the container, quite of their own accord. It *could* happen as a result of the random movement of the molecules, purely by chance. Note that we said it would be *possible.* But *probable?* No. The *probability* that any such thing should ever happen anywhere, even once, in all the history of the universe, is so infinitesimally small that it can as well be ignored. And if Maxwell's demon were on hand and able to sort molecules and open and shut the hole in the partition, it would be violating mathematical probability by making hot molecules flow to the hot side and cold molecules flow to the cold side *every time,* whereas the "natural" flow of heat energy from an area of organization to an area of disorganization is a matter of probability so high that the possibility of the opposite sometime coming to pass under some natural circumstances can simply be dismissed from consideration altogether.

We can see the distinction between *possibility and probability* with a simple experiment. Take a roll of 100 pennies, shake them in your hand, and then toss them freely on the rug. Since each one of the pennies spins freely in the air and may fall either heads up or tails up, there is a possibility that all 100 pennies will land heads up, without a single tail showing. The probability that this will occur, however, is extremely remote—so remote that we would be tempted to say that it "never" would happen—except that we know of no law of nature that says it *can't* happen.

The *probability* is that any time the pennies are so thrown the result will be within a few pennies one way or the other of a perfectly equal distribution, and this will be the case even if all the pennies in the roll have been stacked heads up to start with. A deviation of 10 per cent either way from a perfectly even distribution of heads and tails would be unusual. A deviation of 20 per cent either way would be remarkable, and a deviation of 30 per cent either way would be so extraordinary that it would be unlikely to occur a second time if you kept tossing the pennies in this fashion over and over again all year—although it would be just as *possible* for it to happen on the second throw as on the first.

Exactly the same situation obtains in the case of hot and cold gas molecules separating themselves out by means of their random motion into opposite sides of a container: possible but highly improbable, with the improbability all the greater because there are billions and billions of gas molecules to take into consideration while there were only 100 pennies.

The study of the behavior or movement of heat from an area with a certain heat content to another area with slightly lower heat content (or vice versa) is known to physicists as *thermodynamics,* a term that seems carefully calculated to scare nonscientists away in a panic. All the term means, of course, is the study of the dynamics (i.e., of the movement and/or behavior) of heat or heat energy. In thermodynamics a situation in which a container containing only hot (that is, fast-moving) gas molecules is adjacent to and connected with another container containing only cold or slow-moving gas molecules is said to be an "ordered" or "organized" system simply because we know a great deal about the individual molecules in such a system. We know that most of the molecules on the hot side are going to be moving around swiftly while those on the cold side will, on the average, be moving more slowly, and that the probability that any given molecule in the system would be found to be moving swiftly would be far greater on the hot side than on the cold side. As the hot molecules from the one side intermingle and mix with the cold molecules on the other side, the heat differential between the two sides decreases—and so does the accuracy of our knowledge about any given molecule in the system. By the time equilibrium has been reached between the two sides, we know virtually nothing about any given molecule's movement—the probability of finding any given molecule moving faster or slower on either side would be precisely the same.

Thus we could equally well say that the entropy of a given system is a direct measurement of *how little we know about the individual molecules in the system.* When the system is highly organized, we know a great deal and the system has low entropy. If the molecules diffuse from side to side and the heat content of the two sides begins to equalize, the entropy of the system increases and by the time perfect equilibrium is reached, entropy (i.e., disorder or disorganization) of that system has increased to its maximum. *But then there is no way for the entropy of the system to decrease again without outside help.* The heat content in the ordered system with low entropy has been dissipated and diluted into a disordered or disorganized system with much higher entropy, and the process does not reverse itself naturally. When the opposite thing occurs, as when our open freezer is forced to become colder even when open to a warm room, the entropy of the system of cool freezer and warm room has decreased, but we had to expend a great deal of electrical energy and mechanical energy to bring this about. In fact, in any system in which a form of energy higher on the ladder than heat is converted or "lost" to heat energy, the entropy of the system increases.

As we have seen, physicists of the early 1800s who were studying heat energy in relation to other forms of energy became so convinced of the natural tendency of heat to move from an organized (hot-cold) system to a disorganized (uniformly lukewarm) system that they began to consider this pattern of behavior as a fundamental law of nature and described it in two simple rules which have come to be known as the *laws of thermodynamics*. Today both are considered basic laws of nature. The first law of thermodynamics is simply a restatement of the law of conservation of energy: *Energy is always conserved, and can never be destroyed.* (This implies, among other things, that no perpetual motion machine such as Maxwell's demon may be built.) The second law of thermodynamics states: *In any closed system there is a tendency for the individual units making up the system to change from a state of order into a state of disorder, so that the amount of information we have about the system becomes smaller as time goes on.* A simpler way of stating this law, providing we understand what is meant by "entropy," would be to say: *In any closed system the entropy must always either remain unchanged or increase.*

This second law of thermodynamics is really nothing more than a restatement of Lord Kelvin's principle that we quoted earlier: "*Heat cannot be transferred from a cold body to a warmer body without the aid of some outside source of energy.*" But the fascinating thing about this second law of thermodynamics is that it is the first fundamental natural law we have encountered which depends upon probability rather than invariable behavior. If we wanted to be absolutely precise, we would have to rephrase the second law to say: "While it is not *impossible* for heat to move from a cold body to a warmer body without the aid of some outside source of energy, such an occurrence is so incredibly *improbable* that we can never reasonably expect it to happen."

Assuming that these laws of thermodynamics were really valid everywhere in the universe under all circumstances, not just in the small part of it they had been able to observe personally, those physicists of the mid-1800s became exceedingly gloomy about the likelihood of anything occurring in the universe to "change the monetary system"—to make it suddenly possible for the entropy of any closed system (including the entire universe as a closed system) to begin to increase again. Certainly an extensive survey of known examples of energy transformations within natural systems seemed to indicate that the entropy of any given system would almost invariably remain unchanged or would increase, with only a rare system occurring in nature in which entropy decreased even slightly. Actually, there *are* situations in which decreases in entropy seem to occur under natural circumstances—situations, for example, involving the exchange of energy in the cells of living organisms, in which heat is converted back into other forms of energy, or moves from areas of low heat content to areas of higher heat content. But whenever this happens, even in the metabolism of living cells, it is *forced* to happen in the course of chains of one-way, irreversible chemical reactions which require the addition of significant amounts of new energy in order to proceed.

Such natural situations, moreover, are rare. For the most part, any systems in which heat is converted into other forms of energy show the fine hand of man at work. These are usually artificial man-made systems such as the household freezer we spoke of which we have ingeniously devised in order to *force* a flow of heat in an "unnatural" direction, with a resulting decrease in entropy, in order to achieve some particular predetermined advantage.

What is more, such "abnormal" systems will continue to operate only at great expense in outside energy input, and only as long as we force them to continue. In operating them we are, in effect, flying in the face of natural law, and though we may gain a temporary advantage it is usually costly. Scientists such as George Gamow have pointed out that if we could create a system in which entropy was easily and cheaply decreased we would in effect be creating a sort of perpetual motion machine. For example, if it were possible to utilize the natural heat of the ocean cheaply as a source of energy, it would be possible to operate a ship almost perpetually simply by drawing the warm sea water into its engine, converting the heat into mechanical energy to turn the screw, and tossing the resultant blocks of ice overboard. If this could be accomplished without using up more energy to convert the heat energy into mechanical energy than the mechanical energy we got out, our ship could continue to travel about scot free until the temperature of the ocean had been reduced to the freezing point of salt water all over the earth—which might mean virtually forever, considering the amount of water there is, the smallness of the ship, and the constant warming action of the seas by the sun.

But such a "perpetual motion machine of the second order" (that is, operating in violation of the second law of thermodynamics) could never be made to work simply because we would always have to use far more energy in some other form in order to convert the heat from the water into useful mechanical energy than we could ever get out in the form of mechanical energy to turn the ship's propeller; far from getting off scot free, we would do far better at far less cost to use more economical fuel in the first place. Maxwell's demon would represent another perpetual motion machine of the second order: The demon, so to speak, would have to be fed in order to keep him working at determining molecular velocities and separating fast molecules from slow molecules, and we would soon find that he had a voracious appetite!

But what do these laws of thermodynamics mean in the long run? If they were really universally valid anywhere under any circumstances, it would appear that in almost all natural systems in the universe there must be a constant dissipation of useful forms of energy into useless heat. The rare instances in which man succeeds by his gadgetry in reversing this process temporarily by means of one device or another would not amount to a hill of beans in terms of the total amount of energy in the universe. Thus, it would seem inevitable that as time passed and more and more energy of all varieties had dissipated into heat, and as more and more matter in the stars and galaxies was consumed to create more energy, that the entire universe bit by bit would increase in entropy, growing steadily hotter and hotter. Indeed, this increase in the heat content of the universe would really only be a function of time; and sooner or later the entire universe would contain nothing whatever but an enormous quantity of heat energy—no living creatures, no planets, no stars, no matter as such, nothing but energy all reduced to an enormous quantity of heat.

That was the dismal forecast for the future that physicists of the late nineteenth century and the early twentieth century were thinking of when they argued and debated about the "entropy increase and heat death of the universe." And as a subject for perpetual fretting, it would be hard to beat, although this gloomy fate was not expected to overtake the universe for another 500 billion years or so. In recent decades, however, modern cosmologists have come to pooh-pooh the whole concept of heat death of the universe, and maintain that it will never actually occur. The assumption that the laws of thermodynamics apply everywhere in the universe under all circumstances may not be entirely valid, they maintain—or at least there is evidence that the laws may not apply quite as they were originally understood. Cosmologists argue that entropy in our universe is not a quantity clearly enough defined under all circumstances for us to make broad generalizations about its increase or decrease throughout the universe—the entropy of a gravitational field, for example, is not defined. Nor do the laws of thermodynamics seem useful in explaining increase or decrease of entropy in an expanding universe, or in an oscillating universe, such as we will discuss in a later chapter. For the moment, the best we can say is that entropy increase is a clearly defined tendency at least in small, localized closed systems—systems here on earth, for example, or systems involving our solar system in its present stage of evolution—but that even our grandchildren need not worry about the heat death of the universe ever occurring unless they really want to.

# CHAPTER 12

## *Electricity, Magnetism, and the Phenomena of Waves*

One of the major problems that we face in understanding the significance of the discoveries and observations that were made by physicists in the seventeenth, eighteenth, and nineteenth centuries is the sheer *number* of physical phenomena that were studied and the number of discoveries made all within a comparatively short period of time. After lying dormant for over a thousand years, interest in the nature of the physical universe suddenly came awake with a start as the work of Galileo, Copernicus, Kepler, and Newton began a veritable chain reaction of experiment and discovery.

As we have seen, much of the work in physics up to Newton's time was concerned with the nature of matter, motion, and energy. But by the time that the laws of motion, the laws of conservation, and the laws of universal gravitation had evolved, a wide variety of other seemingly unrelated natural phenomena were being investigated. As a result, a great many fascinating discoveries were made; physicists broadened their understanding of the universe as they saw it and laid the groundwork for the even more startling and revolutionary theories and discoveries which were to appear at the beginning of our own century. Among the more important of these areas of discovery and investigation were the study of the nature of electricity and magnetism; the formulation of the first really *scientific* theory of the atomic structure of matter; the discovery and investigation of the phenomena of waves; the study of the enigma of light; and the growing awareness that signals of many kinds could be transmitted from one place to another by means of various kinds of waves which, taken together, form a continuous "spectrum" extending from exceedingly long waves to exceedingly short waves—an ordering of waves which we have come to know as the *electromagnetic spectrum.*

Most of these investigations and discoveries were all initiated within the space of a hundred years or so between 1750 and 1850, and dovetailed with each other so intimately that it is impossible to discuss one without discussing all the rest at the same time. But interwoven with them all were two mysterious and inexplicable natural phenomena that were known even to the ancient Greeks: the phenomena of *electricity* and *magnetism.*

## THE PUZZLES OF LODESTONE AND AMBER

The early physicists studying the nature of matter and energy and the behavior of matter in motion were real trail blazers; they had little to go on but what they could see and measure. They had enough difficulty figuring out exactly what was happening when an identifiable force came into observable contact with some object and brought about some kind of change in its motion. It must have been a thousand times more difficult for them to understand what was happening when some force that they could *not* see or understand managed to have an effect upon objects that were clearly located at a distance from the force acting upon them. Yet examples of such "forces acting at a distance" kept appearing with discouraging regularity.

We have already seen one such example in the action of the force of gravity upon objects on or near the earth's surface. Since early investigators could not understand *how* a force could reach out through empty space, as it were, and have an effect on the motion of an object, centuries were wasted in a futile effort to discover *what* gravitational force was, *how* it acted upon objects at a distance, and *why* it behaved as it did. An object dropped from the hand fell to the ground at a constant acceleration. How did it fall? Why did it fall? How could a force be acting on it when nothing was touching it?

These seemed to be perfectly reasonable questions; yet as we have seen they were futile and unanswerable questions. Even today modern physics does not have plausible answers for them. But not until a few brilliant men realized that these questions were futile, and bent their efforts toward observing and describing accurately *what* happened when gravity acted upon an object, was progress finally made in outlining how gravity could be expected to act under any and all circumstances. As for what gravity was, the best that anyone could say was that a "field of force" existed between any two massive objects, and that the strength of this field of force (that is, its ability to influence the motion of some object at a distance) varied according to certain proportions related to the masses of the two objects in question and the distance between them. The more massive the objects, the greater the "field of force" between them would be, and thus the more powerful the action at a distance that the field of force could exert. The farther apart the objects, the weaker that field of force became, and the less capable it was of exerting action at a distance.

But gravity was not by any means the only force that was known to act upon an object at a distance. Even the ancient Greeks knew that a certain kind of rock known as lodestone had the mysterious ability to attract and hold pieces of certain kinds of metal, notably iron, as if the iron and the lodestone were glued together. The effect of a large chunk of lodestone (a variety of iron ore which is known today, significantly, as "magnetite") on a small piece of iron was quite dramatic: The iron would be held so tightly to the lodestone that considerable force was necessary to pry the two apart. This magnetic force between lodestone and iron could even overcome the force of gravity; a small piece of iron on the ground would rise up through the air to meet a piece of lodestone held above it and would then cling tightly even though nothing could be seen to be holding it.

This strange attraction between lodestone and iron was puzzling enough, but even more puzzling phenomena were observed as a result of this mysterious attracting force. When a piece of iron was rubbed on a piece of lodestone for a while, the iron itself seemed to take on some of the lodestone's iron-attracting qualities and retain them when the lodestone was taken away—the iron became a sort of "artificial lodestone." Such a piece of iron was then found to be able to attract and hold another piece of iron.

But this force of magnetism was soon found to be more than just a force of *attraction*. When two pieces of iron had been magnetized by contact with lodestone, their newly acquired magnetism seemed to consist not only of a force of attraction but also of a force of *repulsion*. Two ends of the magnetized pieces of iron seemed to attract and hold each other vigorously, just as lodestone attracted iron; but when one of the two pieces of iron was turned around end for end and brought close to the end of the other piece, the two pieces then actively *repelled* each other and could be held pressed together only by exerting considerable force. Then as soon as the force was released the two pieces of iron literally pushed each other away.

It was also found that when a magnetized iron rod was suspended at the middle by a string, it would always orient itself in a north-south direction relative to the earth, with one end pointing north and the other pointing south. This odd directional orientation always took place automatically as soon as a magnetized iron bar was allowed to swing freely from a string, and always resumed this orientation automatically any time that the bar was turned to point in any other direction. Indeed, the magnetized iron behaved exactly as if the earth itself were a huge magnet, so that one end of a magnetized bar (and always the same end) was attracted to the south end of the earth, while the other end was attracted to the north end of the earth. Any time the bar was turned in the opposite direction, the end that was attracted to the north end originally was actively repelled by the south end of the earth and vice versa. The bar would swing around to its "natural" orientation again.

These strange forces of magnetic attraction and repulsion were known to the ancient Greeks and probably to even earlier civilizations. Nobody understood what these forces were or why they were present. All that could be said was that here was something that was observed to happen: By some mysterious means, one object could act upon another object at a distance and actually cause it to move without touching it, sometimes even against the force of gravity. But the similarity between this magnetic force and the force of gravity was obvious. In both cases, some kind of "field of force" existed between two objects, and whenever that "field of force" was strong enough, it had the power to make objects at a distance move one way or another without actually being touched by anything.

But magnetism was not the only phenomenon known to the ancients that exhibited forces of attraction and repulsion. According to history, the first discovery of static electricity and of positive or negative electrical charges was made when a lump of amber was rubbed with a silk cloth. Some kind of mysterious "fluid" seemed to pass out of the amber and into the silk so that the two seemed to attract and cling to each other. Indeed, our word "electricity" is derived from the Greek word *elektron* meaning "amber," and "electric force" came to mean the force generated by rubbing amber.

Later, other substances were also found capable of generating this strange force which seemed to make some objects attract each other and some repel each other. Silk rubbed on amber would create a force of attraction, but if the same silk were rubbed on a glass rod, the cloth would soon be repelled by the rod, not attracted to it. In our own familiar experience, all of us have seen how human hair is attracted to a comb made of hard rubber or plastic, and we have all seen how nylon or silk garments tend to cling to our skin when we undress on cold winter days.

The ancient Greeks came to recognize clearly that there was a distinct difference between an electric force of attraction and a similar but opposite electric force of repulsion. They also recognized that when two objects were both charged with similar electric force, whether it be positive or negative, those objects would repel each other quite as effectively as the similar poles of two magnets repelled each other, whether the two poles happened to be north poles or south poles. Similarly, two objects charged with opposite kinds of electric force, one positive and one negative, would attract each other just as opposite poles of two magnets attracted each other. Here again was a case of forces acting at a distance to cause objects to move, but in the case of electric forces, it was discovered that certain substances such as strips of copper had the ability to transmit or *conduct* the mysterious "fluid" that caused electrical attraction or repulsion from one place to another. Once again a similarity to gravitational force was obvious: Somehow a "field of force" seemed to come into existence between two electrically charged objects, and when that "field of force" was strong enough it could cause the objects to move together or apart even when nothing was touching them.

## THE ELECTROMAGNETIC CONSPIRACY

We have seen, of course, that the ancient Greeks did little more than to note the existence of the two phenomena of magnetism and electrical attraction and repulsion. Both phenomena seemed related to similar fields of force capable of acting upon objects at a distance, but the Greeks did not recognize the almost conspiratory relationship between the two phenomena. Later, as the concept of energy began to evolve, physicists began to recognize that some kind of energy was involved in each of these fields of force, but little effort was made to find out how the forces of attraction and repulsion behaved in either case. For centuries electricity and magnetism both remained baffling phenomena that no one understood very well and that no one tied together in any way.

Then, in the mid-1700s, a series of singular discoveries were made in rapid succession. Unlikely as it may seem, Benjamin Franklin actually did conduct his famous kite experiment in 1747, proving beyond question that lightning was indeed a natural form of electricity, although how he avoided electrocuting himself in the process remains a mystery to this day. Franklin also recognized that two varieties of static electric charges could be generated, and assigned the terms "positive charge" and "negative charge" more or less arbitrarily.

Somewhat later, Alessandro Volta discovered that chemical energy could be converted into electrical energy; when he placed a carbon rod into a zinc container filled with a mixture of salt and sulphuric acid, and then connected the carbon rod with the zinc container by means of a copper wire, he found that an electrical current was created and passed through the wire—the earliest example of a simple dry cell battery. Other experimenters discovered that when a wire was coiled around and around an iron rod and then connected to the battery so that current passed through the wire, the wire coil would create or *induce* magnetic force in the iron rod so that the rod actually became a magnet as long as the current continued flowing through the surrounding wire. The wire did not even have to be touching the metal core at all; it could be insulated by wrapping the wire with substances that did not conduct electricity, and still magnetism would be induced in the rod. The strength of the rod's magnetic force could be increased by increasing the number of coils around the rod, but the rod would lose its magnetism very quickly the moment the current through the wire coil was disconnected (see Fig. 13).

This sort of device, which we now know as an *electromagnet,* was found to be extremely useful in converting electrical energy into mechanical energy capable of doing work. For example, the electromagnet acts as the main working mechanism of any kind of electric motor, a machine which operates on such a very simple principle that an intelligent child can build one very readily with a few feet of insulated copper wire, a couple of iron

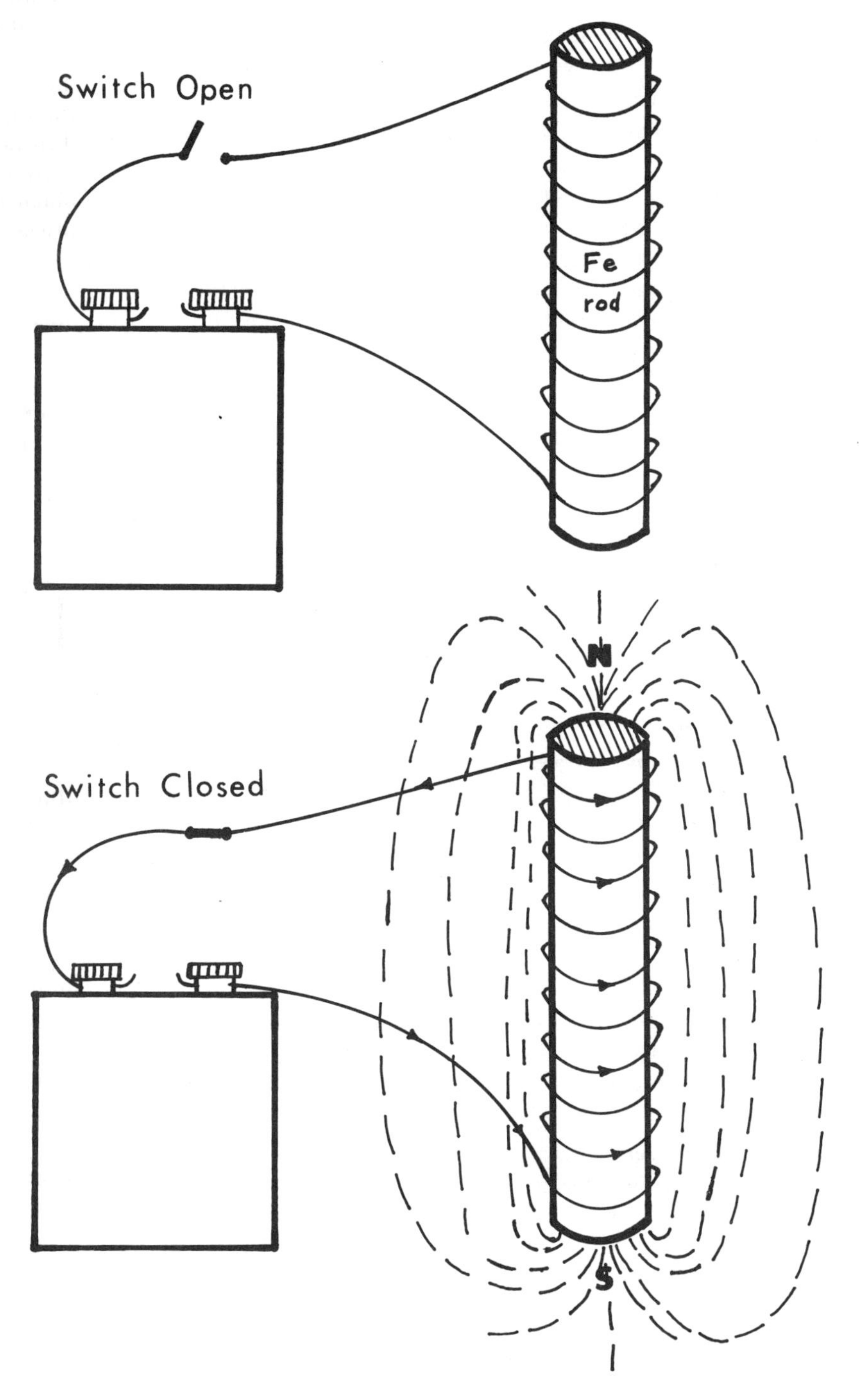


*Fig. 13* Structure and function of an electromagnet.

nails and a dry cell battery. Two electromagnets are arranged on a spindle and equipped with wire "brushes" which make intermittent contact with the electrical current of the battery as the spindle revolves. Another pair of electromagnets set on either side of the pair on the spindle are also connected to the battery as shown in Figure 14. Thus when the magnets on the spindle are approaching the two outside magnets, the wire brushes on the spindle connect the circuit so that the spindle magnets and the outside magnets repel each other, pushing the spindle around until the brushes break their contact again. The spindle with its magnets continues to rotate further around the circle until they have just swung past the outside magnets again, at which point the circuit again is closed by the spindle brushes,

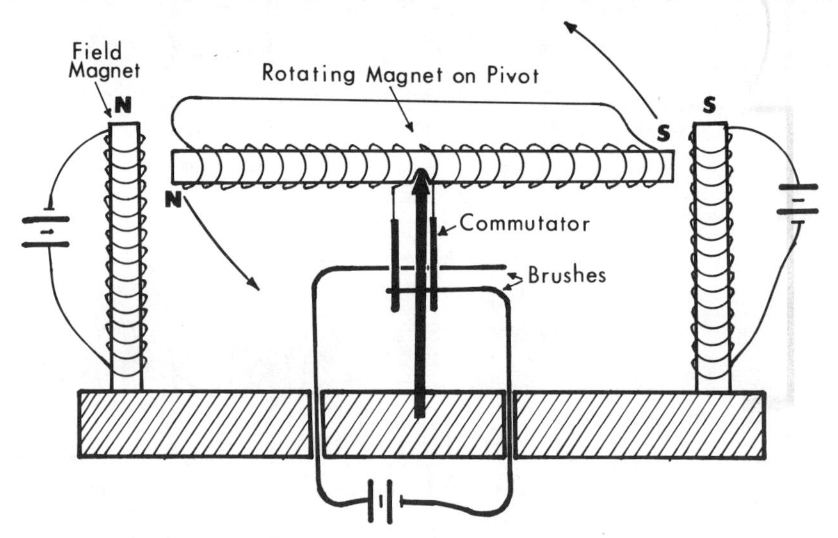


*Fig. 14* Structural diagram of a simple electric motor.

this time reversing the polarity of the spindle magnets and again setting up a repulsive force between spindle magnets and outside magnets which prods the spindle into turning further in the same direction. If the electromagnets are powerful enough and sufficient electric current is provided, the far end of the continuously rotating spindle can be connected by means of pulleys to a set of gears, or even be connected directly to a circular-saw blade so that the pivoting core of the motor can be made to do the mechanical work of sawing through a piece of wood.

The discovery that an electric current in a wire coil could induce magnetism in an iron core around which the wire was wound was startling enough, but it was not the only odd interconnection between electrical energy and magnetic energy. It was further discovered that any time a plain piece of copper wire or any other substance that could conduct electric current was moved through the field of force of a magnet, an electric current was always induced in the conducting wire. Thus, just as electricity moving through a conducting wire induced or created a magnetic field around it, a magnet moving past a conducting wire could induce or create an electric current as long as the wire and the magnet continued to move relative to one another. This phenomenon of "generation" of electric current from a magnetic field soon proved to be of almost as great practical importance as the phenomenon of the electromagnet; for one thing, the principle was the basis for all kinds of dynamos and electricity generators, in which electric currents were generated in coils of electrically conductive wire rotating around powerful magnets. Gradually physicists began to realize that the interrelationship between electrical energy and magnetic energy was so very close that the two forms of energy came to be spoken of together as *electromagnetic energy*.

Once it became clear in the early 1800s that electricity, whatever it was, could be generated at will, and that electricity and magnetism between them had some extremely valuable practical uses, a great many physicists began studying these phenomena intensively, trying to determine *what* electrical energy and magnetic force were, and to define more specifically how they behaved. There was very little progress made in answering the first question, but as for defining how electromagnetic energy behaved, an enormous amount of information was gathered by such men as Volta, Ampère, Coulomb, Michael Faraday, and James Clerk Maxwell. Out of their studies, bit by bit, an increasingly clear picture of the nature of electromagnetic energy began to appear. Whatever electricity was (and it was still frequently considered as some kind of mysterious "fluid" because of its apparent ability to flow through a conducting wire) it seemed to exist in two forms: a positive variety and a negative variety. Once steady and reliable sources of electrical current, such as the early batteries, were discovered it became clear that neither of these varieties of electricity existed alone, but that the two were always associated together, just as the north and south poles of a magnet were always associated together. Soon the idea of positive and negative "varieties" of electricity was dropped altogether, and the idea of positive electrical charge and negative electrical charge substituted. An object carrying a positive electrical charge was considered to have an "excess" of electricity, while another object carrying a negative charge was thought to have a "deficiency" of electricity. Whenever two objects carrying opposite electrical charges were close to each other, there was a tendency for electricity to flow from an area of excessive charge to an area of deficient charge, and whenever the electricity flowed through a conducting substance such as a wire, a one-way stream or "current" of electricity moved through the conductor.

We will have a good deal more to say later about the various types of electromagnetic interactions that were studied and observed in those days, and about the curious interrelationship between electricity and magnetism.

For our immediate purposes here, however, some of the simple principles that emerged from these studies can be listed as follows:

1. Any two objects bearing similar electrical charges (that is, both carrying positive charges or both carrying negative charges) repel each other, with the force of this repulsion becoming increasingly larger the closer together the two charged objects are placed.

2. Any two objects bearing opposite electric charges (that is, one bearing a positive charge and the other a negative charge) attract each other, with the force of attraction becoming increasingly greater the closer together the two charged objects are placed.

3. Two similar magnetic poles (that is, either the north poles of two magnets or the south poles of two magnets) will repel each other, with the force of this repulsion becoming increasingly stronger the closer together the poles are brought. Similarly, two opposite magnetic poles (the north pole of one magnet and the south pole of another) will attract each other, with the force of the attraction becoming increasingly stronger the closer together the poles are brought. (Note here the great similarity between the behavior of electrically charged objects interacting and the behavior of magnetic poles interacting. Indeed, the very same equations that describe the strength of forces of attraction or repulsion between electrically charged objects are also used to describe the forces of attraction or repulsion between magnetic poles. Yet, curiously enough, no force of any kind arises when either a positive or a negative electric charge is brought in close proximity to either a north or south magnetic pole.)

4. When an electric current moves through a conducting wire, the passage of the current induces a magnetic field around the wire, so that an adjacent magnet will orient itself with respect to the wire as if the wire had a north and south pole. If the wire is wound in a coil around an iron rod, the iron rod will become an induced electromagnet as long as the current continues to flow, with its north pole oriented in the direction in which the electrical current is flowing down the coil.

5. When a conducting wire is moved through a magnetic field an electrical current will be induced in the wire which will continue to flow as long as the wire continues to move in the magnetic field. In other words, magnetism induces an electrical current in a conducting wire just as electrical current passing through a wire induces magnetism. Furthermore, when electrical currents are moving through two adjacent wires, the wires themselves will behave as magnets, attracting or repulsing each other depending upon the direction in which the currents are flowing. When the current is running the same direction in both wires, the wires will repel each other, whereas if the current is running in opposite directions in the two wires, the wires attract each other.

## THE BIRTH OF AN AWKWARD THEORY

From the general pattern of these simple rules that we have outlined above, it is clear that both electricity and magnetism behave in specific and orderly ways, and that although they are quite different phenomena, they are nevertheless very suspiciously interrelated, and both generate forces of attraction and forces of repulsion capable of acting upon objects at a distance.

Today this odd capability does not seem so remarkable; even if we are poorly acquainted with the behavior of electricity, we are perfectly familiar and comfortable with the forces of attraction and repulsion that we have observed in toy magnets, compasses, and other magnetic devices. But to the physicists working out the details of the behavior of electricity and magnetism this business of forces acting on objects at a distance was an extremely confusing and frustrating enigma. They couldn't see *how* one magnet could affect the movement of another magnet without even touching it. In the case of other mechanical interactions between objects, one object had to be in actual contact with another before it could exert a force upon it. Frictional force, for example, arose only when one object was rubbing or scraping against another. Similarly, if one billiard ball were to move as a result of an interaction with another, the two billiard balls actually had to strike each other—even the nearest of misses brought about no interaction whatever. Yet in the case of forces of attraction and repulsion associated with electrical or magnetic fields, no actual contact was necessary in order for objects to be moved. Bemused by this undeniable fact, physicists began dreaming up theories to account for it, seeking to find some kind of plausible explanation for how this "action at a distance" could possibly occur.

Among the men studying electricity and magnetism in the century from 1780 to 1880 one of the most brilliant pioneers was an Englishman named Michael Faraday. Faraday did more than anyone else to make orderly sense out of the behavior of these similar but different phenomena. Determined to find some explanation for how electric or magnetic forces could act upon objects at a distance, he came up with a theory that was both plausible and uncomfortably awkward. He theorized that electrical or magnetic fields could act upon objects at a distance because the "empty space" between interacting charged objects or interacting magnetic poles was not really empty at all. He proposed that this space must be filled with a peculiar substance which he referred to as "world ether" and which he believed was responsible for all electrical and magnetic interactions. He proposed that the fields of force existing during these interactions were nothing more than stress and strain lines in this hypothetical world ether: They acted exactly like elastic rubber bands pulling objects together or pulling them apart, except that they were colorless, invisible, and otherwise totally undetectable.

The only real trouble with Faraday's idea of a world ether was simply that because it was colorless, invisible, and otherwise undetectable, there was no way to prove that it existed at all, except to say that, "Well, it must exist because these forces acting upon objects at a distance couldn't do it unless they were acting in contact with something that was in contact with the object they were exerting a force upon." And this was uncomfortably like defining "baby" as "what mothers have" and then defining "mother" as "what babies have."

It was an awkward theory at the very best. Unlike most good scientific theories, which usually arose from observations of nature, this one arose from pure expediency, because of the complete *failure* of anyone to observe any sign of something in nature that physicists thought had to be there. Even worse, it violated a basic requirement of the scientific method: that *any* theory, if it were to command serious attention, had to be subject to experimental or mathematical verification. And this theory, by its very nature, seemed to preclude any possibility of experimental verification.

In spite of all this, the notion of the existence of a world ether did indeed command attention in the world of physics during the middle and late 1800s. For a great many physicists who could not understand how action at a distance could possibly take place, the possibility of an all-pervading world ether seemed to provide a very necessary answer. It even seemed particularly helpful in understanding how gravitational forces were capable of acting upon objects at a distance. For almost a hundred years the theory was accepted—a trifle uneasily—by the majority of physicists and mathematicians studying the behavior of electricity and magnetism, apparently on the basis that an awkward and improbable theory that was scientifically unsupportable was better than no theory at all. It was not until Albert Einstein's work was published at the beginning of our own century that the idea of a world ether was finally discarded and physicists were content to think of electromagnetic fields simply as areas of empty space in which some kind of force was acting. Einstein himself struggled in vain trying to develop some sort of "picture" of what an electromagnetic field itself was, and at one point came to the conclusion that part of an electromagnetic field was *some* kind of material that surrounded electrified and magnetized objects and made interaction between them possible. He thought, for example, that the electromagnetic field around an electrically charged conductor, or around the pole of a magnet, might be a sort of invisible gel-like material forming a cloud around a charged conductor or magnetic pole—a notion that was not really much different from Faraday's "picture" of these fields as local deformations in some kind of all-pervading jellylike medium which filled all space.

Why did these men, many of them among the most brilliant pioneers in the history of physics, feel so utterly certain that some kind of medium had to exist in empty space through which electrical or magnetic energy could be transmitted to affect distant objects? To understand clearly why the world ether theory seemed so attractive and necessary to physicists a century or more ago we must turn our attention to another kind of physical phenomenon which was also commanding intensive investigation at the same time. This phenomenon also was related to the problem of forces acting upon objects at a distance, but seemed always to accomplish this feat by transmission of energy through some solid, liquid, or gaseous medium. Furthermore, in certain ways this phenomenon seemed closely related to the behavior of electromagnetic fields—so closely related, in fact, that it seemed impossible that there was not some fundamental and basic connection between the two. We are speaking of the phenomenon of harmonic vibrations and waves.

## A PROFILE OF VIBRATORY MOTION

We know today that the world around us is full of examples of vibratory motion and wave phenomena which have nothing whatever to do either with electricity or magnetism. The great waves and troughs disturbing the surface of the ocean, the shock waves in the earth radiating out from the epicenter of an earthquake, or the vibration of a reed that produces the clear tone of a clarinet, all are examples of waves or vibrations. In a broad sense we could speak of any repetitive, cyclical, or back-and-forth movement of anything as an example of vibratory motion, whether it be the vibration of a piece of steel struck with a sledge hammer, the vibration of atoms in a heated piece of iron, the vibration of human vocal cords, or the movement of a man bouncing on a diving board. We are so accustomed to encountering vibratory motion of all kinds in our everyday life that we rarely even stop to think about what this kind of motion is; we simply accept such motion and its effects as commonplace in our lives.

Vibratory motion may be very rapid, like the vibration of a tuning fork, or very slow, like the back-and-forth motion of the pendulum on a grandfather's clock. Yet in the broadest sense both are examples of vibratory motion. So also is the vibration of a violin string, the rhythmic tick-tock of the flywheel in a watch, or the motion of a child in a swing. When we drive an automobile it vibrates very gently on account of the repetitive and cyclic firing of fuel in its combustion chambers, the resultant up and down motion of the pistons, and the transmission of part of their repetitive motion to the engine block in the form of vibration. We are so accustomed to this slight vibration as the normal state of things when the car's motor is running that we rarely notice it unless its regularity is disturbed in some way—when the timing of the motor is out of phase, causing an irregular jerkiness to be transmitted to the frame of the car, or when one of the spark plugs fails to fire so that the engine "misses" and the steady vibration is interrupted by a jerk at irregular intervals. In these cases it is not the vibration that we notice but the irregularity.

When physicists began to study and analyze various kinds of vibratory motion, they picked special words to describe certain observable and measurable characteristics that seemed common to all kinds of vibration. For example, they defined the "period" of a vibratory motion as the length of time a vibrating object required to complete one full cycle in its repetitive motion—the time required for one trip to the woods and back, so to speak—whereas the "frequency" of a vibration was the number of complete cycles which occurred in a given unit of time, most commonly given in terms of cycles per second. Thus different kinds of vibratory motion had different periods of vibration: the length of time required for a piston in an automobile engine to move from some arbitrary starting point—say the top of the cylinder—down to the bottom of the cylinder and back up to the original starting point again; or the length of time required for a pendulum to move from its point of release at the extreme right-hand side of its swing through its entire arc to the left-hand side and back to the right-hand side again; or the length of time taken by a vibrating clarinet reed to move down and back up again. The frequency of these vibratory motions might be described respectively as one cycle per second for the piston, two cycles per second for a very fast pendulum or metronome, and perhaps 100 cycles per second for a vibrating clarinet reed.

Vibratory motion can also be described according to its "amplitude," that is, the "distance of swing" of the vibration—how far the object actually moves back and forth, up and down, or around and around, whatever its period and frequency might be. This motion can also be described according to whether it is the sort of straight up-and-down motion that we see in the case of a vibrating clarinet reed, or whether it follows an arc back and forth as a pendulum does, or whether it is a cyclical up-over-and-down-up-over-and-down sort of motion that might be traced through the air by a chalk mark on a bicycle tire while the bicycle was moving. But surprisingly enough it was soon found that the particular *form* of the motion made very little difference, since all of these characteristics of vibratory motion had certain things in common regardless of the form it took. For example, it was possible to plot the movement of any regularly vibrating object on a graph against the passage of time and obtain a characteristic wavy line of motion such as we see in Figure 15—a curve which mathematicians will recognize as a "sine curve" or "sine wave." It also became evident that regardless of the *form* of motion, any object involved in regular vibratory motion would always start from a position of rest or balance at the midpoint of its sine wave, then accelerate to a high midpoint on the curve, then decelerate again to a position of rest or balance, and then undergo a repetition of the process in reverse. Again, as we see in Figure 15, when this kind of oscillating motion is plotted on a graph with one coordinate representing distance traveled up or down and the other coordinate representing time consumed, the result would come out in the form of a sine wave, as long as the vibratory motion was regular.*

Indeed, there is a fancy name for this kind of vibratory motion which follows a regular and repeated pattern and which we see occurring all around us in the swinging of pendulums, the vibrations of tuning forks, the turning of wheels, or the bouncing of rubber balls. Such regularly repetitive motion is called "simple harmonic motion."

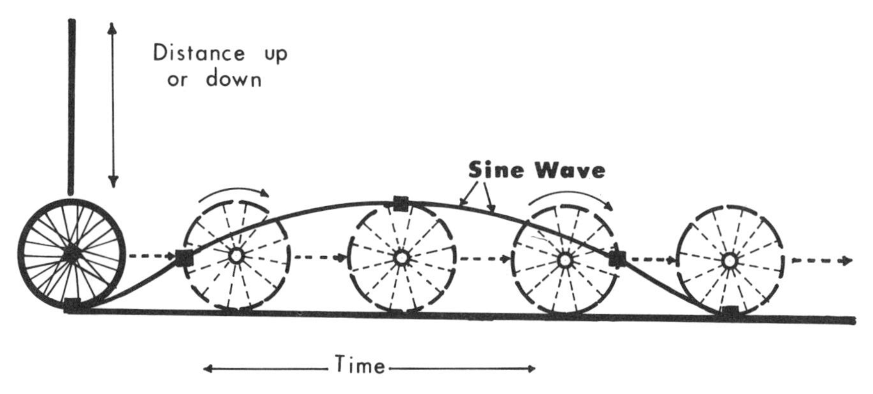


*Fig. 15* An imaginary sine wave generated by the rotation of a bicycle tire. Such a wave is generated as a result of any form of regular vibratory motion.

Commonplace as simple harmonic motion may be in our everyday experience, there is an equally commonplace type of motion that we also see, related to it but not precisely the same. This is the motion of waves. The most obvious difference between repetitive vibratory or cyclical motion on the one hand and wave motion on the other is that vibrating, oscillating, spinning, or resonating objects essentially don't go anywhere—they do not *progress*—whereas wave motion does progress from one place to another, sometimes across great distances.

But what is a *wave*? And what is wave motion? These things are far

* Of course, all vibratory motion is not necessarily regular or periodic, and a graph of irregular or nonperiodic vibratory motion would not produce a sine wave, but rather an irregular and jagged up-and-down type of curve. The vibration of human vocal cords, for example, is nonperiodic and constantly changing because the vocal cords are continually being stretched tight or relaxed, so a graphic analysis of the vocal sounds produced by vocal cord vibration (as seen in optical sound tracks on motion pictures) shows an irregular and often very jagged-looking curve. For our purposes here, however, we need not concern ourselves with nonperiodic vibratory motion other than to recognize parenthetically that such vibratory motion does exist.

easier to illustrate than to define, but for a working definition which also evokes a visual image we might say that a wave is any repetitive succession of moving objects or forms—any repetitive shape or pattern of substance or any regular pulsation of energy—that progresses steadily from one place to another without carrying the rest of the universe along with it.

And if that working definition seems to cover a great deal of ground, it is no surprise, for our universe is rich with examples of waves and of wave motion. For openers, consider the following:

1. We disturb the still surface of a puddle of water by sticking our finger in the middle of it. In doing so, we push some of the water molecules aside with our finger, shoving them against others with a certain force. These molecules, having been pushed, push the molecules next to them, and those push some more molecules. Since the original push we gave to the molecules of water under our finger was in an up and down direction, we now see that the surface of the water begins moving up and down in an expanding ripple originating from the point at which we first dipped our finger in and extending on out to the edge of the puddle. We have created a simple water wave.

But careful observation will show us an interesting thing. Although the wave progresses away from the originating force, our finger, in all directions, the actual molecules of water that were pushed with our finger were not themselves all carried along with the wave. They merely pushed in turn. A chip of wood floating in the water would be raised up as the ridge of the water wave moved out to meet it, and then would slide down into the trough as the wave moved out from under it and beyond, but the chip would not be carried along with the wave. Once past the chip of wood the water wave, consisting of a long succession of pushed molecules which push more molecules which push more molecules, continues moving steadily on out to the edge of the puddle; indeed, it might continue moving indefinitely if there were an infinitely large smooth surface of water for it to move along, except for the fact that other outside forces such as the force of gravity (to mention only one) continually act to flatten out the wave and cause it gradually to dissipate. Given a large enough puddle, the water wave would ultimately become smaller and smaller under the action of such external forces and presently would disappear. But it is important to note that it *progresses* (rather than bouncing back and forth without progress) and continues to progress until it is either dissipated or strikes some rigid, unbending surface which changes its course.

2. We set a tuning fork into vibratory motion by striking it forcefully. As the ends of the tuning fork vibrate, moving swiftly back and forth without going anywhere, air molecules surrounding both vibrating ends are sharply compressed with each cycle of vibration. These compressed

air molecules in turn compress adjacent air molecules, and they compress further air molecules so that a succession of waves of agitated air molecules spreads out from the vibrating ends of the tuning fork evenly in all directions.

Again, the molecules first compressed by contact with the tuning fork were not themselves carried along by these waves; they simply received a succession of shoves from the tuning fork and administered a succession of shoves to the next adjacent air molecules in turn. But now the succession of waves of air molecules suddenly impinges upon a barrier: a delicate membrane known as the human eardrum. This membrane is so tightly stretched, and so elastic, that when the waves of disturbed air molecules strike it it begins vibrating; the mechanical motion of successive clusters of disturbed air molecules press on the eardrum, then relax their pressure, then press again. This mechanical in-and-out motion of the ear drum is then transmitted by way of three tiny movable bones in the middle ear to delicate nerve endings in the inner ear, which in turn transmit this same pattern of disturbance along nerve fibers to certain signal receptors in the brain. In the brain, which is well trained, these patterns of disturbance are interpreted as a sound, with the particular pitch of the sound wholly determined by the frequency of vibration of the ends of the tuning fork.

Thus, although the tuning fork seems to do nothing but agitate air molecules, the pattern of disturbance it causes is transmitted by means of a succession of air waves which in turn brings about a mechanical vibration of the eardrum which is ultimately interpreted as a musical note. This is not to say that the air molecules near the tuning fork themselves travel all the way to the human ear; rather, they simply push other molecules which push other molecules which push still others. But we can also see that the sound waves thus created carry and transmit momentum through the medium in which they are traveling, namely, the air. Obviously they must also transmit energy as well, since energy is required to move the eardrum and that energy could only have come from the sound waves moving through air.

Finally, we can see that if there is a considerable distance between the tuning fork and the eardrum, a measurable period of time will be required for the sound waves generated by the tuning fork to reach the eardrum. If we were to measure this time interval carefully, we would find that the speed of sound traveling through air is always constant (when measured at the same temperature) no matter how loud or soft the sound and no matter how high or low its pitch. The speed of sound will vary somewhat with increase or decrease in temperature, but sound waves traveling through air measured at a comfortable 68 degrees Fahrenheit travel at a comparatively slow 1,130 feet per secound, or approximately 750 miles per hour.

We see some distant similarities between sound waves and water waves: both progress from one place to another at a measurable velocity; both can transmit mechanical energy to some distant object (the chip of wood in one case, the eardrum in the other); and in either case the wave travels through some substantial medium without hauling that medium along with it. We can see a further similarity: As we saw with water waves, sound waves can be dampened or dissipated by the action of external forces upon them as they move. The very motion of air molecules is under the influence of gravitational forces which create a drag or dampening effect that gradually reduces the pattern of disturbance caused by a sound wave. Again, as we shall see more clearly later, when the molecules of any gas are in motion a certain amount of the energy that they carry is steadily converted into heat energy. It might seem far-fetched to imagine that the simple act of shouting at someone across the street serves to warm up the earth's atmosphere a tiny bit; yet this is exactly what happens. The motion of sound waves all over the earth is continuously converting quantities of kinetic energy into heat energy!

3. To create still another kind of wave, suppose we tie one end of a long rope to a tree, then stretch the rope out to full length and hold the other end in our hand. If we then waggle the rope up and down once, we can see a wave pass along it from our hand to the tree. Obviously the part of the rope that we are holding onto does not itself travel along the length of the rope; when we waggled the rope, we merely stretched a bit of the elastic material of which the rope is made, which then in turn stretched an adjacent part of the material, which then stretched the next part, and so on until the energy of our stretching had been transmitted to the tree. If we continued to waggle the rope up and down in a repetitive motion (vibratory motion) we would then create a whole stream of waves moving along the rope from our hand to the tree.

Furthermore, if our movement of the rope is perfectly rhythmic and repetitive, we will be creating a stream of simple harmonic waves which will have a fixed wave length (the distance from one crest of a wave to the next or from one trough of a wave to the next), a measurable period, a measurable amplitude, and a measurable frequency. What is more, if an outside observer looked at the waves traveling down the rope from the side, he would find that the waves looked in profile exactly like a sine curve (see Fig. 16). These simple harmonic waves could also properly be called "periodic waves," since the waves result from the period of our hand movement up and down.

4. Finally, imagine that we have a steel rod held firmly in place by a vise at one end so that it cannot move and that we then strike the rod end-on with a hammer. Although we cannot see it, we have again created a wave by this simple action. We can tell this by touching the rod half

way down and feeling it vibrate as the wave passes under our finger. We would discover that it takes a brief but discrete length of time from the time we whack the end of the rod until the time we feel the vibration with our finger.

What we are actually doing here is compressing the molecules at the end of the rod by hitting them with a hammer—not very much, but a little. The molecules in this rod, like those in many solid substances, have the quality of elasticity which makes them tend to "bounce back" to the rigid form they were in as soon as any distorting force (in this case, the blow from our hammer) ceases to act. But the molecules struck by the hammer and compressed have already compressed the molecules next

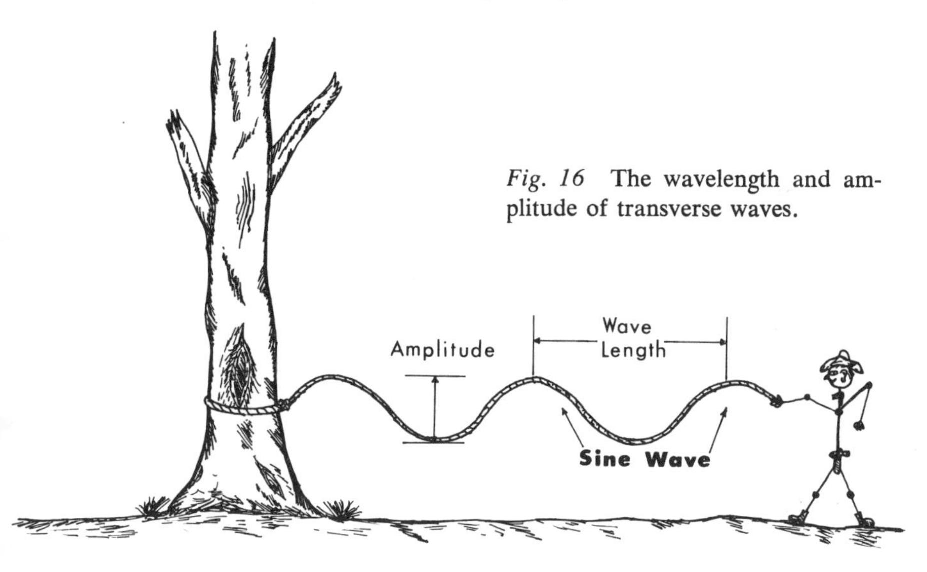


*Fig. 16* The wavelength and amplitude of transverse waves.

to them down the rod; so the compression wave proceeds from molecule to molecule until the end of the rod is reached. Fortunately, in this case we have a rod with enough elasticity to withstand this kind of treatment. If instead of a steel rod we had chosen a rod of glass, a substance which has very little elasticity, our hammer blow might well have exceeded the elastic limit of the glass and shattered it, rather than making a compression wave that traveled down the rod. Similarly, if the temperature of the steel rod happened to be very low—say −60 degrees Fahrenheit—much of the rod's elasticity would have been lost, and the steel rod might have fallen apart from the blow just as if it were made of glass. It is a shattering experience in arctic countries to go out to split firewood in the dead of winter with a sledge hammer and a steel wedge, only to see the wedge fly apart like glass at the first blow—but it happens, as anyone who has lived in the north country can attest.

## THE CURIOUS CHARACTERISTICS OF WAVES

We can see clearly from the above that there is a very close relationship between simple harmonic vibratory motion and the creation of simple harmonic waves. In Figure 15 we saw how a projection of such cyclical movement as the up-and-down cycle of a piston or the turn of a wheel on graph paper traces a perfect sine wave when plotted against the passage of time. In the case of the tuning fork we have seen how simple vibratory motion of an object which does not progress anywhere but merely sits and vibrates can nevertheless produce a moving wave in nature: a succession of compressed air molecules which move from one place to another with a measurable speed. Other kinds of vibrating motion can also create periodic waves, and in all cases the waves produced can be described according to their period, their amplitude, their frequency (number of cycles per second), their wavelength, or their "velocity"—frequency multiplied by wavelength in a given direction.

But we can also see from the above that two rather different *kinds* of waves have been described. In the case of the water wave or the wave created in the rope, the waves we described were moving up and down perpendicularly to the direction they were traveling. If an outside observer could see this kind of wave moving past him he would see that these waves have the familiar sine wave shape, and these are spoken of as "transverse" waves. In the case of the sound waves produced by the tuning fork, or the compression wave that we induced when we whacked the iron bar with the hammer, the waves are called "longitudinal" waves, since the waves of compressed molecules are traveling along with the direction the wave travels instead of perpendicular to it (see Fig.17). As we can see, it is much easier to represent transverse waves in a diagram than it is longitudinal waves, since it is easier to show a vertical rise and fall of the wave than it is to show a horizontal pulsation; therefore longitudinal waves are usually *projected* one way or another in a diagram to look like transverse waves as we have done in Figure 17.

Transverse waves and longitudinal waves are far and away the most common and familiar waves encountered in nature. They are different from each other, as we have seen, but in spite of their differences both transverse waves and longitudinal waves share certain properties in common, both behaving in certain regular and predictable ways.

First, all waves are *reflected* when they strike a suitable reflecting surface, some solid or otherwise inelastic surface which is thick enough that a given wave simply cannot move on through it and continue to progress on the other side. Thus, in the case of any given wave, a "suitable reflecting surface" would have to be thicker than the wavelength of the particular wave, or else present sharp resistance to being moved by the wave or to moving with the wave.

Sound waves, for example, are reflected by a wall, while water waves are reflected by any solid barrier breaking the surface of the water. In the case of water waves, reflection of the waves will be complete, with none passing through a solid or rigid barrier even if it is very thin. If sound waves strike a very thin wall, however, with air on the other side, only a part of the wave will be reflected, while some will pass through the wall and on through the air in the adjoining room. In general only the sound waves with very long wavelengths (low pitch) will be able to pass through the wall, while the shorter wavelength sound waves (higher

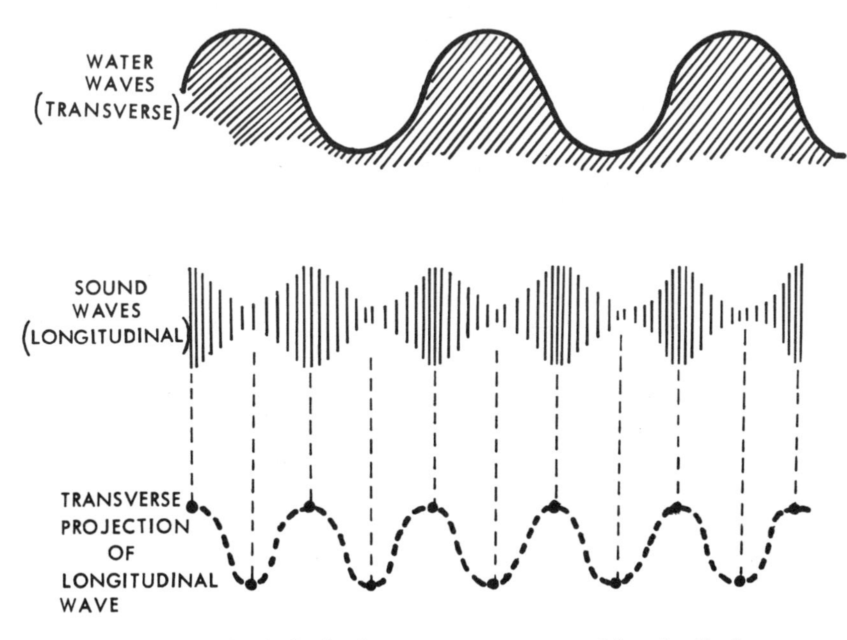


*Fig. 17* The basic similarity between transverse and longitudinal waves.

pitch) are reflected. When we hear music played in the next apartment or the floor below we are particularly aware of the bass tones and rhythms even when we cannot hear any of the higher-pitched melody tones. On the other hand, if a room were divided in half by a partition with the thickness of a sheet of paper, virtually all the sound waves produced on one side of the partition would pass through to the other side and could be heard there very readily, while only a small fraction of the shortest wave-length sound waves would be reflected.

In any case, when *any* kind of wave is reflected from a suitable reflecting surface, it will be reflected in *mirror-image fashion* from the surface, with the angle of the reflection equal in size but opposite in direction from the angle at which the waves strike the surface.

A second universal property of waves is that they are "refracted" any time their velocity changes. This simply means that when waves are traveling at a given velocity through some movable medium such as air, water, or whatnot and then cross into a different medium in which their velocity increases or decreases, the direction that the waves are moving will shift slightly. Sound waves, for example, which can only move through some sort of movable medium, will shift in direction slightly if they move from an area of clear air into densely foggy area, because their velocity in the denser medium will be slightly greater than in the clear, more rarefied medium. The tuning fork which produces sound waves in earth's atmosphere, however, would produce no waves whatever on the airless surface of the moon because there is no medium to carry them. Sound waves traveling through water are sharply refracted when they reach the surface and start moving through air, because their velocity is much greater through water than through air. Oddly enough, sound waves (which are merely molecular compression waves) travel at an even faster velocity through a solid medium than through a liquid such as water.

Probably the most familiar example of wave refraction is the apparent jog or break in continuity that we see when a spoon is placed in a tumbler of water. We know perfectly well that the spoon is not broken, it merely *looks* broken because of the refraction or bending of light waves when they pass from air into water with a consequent reduction in their velocity of travel. We will have more to say about this property of waves later.

A third peculiarity of waves is their property of being "diffracted" or "bent around corners" under certain circumstances. We saw that when water waves encounter a solid barrier breaking the surface of the water, they are completely reflected. But if the barrier they encounter has a gap in it, something different happens. If the gap is very wide (say, wider than several wavelengths of the water waves) the waves encountering the gap will travel straight on through without any evidence of bending or changing direction.

If the gap in the barrier is somewhat narrower than that, perhaps only one or two wavelengths wide, the waves which pass through it will be seen to splay out or bend outward from the gap after passing through it, and if the gap is very narrow (less than one wavelength wide) the waves will seem to splay out in a perfect semicircle from the gap as if it were a completely new disturbance in the water that was causing a new circle of waves (see Fig. 18). This spreading out of waves is called diffraction, which means "a bending away"; the surface waves reaching a narrow gap in a barrier in the water literally "bend around the corner" at the edge of the gap and travel in different directions than they were traveling when they approached the gap.

Finally, all kinds of waves share one other property in common, perhaps the most singular characteristic of all—a property so peculiar to waves and waves alone that it soon became a major criterion for deciding what was indeed a wave and what was not (a question that began turning up with annoying frequency in nineteenth century physics, as we shall see). This odd characteristic of waves is the property of *interference*. Any time two groups of waves of the same wavelength are traveling at the same velocity and happen to arrive in the same region in space at the same time and overlap, one of two rather startling things can happen, depending upon whether the two groups of waves happen to be *in step* with each other or *out of step* with each other at the particular region where they meet. If the two groups of waves are in step or *in phase* with one another they will add together and augment each other, with their crests becoming twice as high and their troughs becoming twice as deep at the region where they

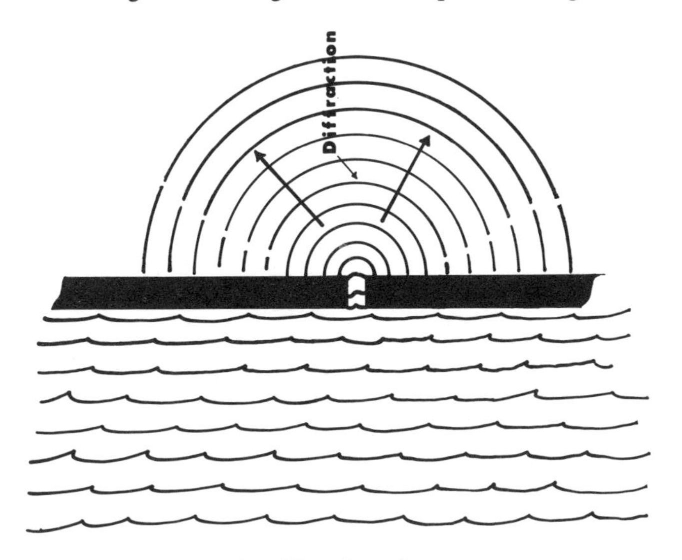


*Fig. 18* The diffraction of water waves.

overlap. On the other hand, if the two sets of waves are out of step or *out of phase* when they meet they will *interfere* with each other: the crests of one group of waves will meet the troughs of the other group of waves and completely cancel each other out. Furthermore, if the two wave fronts are approaching each other at an angle, as they begin to meet and overlap alternating areas will be observed where the waves are in phase and thus augmenting each other and out of phase and thus interfering with each other, so that the end result will be a succession of areas with double wavelength waves alternating with areas of no waves at all.

In Figure 19 we can see this property of wave interference illustrated. We can even produce an example of wave interference experimentally by going back to the puddle of water in which we created surface waves by dipping our finger into the water at regular intervals to create a series of expanding ripples. We saw before that these surface waves will continue moving until they come to the edge of the puddle or until they strike a barrier of some kind. We also saw that if a barrier with one narrow gap in it is placed across the puddle, the waves reaching that gap will be diffracted or "bent around the corners" of the gap so that a new semicircle of waves will originate from the gap at the far side of the barrier.

But now suppose we place a barrier with two narrow gaps in it, fairly close together, across the water puddle. Now when we create water waves with our finger on one side of the barrier, a new set of waves will be created

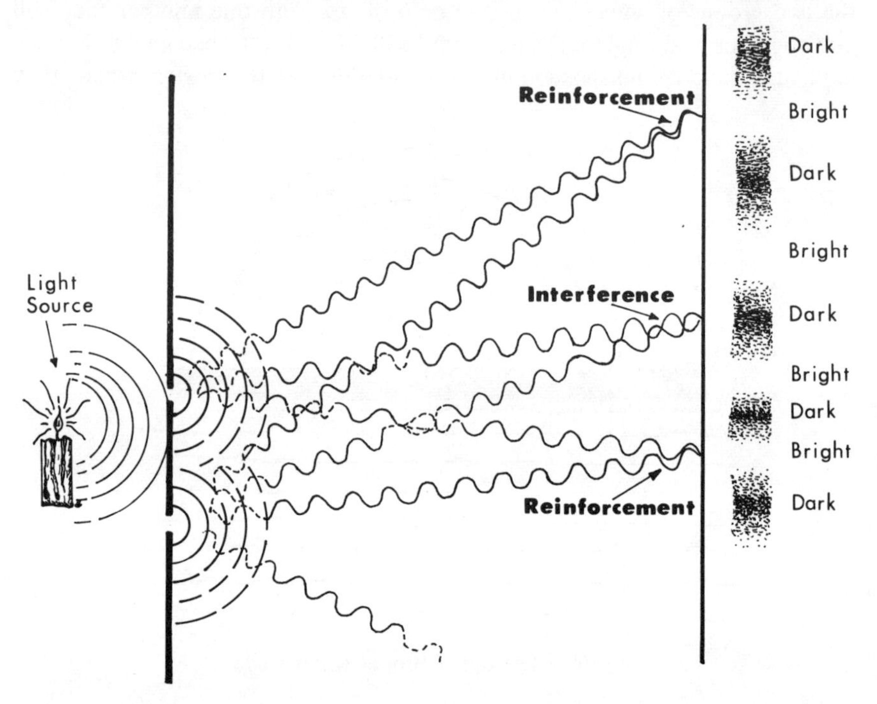


*Fig. 19* The phenomenon of interference, illustrated with light waves. Note the zebra-striped band of light and dark stripes observed on the target card to the right.

at each gap as a result of diffraction. As these new waves spread out and begin to overlap each other, we will see a characteristic mottled pattern of much higher waves regularly alternating with areas of apparently perfectly flat undisturbed water. Those regions where the waves are doubled in amplitude are the regions where two waves have met and overlapped in step (that is, *in phase*) so that a double-high crest and a double-deep trough are formed. The regions where the waves seem to have disappeared altogether are those where the waves have overlapped out of step or out of phase, the crest of one meeting the trough of the other with a net result of zero—no waves. The regions alternate because the two wave fronts are approaching and overlapping at an angle, so the one part of each wave meets and overlaps before the next part has reached the area of overlap.

Furthermore, if we were to place another completely unbroken barrier beyond this region where two sets of water waves are meeting and alternately augmenting each other or interfering with each other, the alternating interference pattern itself would be reflected as it reached the unbroken barrier, so that we could observe areas of high-crested waves being reflected alternately with areas where no waves at all are being reflected.

Of all properties characteristic of waves, this property of interference seems to be the most universal. Physicists have been able to demonstrate interference phenomena with virtually every kind of wave known to man, regardless of its nature or form and of the source from which it originates. In fact, the property of interference seemed so unique to waves and waves alone that physicists of the last three centuries became firmly convinced that any time an interference pattern could be demonstrated there *had* to be some kind of wave phenomenon causing it.

This conclusion seemed to hold true splendidly in the case of such commonly recognized and unarguable wave phenomena as sound waves or water waves. But then, as so often happens with neat and comfortable theories, physicists of Newton's time and later found themselves led into a mare's nest because their certainty that interference had to be a property of waves and *only* of waves. Physicists who thought they had found a bit of firm ground to stand on found themselves face to face with an enigma so thoroughly contradictory and confusing that they had not even gotten themselves fully disentangled by the beginning of our own twentieth century. That enigma, known to the ancients and still one of the great all-time riddles of physics, was the mysterious and frustrating phenomenon of light.

For the physicists trying to make some kind of coherent sense out of electricity, magnetism, and the nature of waves, the behavior of light stood as a great imponderable, because of all the wonders observed in nature, light alone seemed to disobey all of the rules. Here was a phenomenon which on the one hand seemed to exhibit all of the properties of waves, yet on the other hand seemed to have other baffling properties that no wave could possibly have. More than anything else, it was the study of the nature of light and its relationship to electricity, magnetism, and the rest of the world of waves, which led directly to the great revolutionary upheaval in the understanding of the nature of our universe at the beginning of the twentieth century, and which finally threw the door open to the strange and wonderful world of modern physics.

# CHAPTER 13

## *The Baffling Enigma of Light*

There is no way to guess when some subhuman ancestor of man first stared at a dawn or a sunset and wondered about the light that filled the sky: what it was, where it came from, why it was sometimes present and sometimes not, and how it was that it could make rocks or sand too hot too touch, and warm one's back pleasantly yet throw a shadow on the ground ahead. Certainly it was millenniums before the beginning of recorded history; probably it was long before man as we know him today had even appeared on the face of the earth.

We do know that for the centuries and millenniums since the beginning of recorded history students of nature and experimenters in science have constantly been stumbling over this awkward child of the physical universe. Light was obviously present in the universe—one might almost say that it was omnipresent—and could not be ignored; yet it seemed always to defy understanding. Light flooded the earth and all the rest of the solar system, pouring forth endlessly from the sun, blindingly bright during the day, and even reflected back to human eyes from the surface of other planets and the moon at night. Light poured forth from all the other stars as well, filling the night sky with brilliant, sparkling pinpoints, sometimes startling the eyes of spectators as it blazed across the sky from such cosmic accidents as comets or meteorites. It waxed and waned in a regular cycle as night followed day, yet even on the darkest night *some* light was present.

Even the ancients recognized that the only places on earth where no light at all could be found were in the depths of caverns. Anywhere else, with time for the eyes to accommodate, enough light perfused even the "darkness" so that objects could be defined, even if only in shades of gray and black. But deep in the caverns one could wait indefinitely and no accommodation would take place. Of course, men rarely remained deep in the caverns for long without having their own light sources with them, and today we have some idea of the reason why. Recent experiments in sensory deprivation have demonstrated that of all the stimuli upon which we depend for the sane functioning of our minds, light is by far the most important. The man who is accustomed to light cannot tolerate its absence well at all; when experimentally deprived of it, he begins in a matter of hours to create lighted scenes in his mind, and will soon even begin to hallucinate to create the illusion that light is present. It is no accident that in every culture in all human history, congenital blindness has always stirred the deepest compassion, and induced blindness has always been the most dreaded and fearful of all punishments.

Nor is it any wonder that total eclipses of the sun which could produce "darkness at noon" have always been regarded with dread and fear by primitive peoples and considered to be malign acts of gods or spirits. Even today sophisticated people who know precisely what is happening during a total solar eclipse and are prepared weeks in advance for the occurrence still feel a stirring of primitive panic as the noonday light begins to die, a sensation comparable only to the irrational and primitive fear that we feel in the midst of an earthquake, watching an approaching tornado, or in the presence of some other furiously violent accident of nature.

Of course the sun and the stars are not the only sources of light that we know. Early men discovered how to produce light artificially, almost certainly at first as protection against the terrors of darkness. The blazing campfire, the burning torch, the lighted candle, all were controlled means of creating illumination in the darkness, and when early scientists first attempted to describe a given amount of light they seized upon the "candle power" as their unit of description. Oddly enough, that unit survives to this day in spite of the fact that we know that a natural "candle power" of light may vary enormously with the size of the candle or the particular substance that is being burned in it.

Early metallurgists discovered another odd thing, far more significant than they realized: It was not necessary to *burn fuel* in order to produce artificial light. An iron rod thrust into a blacksmith's furnace would, if heated sufficiently, begin to glow dull red, then brighter and brighter red, then yellow, and then (if the furnace were hot enough) finally an incandescent white, emitting enough light to light a room. They also knew that if the rod were allowed to cool slowly it would pass back again through each of the various color stages, emitting less and less light with each one until it faded back to the dull cold gray color of iron again. Thus very early in human history men discovered that some sort of relationship existed between heat and light, although they did not understand precisely what that relationship was. And this is of enormous practical value to us today, in a world in which we depend upon the glowing red of an electric stove burner to cook our food, and upon the heating of a fine tungsten wire filament to incandescence in a glass-enclosed vacuum as our most common means of artificial illumination, very useful in our everyday lives, whether we understand what science has discovered about the relationship between heat and light or not.

## IN SEARCH OF A VELOCITY

From earliest times men have always thought of light as existing in beams or rays emanating from some source or another. Certainly light from any source, from the sun itself to the tiniest Christmas-tree light bulb, seems to spread out in all directions evenly, with its intensity decreasing rapidly as the observer's distance from the light source increases. Early physicists were aware that beams of light could be reflected completely from certain kinds of polished shiny surfaces, and could also be spread out, pulled together, or otherwise manipulated by mirrors, first opaque ones made of bronze, then much better ones made of silvered glass. But as far as they could see, beams of light did not appear to bend around corners; a light ray, emanating from some source, seemed to throw a sharp shadow of anything standing in its path. These things were all easily observed, and a great deal of information about the behavior of light had already been gathered long before Galileo's time.

But the great puzzling question remained unanswered: what, exactly *was* light? No one could say for sure. Was it a substance? Intuitively this seemed to be a ridiculous question. No one could feel it, touch it, or wear it. It had no measurable weight, and seemed to exert no force on objects that it struck; yet even in earliest times fair-skinned men in northern countries knew that cool sunlight on a hazy day could produce a painful sunburn. Later, when it was found that light could be reflected back and forth between mirrors, it seemed reasonable that *something* was being bounced back and forth. Where did light come from? How far could it go, and how fast did it move? When a torch was lighted, the light from it seemed to reach all parts of a room instantaneously and at the same time, while it disappeared just as instantaneously when the torch was blown out. Was light something which could suddenly be created out of nothing one moment and just as suddenly destroyed in another? These questions, it seemed, had no answers.

Much early speculation about the nature of light was hedged about with fears and superstitions, the old security of being able to see and the dread of the terrors of darkness. Thus much early thinking about light was also quite illogical and irrational. When Galileo came along, he approached the problem of light logically and remarkably without superstition. On the one hand he was delighted by the way that light could be manipulated by glass lenses, and excited by the fact that when light was properly manipulated with the proper combination of lenses (as in the early telescopes) it was possible to see greatly magnified images of distant objects. With his primitive telescope Galileo discovered the four largest moons of Jupiter, and clearly recognized that Jupiter was not itself a source of light but merely a lightless heavenly body which reflected the sun's light from its surface back to earth, just as our moon reflects light from its surface back to earth.

On the other hand, Galileo was puzzled and frustrated by the many questions he could not answer about what light was and why it behaved as it did. But even in his confusion, there was one thing of which he was convinced: Light did not suddenly appear out of nowhere when a candle was lit or a lantern unshielded. He was convinced that whatever light was, it *moved* from one place to another, and moved with some finite velocity that could be measured if he could only find the right way to measure it. As we saw earlier, he actually tried to measure the velocity of light and failed, probably without ever understanding why he had failed. As he trudged back down the hill that legendary night after spending hours flicking beams of light from his lamp back and forth to his assistant on a distant hilltop, he must have been a puzzled and bewildered man, confounded by the greatest physical enigma he had ever grappled with.

Logical and faithful as he was to his own observations, Galileo must have realized that he had to believe the evidence of his experiments: that light indeed traveled instantaneously (or with infinite velocity) from one place to another, no matter how much he might like to believe differently. At the same time, he was wise enough to recognize that this could not be true unless light were a totally unique phenomenon in the universe, behaving in its motion differently from any other kind of motion ever encountered, an idea that Galileo was stubbornly unwilling to accept. But fortunately Galileo had a different kind of mind from earlier scientists. It seems unlikely that he concluded that since he did not understand the behavior of light, the behavior of light was therefore not understandable. It seems even less likely that he accepted the conclusion many earlier scientists and philosophers had accepted: that light was a mysterious phenomenon which God did not intend men to understand. Far more likely, Galileo lay awake that night and many others telling himself again and again, "Light *has* to travel from one place to another with a finite measurable velocity. But my method of measuring its velocity didn't work. Why not? How could my experiment have gone wrong?"

And in the long run, Galileo's attempt to measure the velocity of light proved far less of a failure than a challenge to physicists who came after him. One by one scientists who followed Galileo realized that his thinking was correct even if his measurements had failed. More and more scientists became convinced that there really had to be *some* way of measuring the velocity of light and set about rooting it out.

We know today that there actually was nothing whatever wrong with Galileo's experiment except for one simple thing: He grossly underestimated how very great the finite velocity of light really was. The distance between the hilltop where he stood and the hilltop where his assistant stood was far too short for any delay in the passage of light from his lantern to his assistant or back again to have been detected by human senses. The light from his assistant's lamp must have moved from hilltop to hilltop in something less than 1/100,000 of second; it was hardly surprising that it seemed instantaneous! After all, nothing else in all human experience had ever been known to move with anything approaching that velocity.

But although Galileo's experiment failed, later physicists and astronomers began to recognize a necessary implication of that failure. They realized that if light did travel from one place to another with a finite velocity, and if Galileo's method of measuring that velocity failed, then the only possible conclusion was that the velocity of light must be very great indeed, perhaps greater than anyone had even imagined. In the year 1675 a German astronomer named Roemer considered this implication and took the next logical step: He thought of a way of measuring the length of time required for light to travel a distance greater than the distance in Galileo's experiment by a factor of several hundred millions. Like many astronomers of his time, Roemer had studied the moons of Jupiter moving around that planet, observing their eclipses as they entered the broad cone of shadow cast by the great gas giant and studying and recording all aspects of their behavior in minute detail. Roemer reasoned that the observed eclipses of Jupiter's moons ought to occur on a perfectly orderly and completely predictable time scale, just as the time of the eclipses of earth's own moon could be calculated accurately right down to the second. But Roemer discovered from his observations that the eclipses of Jupiter's moons on some occasions occurred as much as seven or eight minutes *earlier* than his predictions indicated they should, while at other times they seemed to occur as much as seven or eight minutes *late*.

He also noticed that the "early" eclipses of Jupiter's moons seemed to occur only at those times when the earth and Jupiter happened to be on the same side of the sun, in *conjunction*, whereas the "delayed" eclipses occurred only when the earth happened to be in *opposition* to Jupiter, on the far side of the sun from the great planet. In other words, the "early" eclipses of Jupiter's moons were observed when the earth and Jupiter were as close together as they ever get, only some 400 million miles apart, while the "delayed" eclipses occurred when the two planets were as far apart in the solar system as they ever get, more than 800 million miles away from each other. It seemed reasonable to Roemer that the differences in the timing of the eclipses of Jupiter's moons that he observed under these two opposite extreme conditions could only be explained on the basis of the difference in time it took light to reach the earth from Jupiter when it was close and when it was far away. Following this reasoning, in 1675 Roemer calculated that light must travel through space with the perfectly staggering velocity of approximately 300,000 kilometers per second—roughly equivalent to 185,000 miles in a single second!

To those early scientists that figure must really have been an eye-opener. If it were valid, it meant that the velocity of light was indeed finite and measurable, but that it was also almost unbelievably great. It meant that Galileo could have placed his assistant on the surface of the moon to flash his lamp and the light would have been delayed only a single second in reaching Galileo's eyes. It meant that if the light of the sun were suddenly to go out, some eight minutes would pass before anyone on earth saw its light disappear because the distance from the sun to the earth is over 93 million miles. Thus in the world of the early physicists light traveled so fast that it was virtually impossible to measure its velocity by any man-made experiments, and even later when the general magnitude of its velocity was already known as a guideline, very special laboratory instruments were required to measure its velocity accurately.

Thus it is only when we think in terms of astronomical distances that the velocity of light begins to mean anything useful to us. But what about light coming from stars and galaxies that are not a mere 93 million miles away, but billions upon billions of miles away? We know now that the lights from those stars and galaxies have reached us only after traveling through space not for a matter of seconds or minutes, but for years, or centuries, or millenniums. Light from the nearest star outside our solar system takes over four years to reach us, so that when we observe that star we are actually seeing not what is happening there *now*, but what *was* happening there over four years ago. When men in the year 1054 first saw the great cosmic superexplosion that later came to be known as the Crab Nebula, that exploding star was so distant from our solar system that those men were seeing light reaching the earth from an event that had actually taken place over four thousand years before. And the distant galaxies observed today with the great reflecting telescope at Mt. Palomar are in fact "after-images" of things that actually occurred long before any man walked on the face of the earth.

Roemer's first measurement of the velocity of light was crude. It answered a great many questions, and it certainly indicated the very great order of magnitude of light's velocity, but in the succeeding centuries accuracy-minded physicists wanted to have precise measurements rather than a general order of magnitude. In 1849 a French physicist named A. H. L. Fizeau devised a fascinating laboratory instrument that permitted a far more accurate measurement of light velocity to be made. Essentially it consisted of a pair of cartwheels equipped with gearlike teeth mounted on opposite ends of a long rotating rod. The teeth of the two cartwheels were so arranged that a tooth of one was exactly opposite a notch between teeth of the other such that a light source set behind the cartwheel could not be seen by the observer looking down the length of the rod no matter what position the wheels were turned to as long as they were rotated slowly.

But Fizeau reasoned that the light passing between the two teeth of the first wheel would take a very tiny but finite length of time to move on to the second wheel, so that if the wheels were rotated *very rapidly* the light source would become visible to the observer, passing through the gap in the first wheel when the gap was uppermost there, and reaching the gap in the second wheel at just the time that *it* was uppermost. Then by calculating the speed at which the wheels were rotating and the distance from one end of the rod to the other, a measure of the velocity of light moving from one wheel to the other should be possible.

It was a good idea, but in practice Fizeau found that he could manage to rotate the wheels no faster than a few thousand revolutions per minute, which was not quite fast enough; light still traveled too fast from the first wheel to the second to escape out the gap between the second wheel's teeth. It was only when Fizeau lengthened the path that the light traveled by means of a fancy system of mirrors that he was able to obtain a measurement of the velocity of light with his gadget. But by doing this he finally obtained a measurement of the velocity of light that agreed very closely with that obtained by Roemer and his astronomical measurement. Subsequent refinements of Fizeau's machine have produced increasingly accurate measurements, but for all practical purposes we need only remember that light's velocity in traveling from one place to another has been repeatedly measured at approximately 186,000 miles per second.

## THE REMARKABLE SPECTRUM OF COLOR

The question of light's velocity was a major problem in understanding the nature of this mysterious entity, but it was not by any means the only property of light that intrigued physicists of Newton's day. Another enigma was the difference in the *color* of light observed in nature. Light from the sun or from campfires and torches seemed yellow or orange, while moonlight appeared silvery-white and a heated iron rod gave off an eerie red light. Brightly colored rainbows had long been observed arching across the sky when the sun came out following a rain storm; obviously there was some relationship between the rainbow and sunlight striking raindrops in the sky, but no one could understand why the rainbow always appeared to have the same succession of brightly colored bands, always in the same order. Light reflected from a mirror seemed to be white or colorless, but when sunlight was passed through a large crystal of quartz it seemed to be transformed into the same sequence of colored bands as seen in a rainbow, and fine diamonds that were cut in certain ways sparkled with the brilliance of rainbow colors.

Isaac Newton became interested in this question of colored light, and devised a number of ingenious experiments to try to discover where colored light came from. In one famous experiment he made a pinhole in the curtain of a darkened room, and passed the narrow band of sunlight that came in through the pinhole through a glass prism. Instead of finding a white pinpoint of light transmitted by the prism as he might have expected, he discovered that the beam of sunlight was broken up into an artificial rainbow with its colors in a series of bands in exactly the same order from one side to the other as they appeared in a natural rainbow. Obviously the white light produced by the sun was not really "pure" light at all, but a mixture of light of a variety of colors. But curiously enough if a second prism were held so that only one of the colored strips of light transmitted by the first prism passed through it, no further breakdown occurred. It seemed that each of the various colored lights making up the white light of the sun was itself indeed pure.

Other investigators seeking to repeat Newton's experiment discovered another curious property of this sunlight-broken-up-by-a-prism. Everyone knew, of course, that sunlight seemed to be associated with a considerable quantity of heat, particularly in desert areas or equatorial countries. Heat also seemed to be a component of campfire light, candlelight, or lantern light. But if sunlight could be broken up into a multitude of "pure" colored bands when passed through a prism, which of the "pure" components of the light carried the heat along with it?

Attempts were made to answer this question by actually holding a thermometer in the different colored bands of light from the prism one by one, but results proved disappointing. None of the colors seemed to be associated with any greater amount of heat than any of the others. But then a very curious thing was discovered: when the thermometer was held *just outside* the top of the rainbow spectrum band, at the very edge of the red, but not actually in the light band, it registered a very marked increase in temperature. Here was something quite unexpected: apparently there was some component of sunlight—an invisible component which had no color—which accounted for virtually all of the sensation of heat in the sun's rays!

Clearly the eight colored bands of light in the spectrum were not all there was to light; another invisible component of white light was identified by the prism, and lay just outside the red-colored band of the rainbow spectrum. This component was spoken of as "infrared" light, and if it had no apparent color, it appeared to carry enormous amounts of heat. Centuries later, when light-sensitive emulsions of silver salts were used to make photographic film, it was found that still another invisible component of light also existed, this one lying at the opposite end of the rainbow spectrum just beyond the violet-colored band. This invisible band of so-called ultraviolet light created no more sensation of heat than any of the colored light bands, but it did have the power to fog and expose photographic film just as visible light did, to cause certain substances to fluoresce with a ghostly white light when exposed to it, or to cause burns on the skin that were quite indistinguishable from very severe sunburns.

## ON THE TRAIL OF LIGHT WAVES

Thus it was apparent as early as Newton's time that there was more to light than met the eye. And as later physicists continued to study various properties of light, a few began to catch the first glimpses of a much larger picture emerging from their studies than anyone had seen before. For one thing, more and more evidence began to gather that light traveled from one place to another in the form of waves, just the way that sound waves or surface waves traveled, and that it possessed a definite wavelength (or wavelengths) and a definite frequency (or frequencies) just the same as any other waves.

One barrier to this idea had already been broken down: it had been established that light did indeed move from one place to another with a definite measurable velocity. Since all other kinds of waves that had ever been studied had definite velocities, it had been difficult to see how light could be a wave phenomenon if it moved from one place to another instantaneously. But the work of Roemer and others had overcome this objection. True enough, light's velocity was enormously greater than the velocity of any other wave known to man, but at least it was a finite velocity. In other words, if light did travel in the form of waves, a light wave then required a certain measurable period of time to travel a specified distance, just as any other kind of wave required a measurable period of time to move a specified distance. The fact that light had been found to travel 186,000 miles in the course of a single second was a bit of a shock—sound waves, by comparison, were known to travel only about 1,100 feet per second—but this was a difference in magnitudes, not a difference in basic patterns of behavior and thus did not really pose any *scientific* problem or objection. It simply meant that if light traveled in the form of waves, these waves traveled enormously faster than any other kind of wave ever encountered, not that they traveled in any different *manner*.

But were there any other properties of light to suggest that it was a wave phenomenon? As a matter of fact, a number of very convincing similarities were found to exist between the properties of light and the properties of known waves. For example, it was known that light was reflected from polished mirror surfaces—from a piece of silvered glass or the smooth surface of a pond. There had been great interest in this property of light from the very earliest of times, probably a direct result of the fact that approximately half the people in the world were women. If a flat mirror was used, the light reflected from it would give a true and proportionate but "backwards" image of the observer. Furthermore, if the position of the source of the light beam was changed with respect to the mirror, the beam of light reflected from the mirror would also change in direction: no matter at what angle the light struck a flat mirror, it would always be reflected away from it at the same angle but in the opposite direction.

This was not to say that light could be reflected *only* from polished or shiny flat surfaces. Almost any kind of flat light-colored surface could serve as a reflector of light. Thus, a large sheet of white cardboard, or a wall freshly coated with flat white paint could reflect light, but from such surfaces light would be somewhat scattered, rather than reflected in a perfect beam like a mirror reflection, because of the many tiny irregularities of the reflecting surface.

What about the reflection of light from mirrors which are *not* flat—from concave or convex mirrors, for example? Here again it was found that light was always reflected away from the mirror at an angle exactly equal to the angle at which it strikes the mirror, but in the opposite direction. But if the mirror surface itself is curved, some strange things can happen to the image of the observer that is reflected. Anyone who has ever studied his face in the shiny curved surface of an electric toaster, or in the concave surface of a shaving mirror has encountered this kind of distortion. This is also the principle of the full-length mirrors found in amusement park fun houses in which the observer sees his image grotesquely elongated or flattened or otherwise distorted because of various curvatures of a mirror surface.

Certain types of curved mirrors, however, were known to reflect light in especially useful ways. For example, a concave mirror ground to take the shape of a segment of a sphere, or of a parabola, could collect light and by means of reflection *focus* it at a given point in front of the mirror.

Thus concave mirrors of this sort became extremely useful in building powerful reflecting-type telescopes; as shown in the diagram, light reaching the reflecting mirror of a reflector telescope in straight lines from outer space will be focused at a point close to the mirror's surface where a very sharp image can be photographed or observed. The reverse is also true; when a light source is placed in front of a concave mirror at the focal point, the light from the source will be reflected away from the mirror in an almost perfectly cylindrical column, thus carrying the light for long distances—the principle of the spotlight or the automobile headlight. As we will see later, this same principle of using a spherical or parabolic "reflecting dish" has made the construction of radio-radar telescopes possible. One such telescope built near the town of Aracibo in Puerto Rico uses a spherical reflecting dish made of wire mesh to send radar signals out to distant objects such as the moon or the sun and then catch those same signals as they are bounced back again to the wire mesh "mirror" of the reflecting dish.

Another property which light seemed to share with other forms of waves was that of "refraction": Light was found to travel at slightly different velocities in different media, so that when it passed from one medium to another in which its velocity was different, its rays would be bent slightly from the original direction they were traveling. We mentioned before the odd "jog" or apparent break in continuity of a long-handled spoon sitting in a tumbler of water. When we look at it, it appears as if the spoon handle has been broken sharply at the interface between the air and the water. Of course, when we use the spoon to stir the water, this "jog" remains there and moves along with the rest of the spoon; obviously the spoon is not actually broken at all. Small children are often amused and puzzled when they first notice this phenomenon, but we are so accustomed to it that we hardly notice it. The light rays illuminating the room are simply bent when they pass from air into water so that the *reflected image* of the spoon that reaches our eye seems distorted.

We could even observe and measure the amount of distortion in the direction the light waves are moving if we used a pinpoint light source in a

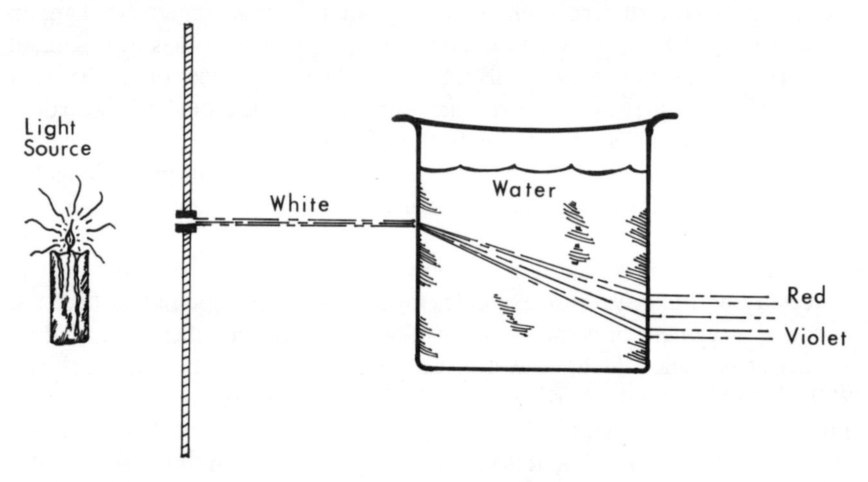


*Fig. 20* The refraction of white light passing through water. The different wavelengths of the components of white light travel through water at differing velocities, so that the light is separated into its colored components by refraction.

very dark room to shine a tiny horizontal beam of light through the tumbler of water. As in Figure 20, we would observe that the beam of light traveled horizontally until it reached the glass, then bent downward as it entered the water, then emerged from the glass at a slightly lower level than it went in, and proceeded on a horizontal course until it reached the vertical screen. But we would notice that the image formed on the screen would not be a *pinpoint* of light; it would be a slightly larger, fuzzy, and blurred image with a tiny fringe of rainbow colors at the top and bottom. Indeed we would see much the same thing that Isaac Newton saw the first time he punched a pinhole in his windowshade and allowed a beam of light to stream through a prism onto a sheet of white paper. If we used a glass full of mineral oil instead of water, we would find that the amount of bending of the light upon moving from air into mineral oil would be even greater than when the light passed from air into water, and if we passed it through a solid glass cylinder the amount the light bent would be greater yet.

This bending of light as it passes from one medium into another is called refraction, and the amount that the light is bent in passing into any given substance is the "refractive index" of that substance. Physicists of Newton's day were familiar enough with this phenomenon, but they could not agree as to *why* light was refracted. In fact, two quite different theories that were proposed to explain this phenomenon raised such a storm of controversy in Newton's time that scientists for the first time were led to a really careful and scientific investigation of exactly what light consisted of.

It was a remarkable controversy, if for no other reason than that it turned out to be one of the rare cases in which Newton was wrong about something. Newton was dissatisfied with the commonly held notion that light was a wave phenomenon, and sought to explain its property of refraction on some other basis. He believed that light had to have *some* material quality to it; after all, it traveled from stars to the earth across millions of miles of empty space. If light was nothing but a series of waves, then what transmitted the waves? If there was nothing out there for light to travel through, Newton insisted, then light waves could never reach the earth even from our own sun.

Newton proposed, rather, that light was really made up of a multitude of tiny material particles that were thrown off from light sources and could travel at very high speeds through space. He contended that the reason a stream of these tiny pellets could pass through transparent substances such as glass or water was simply that the pellets were so very tiny and traveling at such a high speed that they simply bored their way through unimpeded, or else were drawn through by some sort of attractive force such as surface tension. Furthermore, Newton contended that as a stream of tiny bulletlike pellets, light had to have weight, and to strike objects with a certain force. This, he felt, helped explain why light could be reflected; if you blow a stream of dried peas through a pea shooter at a glass mirror, the peas will bounce off or be reflected just as light is. Of course, if you happen to hit someone with a stream of dried peas from a pea shooter, he would feel them hitting, whereas no one felt light pellets striking them. But Newton argued that a ray of light was made up of such very tiny particles traveling so fast that there were simply no instruments or sensory apparatus sufficiently sensitive to detect them hitting.

Newton believed that the refraction of light was the proof of his theory that light rays were made up of a stream of material particles. He argued that when light which had been traveling through the air struck water there was some kind of an attractive force similar to ordinary surface tension which acted upon the particles of light to pull them into the denser medium. Thus he argued that a light ray moving from air into water was tugged slightly at the water's surface and thus was slightly speeded up as it moved through the water, causing the light ray to bend. The more dense the material through which it was traveling, the greater this attractive force and the faster the light rays were forced to move. Then when they moved out of the water glass again, the light particles slowed down to their original speed again. According to Newton's idea, light rays from the sun would be traveling at their lowest speed through the emptiness of space, and then would be speeded up upon entering an atmosphere of oxygen and nitrogen gas on the surface of the earth.

A Dutch physicist named Christian Huygens took sharp issue with this notion of Newton's. Huygens felt that there was far too much evidence that light behaved as a wave phenomenon to take Newton's idea of light as a stream of tiny pellets or bullets seriously. He argued that light waves reaching a water surface after traveling through the air were bent or refracted because their velocity *decreased* while passing through the more dense material, and then increased again upon leaving the water to reenter the air. In fact, Huygens argued, the more dense the transparent medium through which light was moving, the more its velocity had to be slowed down.

Of course, today we can think immediately of the obvious "crucial experiment" which might have been performed to prove whether the "wave theory" of Huygens or the "pellet theory" of Newton was correct. According to the pellet theory, light speeded up in traveling through a denser medium such as water; according to the wave theory, light should be found to slow down in traveling through such a denser medium. The obvious way to settle the controversy would be to *measure* carefully the velocity of light as it traveled through the denser medium, and then compare the result with the known velocity of light as it traveled through air.

But unfortunately this controversy was going on in the mid-1600s, long before anyone had devised any method for actually *measuring* the speed of light under experimental conditions. Indeed, this crucial experiment was not successfully performed until 1850, some two hundred years later, when laboratory measurements proved conclusively that light did indeed travel more slowly in water than in air, providing what seemed to be conclusive experimental proof that the Huygens wave theory of light was correct and that Newton's was wrong.

But before this evidence was finally available, physicists had studied certain other properties of light which lent firm support to the wave theory, and indeed seemed to rule out Newton's pellet theory completely. First, it was shown beyond doubt that light, like other kinds of waves previously investigated, could be *diffracted* when an interrupted barrier was placed in its path. In addition, light displayed the unmistakable wavelike property of *interference*. Finally, it was found that light could be *polarized* (as we will describe later), a property that it could only have if it were truly a wave phenomenon.

We mentioned the phenomenon of interference earlier when we were discussing water waves, but the concept is so important to understanding the enigma of light that it is worth reviewing. As we saw, when a finger was stuck in a puddle of water, waves spread out in circles from the point of disturbance. If a barrier with a gap in it was set across the surface of the water, we saw some interesting changes in the pattern of those waves when they reached the barrier. If the gap in the barrier was very wide (as wide as several wavelengths of the water waves) those waves that came to the gap traveled straight on through without any disturbance, while the waves striking the barrier on either side of the gap were simply reflected back (see Fig. 18, p. 213). If the gap was made narrower, but still not as narrow as a single wavelength of the water waves, the waves that reached this narrower gap still passed through, but were seen to splay out or bend somewhat beyond the barrier at either edge of the gap. We saw that this "bending around the corner" of the water wave was called diffraction. Finally, if the gap in the barrier was made so narrow that it was less than a single water wavelength wide, the water waves reaching the gap would be stopped, and a new set of waves would be seen splaying out in a semicircle beyond the gap just as if the wave striking this narrow gap itself created a new disturbance in the water at that point.

Now this property of diffraction of a water wave from a gap in an obstructing barrier is not in itself particularly exciting. The diffracted waves go on traveling beyond the gap in the barrier much as they did before they struck the barrier, although bent in different directions. But if we removed the barrier with the single gap and substituted a barrier with two or more very narrow gaps, each only slightly wider than a single water wavelength, we saw that when the water waves struck such a "diffraction grating," a "beam" of water waves would pass through each of the gaps, each diffracted, and presently these diffracted waves would meet each other on the surface of the water somewhere beyond the grating and overlap. Wherever those water waves overlapped with two crests coinciding or two troughs coinciding, the height of the crest or the depth of the trough would be accentuated, whereas at those places where the crest of one wave met the trough of another, the crest and trough would cancel each other out, leaving a patch of calm water. Thus in the first case we saw that where the waves met each other *in phase* they augmented one another. In areas where they met *out of phase* with one another, they interfered with one another. This we have shown diagramatically in Figure 19, p. 214.

By the time physicists were seriously investigating the nature of light, it was commonly known that the phenomena of diffraction and interference were properties unique to waves and only waves. If a stream of pellets were launched at a diffraction grating (a barrier with a series of narrow gaps in it) the pellets could not possibly be diffracted. If they were narrower than the gaps, all those pellets that came to a gap would pass straight on through without any change in their course. If they were wider than the gaps, *none* of the pellets would go through. And if a pellet struck the edge of a gap it would be reflected back. Thus, the scientists investigating the nature of light reasoned that if light really were made up of a stream of tiny pellets, then light passed through a diffraction grating should show no evidence of diffraction or bending at all, whereas if light were really a wave phenomenon, and if the gaps in the diffraction grating were narrow enough and close enough together, then light waves should be diffracted just like any other kind of waves, and should show an interference pattern of some sort on a screen set behind the grating.

When this thesis was experimentally tested, the results seemed a clear victory for the wave theory: Light was shown unmistakably to demonstrate both diffraction and interference. A physicist named Thomas Young was the first to devise a crucial testing experiment, about the year 1800. In his laboratory Young directed a beam of light from a pinhole source so that it struck an opaque barrier that had two very thin transparent slits scratched on its surface very close together. Young reasoned that as light traveled in the form of waves, the light that passed through those slits should be diffracted—bent outward from each slit—and that as the diffracted light waves from the two slits met each other on the screen he had placed beyond, there ought to be some areas where the waves or the troughs of the two series of diffracted light waves would augment each other and cause a bright line of light to appear, and other areas where the two streams of light waves would interfere with each other and cancel out any visible light at all. Sure enough, when Young tested his light-diffracting device, a zebra-striped pattern of light appeared on the screen behind the grating. At the areas where the bright stripes were seen, light waves had arrived at the screen in phase, thus increasing the illumination on the screen at those areas. In the areas of darkness on the screen the light waves had arrived out of phase, thus canceling each other out (see Fig. 19, p. 214).

This simple experiment at the particular time it was done proved to be doubly valuable. First, it helped clear the air of controversy by demonstrating beyond any doubt that light traveled in the form of waves. Second, by means of geometric analysis of the width of the slits in Young's defraction gratings, their distance apart, and the distance from one bright stripe to the next on the screen it was actually possible to determine the precise wavelength of light. But Young's experiment was only the first in a rapid-fire series of studies of light by means of diffraction gratings. In 1814 a French physicist named Augustin Fresnel repeated Young's experiment using different kinds of light sources and using the reflection of a beam of light from two mirrors placed close together at a very slight angle to create an interference pattern on the screen. Again the screen revealed a zebra-striped pattern of light and dark stripes, with the stripes rather fuzzy and blurred as in Young's experiment. But this time Fresnel noticed that at the fringes of the interference pattern on either side, the stripes of light were not white but took on rainbow colors. Both the blurriness of the interference pattern and the rainbow colors could be eliminated, however, simply by placing a piece of colored glass in front of the light source which would absorb all of the light excepting light of a single color. By careful calculation from the sharp and distinct lines of colored light that appeared on the screen under these conditions, Fresnel not only confirmed that light was propagated as waves, but demonstrated that each separate color of light had its own wavelength different from all the other colors. Pure red light, for example, was found to have the longest wavelength of all the visible colors, while pure violet light had the shortest wavelength of all colors that could be detected by the eye. Infrared waves were shown to have a somewhat longer wavelength than pure visible red light, and ultraviolet waves had slightly shorter wavelengths than visible violet light at the other end of the spectrum.

Of course the truly amazing thing about these experiments was that it was possible even with such crude instruments as these for physicists to measure differences in the wavelengths of light of various colors at all, because those wavelengths were exceedingly small indeed. The wavelength of red light, for example, was shown to be approximately 7/100,000 (0.00007) centimeter, while the much shorter wavelength of violet light was only 4/100,000 (0.00004) centimeter—considerably less than the width of the tiniest bacteria known at the time.

Obviously, trying to record measurements in terms of such clumsy fractions was soon found to be unmanageable; measuring the wavelengths of light of different colors in fractions of centimeters was roughly equivalent to measuring the width of eyelashes in fractions of miles. Physicists soon agreed upon a new unit of measurement for describing the wavelength of light: the *angstrom unit,* defined as 1/10,000,000,000 (0.0000000001) meter, or 1/100,000,000 (0.00000001) centimeter. Thus in modern physics red light is said to have a wavelength of approximately 7,000 angstrom units while the wavelength of violet light is approximately 4,000 angstrom units. When we consider that audible sound waves have wavelengths ranging from 1½ *centimeters* for high-pitched squeaks to as long as 15 *meters* for very low-pitched rumbles, we can begin to appreciate what an achievement the measurement of the wavelength of light by means of the geometry of diffraction gratings really was.

The discovery that light waves demonstrated interference patterns, and the subsequent measurement of the wavelengths of light, all seemed to refute Newton's concept of light as a stream of pellets quite completely and to confirm Huygens's wave theory. The discovery of still another wavelike property of light further added to the almost overwhelming preponderance of evidence. This was the discovery that light could be *polarized.*

Earlier we discussed the difference between longitudinal waves, in which the motion of the waves is along the line of propagation (sound waves, compression waves traveling down a metal rod, etc.) and transverse waves in which the motion of the wave is up and down or perpendicular to the line of propagation of the wave. Water waves, of course, are the most familiar example of transverse waves we have discussed. It seemed very likely to the classical physicists that if light was a wave phenomenon, then it must be propagated as one kind of wave or the other—but which kind? In the case of water waves, it was easy to tell that they were transverse waves from the way their crests and troughs were seen to ripple in one direction only, perpendicular to the earth. If a barrier with a perpendicular gap in it is placed in their way, such one-direction transverse waves will pass through the gap unimpeded, whereas if a barrier with a very shallow horizontal gap at the water's surface is placed in the way of water waves, the waves will be stopped.

It seemed probable that light also traveled in the form of transverse waves, but there was no way to guess in which direction along the line of propagation the crests and troughs of those light waves might be oriented. Then Augustin Fresnel discovered that certain kinds of transparent crystals acted like one-way gratings when placed in the path of a beam of light, so that only those light waves which were moving in one direction along the line of propagation—for example, those light waves whose crests and troughs were directly perpendicular to the earth—could pass through while light waves with any other kind of orientation at an angle from the perpendicular were blocked and not permitted to pass. In other words, the internal structure of these crystals was such that they acted like the slats in a picket fence, so that only light waves oriented in the perpendicular direction of the slats could pass through, and the particular perpendicular direction through which light could pass was spoken of as the "optical axis" of the particular crystal.

One such polarizing crystal by itself would always allow *some* of a beam of light to pass through no matter how much the crystal was rotated in either direction. When two such crystals were placed together and rotated so that the optical axis of one crystal was turned in the same direction as the optical axis of the other, the part of a beam of light that was composed of waves matching the common optical axis of the crystals could pass through both first and second crystal, but if the second crystal were rotated 90 degrees with respect to the first, the "polarized" or direction-filtered light that passed through the first crystal could not pass through the second at all, so that *none* of the light waves could be seen passing through both crystals (see Fig. 21). Since polarizing crystals of this sort could only differentially filter light waves out of a beam of light if the light waves were traveling in crests and troughs perpendicular to the direction they were moving—that is, if light waves were transverse waves—Fresnel's experiments with the polarization of light not only added further evidence to the theory that light was indeed a wave phenomenon, but demonstrated quite conclusively that light waves were transverse waves, not longitudinal or compression waves.

## THE SEARCH FOR A TRANSMITTING MEDIUM

Thus we see that in the early 1800s, at the same time that physicists were busily investigating the phenomena of electricity and magnetism they were also piling up bit by bit a convincing body of evidence that light was not some kind of material substance that traveled as a stream of pellets or tiny bullets through space but, rather, a wave phenomenon. There was still no clear idea of what precisely light was, but whatever it was it clearly traveled

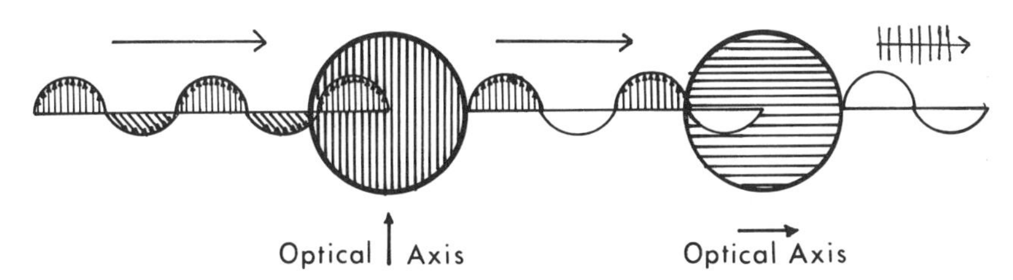


*Fig. 21* Polarization of light. Light waves in horizontal axis are screened out by first lens with crystals oriented vertically; light waves in vertical axis are screened out by second lens with crystals oriented horizontally. Result: *No* light passes beyond second lens.

or "propagated" from one place to another in the form of transverse waves, very much the way water waves travel or propagate, but at an enormously higher velocity.

Considering the many centuries during which the nature of light had remained a total enigma, this knowledge represented a great deal of progress. But there was one nagging problem with this wave theory of light that continued to bother the scientists who supported it most strongly. If light did indeed travel in the form of waves, as it certainly appeared to do, then it seemed to follow logically that some kind of medium had to exist to carry those waves, and nobody could figure out what that medium might be. Once light traveling from the stars or from the sun reached the earth's atmosphere, it was easy: Surrounding the earth was a blanket of atmospheric gas which could act as a medium to carry light waves, just as water or glass also seemed able to carry light waves, with those waves traveling more slowly the more dense the medium.

But what about the vast empty spaces between the stars? How could light travel through *those* vast regions without traveling *through* something? To the classical physicists it seemed obvious that it couldn't. There had to be *some* medium up there to carry light waves. Any other answer seemed akin to launching a wagonload of groceries off into space from one place and having it arrive at some distant place without benefit of any wagon to carry it. Clearly, if the groceries got to their destination, there must have been a wagon in which they were carted—even if one were forced to concede that the wagon was completely invisible, weightless, and incapable of being detected in any way!

Obviously this was not a very comfortable solution, yet we can see how those investigators felt compelled to accept it. Sound waves had to have some kind of medium through which to be transmitted as one molecule bumped another to carry the wave along. Water waves had to have the fluid medium of water to ripple through. So nineteenth-century physicists concluded that there *was* a medium which carried light waves between the stars and across the empty reaches of space—some kind of an all-pervading "fluid" through which those waves could travel. But this mysterious "fluid" medium was quite invisible; it apparently had no weight nor substance; as far as those scientists could tell, no instruments existed that could possibly detect its presence. To be entirely logical about it, these men had to assume that this strange invisible "fluid" which carried light waves filled the entire universe that was not taken up by matter, filled all of the space between the stars, and was even intermixed with the air, with water, and with the solid material of the rock on earth itself.

Since this "fluid" was unlike any other substance known to man, scientists were hard put to describe it or even to give it a name. It was believed to exist only because it *had* to exist if the things that had been observed about the wave nature of light were really true. Furthermore, as we have seen, the existence of this all-pervading "fluid" in the universe was handy in explaining not only how light waves could be transmitted from place to place, but also in explaining how gravitational fields, magnetic fields, or electrical fields could manage to act upon objects at a distance without being in physical contact with them. Reluctantly, the most brilliant and hardheaded physicists of the day grudgingly came to accept the necessity for such an invisible "fluid" to fill the universe, but the insubstantial nature of this stuff was wryly reflected in the word that they used to describe it. This invisible "fluid," as we mentioned earlier, was called "the ether" or "the world-ether," and from the very first the concept was an uncomfortable bedfellow.

It is important to understand that this notion of a "world-ether" was not some facile explanation that scientists cooked up on philosophic or esthetic grounds as the Greeks might have done. It evolved most painfully and was accepted by scientists of the nineteenth century only because their backs were to the wall; they could see no alternative to accepting it. Unless light waves were in some way totally unique and completely different from any other kind of wave phenomenon known (a notion that *no one* was willing to accept) then there had to be a world-ether through which light waves could travel.

But if this idea was uncomfortable and awkward, certain logical implications that arose from it were even more awkward and uncomfortable. Isaac Newton had proved experimentally and demonstrated mathematically that all of the universe as far as men could examine it was composed of matter in motion. He had outlined and defined the great laws of motion describing the way matter could and did move in the universe, and describing what forces could bring about what kinds of motion in what ways. Objects on the earth's surface moved under the influence of earth's gravitational field. The moon and the planets moved about each other and the sun in precisely calculable paths under the influence of similar gravitational fields.

The law of inertia played its part in the behavior of moving objects and the laws of conservation of momentum and conservation of energy and matter had all been repeatedly challenged and seemed invariably to hold true. But if the entire universe was filled to the scuppers with an invisible and substanceless world-ether, then that ether had to represent a completely motionless substance through which every material object in the universe, as well as light waves, must be moving. Insubstantial as the ether had to be, every object in motion anywhere had to be plowing a furrow through it. The earth in its path around the sun, for example, had to be shouldering its way steadily through the ether, pushing it aside as it passed and leaving behind whorls and eddies of ether in its wake. Even if the ether permeated all solid matter as well as all empty space, solid matter took up space and must create *some* kind of disturbance in the ether as it passed through.

Well and good; but if all this were so, then surely such a disturbance in the ether as would be caused by the earth passing through it ought to be detectable in *some* way. If a man moved his hand in an arc through the air, he created a "wind" by pushing the air molecules aside. Surely, then, the earth in its orbit must be creating some kind of "ether wind."

But if so, then why couldn't that "ether wind" be detected?

More and more uncomfortable all the time.

It is fortunate, therefore, that scientific investigation, for all its rigid rules and strict disciplines, is as dynamic a process of discovery of the unknown as it is. It is also fortunate that new scientific discoveries so consistently and characteristically raise more questions than the questions that are answered. In the mid-1800s the theory of the ether wind was a blind alley that offered no apparent escape. Physicists of the day were trying desperately to think their way out, and in the course of their efforts made another discovery which answered many questions about electricity, magnetism, and the nature of light, and at the same time forced scientists to consider the nature of light in a completely new and startling way. Just as the theory of the world-ether was beginning to become well entrenched, with all its frustrations, someone discovered that light waves were *not* the only kind of waves that were capable of traveling at staggering velocities across empty reaches of space. In fact, it appeared that there was another totally unsuspected kind of wave that was not light but that behaved in many ways precisely the way light did—a discovery that could only mean that light had to be something quite different indeed from what it had been thought to be. And because these startling newcomers to the world of waves seemed consistently to be produced by the interaction of electrical and magnetic fields, they came to be known simply as "electromagnetic waves."

## THE SEARCH FOR UNITY IN THE WORLD OF WAVES

To understand more clearly what electromagnetic waves are, how they are produced, why they were not discovered far sooner than they were, and why their discovery had such significance to physicists who were studying wave phenomena in general and light waves in particular, we must backtrack a bit and review some of the discoveries that had been made with regard to electricity and magnetism in the early part of the nineteenth century.

It is difficult to outline and summarize the history of the discovery of electromagnetic waves in a coherent and orderly fashion for the simple reason that electricity and magnetism were only two of a wide variety of phenomena that were being intensely investigated during the same few decades. Some physicists were investigating the basic rules of wave phenomena; some were investigating the nature, velocity, and wave theory of light; chemists during the same era were making enormous strides in their investigation of the physical and chemical properties of inorganic and organic compounds (chemical substances which respectively did not contain carbon atoms or carbon-hydrogen linkages, and those which did), discovering a great deal about the nature of chemical reactions of all sorts, and beating together the first really scientific foundation for an atomic theory of the structure of matter. Some of these investigators were working in two or more of these areas; others were working in all of them; and in the early and middle 1800s the work of all these scientists began to dovetail as interrelationships were discovered among all these various fields.

We saw in an earlier chapter that the physicists studying electricity and magnetism had discovered that invisible fields of magnetic force existed in the area around the magnetized ends of an iron bar, and that equally invisible electrical fields of force existed in the empty space between two electrically charged objects. In either case, invisible "lines of force" seemed to exist in the air or space around the object, and in either case these force fields could exert two different kinds of force: a force of attraction under certain circumstances, and a force of repulsion under other circumstances.

As we saw, these fields of magnetic or electrical force were real forces, just as real as the force exerted by one object striking another in collision. But they were "forces acting at a distance." The fact that there was no material physical contact between objects, nor any apparent kind of substance through which these forces could act did not seem to matter. Furthermore, we saw that the force of attraction or of repulsion created when two magnets were brought together depended upon the polarity of the magnet: Similar magnetic poles repelled each other with the force of repulsion increasing sharply the closer similar poles were brought together, while unlike poles attracted each other with a force of attraction which increased sharply the closer the dissimilar poles were brought together. By the same token we saw that there were "two kinds of electricity"—or, more accurately, two kinds of electrical charge, one called a positive charge and the other a negative charge. At first no one understood precisely what these forms of electrical charge were, nor why they were dissimilar, but early investigators such as Charles de Coulomb in the year 1785 were actually able to measure the magnitude of the electric force which existed between two small electrically charged bodies. Coulomb found that the strength of the electrical force between two charged bodies was directly proportional in magnitude to the amount of electrical charge each of the two bodies possessed, and inversely proportional to the square of the distance which existed between them. This relationship seemed strangely similar to Newton's description of the magnitude of gravitational forces between two objects, but the force of gravity was always a force of attraction, while electrical force could be a force either of attraction or repulsion.

In the early 1800s a whole squadron of physicists experimented intensely with magnetic fields and electrical forces. They learned very early in the game that electricity and magnetism were by no means identical, but they certainly bore some very striking resemblances to each other, and their behavior actually seemed to be linked together in a very close and puzzling interrelationship. As we saw, when an electric current moved through a wire, a magnetic field was created around the wire—the principle of the electromagnet. Reciprocally, when a copper wire carrying no electric current at all was moved through a magnetic field, an electric current was created in the wire as long as the motion continued.

Of all the physicists studying these phenomena, Michael Faraday was among the most painstaking. He discovered that a current induced in such a wire being moved past a magnet rose and fell according to the strength of the magnetic field through which the wire was passing. He also found that the closer the wire was to the magnetic field, the greater the induced current, and the slower the wire was moved past the magnet, the lower the current.

In fact, it was Michael Faraday who gathered together the multitudes of observations that had been made with regard to electricity and magnetism and recognized for the first time the intimate relationship between electrical fields of force and magnetic fields of force. He pointed out that although both were separate physical "entities," nevertheless the effect of magnetism on electricity and the effect of electricity on magnetism resulted from *interaction* between the two different invisible fields of force, and it was he who first began speaking of the two fields of force working together and interacting as an "electromagnetic field."

It was clear that an electromagnetic field was not visible, so it could not be observed and measured directly. It could be observed only in terms of things that happened on account of it, in other words, in terms of changes that it brought about in observable and measurable objects. This was the case with gravitational forces too, you recall, but in the case of electromagnetic fields there were not even any instruments capable of measuring the changes electromagnetic force could bring about. Consequently, Faraday had to invent the instruments he needed himself. More of an engineer than a mathematician, Faraday first sought to demonstrate by repeatable experiments that under certain circumstances when electrical fields and magnetic fields were brought close to each other certain forces were created which seemed capable of acting along invisible lines of force to influence the motion of objects even though there was no physical contact between the objects and the sources of the electromagnetic fields. Thus Faraday confronted science with the enigma of two "somethings" which had no material substance or weight, which could react with each other without destroying each other, and could only be detected by the effect they had on material objects as a result of their interaction.

If Faraday had tried to produce a physical description of what these interacting fields of force were, or to describe in physical or geometrical terms how they operated, he would have been in trouble. If he had even attempted to describe what these fields of force were, where they came from, or why they behaved as they did, his work might have met with grave doubt or flat rejection by other scientists of his day. Fortunately, he did not attempt any of these things. All he did was to conduct experiments, observe how electromagnetic fields behaved, and then describe what he observed happening when electromagnetic fields interacted under certain circumstances.

In short, Faraday's great contribution in this area was that he introduced the *concept* of electromagnetic fields and described in physical terms the things such fields could cause to happen. It remained for his student and assistant, James Clerk Maxwell, to find a way to describe in terms of a series of mathematical equations the physical phenomena Faraday had discovered and investigated—in other words, to define Faraday's concept of electromagnetic fields in the language of mathematics, so that other physicists could not only attempt to duplicate Faraday's experiments and observations, but could verify their findings by means of mathematical analysis, using Maxwell's equations.

This use of mathematics as a descriptive language and a tool was by no means a new thing. Galileo had used it, as had Newton and a multitude of other classical physicists. But the use of mathematical terms to describe natural phenomena was becoming more and more important the more complex the concepts in physics became. We would have to search far and wide for a better example of the enormous value of being able to translate observations of nature into the language of mathematics than in this example of Maxwell's "translation" of Faraday's concepts into mathematical terms.

How could such a process be so useful and important to the clear understanding of natural phenomena? Consider the following example. Suppose that we have a magnet on the table, and move a loop of copper wire through the magnet's magnetic field. Both ends of the wire are attached to a galvanometer, an instrument used to detect the presence of an electric current and to show its magnitude and the direction it is traveling through the wire by means of deflections of a delicate needle on a dial. Even a crude galvanometer such as the one Faraday used was very sensitive; it could detect the tiniest trickle of current flowing through the wire, and the needle on its dial was deflected more and more from its rest-point at the center of the dial the stronger the current was.

With such a gadget connected to our loop of wire, we would observe the same thing that Faraday observed when he moved a loop of copper wire through a magnetic field so that it cut transversely across the magnetic lines of force. We would see that the moment the loop of wire entered the magnetic field the galvanometer needle would begin to deflect from rest-point. The closer we brought the wire toward the center of the magnetic field the more the needle would be deflected, indicating an increasingly strong current appearing in the wire. Then when the wire reached the center of the strongest area of the magnetic field and started moving away from the magnet, we would see the needle of the galvanometer abruptly shift over to the far side of its rest-point, indicating a reversal in the direction in which the electric current in the wire was moving. Finally, if the wire loop were gradually brought farther and farther away from the magnet we would see the galvanometer needle gradually subside back to its rest-point, indicating that no current was flowing (see Fig. 22).

Notice from what we can see happening to the galvanometer needle that the current that was created in the wire by moving the wire through the magnetic field did not appear instantaneously at full strength, and then disappear instantaneously. Rather, it appeared as a very weak current which grew steadily stronger as the wire went deeper and deeper into the magnetic field, then reversed direction at the point of maximum strength as the wire began moving away from the magnet and out of the magnetic field

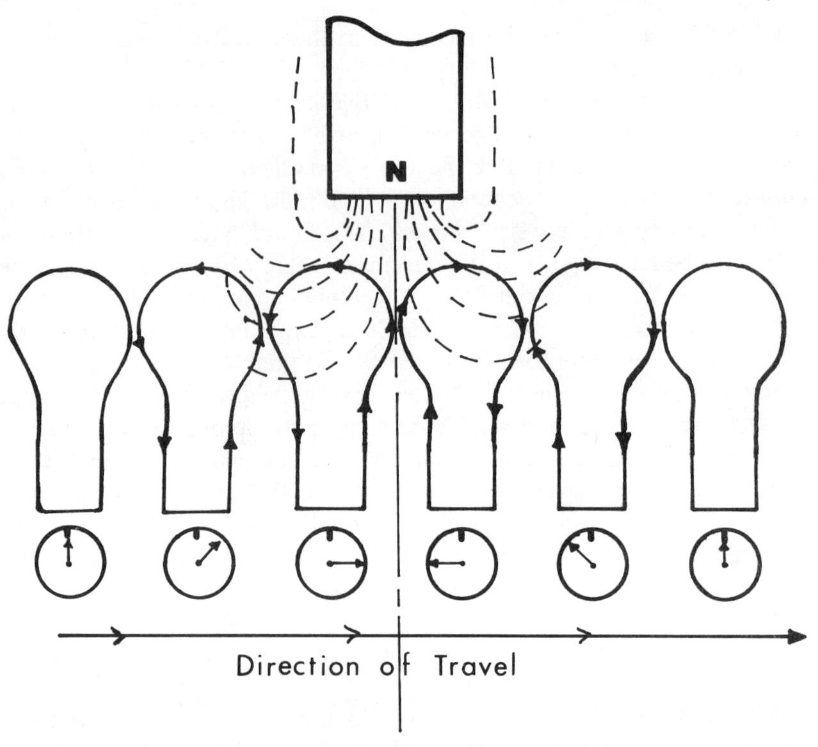


*Fig. 22* The function of a simple galvanometer. As the wire loop is moved from left to right through a stationary magnetic field, a current begins to flow through the wire, indicated by the deflection of the needle on the galvanometer dial. As the wire reaches the point of greatest magnetic field strength and begins to move out of the field, the direction of the current reverses and the galvanometer needle deflects to the opposite side of the dial. The strength of the current flowing in the wire is reflected by the degree of deflection of the needle (in either direction) from mid-point on the dial. Note that the wire loop has *no connection to any power source;* the current flowing through it is *induced* by its movement through a magnetic field.

again, finally dropping gradually away to "no current" again. In other words, the induction of a current in a wire cutting through a magnetic field is not a broken and discontinuous "now it's here and now it's not" phenomenon. Instead, it is a *constantly changing* phenomenon, a continuous and unbroken process with characteristics that depend at any given instant on where the wire is located in the magnetic field, what direction it is moving with regard to the magnetic field at any given instant, and how close to the center or strongest part of the magnetic field it is at any given instant.

Now consider that we have just produced a lengthy and rather vague verbal description of a very simple phenomenon. Even selecting our words with great care, we still do not succeed in presenting a *precise* description of what happens as we move a loop of wire into, through, and out of a magnetic field. Certainly we have not even begun to describe precisely what is happening at any given instant during this experimental procedure. Of course, we can imagine this interaction between a loop of wire and a magnetic field as occurring through a series of tiny instants of time and plot what is happening at each such instant on a piece of graph paper. We could then find out from the graph more or less what was happening at any given instant we wanted. But unfortunately, even this laborious process would still not show us *precisely* what was happening at any given instant, because the interaction between the wire and the magnetic field did *not* occur as a series of tiny stepwise changes occurring instant after instant until the process was over. The changes that occurred were continuous, smooth, and unbroken.

But then, is there *any* way we could describe such a phenomenon with absolute precision? Indeed there is a way, and Maxwell used it. He described what was happening by using a form of mathematical language that had been expressly invented in order to describe continuous unbroken motion or action—a type of mathematical expression that is known as the "differential calculus."

This particular form of mathematical language is particularly fascinating because it was literally *invented*—made up out of whole cloth—at almost precisely the same time by Isaac Newton in England and Gottfried Wilhelm Leibnitz in Germany, working quite independently, in order to be able to describe a type of motion that could not be accurately described in any other way. Newton invented this form of mathematical expression for precisely the same reason that Maxwell needed it to describe Faraday's concept of unbroken changes arising from interaction of electromagnetic fields. Newton needed some way of describing in mathematical terms the continuous unbroken motion of physical objects; for example, the motion of an artillery shell fired from the mouth of a cannon and moving horizontally through a gravitational field. When he discovered that there was no mathematical language available for such description, he cooked up his own, and his differential calculus proved invaluable for describing the rate of change of any number of smoothly variable quantities.* Maxwell used the differential calculus to describe the rate of change of another variable quantity: the developing, cresting, changing of direction and then receding

* Interestingly enough, although Newton and Leibnitz invented the same differential calculus independently of one another, they used two entirely different sets of mathematical symbols and notations. At first, of course, Newton's notations were used in England and Leibnitz's in Germany, but later on at Cambridge two separate groups of mathematicians formed in strong opposition to each other, fighting bitterly about what notations to use. They ended up coming much closer to Leibnitz's notations (which are approximately those we use today) than to Newton's, which were much more complex and more difficult to follow. But this was essentially a quibble about what mathematical symbols to use; the basic method of the calculus that Leibnitz invented (as well as the basic application of it) was identical to Newton's.

of an electrical current in a wire that was moving through a magnetic field.

Maxwell's equations were a triumph in two ways. First, they demonstrated beyond any question that there were certain phenomena of nature in the universe which could be better described in the language of mathematics than in any other way—indeed, phenomena which could not be described *at all* except in the language of mathematics. Second, his equations demonstrated clearly that electromagnetic interactions, at least, could in fact be described by a set of mathematical equations based on the differential calculus, and that this procedure did a better job of describing the stresses and strains in this mysterious, invisible yet apparently elastic medium known only as a "field of force" than anything else. Maxwell's equations made it possible to determine the strength of electric and magnetic force fields during any given interaction.

But even more significant and more amazing, those same equations described perfectly the *shape of the waves* that were known to be formed by a vibrating violin string, or by a sound wave traveling through air. In fact, Maxwell's equations indicated that comparable waves ought also to be formed any time an electric field of force interacted with a magnetic field of force. They indicated that the potential energy of an electromagnetic field (that is, energy stored in such a field) should be capable of conversion into kinetic energy in the form of high-velocity waves and then be reconvertible into potential energy stored in the electromagnetic field again whenever a loop of conducting wire was moved through a magnetic field.

In short, his equations indicated that an interaction between an electric field of force and a magnetic field of force ought to create an oscillating shift of energy from potential energy to kinetic energy and back to potential energy again—a shift of energy which behaves in a regular repetitive manner virtually identical to the behavior of a pendulum or any other vibrating or oscillating system.

We do not intend to go into the mathematical expressions that Maxwell worked out; there is no point to our doing so here. The mathematical analysis of the results of this kind of interaction in an electromagnetic field can be found in any textbook of physics by any mathematically inclined reader who wants to look them up. From our point of view, it is enough to recognize that Maxwell, by means of his equations, predicted unequivocally that the interaction of electrical and magnetic fields ought to produce "electromagnetic waves" that would radiate out into space any time an interaction between such fields occurred. In other words, Maxwell predicted that electromagnetic forces not only interacted at their source, but also inevitably emitted energy in the form of electromagnetic waves any time such an interaction took place. What was more, those electromagnetic waves, according to Maxwell's calculations, had to travel from one place to another at the same staggering speed that light traveled from one place to another, and that like light waves, these electromagnetic waves could travel freely not only through air and water but through totally empty space.

When we stop to consider the implications of this notion, we can begin to see what a truly revolutionary idea it was. For the first time in history, a scientist was using theoretical mathematical calculations to predict that something existed in the universe which no one had ever detected, or even thought of before. As we might expect, Maxwell's equations were met with skepticism if not flat disbelief. Physicists all over Europe said, in effect: "These equations are very nice, and they suggest that under certain circumstances something will happen that we've never heard of before—but show us. Where is the experimental evidence? Where is the proof?" Maxwell's equations stirred a furor, even in the minds of those investigators who had deeply admired the experimental work of Faraday, and had been working to duplicate it in their own laboratories. They did not realize that James Clerk Maxwell had in fact pioneered one of the mightiest and most revolutionary concepts of modern physics.

It is not hard to see why Maxwell's equations made other scientists uneasy. He was saying, in effect, that by pursuing mathematical calculations to their logical conclusions a scientist might predict a phenomenon, or a series of events, or some other kind of occurrence in the physical world before any such phenomenon, series of events, or occurrences had ever been observed under experimental conditions. In the world of modern physics this is not merely an accepted idea which manifests itself on rare occasions; it is one of the most well accepted and basic concepts that modern physicists count upon: that mathematics can predict reality, that mathematics indeed is the guide that shows experimental physicists where to experiment; in fact that if mathematics predicts one thing and experiment suggests another, it is wiser to accept the predictions of mathematics than the experimental results in the laboratory.

Today this is a commonplace and accepted principle, but for the physicists of Maxwell's day, it was a most uneasy notion. Not that those men could find fault with Maxwell's mathematics; he was far too good a mathematician for that. But they could indeed question sharply whether Maxwell's equations and the predictions that they implied had any real *meaning* in terms of observable phenomena in the physical world. If, as Maxwell's equations predicted, there really were "electromagnetic waves" which were created and sent moving out through space every time an electric field interacted with a magnetic field then, they said, *it ought to be possible deliberately to produce such "electromagnetic waves" experimentally in the laboratory*. They also insisted that it ought to be possible to *detect* such waves, and that when detected they ought to be found to behave precisely in the fashion that Maxwell's equations predicted they would behave.

As with most complaints, scientific or otherwise, it proved far more difficult to produce the experimental proof than it was to make the prediction. In this case, more than twenty years elapsed before a German physicist named Heinrich Hertz succeeded in producing a beam of electromagnetic waves by driving electrical charges rapidly back and forth through a wire connected to two electrically charged conductors bearing opposite charges. In doing so, in 1888, he showed that any vibrating or oscillating electrical charge sent out a train of electromagnetic waves which did indeed rush off into space in all directions with an enormous velocity.

This was very strange, something which never before had been accomplished or observed. Even more strange was the discovery that the velocity of these electromagnetic waves was *precisely* the same as the velocity of light, and the fact that these waves could move indefinitely through empty space *precisely* as light traveled through empty space. The discovery of Heinrich Hertz did not go unnoticed very long; in the year 1896 an Italian engineer named Guglielmo Marconi discovered that these waves arising from oscillating electric charges could be transmitted from one place and received or picked up at a great distance, and at one fell swoop Marconi laid the whole basis for modern radio and wireless telegraph communication systems.

The discovery that electromagnetic waves could be produced was not an isolated scientific achievement; it dovetailed neatly with the studies of wave phenomena of all sorts, and with the investigation of the nature of light. Like the key piece in an immense jigsaw puzzle, this discovery suddenly fitted into place and made sense of a multitude of other pieces which fell into place bit by bit with increasing rapidity. The phenomenon of light, which was so long considered to be a completely unique phenomenon of the universe, began to appear more and more like a small isolated fragment of a much greater picture. It became more and more clear that light waves were merely a small and perhaps even relatively insignificant branch of a much larger family, so to speak. It was already known that in addition to visible light in the rainbow spectrum, there were also two varieties of "invisible light," one with a longer wavelength (infrared) than visible light and one with a shorter wavelength (ultraviolet). The electromagnetic waves which Hertz produced experimentally had a much longer wavelength even than infrared radiation, but also traveled through space at the speed of light. Much later, in the early 1900s, yet another kind of radiation or stream of electromagnetic waves was produced by bombarding a target of tungsten with a beam of electrically charged particles driven with enormous energy. These waves had wavelengths that were far shorter than the already identified "short" ultraviolet waves, and for reasons that we will explore later, became known as X-rays.

In fact, it soon became obvious to physicists that a wide band or "spectrum" of electromagnetic waves existed, ranging from exceedingly short wavelengths to extremely long wavelengths, and that these waves were invariably produced by oscillations or vibrations of electrically charged objects or particles in electromagnetic fields. Only a very narrow band of this broad spectrum of electromagnetic waves was made up of waves of just the right wavelength that the human eye could detect them as visible light—but the implication of the idea was clear. Visible light itself also consisted of electromagnetic waves, and visible light, like all other electromagnetic waves, always originated from the movement of charged particles in an electromagnetic field. Later, it was found that light waves had their origin in the movement of charged elementary particles moving in the interior of atoms themselves. But even before this was recognized, physicists at last felt that they had encountered the beginning of an answer to the ancient questions of what light really is and where it really comes from.

## IN QUEST OF THE ELECTRICAL CHARGE

It is hardly surprising that at the time these discoveries were made, from 1850 to 1880, there was a great concern among scientists to determine what exactly an electrical charge was. As we have seen, even the ancients knew electrical charges could be created on various objects. We are just as aware of that today. Whenever we run a hard rubber comb through our hair, or stroke a piece of amber with a bit of silk, an electrical charge is created, often of considerable magnitude. When a person rubs the soles of his shoes against a woolen rug, the charge of electricity that he accumulates in his own body can be discharged by touching some uncharged object. In a darkened room, you can even see the discharging of static electricity occurring in the form of an electric spark between two objects.

But the classical physicists had more significant questions to ask. Static electricity could be generated, yes, and stored as an electrical charge on various objects; but was there some smallest possible unit of electrical charge that could be detected? For that matter, what was it that actually moved when electricity was conducted through a copper wire from one point to another? Was there some tiny particle of matter, too small to be observed in any microscope, which embodied an electrical charge and could itself move through certain substances (such as metals) carrying the charge along with it? Or was the electric charge doomed to remain an intangible, invisible, and abstract concept just as the electromagnetic field remained an intangible and abstract concept?

Michael Faraday, who first investigated the continuous or flowing motion of electric currents generated in electromagnetic fields, came upon the first clue to the answer while studying the phenomenon of "chemical electrolysis." Faraday and others had earlier discovered that pure, freshly distilled water was an exceptionally poor conductor of electric currents. Wire made of copper or silver would carry an electric current from one point to another readily, but pure distilled water would not. On the other hand, when that same water had certain kinds of chemical salts dissolved in it, even in very small quantities, it suddenly became an exceptionally good conductor of electric currents, in many cases even better than copper wire!

How could this be? The answer lay not in the water, but in the chemical substances dissolved in it. When crystals of common table salt (a chemical compound formed of metallic sodium and chlorine gas) were dissolved in a quantity of water, it seemed that this chemical compound was broken up or *dissociated* into two kinds of charged particles: positively charged sodium atoms and negatively charged chlorine atoms. If a battery with the potential for creating an electric current was connected by conducting wires to two carbon rods (excellent electrical conductors) dipped into a solution of sodium chloride and water, the salt solution suddenly proved to be a splendid conductor of electrical current.

The reason was not hard to see: When a "potential difference" existed between the two poles, the rule of "like charges repel and unlike charges attract" prevailed. The positively charged sodium atoms began rushing in a stream toward the negatively charged pole or "cathode," while all of the negatively charged chlorine atoms rushed with equal haste toward the positively charged pole or "anode." As long as positively charged sodium atoms and negatively charged chlorine atoms still remained in solution, the water containing the solution was thus an excellent conductor of electrical current.

Of course, when the charged sodium atoms reached the carbon rod bearing the negative charge, their positive charges were neutralized and free metallic sodium collected on that rod; similarly, as the negatively charged chlorine atoms reached the positively charged pole or anode, their negative charges were neutralized by the positive charge of the anode and free atomic chlorine gas atoms without any charge were released. Again, when some acid was dissolved in water and a current passed through the solution, the water itself broke up into its elements, hydrogen and oxygen, with hydrogen atoms carrying a positive charge and oxygen atoms carrying a negative charge. In the course of his experiments, Faraday discovered that in a given length of time a given amount of electrical current would always deposit the same amount of these elements on either the anode or the cathode, depending on the charge they carried while in solution. The amount of an element liberated from its compound by this process of *electrolysis* was always proportional to the quantity of electricity that passed through the solution.

But what really was passing through the solution? As he experimented, Faraday began to realize the answer: *nothing* was "passing through" the solution. All that was happening when a battery was connected by wires to carbon rods dipped into a solution that contained positively charged and negatively charged atoms was simply that the old rule of "like repels, unlike attracts" was obtaining: Positively charged atoms were repelled from the positively charged anode and were attracted to the negative cathode and vice versa. These electrically charged atoms in solution were called "ions." But in the course of his studies, Faraday discovered that the total *amount* of charge carried by a given ion was always equivalent to a certain tiny amount of electricity, or to some even multiple of this tiny amount. An ion might carry an "electricity-equivalent" of 1, or of 2, or of 3, but never of 1½ or 3⅔. The smallest electricity-equivalent was that equal to the electricity-equivalent of a single hydrogen atom in solution carrying a positive charge. Oxygen ions carried an electricity-equivalent equal to twice that of a hydrogen ion, while ions of aluminum carried an electricity-equivalent three times greater than that of a hydrogen ion. Thus the electricity-equivalent of a charged hydrogen atom was taken as the basic smallest unit of electrical charge, and because hydrogen ions moved to the negative pole or cathode in an electrolysis setup, its charge was taken as the basic unit of one positive electrical charge or +1.

This discovery that substances dissolved in water broke up into charged atoms, or that even water itself might break up into charged atoms, had several implications. For one thing, it implied that atoms of hydrogen in some way must be basic building stones or primary elements in nature; in other words, that atoms of all other elements might conceivably be made up of varying numbers of hydrogen atoms bound together. It also implied that the bonds which held various elements together in chemical combination must somehow be electrical in nature.

Most important of all, it implied that any electric current was not a continuous flowing entity, utterly unique to itself and unlike anything else that existed in the universe, but rather was made up of nothing more than a stream of tiny distinct units of definite size moving to one pole or the other of an electrolysis setup purely on the basis of "like charges repel, unlike charges attract." Half a century later another Englishman, J. J. Thompson, was able to demonstrate that these tiny "basic units" of electricity could actually be detached from the atoms with which they were associated, and that when detached these tiny units of electricity had to have masses far smaller than the mass of any individual atom. These tiny particles were called *electrons*. Another half-century later, the American physicist Robert Millikan actually measured the magnitude of the electrostatic charge of an electron, and then by using this value in Thompson's earlier measurement of the ratio of an electron's charge to its mass, demonstrated that the electron had a mass approximately 1/1,800 of the mass of a hydrogen ion. Furthermore, it was found that a single hydrogen ion, in spite of the fact that it is some 1,800 times more massive than an electron, carries exactly the same magnitude of electrical charge, excepting that the charge on the hydrogen ion is positive, whereas the charge on the electron is negative.

Thus, bit by bit, it became clear that a current of electricity was nothing more than a continuous stream of exceedingly small, negatively charged particles—particles so very tiny that they could stream freely in between the atoms of even a dense metallic substance such as copper, as readily as dust in a dust storm can stream through a chicken-wire fence. It also became clear that these tiny electrons were somehow an integral part of the interior structure of atoms, those elusive submicroscopic building blocks of all solid matter in the universe. Furthermore, as "charged particles," it followed logically that whenever electrons moved from one place to another, whether they formed an electric current in some conductor or not, those electrons inevitably created magnetic fields around themselves, and thus it was the *movement of electrons* which generated electromagnetic waves. Indeed, just as a musician might jar or disturb a guitar string with a pick and find that the disturbance spreads along the string in both directions with the speed of sound, it was possible to jar or disturb an electromagnetic field by moving an electron through it, with the result that this disturbance in the electronic field would create and transmit electromagnetic waves which would travel in all directions and with the speed of light.

Although no one knew it at the time, the long era of classical physics was coming to an end with these investigations and discoveries. Galileo had established the basic technique of the scientific method of investigation, and had himself built a solid foundation upon which subsequent physicists could work. Copernicus had discarded the ancient idea that the earth was the center of revolution of the universe, and had offered the first scientifically plausible hypothesis to explain the motion of the earth and the other planets around the sun. Tycho Brahe and Johannes Kepler had carried the labors of Copernicus still farther, working out more precise descriptions of the kind of orbits the planets followed around the sun. Sir Isaac Newton dominated the investigative work of physics completely in his day, working out his laws of motion, his law of universal gravitation, his law of conservation of momentum, and by implication, the laws of conservation of matter and conservation of energy.

Following Newton's work there was an enormous surge of scientific investigation done by a multitude of men. Among these Michael Faraday stood as a giant not only for his splendidly accurate observation of the behavior of electromagnetic fields, but for proposing the apparently necessary theory of an all-pervading "world-ether" as a medium through which light waves and other electromagnetic waves could travel from star to star.

By the late 1800s, many physicists felt that all the really important phenomena of nature had been described in these classical laws of physics, and all that was required further was a matter of mopping up, tying together loose ends.

But in fact, those optimistic scientists who held this view were in for a rude awakening. The question of how light and other electromagnetic waves could be propagated through space at their enormous velocities remained unsolved. The theory of a "world-ether" which acted as a medium through which they could be transmitted was an uneasy hypothesis at best. The flaw in this hypothesis was soon to be revealed, and a totally revolutionary view of the behavior of light and electromagnetic waves was soon to be presented to the world of science by an obscure, mild-mannered Swiss patent clerk who almost single-handedly squashed the idea that everything of importance in physics had already been discovered, and who pioneered a revolution in thinking about the nature of the physical world which is still reverberating today.

# Part III

## *The Einstein Revoution*

# CHAPTER 14

## *The Riddle of the Ether Wind*

For all of the aura of mystery that surrounds physics and the work of modern physicists, there are a few names in the history of physics which are familiar to all of us. Almost everyone, physicist or not, has heard of such men as Galileo, Isaac Newton, or Michael Faraday; most people even have at least a vague idea of what kind of contribution those men made to our understanding of physics.

Albert Einstein's name is even more widely familiar, but in this case the nature of the work that lies behind the name remains an enigma to most people. They are aware that he was an eccentric little man, and a brilliant man; they know that he somehow originated the idea that matter could be transformed into energy, and that this idea was in some way basically related to atom bombs and thermonuclear weapons; they know also that he was the originator of certain bizarre and incomprehensible theories about the nature of the universe, a group of concepts that became known as the "theories of relativity"—but there, for most people, familiarity comes to an end.

There have been many books written explaining Einstein's views of the universe and his theories of relativity (always a vague word to non-scientists, without any very distinct meaning). One of the main problems in trying to make comprehensible sense out of Einstein's work has always been the temptation to deal with the relativity theories as though they were ideas that sprang full-bodied and unannounced upon the world in the early decades of the twentieth century, bearing little or no relationship to anything that had gone before in physics. And this, of course, was not what happened at all. Einstein's work followed logically from a growing need to find answers to certain questions that had perplexed and confounded scientists for decades and centuries. His work cannot possibly be understood without first understanding the rich background of discovery and inquiry that came before it. It did not just miraculously appear; it evolved by necessity from the work of earlier men.

In the previous section of this book we have discussed a great deal of this background of discovery. Often the discussion has seemed to meander from one point to another without following any plot or pattern. But as we have seen, many of these apparently "isolated" discoveries and theories were indeed interrelated and began, bit by bit, to dovetail into a coherent shape. And in fact, even if it appeared that those many paths of inquiry seemed to wander vaguely along without going any place in particular, they were all leading slowly but inevitably in a very definite direction, ultimately to converge upon and contribute to the work of one brilliant man.

Not that these paths did not lead somewhere in themselves. They did. By the late 1800s, practically everyone in physics was firmly convinced that light traveled from one place to another as a stream of waves. By then it was also recognized that light was just one of a whole family of electromagnetic waves which had wavelengths varying all the way from the very short to the very, very long. The only thing that really seemed to distinguish light from any other kind of electromagnetic wave was the simple fact that human eyes were somehow physiologically capable of responding to the particular electromagnetic waves that fell within a certain very limited band of wavelengths ranging from 4,000 to 7,000 angstrom units, and that the human brain interpreted these particular electromagnetic waves as "light." Furthermore, it was recognized that these and all other electromagnetic waves travel from one place to another with the same enormous speed—the speed of light.

But this enlarged concept of what light really was did not help resolve the nagging question which physicists had been struggling to answer ever since the time of Newton: the question of *how*—through what medium—electromagnetic waves of any sort could be propagated through the empty space that lay between the stars. We have seen that for lack of any better answer, physicists had at least temporarily agreed to an explanation of expediency: that these waves traveled through some kind of invisible, weightless, substanceless, rigid yet completely frictionless medium known as "world-ether" which filled all space between the stars and planets and even permeated planetary atmospheres and the solid matter of which planets were composed. Awkward as this theory was and difficult as it was to understand how such a world-ether could be what it was supposed to be and do what it was supposed to do, there seemed to be no other explanation for the transmission of electromagnetic waves, and by the mid-1800s most physicists had made some kind of peace in their minds with this uncomfortable bedfellow. Even so, recognizing the paradox that they were grappling with, most of them wished fervently that some better solution could be found. And as it happened, within a matter of a few decades their wish was granted; the only trouble was that that "better solution" proved to be a far more awkward bedfellow than the theory of a world-ether ever was.

As we have seen, one of the most uncomfortable things about the ether concept was that there seemed to be no possible way of identifying the stuff experimentally or demonstrating its presence, and physicists have always been much happier with ideas that they could get their hands on in the laboratory than with ideas that were by definition experimentally unprovable. Obviously, things would be far more comfortable all around if only *some* way could be found to "trap" the world-ether into revealing itself or demonstrating its presence. And as physicists became more and more convinced that light and other electromagnetic waves had to travel from place to place through some sort of a world-ether, a number of men came up with ingenious ideas for actually *proving* its existence.

The reasoning behind most of these ideas seemed entirely logical. All other known waves (such as sound waves) had certain characteristics in common. They all had identifiable wavelengths, identifiable frequencies, and traveled at identifiable speeds. The same could be said about light and other electromagnetic waves, excepting that the speed with which these waves traveled was far, far greater than the speed with which sound waves, for example, traveled. It was therefore only common sense to assume that the speed of light waves ought to vary in some degree according to the motion of their source, just as the speed of sound waves was known to vary according to the motion of their source.

## A TALE OF TWO TRAINS

Earlier we pointed out that sound waves are propagated through the air at a speed of approximately 1,100 feet per second (when the temperature of the air is about 70 degrees Fahrenheit), implying that as long as the temperature did not vary, the speed of the sound waves would remain constant. But this is not the whole story. The speed at which sound waves travel *does* vary, often by a great deal, according to the motion of the source of those sound waves. Whether we know it or not, we have all seen this demonstrated innumerable times, and we could prove it by any number of experiments. To understand more clearly what this means, consider a simple imaginary example of two trains tooting their whistles at each other.

Suppose that we have two trains sitting on the same set of tracks with the engine of train A 1,100 feet ahead of the engine of train B. Suppose each of the trains is equipped with a red signal flag and a stop watch, and each is capable of traveling very fast indeed when necessary. The engineers of the two trains have agreed to conduct an experiment to measure the velocity of sound traveling ahead from train B to train A under varying conditions. The sound signal to be measured is a toot of a whistle; it is agreed that when train B toots its whistle at train A, the engineer on train B will punch a stop watch the instant he pulls the whistle cord. The engineer in train A agrees to wave a red flag the instant he hears train B's whistle sound, and when the engineer in train B sees the flag waving up ahead he will again punch the stop watch, thus measuring the time that the sound wave has taken to travel from the engine of train B up ahead to the engine of train A.

First of all the test is conducted with the two trains sitting at rest on the same set of tracks with train A 1,100 feet ahead of train B. (We will assume that there is no wind blowing, so that the air is perfectly still, and that all the experiments are conducted at sea level at the same temperature so as to rule out any extraneous factors that might influence the results.) With the two trains sitting at rest, engineer B toots his whistle, starts his stop watch going, and then sees the red flag wave from train A almost exactly one second after the signal was sounded. Since the two trains are at rest, the conclusion is obvious: the sound waves traveled from train B to train A with a speed of approximately 1,100 feet in one second, and repeated tests under these same conditions would reveal the same speed for sound every time.

Does this then mean that sound waves *always* travel at a speed of 1,100 feet per second no matter what the circumstances? It does not. Suppose the same experiment were conducted while both trains, still 1,100 feet apart on the same set of tracks, were *moving* down the track with equal speeds of 70 miles per hour (approximately 100 feet per second). Under such circumstances, when engineer B toots his whistle, the sound signal would still travel through the air toward train A at a velocity of 1,100 feet per second *relative to the ground*. However, in this case the engineer in train A listening to receive the sound signal is himself moving forward at a speed of 100 feet per second in the same direction that the sound waves are moving. Consequently, the sound waves will take longer to catch up with him than they did when the two trains were standing still. Not a great deal longer, of course, but enough longer that the difference would be measurable with a very accurate stop watch.

In fact, it would seem to engineer A that the speed of the sound waves traveling from train B behind him was slightly slower than when the two trains were standing still. By the same token, if engineer A sent a sound signal back to train B by tooting his whistle while the two trains were still moving at the same speed of 70 miles per hour, the sound waves from train A (which still move through the air at a velocity of 1,100 feet per second *relative to the ground*) would reach the oncoming train B in *slightly less* than one second, simply because while sound waves were traveling back to train B, train B was simultaneously rushing forward to meet the on-coming sound waves.

What is more, if these two trains, still 1,100 feet apart on the same set of tracks, could simultaneously increase their speed to some 740 miles per hour (approximately 1,100 feet per second) the sound signal from train B would *never* catch up to train A in order for engineer A to hear it, since train A would be traveling ahead just as fast as the sound waves that were pursuing it were moving. On the other hand, under such circumstances, a signal from train A would be heard by engineer B at the very instant that it was sounded.

This, of course, is not to say that sound travels at different speeds *relative to the ground* under the varying conditions of these different train speeds. It is merely to say that the speed of sound, *as far as the observer is concerned* and *as best he can measure it,* depends very much upon the speed at which the source of the sound is traveling and upon the speed at which the observer himself is traveling relative to the sound source!

Now let's change the conditions of the experiment a bit. Suppose another set of experiments were done with the two trains running not on the same set of tracks with one ahead of the other, but running on two separate sets of tracks parallel to each other but 1,100 feet apart. Suppose then that both trains were traveling in the same direction and at the same high rate of speed, so that they remained neck and neck with each other as they traveled along. Suppose the same two engineers used the same red flag and stop watch technique for measuring how long it took a sound signal from one train to reach the other and vice versa under these conditions. We might initially assume that a sound wave from train B would require only one second to reach train A, or vice versa, since both trains are moving at the same speed and only 1,100 feet separate the two moving trains. But when the experiment is conducted, each engineer would find that the signal from the other train traveled far more slowly than 1,100 feet per second. With a little thought we can see why it would have to seem that way to the engineers doing the measuring. The sound waves in each case would not only have to cross the 1,100 feet between the trains, but would also have to travel quite considerably more than 1,100 feet in the direction the trains were moving before the sound wave from train A could catch up with the engineer in train B or vice versa.

We can also see that the faster the trains are moving, even though they are both traveling at the same speed, the longer the time lag necessary for the sound signal to reach from one to the other, thus the longer the *apparent* velocity of sound as measured by the moving observers in this experiment. In other words, in this case the measured speed of sound would depend not only on the speed at which the observer was traveling and the speed at which the source of the sound was traveling, but also upon the direction in which the measurement was made.

Again we must emphasize that these measurements of the speed of sound made by engineer A and engineer B have nothing to do with the speed of sound *relative to the ground*—only with the speed of sound waves as measured by observers who are themselves in rapid motion. We can also imagine something very interesting indeed: If neither of our engineers had any knowledge of the speed at which sound waves travel through the air relative to the ground, and if their respective trains were traveling through totally unknown country in a fog so thick that they could not see the ground nor any passing "reference points" that were attached to the ground, and if their respective speedometers were not functioning so that they had no knowledge of how fast either of the trains was traveling relative to the ground, then these two engineers would have *no way in the world* to tell that the speed of sound waves that they were able to measure under these conditions was not the absolute speed of sound under any circumstances whatever. The engineer in one train could only measure the speed of sound *relative to the other train.*

This is not nearly as complicated or difficult a concept as it might seem at first glance; nor does it add up to a particularly new or unfamiliar idea. All that we are really saying is that in order to describe the motion of sound waves, or of any other moving objects, it is necessary to relate their position and speed to some fixed reference point. And each of us, whether we realize it or not, is constantly doing precisely that innumerable times in the course of his everyday life. In most cases in our own commonplace experience, we assume the surface of the earth as our fixed reference point. Thus when we say that an automobile is moving at a speed of sixty miles an hour, we are actually making an incomplete and meaningless statement, even though it is comprehended perfectly well. Sixty miles an hour with reference to what? Obviously, we mean with reference to the ground. We merely neglect to state this because we know other people will assume (just as we do) that that is what we mean. Our statement would be quite meaningless except that we *imply* the "with reference to the ground" part of it.

Similarly, when we say that the earth is moving through space with a velocity of 18 miles per second, what we really mean (although we do not say it) is that the earth travels at a velocity of 18 miles per second *with reference to a specific, well-defined orbital pathway around the sun.* Again, without including such a frame of reference, our statement is incomplete and meaningless, even though we understand what frame of reference is implied.

The fact that the motion of an object, or of a wave such as a sound wave, could accurately be described only by reference to—or "relative to"—some other object serving as a fixed reference point was perfectly well known long before the mid-1800s. Galileo had recognized this in his studies of gravitational forces. Newton also recognized it, and it had been expressed repeatedly by such eighteenth-century philosophers and scientists as Bishop Berkeley, and was restated again in the nineteenth century by such men as the German physicist Ernst Mach.

And indeed, if we fail to describe motion in terms of a fixed reference point, we can get into all kinds of confusing trouble. Every day in our ordinary life we see numerous examples of the sort of confusion that can arise when two observers are moving relative to each other but have no fixed reference point upon which both can agree. For example, we have all experienced momentary confusion sitting in a train looking out a window at another train on an adjacent track and wondering, "Is *that* train moving backward, or is *this* train moving forward?" The reason we are puzzled is that there is no way of answering that question until we find some fixed reference point—one of the supporting pillars of the train station, for instance, or an advertising billboard on the wall—so that we can "orient ourselves" not only relative to the adjacent train but to the ground as well.

Similarly, all of us have watched the moon "rise" over the horizon and move up across the sky at night while the earth apparently stood still, yet we know that the first man to stand on the surface of the moon will see the earth "rise" above the horizon and move up across the sky while the moon apparently stands still. Which observation will be correct? If we knew nothing about the nature of the orbits of the earth and the moon we would have to conclude that both observations are correct, from the differing viewpoints of the individual observers. Of course, we know from multitudes of other observations that the earth is the "fixed reference point" in the earth-moon system, and that it is the moon that does the moving around the earth as its satellite, not vice versa. But without that prior knowledge at our disposal we would have no way to tell which observer was right without using the sun and the position of the stars in the heavens as fixed reference points. It would be absolutely necessary to have some stationary or fixed point of reference—some "inertial frame," as physicists would call it—in order to tell.

## THE CASE OF THE SPEEDING CAR

If this is true, then how can any observer of motion describe what he observes and have it mean anything in terms of what is really happening? Consider another common example of the relative motion of two moving objects—the very familiar example of automobiles moving on a fast superhighway. We might well travel for miles behind another car without either moving up to pass it or seeing it draw farther ahead of us. Obviously, in such a case, that car is not actually moving at all relative to our own state of motion. Then suppose that that car slowly began moving away from us on the highway. How could we interpret this event? It might mean one of two things: Either the velocity of the car ahead of us has begun to increase, or else we have begun to decrease the velocity of the car we are driving. *We would see precisely the same thing* in either case.

If we then increased our velocity and drew up to pass the car ahead, we would seem to be moving very slowly in reference to that car, barely creeping along past it, even though our speedometer tells us that we are traveling 70 miles per hour with reference to the ground. At the same time a car on the opposite side of the highway approaches us at what seems an extremely high velocity and seems to pass us *swish* as if its driver were going to a fire. We might say to ourselves: "That driver is going too fast for comfort, and he ought to be reported," at the very same time that the driver of the car going in the opposite direction is saying to himself about us, "That driver is going too fast for comfort, and he ought to be reported." But an outside observer flying overhead in an airplane would see quite a different situation: He would see one car traveling 70 miles per hour relative to the ground passing a car going in the same direction at only 60 miles per hour relative to the ground, while a third car traveling 70 miles per hour is approaching and passing from the opposite direction.

So who is observing correctly? You? The observer in the airplane? The driver of the car you are passing? Or the driver of the oncoming car? In point of fact, each of you is observing correctly, but only the observer in the airplane is able to observe and measure the speeds of all three cars in a common frame of reference relative to the ground. This does not mean that the observations made by you or either of the other two drivers is invalid; it simply means that each of you is observing in a different frame of reference, and must find a way to *convert* or *transform* your individual relative observations into some frame of reference common to all three of you. In this case, we can easily see that the difference between what you observe and what the observer in the airplane observes amounts to nothing more than adding or subtracting velocities of the various cars. As you move up and pass the car going in your direction, you and he observe only the difference between your velocity and his somewhat slower velocity in the same direction. When the oncoming car passes you at (apparently) very high speed, what you observe is no more than the sum of your velocity going in one direction and his velocity going in the other. The time intervals involved, of course, would appear the same to all observers.

We might simplify the highway problem we are discussing by saying that each of the drivers is observing the speed and direction of motion of the other cars only relative to his own car, whereas the airplane observer measured the "absolute" speed of each of the cars relative to the ground. Thus, although your observation is perfectly correct, as far as it goes, in order for you to *interpret* your observation correctly you must convert or transform what you observe from a frame of reference relative to your own car to the "absolute" frame of reference of the airplane observer—in this case, a frame of reference using the ground as the fixed reference point. You could so "transform" your observations into the airplane observer's frame of reference simply by using an algebraic formula that would relate your "relative" measurement of the ground speeds and directions of the other cars to the airplane observer's "absolute" frame of reference, essentially by means of adding or subtracting the various ground speeds and directions you observe. Such a mathematical formula would tell you how much you would have to add or subtract from the apparent ground speeds of the other cars in order to account for your own direction and ground speed and bring your observations into the same frame of reference used by the airplane observer.

Such a mathematical formula would be useful not only in the specific case in point, but could equally well be used in *any* case in which two observers of the same events came up with differing measurements of velocity or direction because they themselves were moving when the observations were being made—in other words, because they were measuring speeds and directions from different frames of reference. Precisely the same mathematical formulas could be applied not only to any problem of cars moving on a highway, but equally well to problems of the movement of sound waves as measured by moving observers such as our two train engineers.

As we pointed out before, there is nothing the slightest bit new about this idea. Even Galileo realized that any time observations were made of the velocities of moving objects, the resulting measurements could vary a great deal depending upon the motion of the individuals doing the measuring. He recognized that mathematical equations were necessary to correct or *transform* such experimental results to conform with some widely acceptable and generally unmoving frame of reference such as the surface of the earth, in order to make allowance for the effect of the experimenter's own movement. He even worked out the necessary equations. More than a century later Isaac Newton derived his own "transformation equations" and found that his boiled down to precisely the same thing as those Galileo had derived.

Because of their origin, scientists came to speak of these equations as "the Galilean transformation" or "the Galilean transformation equations," but they were not nearly as frightful as the fancy name seems to suggest. Essentially, they were merely equations telling any observer how to add (or subtract) his own speed of motion to (or from) the speed of motion of the object he was observing (depending upon whether he was moving away from the object he was observing or toward it) in order to bring both its motion and his "down to earth," to a fixed common frame of reference. Nor was there anything in these Galilean transformation equations that in any way contradicted Newton's laws of motion. In fact, just the opposite was the case: The Galilean transformation proved to be a very common-sense mathematical tool which was very useful in confirming that Newton's laws of motion really *did* prevail in all cases, even when measurements made by a moving observer seemed to contradict them.

The Galilean transformation, far from being mysterious, is so commonplace that all of us use it without even thinking about it in our everyday lives. We use it any time we judge the relative speeds of other vehicles on the highway when we are entering traffic, pulling off the road, or trying to decide whether to pass another car or not. The highway patrolman has to apply the Galilean transformation equations instinctively but quite accurately in order to hand out speeding tickets. We must make the same sort of instinctive calculations any time we run to catch a fly baseball, fade back to pass a football, or try to evade a barrage of snowballs.

And if the Galilean transformation equations were known to be so useful for interpreting observations of objects in motion made by moving observers, or for interpreting the measurements of the speed of motion of waves such as sound waves, we can see why physicists in the late 1800s, searching desperately for some way to trap the world-ether into revealing itself, began to think of using the Galilean transformation equations in measuring the speed of light traveling in different directions through the ether as a crafty and indirect but valid scientific way of demonstrating that there really was such a thing as the ether in the first place.

## MICHELSON, MORLEY, AND THE ETHER-WIND FIASCO

By the middle decades of the 1800s a truly impressive body of evidence had been collected indicating that light traveled through air, through water, through glass, or even through apparently empty space in the form of various swift-moving waves. Furthermore, visible light had been found to be merely a tiny part of a great family of waves, all of which were electromagnetic in nature and all of which traveled at the rather staggering velocity of 186,000 miles per second—almost 670 million miles per hour. We have also seen that physicists felt obliged to assume that light and other electromagnetic waves had to be traveling through some kind of invisible medium between the stars, uncomfortable as the notion might be, since all other known waves required some kind of medium through which to travel.

But if the idea of the world-ether was scientifically awkward, the ingenious experiments that were devised to demonstrate its presence were not. They were based upon logical scientific speculation about certain phenomena which ought to be present if it were assumed as a basic premise that the world-ether did exist. For one thing, scientists assumed that to act as a medium for the transmission of light and other electromagnetic waves, the world-ether had to be an absolutely motionless substance through which the planets, the satellites, and the stars were all moving. They also speculated that if the measured velocity of sound waves traveling through air could be observed to vary according to the motion of the observer (as in our two-trains experiment) then the velocity of light traveling through the world-ether should also vary according to the motion of the observer.

But if the world-ether were indeed an absolutely motionless medium, then clearly the earth itself had to be plowing through this ether at a fairly respectable clip as it traveled in its orbit around the sun, so that any observer measuring the velocity of light on the surface of the earth would automatically qualify as a "moving observer." Furthermore, if the earth were indeed moving through absolutely unmoving world-ether like a minnow through pond water, the surface of the earth must be constantly brushed by a "wind" of ether passing by in the direction opposite the earth's direction of motion. Thus, the reasoning went, an observer on earth seeking to measure the speed of light through the ether should get one value if he measured the speed of light in the direction of this ether wind and a different value for the speed of light if he measured it traveling against the ether wind.

At first, it was hard to guess how great these differences in the speed of light measured with and against the ether wind might be, since no one knew exactly how fast the earth was moving in its orbit. Conceivably, the difference might be immeasurably small. But soon after Isaac Newton's death an astronomer named James Bradley began measuring the tiny annual to-and-fro motion of the stars, an observation long assumed to be a result of earth's motion in its orbit around the sun. From a careful analysis of his measurements, Bradley estimated that the earth was actually traveling through the world-ether with a velocity of almost twenty miles per second as it made its annual elliptical tour around the sun. Thus, it seemed that such a very rapid speed of the earth traveling through the ether must be creating not just an ether wind but a veritable gale across the surface of the earth, even though nobody was able to feel it.

Of course this velocity of the earth pushing through the ether was not great compared to the extremely great velocity of light—but it seemed reasonable that light waves traveling *with* the ether wind, or in the same direction as the ether wind, would have to appear to be moving measurably faster than light waves traveling *against* the ether wind, and that the speed of light waves traveling crosswind, so to speak, would appear to an earthly observer as measurably different from the speed of light traveling either upwind or downwind. By use of the Galilean transformation equations, physicists calculated that light waves traveling with the ether wind would appear to an earthly observer to have a speed equal to the speed of light *plus* the speed of earth's movement through the ether, whereas light waves traveling against the ether wind would appear to have a speed equivalent to the true speed of light *minus* the speed at which earth was traveling through the ether.

Granted that no one expected the difference in the two measurements of light velocity to be very great, but by the late 1870s physicists had learned to measure the speed of light with considerable accuracy. It followed that if someone could just think up a device by means of which the speed of light traveling with the ether wind could be compared simultaneously with the speed of light traveling against it, the difference ought at least to be measurable, however slight it might be.

In physics, as in most other things, necessity has always been the mother of invention. In the year 1886, two American physicists named A. A. Michelson and E. W. Morley dreamed up just such a measuring device, and quite a device it was, too. As we discussed in Chapter 1, the Michelson-Morley experiment depended upon the use of a semitransparent mirror; that is, a mirror which would reflect part of a beam of light striking it but would permit another part of the same light beam to pass freely on through. If such a mirror was placed at a 45-degree angle with

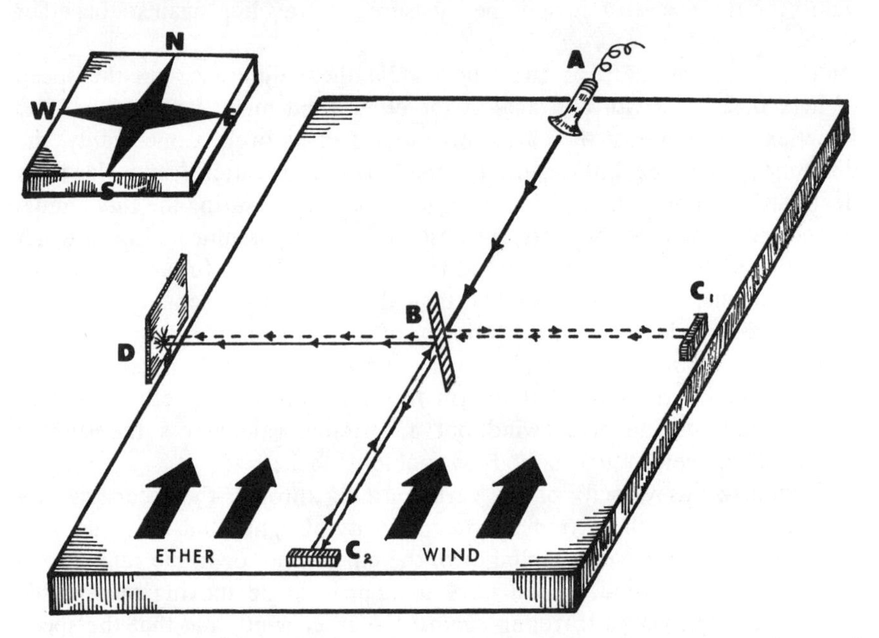


*Fig. 23* A simplified diagram of the Michelson-Morley experiment. A beam of light from source A was divided by semi-reflective mirror B so that part of the beam (unbroken arrows) went straight through to mirror $C_2$ while part (broken arrows) was diverted to mirror $C_1$. Both beams traveled equal distances against the "ether wind" to reach target card D, but one beam (broken arrows) had to travel three times as far *across* the "ether wind" as the other, and was thus expected to reach card D slightly *later* than the other beam, creating an interference pattern.

respect to a pinpoint light source and a beam of light was then aimed at the mirror, the beam would be divided into two beams, one moving straight through the mirror in a north-south direction, for example, while the other would be reflected at a 90-degree angle and travel at right angles (that is, in an east-west direction) to the north-south traveling beam (see Fig. 23). Both segments of the original light beam were then reflected back by fully reflecting mirrors and rejoined to form a bright spot on a receiving screen. (For clarity, in the diagram we have shown the original light beam as a heavy black line, and its two components leaving the semireflecting mirror as, respectively, a thin unbroken line and a thin dotted line.)

Michelson and Morley reasoned that the half of the light beam that was traveling in the direction of the ether wind (say the north-south beam) ought to return to the screen slightly sooner than the half of the beam which was traveling *across* the ether wind (that is, east-west), so that the two halves of the light beam finally reaching the screen ought to be slightly "out of phase" with each other by the time they struck the screen and thus ought to form a clearly visible interference pattern of light and dark stripes on the screen.

Obviously, since light waves were known to have such very short wavelengths, Michelson and Morley did not expect a very great difference in "return time" of the separated components of the light beam, but they expected enough difference for the two beams to be out of phase by the time they reached the screen. Since nobody knew for sure the precise direction the ether wind would be blowing, their whole apparatus—light source, mirrors, screen and all—was set on a huge block of stone floating on a pan of mercury so that it could be rotated in any direction. With this arrangement, the experimenters reasoned that they could turn the entire apparatus through the whole 360 degrees of a circle if necessary, watching the screen all the time. They recognized that if the apparatus were so oriented in direction that *both* components of the light beam were traveling equally "crosswind" to the ether wind, no interference pattern would be seen, and they expected that as they slowly rotated the apparatus the telltale interference pattern would disappear when the apparatus was turned in such a fashion, only to reappear again as the device was rotated further.

The experiment was certainly ingenious, and the reasoning behind it seemed flawless. Michelson and Morley (as well as the vast majority of physicists of the day) fully expected their experiment not only to demonstrate the effect of the ether wind on the velocity of light, but also to reveal once and for all both the *direction* and the *magnitude* of that ether wind. In addition, both men confidently expected their experimental results to confirm Bradley's earlier estimate of the velocity of the earth in her orbit by revealing how fast the ether wind was passing the earth and in what direction.

With the use of their cleverly conceived device, Michelson and Morley proceeded to perform one of the most famous and decisive laboratory experiments in all the history of physics. The experiment was carefully thought out. There were no logical loopholes, nothing that could conceivably go wrong. All the scientific community throughout the world waited eagerly for the experimental results that would pin down the elusive world-ether once and for all. And when the experiment was finally performed, it proved to be a scientific triumph, a major turning point in the history of man's understanding of the universe—for the simple reason that it failed completely.

What happened when these men turned their pinpoint beam of light on their semitransparent mirror? *Nothing* happened. The part of the light beam that traveled *across* the ether wind returned perfectly in phase with the part that traveled *with* the ether wind no matter in what direction the apparatus was oriented. Hardly able to believe their own eyes, Michelson and Morley repeated the experiment again and again, always with the same result: They could demonstrate no evidence whatever that light traveled crosswind or even against the ether wind at any different speed than it traveled with the ether wind. Other physicists repeated the same experiment at different seasons of the year and with innumerable modifications of the apparatus and still always came up with the same answer: nothing happened, no difference in light speed measured.

The experiment came as a bombshell in the world of physics. Michelson and Morley's experimental results seemed to be saying that the speed of light was always the same, under all circumstances, regardless of what direction the light beam was traveling and—most devastating of all—*regardless of the speed or direction of motion of the observer!* The Galilean transformation equations which applied so splendidly to the measurement of the relative velocity of sound waves as measured by moving observers, or to the relative motion and direction of moving objects as described in Newton's laws of motion, did not seem to apply at all to the relative motion of light waves: An observer moving swiftly *toward* a given light source (say, moving toward a star in a high-velocity rocket ship) would obtain the same measurement for the speed of light as another observer moving equally swiftly *away* from the light source when he measured the speed of light. It seemed impossible; it simply didn't make sense; yet this was precisely what that famous experiment indicated was indeed true.

## THE FITZGERALD CONTRACTION

For centuries scientific knowledge had been expanding in explosive fashion in all directions on the basis of a single unshakable conviction: that whenever the results of actual observation and experiment contradicted theory, the theory was considered invalid and not the experimental results. Now that conviction was put to its gravest test. Michelson and Morley were not the only scientists who were staggered and chagrined at the abysmal failure of their experiment to detect the direction and magnitude of the ether wind. So also was virtually every physicist and mathematician alive.

Obviously something was drastically wrong here. This was not a matter of an experiment producing slightly different results from those anticipated; this was a matter of an experiment which practically *had* to produce *some* kind of results producing no results whatever. Of course, as in any other case in which experimental results fail to confirm reasonable predictions, the technique of the experiment came under scrutiny. How sensitive was the apparatus that Michelson and Morley had used? What possible flaw in the experimental setup could have accounted for their failure? If these men had been clumsy or sloppy experimenters, using badly engineered apparatus, following poorly conceived techniques or pursuing a faulty line of reasoning, the world of physics might have been able to ignore their results, at least for a while.

But this was not the case. The men were reputable investigators, their experiment flawlessly conceived and executed in such a way that it could readily be duplicated by other experimenters anywhere in the world. Indeed, their experimental apparatus was so sensitive that it should have revealed the earth's motion through the ether even if its velocity in orbit had been only a quarter of the velocity that Bradley had estimated earlier.

In short, physicists of the day faced an irresolvable contradiction. There appeared only two ways to explain the negative results of the Michelson-Morley experiment, with each explanation as untenable as the other. Either the earth was *carrying its own cloud of ether along with it* as it moved in its orbit (a notion no one could seriously accept) or else the earth *was not moving through any ether at all.* As the Michelson-Morley experiment was repeated again and again by multitudes of other physicists, with the same monotonously negative results as before, it began to look more and more suspiciously as if no such thing as the world-ether even existed. More and more it began to appear that light waves not only traveled through completely empty space without the aid of any medium whatsoever through which to move, but that these strange waves moved at a velocity which remained constant no matter how fast or in what direction the individual doing the measuring might be moving.*

Faced with such a conundrum, physicists naturally began searching for explanations. They were desperate enough that the explanations didn't even have to be particularly plausible in order to command attention, providing that they somehow explained how the Michelson-Morley experiment could have failed. An Irishman named George Fitzgerald hit upon one way out of the dilemma. Fitzgerald suggested the possibility that all

* This is not to say that attempts to refine and improve on the Michelson-Morley experiment were soon abandoned; in the very face of the accumulation of a wealth of evidence that light waves and other electromagnetic waves needed no medium through which to travel, physicists have continued repeating that famous experiment in various forms with various kinds of electromagnetic waves under various conditions right up to the present day. As late as 1958 a physicist named Towne repeated the experiment with negative results although his equipment was so highly refined that it should have shown a difference in the velocity of light traveling with the ether as compared with its velocity traveling against the ether even if the earth's orbital speed were as little as 1/1,000 what it actually is. If nothing else, physicists are tenacious.

moving objects, including rulers and measuring devices of every kind, actually became a trifle shorter or flattened out as they moved in the direction that they were moving, and that the faster they moved, the more they "contracted" or flattened out. Thus, he suggested, when physicists worked in laboratories on the surface of the earth which was known to be moving some eighteen miles per second in its orbit, the physicists themselves together with all of their measuring instruments would be flattened out or contracted slightly in the direction of the earth's motion from what they would be if the earth were absolutely motionless in space.

Furthermore, Fitzgerald postulated that the *amount* of this flattening out or contraction that occurs to objects in motion conveniently happens to be *just the right amount* so that any measurement physicists might make of the speed of light would always come out to be the same no matter how fast the observers might be moving or in what direction. If this were the case, it followed logically that the Michelson-Morley experiment would have inevitably failed since Michelson, Morley, and all of their measuring instruments were squeezed down and contracted in the direction of earth's motion just enough to cancel out perfectly the difference in the speed of light through the ether that they would have been able to detect if they and their measuring instruments were *not* squeezed down in this fashion.

Now, granted, this train of logic might well seem uncommonly crafty and peculiar even to the untrained layman, particularly since the very nature of the notion automatically rules out any possibility of ever proving it right or wrong. In fact, it sounds very much like the carnival peddler of a remedy for dragon bites saying, "The world around us is really filled with dangerous dragons, *but* any time one of us tries to detect a dragon it automatically becomes invisible and otherwise undetectable until we quit trying to detect it, so there isn't any way that anyone can actually *prove* that the world is full of dragons. You'll just have to take my word for it." This is a tough one to argue; of course we could ask such a person, "Who told you?", but then he might merely look inscrutable and say, "I just happen to know, but *you* are the one in danger of dragon bites; I've already taken some of my remedy." And there you would be.

In spite of the logical awkwardness of Fitzgerald's hypothesis (which came to be known as "the Fitzgerald contraction"), it gained a great deal of attention among physicists simply because it provided *some* kind of explanation for a phenomenon that simply didn't make sense any other way. Unfortunately, later experimenters trying various modifications of the Michelson-Morley experiment began to accumulate evidence that the Fitzgerald contraction was not quite as good an explanation as it seemed at first. For example, when the Michelson-Morley experiment was repeated using unequal distances for the two segments of the divided light beam, the results were still entirely negative. This suggested that Fitzgerald's idea of shrinkage or contraction of length of a moving object was not sufficient to explain the paradox of the behavior of the divided light beam. In fact, some experimental results implied rather strongly that not only did objects contract in length in the direction in which they were moving, but that the time scale against which time intervals were measured would also have to change if the Fitzgerald contraction were correct!

Now this was a very sticky idea indeed. As we have seen, for centuries physicists had recognized that the measurement of the motion of objects and the direction of that motion had to be relative to the motion of the person doing the measuring; the same was true about the motion of sound waves, and there had been no reason to suspect that light waves would be any different. Of course, in the case of moving objects or moving sound waves there was always a fixed frame of reference against which motion could be measured: the surface of the earth. As far as light waves and other electromagnetic waves were concerned, the world-ether was considered to be the fixed frame of reference against which the motion of those waves could be measured, but now physicists were seriously beginning to question whether that world-ether even existed or not.

In fact, it was beginning to appear that nature provided no fixed frame of reference *anywhere* for such measurements, no place or thing in the entire universe to which one could refer and say, "This point is immutably fixed and unmoving even though everything else in the universe is moving relative to it." Unsettling as such an idea was, physicists were beginning to consider this a distinct possibility, however uneasily—but never before had anyone ever seriously suggested that time itself might be a changeable or variable entity. Throughout all history, time had always been considered as an absolute and constant entity, moving or flowing (if it could be said to move or flow) from the past to the present to the future at a fixed rate, always precisely measurable on the same scale for any observer anywhere in the universe whether he were moving or stationary, and no matter what he might be doing. How could anything else even be conceivable? How could it be possible that any object, person, or particle in the universe could *move through time* at any different rate than any other object, person, or particle in the universe was moving through time?

Oddly enough, during this unsettled period in the history of physics, certain mathematicians had already suggested that such a thing not only was possible but in fact *must be true* under certain theoretical circumstances. At the same time that the world of physics was thrown into confusion by the failure of the Michelson-Morley experiment with light waves (one form of electromagnetic wave, we must remember) a Dutch physicist and mathematician named Hendrik Antoon Lorentz had been struggling with some imaginary problems that had come to his mind as he analyzed Maxwell's equations describing the interrelationships between electrical and magnetic fields.

These equations (and the actual experimental behavior of electrical and magnetic fields as observed in the laboratory) had been taking on greater and greater importance in the world of physics as the basic discoveries about the nature of electromagnetic waves were made. Just as a highway patrolman needs to translate what he observes automobiles on the highway to be doing in terms of actual automobile velocities and directions with reference to the ground, so Lorentz was trying to translate the behavior of moving electrical currents and electromagnetic fields as measured by some moving observer in a moving laboratory into some absolute fixed frame of reference such as the world-ether. To his chagrin, he began discovering that it could not be done. Although Newton's laws of motion were not violated when the Galilean transformation equations were used to relate the behavior of moving objects to a fixed frame of reference such as the ground, Lorentz found that if he tried to apply these same transformation equations to the behavior of moving electric currents and electromagnetic fields in order to relate them to some fixed frame of reference, Maxwell's equations ended up altered and distorted.

Accordingly, Lorentz had revised and modified the Galilean transformation equations in such a way that they could be applied to the behavior of moving electromagnetic fields and electrical currents without altering Maxwell's equations. What Lorentz came up with was a new set of transformation equations involving length, width, height, *and time*—equations that accurately described what happened to electrical currents, magnetic fields, and electromagnetic waves no matter what the velocity of the observer and measurer might be, and without altering Maxwell's equations at all.

These new transformation equations of Lorentz were very similar in form to Galileo's transformation equations, but they were not precisely identical. When they were applied to the motion of objects or waves moving at relatively low velocities—to the behavior of sound waves, for example, or to the movement of vehicles on a highway—the Lorentz transformation equations produced results so closely similar to the Galilean transformation equations that there was no way of measuring or even detecting any difference between them. It was only in cases in which objects (or observers) were moving at *extremely high velocities* that any difference in the results of these two transformation systems became measurable—but in such cases of very high velocity motion the Lorentz transformation equations seemed to twist the classical laws of motion into forms that flatly contradicted Galileo's common-sense relativity and Newton's laws of motion.

The uncomfortable thing about Lorentz's equations was that when they were applied to the question of how the speed of light could remain constant regardless of the relative motion of the observer, as the Michelson-Morley experiments indicated, they provided a very good theoretical, mathematical explanation. They also explained why George Fitzgerald's "contraction" hypothesis did not hold up well under close scrutiny. For one thing, Fitzgerald's contraction was a *visual* explanation of something that no one could visualize. For another thing, it was incomplete. The Lorentz transformation equations indicated that moving objects did indeed undergo a flattening or "contraction" as Fitzgerald had hypothesized, but that the amount of contraction that occurred in the case of objects moving at normal everyday velocities was so extremely tiny that there was no possible way to detect it. According to Lorentz's equations it was only when objects (or waves) were moving at extremely high velocities—velocities approaching the speed of light, for example—that a significant and measurable contraction took place.

Furthermore, these equations indicated that the contraction or flattening out of an object under these conditions of extremely high velocity occurred not only in the object's physical linear dimensions, *but also in its time scale,* an element that Fitzgerald had ignored. In short, it seemed that when an object was moving at a very high velocity Lorentz's equations indicated that it ought not only to "shrink" or flatten out in the direction that it was moving, shrinking increasingly more the faster it moved, but that the time scale in which it moved ought also to slow down more and more the faster it moved. Not only should the measuring yardstick shrink in such a high-velocity system; the clock should actually tick more slowly!

## ON THE HORNS OF A DILEMMA

It was clear that things were getting worse rather than better as a result of these lines of reasoning. Here, it seemed, was an even worse paradox than the paradox of the ever-constant velocity of light. Lorentz's transformation equations were mathematical expressions, not observed experimental results, but mathematics had long since proved itself to be such an extremely reliable aid to scientific investigations that investigators could not ignore these equations. Physicists had long used theoretical mathematics as a means of predicting what experimental results they might expect to obtain in the laboratory, and this use of mathematics had never failed them. Now Lorentz's transformation equations were doing no more than theoretical mathematics had always done: They were merely predicting that if experimental observations were made under certain circumstances, certain results would be obtained. Yet the results that these equations predicted were obviously so completely bizarre and unimaginable that no physicist in his right mind could seriously believe them.

What, then, was to be done? By odd coincidence, a great fictional detective of the day put his finger squarely on the only possible answer. In one of his popular short stories, Conan Doyle's detective, Sherlock Holmes, said, "When you have eliminated the impossible, whatever remains, however improbable, must be the truth."

And this, in fact, was what had happened. The Michelson-Morley experiment had eliminated the idea of the existence of an ether wind as impossible. What remained, the fact that an observer in a moving system such as the earth traveling in its orbit would always come up with the same measurement of the speed of light no matter in what direction that light was traveling relative to the observer—however improbable it seemed—had to be the truth. Physicists were faced with several possible solutions to this dilemma, each as highly improbable as the other:

1. *The behavior of light involved some kind of magic completely outside the control of any law of nature.* Improbable, indeed, and scientists of all people were simply unable to settle for "magic" as an acceptable solution. To do so would be to discard all hope of ever learning how things worked in an orderly universe governed by natural law. It would mean the bitter end of scientific discovery.

2. *Light behaves as it does because some special, unseen, unknown, and unknowable mechanism is at work.* Again improbable, and really just another way of saying, "It involves magic."

3. *Because of what we have observed of the behavior of light, we are forced to conclude that some of the "laws of nature" that we have relied upon for so long simply aren't valid—at least not all the time, absolutely, under all circumstances—or else they are not valid the way we have worked them out. Therefore we must either work out new "laws of nature" or else modify the old ones so that all our observations are consistently explained by them.*

When we consider that this was a time when a great many brilliant physicists had begun to think that virtually all of the great natural laws of the universe (including among others Newton's laws of motion, the law of universal gravitation, and the laws of conservation of mass, energy, and momentum) were already completely worked out, or very nearly so, we can begin to appreciate what a dreadful prospect it must have been to consider that all of those laws might be invalid, or require modification. Some simply could not or would not face this prospect; they turned their backs on it and tried again and again to figure out what could possibly have gone wrong to lead their scientific colleagues down the path to this intolerable conclusion.

Others realized that they *had* to face this prospect, somehow, or else discard their most deeply entrenched conviction—that the scientific method of observation, hypothesis, and experimentation was the only valid way to investigate the universe around them. There were only two choices. Either they had to accept that the great laws which they had thought perfect and complete were in fact not complete at all but required serious modification, or else they had to give up scientific investigation altogether and admit that the work to which they had devoted their lives was a fraud.

It took courageous men to see the whole structure of natural law as they had always understood it begin to crumble around their ears and still step back and say, "All right, what can we salvage? What must we modify, and how? What must we throw out completely and start building up again from scratch? And how can we go about doing it?"

Fortunately there were men with sufficiently great courage and sufficiently rigid scientific discipline to do just that. Perhaps some of them sensed that they stood on the brink of revolutionary scientific discovery such as the world had never seen. Undoubtedly many were men and women with such an enormous reserve of built-in insatiable curiosity that they just couldn't bear to walk away from an unsolved enigma. Doubtless others were just too stubborn to admit defeat.

All of them realized, to one degree or another, the horrendous task that faced them. They knew what had to be done, but it took a genius to show them the way.

Albert Einstein was that genius.

# CHAPTER 15

## *The House That Einstein Built: Special Relativity*

One of the most fascinating ideas in all the study of history is the concept of historical watersheds, the notion that seemingly minor incidents and eventualities, or apparently insignificant occurrences, can in retrospect prove to be the critical forks in the road that have decided the entire future course of human events.

Historians are great believers in the principle of cause and effect, and certainly a multitude of the great decisive changes in human affairs have arisen from the simplest and humblest of nondescript events. Some time in history a caveman, terrified and trapped away from his cave one night by the magnificent violence of a thunderstorm, must have found that a lightning-struck log, still burning, could not only keep him warm and dry throughout the night but could also keep away the predators that would otherwise have been very happy to eat him. Certainly the harried decision of a Roman governor in a remote eastern province two thousand years ago changed the course of history, as did the bullet of an anarchist assassin in central Europe two thousand years later. It has even been suggested that Napoleon may have lost the battle of Waterloo simply because he had become too fat to go out and properly inspect the state of his troops and resources on the eve of the battle.

Similarly, although it was not entirely apparent at the time, there is little question today that the course of history was changed half a century ago when a young Swiss patent clerk named Albert Einstein quietly but firmly showed a multitude of chagrined and chastened physicists the direction they would have to take in order to achieve the necessary modification of the great natural laws of physics which had somehow to be accomplished. Even today many nonscientists cling to the popular and pernicious idea that Einstein's work was far too complex and difficult for anyone excepting other scientific geniuses to understand. Nothing could be farther from the truth; characteristically, the route that Einstein charted through the no man's land that physicists had gotten themselves into was no more complicated or sophisticated than the shortest distance between two points. And contrary to the popular notion that this man was an impractical theoretical dreamer, the course that he charted was at first challenged not because it was "impractical" but precisely because it was so completely and mind-bendingly practical that his colleagues in physics couldn't stand it!

When all the mystery and misconception is cut away, the revolutionary proposals of Albert Einstein rested upon two amazingly simple foundations: *an unflinchingly honest (if uncomfortable) point of view,* and *a single simple but unsupported hypothesis.* Einstein did not say, "Come, let us reason together," but he might as well have. "Accept these two foundations," Einstein said, "and there is a solution—the only possible solution."

The honest (if uncomfortable) point of view was really very simple. In effect, Einstein said: "Base your thinking only upon those things which can be observed in real experiments. Forget questions about things which cannot be observed. Ignore them—they are not only unanswerable, they have no place whatever in scientific thinking. Thus, if you are convinced by multitudes of experiments that there is no way to *observe* something such as 'the ether' nor to measure earth's motion through it, then forget it. Simply throw it out, bag and baggage, and assume for scientific purposes *that there is no such thing as 'the ether.'* "

In other words, as one writer expressed it, Einstein was merely saying, "Let's be realistic" on a ruthless—and cosmic—scale. It seems very simple and obvious, expressed this way; but at the time Einstein first introduced it, this was a perfectly staggering concept in the world of physics, totally unlike anything anyone had ever thought of before. It is hard to exaggerate the impact of this philosophy of an "operational approach" to scientific investigation, for it became the open door to virtually all the revolutionary concepts of modern physics.

So much for the honest (if uncomfortable) point of view. The single hypothesis Einstein proposed, also uncomfortable, was equally simple. In effect he said: *"Let us take as a basic assumption that the measured speed of light (or of any electromagnetic waves) will always be the same anywhere in the universe, no matter what the motion of the observer who is making the measurement, and no matter what the motion of the light source."*

Understand that Einstein was not playing meaningless word games. He recognized perfectly well that this hypothesis was contrary to the great classical laws of nature. It was contrary to common sense. But at least it was consistent with the coldly practical viewpoint that Einstein had suggested. *Base your thinking only upon things that can be observed in real experiments; forget questions about things which cannot be observed.* The world-ether, conceived by earlier physicists as an absolutely motionless substance filling the universe through which objects and waves moved according to Newton's laws of motion, had never actually been observed in real experiments. In fact, its existence had never been demonstrated in any way whatsoever, in spite of some highly ingenious and imaginative experimental attempts to pin it down. All this effort had been expended because physicists felt that this world-ether *had* to exist in order to fill a necessary hole in their understanding of the way things worked. Einstein merely said that since the ether had never been observed, and since in all probability it never *could* be observed or demonstrated in any way, it should therefore be ignored, thrown out wholesale. In Einstein's view, this elusive world-ether had no place in scientific thinking.

On the other hand, however, the behavior of light relative to a moving observer—behavior that seemed to violate common sense—*had* been observed in real experiments such as the Michelson-Morley experiment. Indeed, every single experiment in which measurement of the velocity of light was involved, under all circumstances, had always suggested the same basic fact: that light always traveled at the same velocity no matter what the motion of the individual doing the measuring, and no matter what the motion of the light source. For example, astronomers in their search of the heavens had discovered a great many binary stars—star systems in which two stars relatively close to one another revolved perpetually around a common center of gravity. In the case of some such binary star systems, it was obvious that the light traveling to the earth from one of the two stars came from a source that was moving toward the earth at the same time the light coming from the other star in the system was transmitted from a source moving *away* from the earth. Thus if the measured velocity of light from any given source could vary according to whether the light source was moving away from the earth or toward the earth, then the measured velocity of light coming from one star of a binary system should be cyclically increasing and decreasing as the star moved in its orbit.

Furthermore, since the orbital velocity of such stars had to be at least as great as the orbital velocity of earth around the sun, and more likely much greater, common sense would dictate that the light transmitted from one of the stars moving *toward* the earth would reach the earth sooner by a measurable amount than light transmitted from the other star moving *away* from the earth at the same time. Over a distance of scores of light years (not uncommon for observed binary stars) this "small" difference in the velocity of light transmitted by the two stars should, according to common sense, result in a difference of arrival time of that light of days or even weeks, with this difference reversed every half revolution. Yet no such difference in arrival time of light had ever been observed by astronomers studying the motion of these stars.

Here was merely another observed situation in which the velocity of light apparently was not affected by the motion of its source: precisely what Einstein's hypothesis said, however contrary to common sense it might have been. And Einstein's expressed viewpoint was uncompromising: *Believe what you observe, and ignore what you think you ought to observe but fail to observe.*

## THE RIDDLES OF RELATIVITY

The implication of these two points were frightening indeed. If there were no motionless fixed and immovable ether, no absolute frame of reference, anywhere in nature, this implied that *everything in the universe had to be moving*. But if that were true, then there could be no possible way to describe the motion of one moving system except by comparing it with another moving system—there could be no *absolute* fixed frame of reference. And this was precisely the conclusion that Einstein reached. He pointed out that the whole concept that an absolute, motionless point of reference existed anywhere in the universe had to be abandoned completely, and along with it the concept of any kind of absolute motion. Einstein's incomparable contribution to physics was to explore in the minutest detail, with the most rigid and unyielding logic, just precisely what this basic postulate *had to mean or imply,* and then to develop from this basic postulate a theory to describe how moving objects in the universe behave relative to each other at all velocities from the very lowest velocities of normal experience to the inconceivably high velocities of light traveling from one star to another.

Today the end results of this man's logical thinking are commonly known as the "theories of relativity." His work on these theories was to occupy the remaining forty-five years of his life, and to create a totally new concept of what the physical universe really was like and how it really behaved. Because the implications of Einstein's basic postulate were so staggering, and because the questions that it raised were so extremely complex. Einstein first confined his thinking about the behavior of matter in motion relative to other matter in motion to a very special and restrictive form of motion: *the motion of objects or waves traveling at constant velocities without acceleration or deceleration*. For this reason, his earliest exploration of relativity theory is referred to as "the theory of special relativity," since he first worked out new natural laws describing the relative motion of objects under *special theoretical circumstances* in which the motion of those objects is straight-line uniform motion at constant velocities.

Such circumstances, of course, were unrealistic, as Einstein recognized perfectly well. After all, in the *real* universe, material objects rarely if ever actually move with constant velocity. In the *real* universe, material objects are constantly speeding up or slowing down (that is, accelerating or decelerating) or changing in direction under the influence of any number of outside forces such as gravitational force, for example, or the force of electromagnetic fields.

Thus at first Einstein deliberately limited the scope of his thinking in order to build some kind of solid platform upon which to stand. Later he extended his laws of special relativity to encompass all other kinds of motion—accelerated motion, decelerated motion, changes in direction of objects, etc. This far more complex and time-consuming job finally reached an end in the year 1915 when Einstein at last published his "theory of general relativity."

Albert Einstein charted a course through a no-man's land, and then proceeded to follow it to the bitter end. But what were the implications and conclusions that had to be drawn if physicists were to accept Einstein's very practical viewpoint and his basic postulate that the velocity of light remained constant under all circumstances? A great deal has been made of the complexity of Einstein's ideas, of the "paradoxical" nature of his conclusions and of their apparent "impossibility" or "meaninglessness." Certainly a number of the conclusions we will discuss were greeted with incredulity by the layman. They violated common sense. They violated everyday observation, and seemed to be quite unprovable in any way. Indeed, many of these conclusions seemed to have no realistic point to them at all. And it cannot be denied that many of Einstein's hypotheses and proposals do indeed have very little realistic meaning in our everyday world in which objects move at low velocities and the laws of Galileo and Newton apply perfectly well. To a great many people, Einstein's world of objects moving at extremely high velocities seemed nothing more than a fanciful wonderland with no relationship whatever to anything that occurs in the world of our everyday experience.

And we might well wish today that the "wonderland" of Einstein's relativity theories had never come close to our "real" world of experience. Unfortunately, things did not work out that way. For better or for worse, Einstein's "wonderland" of relativity theory was found to be very real and very valid, and inevitably it began, bit by bit, to impinge on our "real" world and influence what happened in it. Time after time the most "impossible" or "paradoxical" predictions of Einstein's relativity theory have been verified by laboratory experiment and observation. And when a theoretical prediction, however improbable it may sound, is repeatedly confirmed by controlled scientific experiment, and when the implications of those predictions can control our very lives and destinies, it is perhaps time to begin paying attention to them.

The fact of history is that Einstein's work led to a long sequence of experiments and discoveries which have not only changed our entire concept of the universe, but have also changed our everyday lives in a multitude of ways. Einstein's work has led to an incredible explosion of technological development in a world which not only thrives upon technological development but depends upon it for its very survival. His work led to a massive revolution in the techniques and concepts of warfare, and the world in which we live today has been shaped by that revolution. Einstein's work has literally molded the modern society in which we live, and has determined the rise and fall of nations. It has led to the present-day exploration of our solar system, and has ultimately offered mankind two alternative paths to follow. One path can lead to the total obliteration of all life on earth by means of uncontrolled exploitation of thermonuclear bombs, purposely or even inadvertently, in the interests of war. The other path can lead to the peaceful control and utilization of unimaginable quantities of raw power which could, if adequately controlled and compassionately used, enrich the lives of every living human being from now until the end of time.

## A NEW LOOK AT MATTER AND ENERGY

What was there about Einstein's practical viewpoint toward scientific investigation, and about his basic hypothesis that the velocity of light would always remain constant regardless of the motion of the observer, that caused such a furor in the world of physics? The overwhelming importance of Einstein's work lay not in these simple basic premises themselves but in the inescapable consequences of these premises if they were followed out to their logical conclusions. And this, of course, was precisely what physicists of the world, including Einstein himself, did: They pursued the logical conclusions that arose from Einstein's premises to the bitter end, no matter how odd or incredible those conclusions were.

One of the first things which became apparent was that if Einstein's premises were true, then the old view of the nature of electromagnetic interactions, even the old view of the nature of light itself, had to be thrown out wholesale and replaced by quite a different view. As we have seen, physicists had originally regarded light waves and other electromagnetic waves as nothing more than disturbances in a fixed and motionless world-ether that filled all empty space and served as a medium through which those electromagnetic waves could travel from one place to another. In other words, they thought of light waves and other electromagnetic waves as entirely comparable to sound waves or water waves in their mode of travel. We have also seen that physicists previously had regarded electric and magnetic fields as nothing more than localized areas of stress or deformation in the all-surrounding world-ether.

But if, as Einstein insisted, physicists could accept as real only those things which could be observed and measured, it was therefore necessary to discard the whole idea of a world-ether—to assume that there was no such thing, since it could not in any way be observed or measured. Fine—but then what was it that caused the poles of two magnets to attract or repel each other? And *what was traveling through space* when light or electromagnetic waves moved from one side of the room to the other, or from one star to the other? If these phenomena were *not* disturbances in some all-surrounding medium, and yet were indeed real because they could be observed and measured, then there was only one conclusion possible. It was necessary to assume that electromagnetic fields, or light waves, or radio waves, or other electromagnetic waves were themselves real physical entities. It was necessary to think of them as definite physical "things" which had just as much physical reality as any other material "things" in the universe from atoms to galaxies.

But where did that assumption lead? Followed out with rigid logic, it led necessarily to a thoroughly startling and unexpected conclusion. If a magnetic field in itself was as much a real physical entity as, say, a rubber ball was, then it followed that that magnetic field itself *had to have a certain amount of mass.* The conclusion didn't indicate *how much* mass a given magnetic field had to have—it might be a lot or it might be a little—but if a magnetic field's mass were great enough to be measurable by laboratory instruments, it could be measured in grams, pounds, or tons, precisely like a sack of corn, and would possess inertia—the tendency of any mass to resist any change in its state of rest or motion—just as a sack of corn possesses inertia.

Similarly, following the same reasoning, a beam of light or a stream of radio waves would also have to possess a certain amount of mass. In the case of propagating light waves or electromagnetic waves, the mass of those waves would have to be carried along with the light waves as they traveled. Thus, when a flashlight bulb is lit up by the flashlight batteries, the batteries would slowly but steadily decrease in mass as long as the bulb was emitting light, since the light waves traveling away would constantly be carrying a tiny amount of mass away with them. Similarly, the magnetic field around an ordinary horseshoe magnet has a tiny amount of mass. In fact, if this line of reasoning were carried still further, we would be forced to conclude that *any* form of energy must have a certain amount of mass associated with it, and that, conversely, *any* solid massive object or particle has to have a certain amount of energy associated with it.

Now this conclusion was certainly confusing. Physicists had long regarded solid matter on the one hand and any of the various forms of energy on the other hand as two totally different entities, quite separate and distinct from each other. Hundreds upon hundreds of experiments performed over a period of centuries had invariably confirmed the law of conservation of mass (applying of course to solid matter) and the law of conservation of energy (applying to energy in one or another of its forms). Why had these experiments never gone awry? If energy inevitably possessed a certain amount of mass, or if matter inevitably was associated with a certain quantity of energy, why were no discrepancies ever discovered in testing the conservation laws? If a part of the mass of a glowing light bulb was constantly being carried off by the light waves it emitted, why was it that the most delicate balances ever devised by man could detect no change in that light bulb's weight after it had been lighting a room for a while? Why was it that the same delicate balance could detect no increase whatsoever in the weight in a piece of iron after it had been magnetized? The answer was both simple and frustrating: The amount of mass carried away from the light bulb by light waves emanating from it was so unimaginably tiny that the most delicate balances devised by man could not even come close to measuring it; and the mass of the magnetic field created when the iron bar was magnetized was likewise so incredibly tiny that no instruments existed delicate enough to detect it.

But how, then, could we be sure there was *any* mass carried away by the light waves, or that the magnetic field of the magnet contained *any* mass? Einstein provided the answer. He even demonstrated how one could calculate *precisely how much* energy was associated with a given quantity of matter and *how much* mass was possessed by a given quantity of energy. These calculations showed clearly that in the ordinary everyday world these quantities were so exceedingly tiny that they rarely if ever were measurable, but that in the vast realm of stars, galaxies, and uncharted reaches of empty space those quantities became very measurable and very significant indeed.

This idea of the relationship between mass and energy is vitally important to our understanding of the explorations of modern physics, but it is very often misinterpreted. Einstein was not saying that matter and energy are two separate things which can, under the right conditions, be *converted* from one to the other, much as one might convert cornstarch, water, sugar, and flavoring into a bowl of butterscotch pudding. What he was saying was that matter and energy are two differing manifestations—two completely interchangeable forms—of the same physical entity, an entity which we might for convenience call *mass-energy*. We might consider it as comparable to the two sides of the same coin: Turned one way we see heads showing, turned the other way we see tails, but whichever side we are looking at, it is still just one coin. What is more, that one coin *always possesses both heads and tails* no matter what side we happen to be seeing at the moment.

Alternatively, we might visualize solid matter as a highly concentrated and compact form of this mass-energy entity, while heat energy, electrical energy, or radiant energy are precisely the same mass-energy entity in an extremely diffuse and insubstantial form. For an analogy we need look no further than the different forms a common chemical compound, carbon dioxide, takes under different temperature conditions. At very low temperatures carbon dioxide (analogous to the mass-energy entity we have been discussing) assumes the solid compact form of dry ice (solid matter). When the temperature rises, the solid dry ice evaporates into the diffuse and insubstantial form of carbon dioxide gas (comparable to energy). Yet whichever of the two forms may exist at a given time under given conditions, both forms are exactly the same compound—carbon dioxide (mass-energy), whether in the concentrated form of dry ice or in the diffuse form of carbon dioxide gas.

When we regard matter and energy in this fashion, as two dissimilar manifestations of the same entity rather than as two separate entities, we can also see that matter and energy must be equivalent to each other. A given measured amount of energy in whatever form we choose must be equivalent to a certain amount of mass which is measurable in grams or any other units that mass can be measured in, just as a given volume of carbon dioxide gas with a measurable density at a given temperature and pressure must be equivalent to the same amount of carbon dioxide in the solid form of dry ice under other conditions—a quantity of dry ice that can be measured in grams or any other units mass is measured by.

The reason that no known measuring instruments are sensitive enough to detect a decrease in the mass of a flashlight battery after the flashlight has been on for a while is simply that the equivalence of matter and energy is exceedingly lopsided. A very tiny amount of mass-energy in the form of solid matter is equivalent to a perfectly enormous amount of energy, while the same amount of mass-energy in the form of a great amount of energy is equivalent only to an exceedingly small amount of solid matter. Indeed, Einstein calculated that the mass which is equivalent to a given amount of energy is equal to that amount of energy divided by the *square of the velocity of light.* Now the square of the velocity of light—186,000 miles per second times 186,000 miles per second—is a pretty large number, something on the order of 34,596,000,000 miles per second each second, but at least it is an *imaginable* number, roughly equivalent to eight times the total number of people living on the earth today. If we take such a large number, however, and make it the divisor in a fraction—for example, taking 1 as the unit of energy for which we want to calculate the mass equivalent, and then divide it by roughly 34,500,000,000, we find that the equivalent amount of mass is exceedingly tiny, so small that we have no practical way to measure it at all in ordinary earthly experiments. In fact, if we calculate the equivalent mass of any manageable amount of energy in our everyday experience, such as the amount of energy in the magnetic field of a horseshoe magnet, we would find that the same mass-energy manifested as solid matter would almost always be so small as to be immeasurable.

On the other hand, today we know of circumstances in which huge enough quantities of energy are present and which have very significant and measurable mass-equivalents. When deuterium and tritium are fused together under conditions of extremely high temperature and pressure to form helium atoms, as occurs in modern hydrogen bombs, a perfectly staggering amount of energy is released, while the total mass of the helium atoms and free neutrons formed by the fusion can be found by indirect measurement to be slightly but measurably less than the total mass of the deuterium and tritium atoms that entered into the fusion reaction. In such a case we can say that a tiny bit of matter has been *transformed* into its energy-equivalent—a very huge amount of energy—and repeated experiments and measurements have proven that the transformation of this mass-energy from the form of solid matter into the form of energy follows precisely the equivalence formula Einstein worked out.

The fact is that for all practical purposes we must look to places and circumstances in the universe in which enormous amounts of energy are being released in order to detect significant enough changes in mass for measurement to be possible. One place in which such conditions prevail is in the interior of stars, which are radiating huge amounts of energy constantly in the form of heat, light, and other electromagnetic waves. We now know that the source of this energy is the very matter of which the star is composed: Part of the matter making up the star is being transformed into energy by the same fusion reactions that take place in a hydrogen bomb, except on a far more massive scale. Other kinds of fusion reactions may also be taking place in the same stars, transforming still more matter into energy which is radiated away. Under such circumstances the amounts of energy released are staggering indeed, and the amount of mass-energy in the form of matter which is being transformed into such huge amounts of energy is very significant.

Physicists have calculated, for example, that the amount of energy pouring from our own sun in the form of heat, light, and electromagnetic waves in a single *day* is equivalent to a mass of some 400 billion tons—well over 350 times the total tonnage of steel produced in the United States in the last ten *years!*

Einstein's formula describing just how much mass in the form of matter was equivalent to how much energy was stated as a simple equation: Mass ($m$) equals energy ($E$) divided by the square of the velocity of light. Since physicists all over the world agreed to use the small letter $c$ as the symbol for the velocity of light, Einstein's equation could be written as follows:

$$m = \frac{E}{c^2}$$

The same equation is often seen in its more familiar transposed form:

$$E = mc^2$$

No matter how the equation is stated, however, it always has the same meaning. Either way, it simply states that an exceedingly tiny amount of matter is equivalent to an enormous amount of energy, or inversely, that a huge amount of energy is equivalent to only a very small amount of matter. Thus it implies that the amount of mass associated with quantities of energy that we might normally expect to encounter in our everyday lives is far too tiny for us to measure or even to pay much attention to; physicists would say that it is "negligible." We could never hope to measure the mass of the magnetic field of a horseshoe magnet, for example, or the mass associated with the heat energy thrown off by a huge bonfire. On the other hand, as we will see later, the mass of a relatively small amount of lithium, under the right conditions of temperature, pressure, and neutron bombardment, can be transformed explosively into horrendous quantities of energy released by a thermonuclear bomb.

## THE COSMIC SPEED LIMIT

But the fact that matter and energy are merely different manifestations of the same entity—two sides of the same coin—with their equivalence easily calculable by means of a simple mathematical equation was not the only unexpected conclusion that Einstein's basic premises forced upon the world of physics, nor was it any accident that the velocity of light was a critical factor in Einstein's mass-energy equations. In fact, it soon became clear that the velocity of light was intimately related both to matter and to energy, especially in cases in which objects of any size down to the tiniest elemental particle were moving at very high velocities.

Consider that when Einstein pointed out that the idea of the world-ether had no place in scientific thinking and had to be discarded, he was actually pointing out that *nothing* in the universe is absolutely motionless or fixed, nor is there any such thing as absolute motion in the universe—only the motion of one object relative to another. In spite of all this, Einstein still insisted that there was one single constant and unchanging property of the universe: the velocity with which light waves or other electromagnetic waves travel through empty space. Remember that one of Einstein's basic premises was that the measured velocity of light is always constant, anywhere in the universe, no matter what the relative motion of the source of that light, and no matter what the relative velocity of the observer might be. If that basic premise were accepted and its implications followed out logically to the bitter end, Einstein reasoned, a startling conclusion must inevitably be reached: that all motion in the universe had to occur under an absolute speed limit—the velocity of light through empty space—and that no material object anywhere in the universe could ever be induced or even forced to move faster than that speed limit, no matter what means of propulsion was used.

We will go into the question of *why* Einstein regarded the speed of light as the maximum possible speed a bit later. This idea, of course, was quite inconsistent with Newton's laws of motion, as Einstein fully recognized. According to Newton's laws, the velocity of one moving object could simply be added to or subtracted from the velocity of another moving object (depending upon whether the objects were moving toward each other or away from each other) and a new resultant velocity could thus be calculated for either object relative to the other. For example, a highway patrolman could add the velocity of his patrol car with reference to the ground (say 60 miles per hour) to the *apparent* velocity of a car passing him going in the same direction (say 20 miles per hour) and conclude quite accurately that the velocity of that passing car with reference to the ground was 80 miles per hour, fast enough to deserve a speeding ticket. This form of translation of the relative velocities of moving objects into a common and comprehensible frame of reference (such as the ground) by means of simple addition or subtraction of the relative velocities of those objects had been tested repeatedly since the time of Galileo and had always been found to be valid.

But now Einstein was insisting that if an imaginary space ship were speeding through space with a velocity of 150,000 miles per second and then fired an artillery shell with a velocity of 100,000 miles per second at some enemy ship far ahead, that shell would *not* be traveling at a velocity of 250,000 miles per second. According to Einstein, the sum of the velocity of the ship and the velocity of the artillery shell fired in the direction of the ship's motion could never in any case exceed 186,000 miles per second, the "limiting velocity" of light. Similarly, he pointed out that if a space ship were traveling with a velocity of 150,000 miles per second through space toward a distant galaxy and then transmitted a radio signal toward that galaxy at light speed of 186,000 miles per second, that radio signal would travel ahead and reach the distant galaxy at a resultant velocity of only 186,000 miles per second and no more. On the other hand, if the ship were traveling at 150,00 miles per second *away* from that distant galaxy and then transmitted a radio signal back toward it at light speed, the radio signal would *still* travel to the galaxy with a velocity of 186,000 miles per second, no more and no less. In either case, the signal would travel toward its goal with precisely the same velocity *as if the ship were not moving at all* when the signal was sent (see Fig. 24).

But how could this be? Newton's laws of motion had been tested innumerable times and always had been found valid. Galileo's transformation equations for translating the relative motion of two moving objects into a fixed common frame of reference had proved to hold true over and over again. Yet this new "cosmic speed limit"—the limiting velocity of light—was inconsistent with both. Einstein's basic postulates seemed to create an insoluble paradox, until physicists began to realize that Newton's laws of motion and Galileo's relativity equations had *not* actually been tested in all parts of the universe under all conceivable conditions. The truth was that they had been tested *only in the case of objects moving at normal (i.e., comparatively low) velocities, either on or near the earth's surface or at best only within the comparatively small neighborhood of our own solar system.* No one had ever even attempted to apply them to objects in remote reaches of the universe, nor to objects moving at velocities even beginning to approach the velocity of light.

This is not to say that Einstein or any other physicist was seriously proposing that one set of natural laws applied to the motion of objects under one set of circumstances and that a completely different set of natural laws applied to the motion of objects under other circumstances. There was probably no scientist in history more firmly convinced than Einstein that the great laws of nature were *universal,* applying in the same way under all circumstances in any part of the universe. Indeed, Einstein was

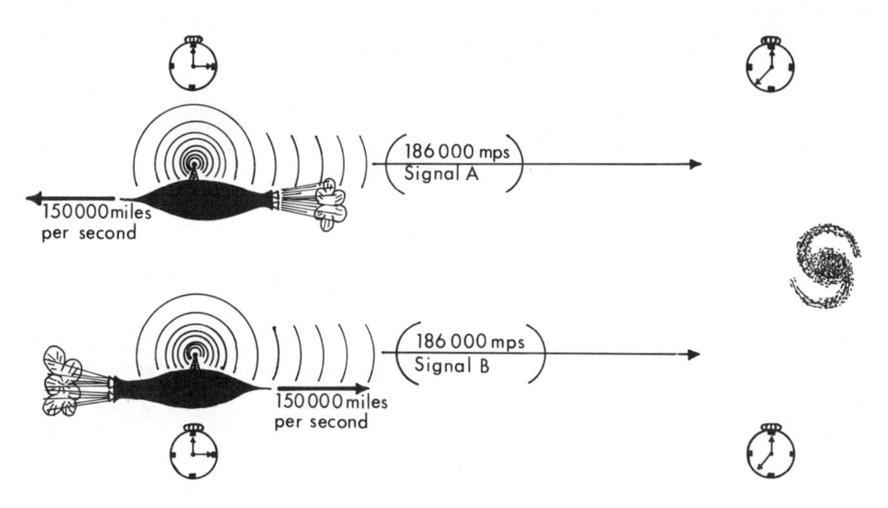


*Fig. 24* Einstein's light-speed paradox. Radio signal A transmitted by ship going *away* from a distant galaxy at 150,000 miles per second will reach the galaxy at the same time as signal B sent by ship traveling *toward* the galaxy. No Doppler effect here!

really contending that Newton's laws of motion and the Galilean transformation equations, while describing the behavior of nature accurately enough under certain limited circumstances, simply didn't go far enough. It was precisely because they were *not* universal—because they demonstrably did *not* accurately describe the observed behavior of nature in the case of objects (or waves) moving at very high velocities—that those classical laws needed to be modified.

But any modification of those laws had to remain consistent with the old laws in the areas of experience in which the old laws were known to be valid. Thus, in the case of objects on or near the surface of the earth or in the "small-scale" neighborhood of our own solar system, with objects of commonly encountered size and mass (rifle bullets, trains, billiard balls, or what have you) moving at very low velocities in comparison to the velocity of light, Newton's laws of motion held up splendidly, and any modification of those laws would have to be such that the modified laws *boiled down to the same thing* as Newton's laws of motion when they were applied to objects moving at such low velocities. The predictions of these modified laws of motion would only begin to deviate from the predictions of Newton's classical laws of motion (or at least to deviate noticeably) when they were applied to the motion of objects (or waves) traveling through vast reaches of space at very high velocities—velocities approaching or equal to the velocity of light, for example.

But how could those classical laws of motion possibly be modified according to such specifications as these? In searching for a logical basis for such modifications Einstein took a new look at an earlier idea which had previously seemed too fantastic to be credible. He adopted George Fitzgerald's idea that moving objects actually shrink slightly in the direction they are moving—in fact, that all the linear dimensions of any material object change slightly when the object is moving, with the change becoming more and more noticeable to an outside observer the closer the object approaches the velocity of light. Einstein further proposed that even the form of the mass-energy of a moving object changes as its velocity approaches light speed, with more and more of its kinetic energy or energy of motion assuming the form of mass rather than energy, so that the closer to the speed of light it approaches, the more massive the object becomes, the greater its inertia and therefore the greater its resistance to any further increase in its speed. Finally, he contended that even the time-scale in which the object moved is progressively more and more altered as its velocity approaches the velocity of light, so that an outside observer (given a motionless platform on which to stand and make his observations) would actually see that a clock carried by the moving object ticked more and more slowly as the object approached light velocity.

In short, Einstein carried the idea of Fitzgerald's contraction to its logical conclusion and found the idea not incorrect but merely incompletely thought out. Then, in order to find a mathematical description of these dimensions, mass-energy, and time changes so that he could calculate to precisely *what degree* they came about at what velocity, Einstein adapted the transformation equations that the Dutchman Lorentz had previously worked out. As we recall, Lorentz had developed his transformation equations in order to relate the measurements of electromagnetic interactions by one observer in one frame of reference to those measurements of another observer in another frame of reference. As we saw, those transformation equations were nothing but modifications or extensions of the Galilean transformation equations applicable to the relative motion of comparatively slow-moving objects on the earth's surface, and actually boiled down to the same thing as Galileo's equations (or so nearly the same thing that the difference couldn't be measured) in any case of commonplace low-velocity motion. The difference was that Lorentz's equations also applied to the motion of electromagnetic waves traveling at the velocity of light, and were consistent with Maxwell's equations describing electromagnetic fields—areas where Newton's laws of motion and Galileo's simple concepts of relativity no longer seemed to apply at all.

It is important here for us to understand very clearly that the "new laws of motion" that Einstein proposed were not a total overthrow of Newton's classical laws of motion, any more than Einstein's "new laws" of mass-energy equivalence constituted a total overthrow of the laws of conservation of mass and of conservation of energy. On the contrary, in each case Einstein's "new laws" were no more than *modifications* of those classical laws of physics that Newton and the other earlier physicists had so painfully discovered and described. Einstein was *not* throwing out the work of Galileo and Newton and all the others and substituting his own ideas. Rather, we could imagine him saying to Galileo and Newton, "You did not look far enough. For that matter, at the time you worked out your laws there was probably no way you *could* have looked far enough. The rules that you worked out to describe and predict occurrences in nature were perfectly valid for the part of nature which you were able to observe and study: objects moving at comparatively low velocities on or near the earth or in a nearby solar system. Your laws worked splendidly in relating the measurements made by one observer in one frame of reference to those of another observer in another frame of reference, *as long as there was some fixed reference point such as the surface of the earth that was common to both.*

But now we realize that those laws you worked out, valid as they were when applied to the world of everyday experience, were simply too limited. They do not accurately describe the behavior of objects moving at very high velocities—velocities which you could not even have imagined. They do not accurately describe how to relate measurements made by one observer in one frame of reference to measurements made by another observer in another frame of reference in a universe in which there is no fixed reference point. And since we cannot have one set of natural laws applying to one set of circumstances and a different set of laws applying to other circumstances, we now have no choice but to modify your laws in such a way that the new modified laws apply to *both* sets of conditions. Naturally, we hate to do this because your classical laws of motion were splendid and elegant achievements of scientific thinking—but great scientists that you were, you yourselves would have sought to modify those laws, just as we are, if you had discovered circumstances in which those laws proved invalid without modification, as we have."

By the same token, it is important to understand that Einstein was not throwing out Galileo's concept of relativity and Galileo's transformation equations and substituting something totally different in their place.

Essentially Lorentz's transformation equations had the same purpose and goal as Galileo's: to relate measurements taken by one observer in one frame of reference to the same measurements taken by another observer in another frame of reference. The difference between the two sets of equations was simply that Galileo's equations depended upon some fixed frame of reference common to both moving observers, while Einstein's adaptation of Lorentz's equations recognized that no absolutely fixed frame of reference existed anywhere in the universe, only a constant and absolute limiting velocity beyond which no material object can move, namely, the velocity of light. And Einstein's adaptation of the Lorentz equations amounted to nothing more than the introduction of the velocity of light as an unchanging and limiting factor in all calculations.

Of course these modified natural laws did indeed differ from the classical laws of Newton and Galileo, but the difference only became apparent when the laws were applied to objects moving at very high velocities. They were also inconsistent with Newton's quite understandable assumption that physical measurements such as length, mass, and time are absolute entities that are independent of any observer. In fact, the modified laws predicted that measurements of length, time, and mass *had* to be different for different observers moving at different velocities, no matter how high or low those velocities were. It was just that at the low velocities we normally encounter in our everyday world the difference is so small as to be negligible.

Thus, Einstein contended that the track star running the 3½-minute mile would indeed be shrinking in the direction he was running; his body would indeed be increasing in mass, and his watch would indeed be ticking more slowly as he ran. But in each case the change, whether it be contraction of length, increase in mass, or slowing down of time and body metabolism, would be so incredibly tiny at the low velocity the track star was running that it would not only be unmeasurable but negligible. For predicting this track star's motion, Newton's laws would be quite close enough in approximation; but send him racing through space at 99 per cent of the velocity of light and the contraction in his length from front to back as measured by another observer traveling at a different and slower speed would be *very* measurable. His mass would increase to a very significant degree, and if his race were long enough, the slowing down of his time scale and metabolism would be great enough to be very awkward when he got home; the rest of the world would have aged much more than he had in the course of his cosmic race, perhaps so much so that his wife would appear to him to be thirty or forty years older than she was when he started the race at what seemed to him only a short time before. At the same time, he might appear to her not to have aged in the slightest throughout the whole thirty or forty years' time that it seemed to her his race through space had taken!

## A VOYAGE TO THE LIGHT LIMIT: VARIATION I

If there seems to be something vaguely eerie and distorted about this whole concept, it is hardly any wonder. Certainly this is not the view of time, space, and dimension to which we are accustomed in our everyday experience. It was a mind-wrenching concept to physicists themselves when Einstein first introduced it at the beginning of this century. It was, of course, based upon his fundamental postulate that the velocity of light was an unchanging constant in the universe regardless of any motion of the light source and regardless of any motion of the observer. It was also based on Einstein's conclusion that the velocity of light was a sort of cosmic speed limit—a velocity which no material object in the universe could ever exceed.

For ordinary people who enjoy the thrill of riding in a high-speed sports car and like to daydream about space voyages to distant stars, this idea of a universal speed limit is perhaps the most irritating and unpalatable of all aspects of Einstein's relativity theory. *Why can't a spaceship, for example, accelerate beyond the velocity of light?* Even if we concede, somewhat grudgingly, that objects moving at very high velocities may undergo an increase in mass or changes in dimension or in time-scale (as measured by some outside observer), why does this mean that the velocity of light has to be the *absolutely maximum* velocity that any material object can move? Suppose we had a spaceship capable of traveling through limitless reaches of space unimpeded, with a limitless amount of time for the voyage and a limitless fuel supply—why then could we not continue to accelerate until we had reached the speed of light or even exceeded it if we felt like continuing?

The answer is very simple. It lies in the fact that mass and energy are interconvertible and equivalent, two manifestations of the same entity, and that the entire universe contains only so much of this mass-energy entity and no more, even if the universe extends infinitely in all directions. The answer is that if we could fulfill the conditions we have set forth for such an imaginary spaceship flight, we could *try* to exceed the velocity of light, but unfortunately our ship would inevitably run out of fuel before it could even attain the velocity of light, much less exceed it.

But *why*? Suppose we were aboard a spaceship making this historic attempt—what would we observe happening as we moved on toward our goal?

What we would observe happening is easy to envision but rather startling to contemplate.

In fact, as crew members on this imaginary spaceship we would observe *nothing* very much out of the ordinary happening except for two curious and frustrating things. First, we would discover (after our accelerating journey had gotten well under way) that our ship would seem to be increasing its velocity at a progressively slower and slower rate, even though we were feeding the engines progressively more and more fuel. Second, as we went along, we would discover that the increase in our ship's velocity continued to slow down progressively the closer we approached the speed of light, even as our fuel consumption went up. And ultimately we would find that we *never could* achieve light velocity, no matter how long we took nor how hard we tried.

Let's see how this would work in the course of our imaginary flight. Suppose we first carefully calculated how much increase in velocity—that is, how much forward acceleration—we ought to be able to expect for each tank of fuel that we used to produce the thrust that keeps accelerating our ship forward. Trusting implicitly in Newton's laws of motion, and basing our calculations upon them, we stored aboard the ship the proper number of tanks of fuel to increase the ship's velocity to the velocity of light. To be extra safe we even loaded aboard a large *surplus* of fuel to guard against any unexpected shortages. Then, with caps waving, we board the ship and begin our flight.

At first, as the ship begins accelerating from zero, everything seems to go according to plan. By the time our first tank of fuel is expended, the ship's velocity has increased from rest to 18,600 miles per second, approximately one-tenth of the speed of light. As the second tank of fuel is emptied, things still seem to be going according to plan, although we hear the captain grumbling that somebody must have failed to fill the second fuel tank completely full, because the ship has increased in velocity only an additional 18,500 or so miles per second on the second tank, somewhat less than on the first. Still, at this point the ship has a total velocity of slightly over 37,000 miles per second according to its instruments—almost one-fifth the speed of light—and is still accelerating.

On emptying the third tank of fuel, however, we find that something clearly is wrong, and the captain begins to wonder if someone is sabotaging the expedition. Tests indicated that this third fuel tank was full to begin with and contained only the highest-grade fuel, yet by the time it has been totally consumed the ship's velocity has increased only another 17,800 miles per second—a total so far of only 54,900 miles per second instead of the 55,800 expected.

Puzzled and angry now, the captain orders a second auxiliary engine into operation, and starts pouring fuel into both engines simultaneously. This seems to help at first; the ship accelerates more rapidly again for a while. But very soon its increase in velocity from *both engines working at once* is lagging far behind the expected velocity per unit of fuel consumed, according to our calculations. In fact, according to those calculations, if one tank of fuel increased the ship's velocity from zero to 18,600 miles per second, only ten tanks of fuel should have been necessary to achieve light velocity of 186,000 miles per second, yet soon all those ten tanks of fuel have been expended and we are dipping into our reserve supply, and the ship has still only reached a velocity of about 185,000 miles per second, a thousand miles per second short of the goal. And even though that ample reserve supply of fuel is dumped in faster and faster, we notice that the increase in the ship's velocity for each additional tank of fuel is becoming less and less.

Finally, in desperation, our captain plays his last trump. Since this is an imaginary ship with imaginary equipment aboard, let's imagine that it is carrying with it a fanciful machine that will permit additional fuel to be transported to the ship instantaneously from any place in the universe. The captain sends out an emergency call for more fuel, and immediately huge reserves of fuel begin pouring into the ship's receiving machine to be fed into the ship's engines. First a river and then a veritable flood of fuel pours into the ship and out the engines just as fast as it comes in, yet it seems that with each increase in the amount of fuel used, the increase in the ship's velocity falls off more sharply, dropping first from a run to a walk and then to a bare crawl.

But the captain is a determined and stubborn man. He has all the time in the world, and the ship *is* still accelerating, even though more and more slowly all the time. Months later, perhaps even years later, with the ship's velocity inching up more and more slowly all the time as more and more fuel pours in, it actually appears that the goal may be in sight. Even though the ship's velocity has been increasing, toward the end of this time, first just one mile per second at a time, then only a *half* mile per second at a time, then only a *quarter* of a mile per second at a time, as more and more fuel is poured into the engines, it has by now attained a velocity of 185,500 miles per second—almost the speed of light. Flushed and excited by the imminent achievement of his gallant goal, our captain doubtless imagines that the ship is being cheered on from all sides throughout the universe. But then, suddenly, a frantic message is received from outside:

"*Cease and desist; you must abandon this flight.*"

"Abandon it!" the captain cries. "Why? We are finally almost within reach of our goal!"

"Not so. You are not significantly closer to your goal now then you were when you began your journey," the answer comes back. "Now you must stop at once; *the entire universe is running out of fuel!*"

To those of us on board this seems to be extraordinary news indeed. As we said before, nothing whatsoever out of the ordinary seems to have happened in the course of this remarkable flight except that our ship has seemed to increase its velocity less and less as progressively more and more fuel has been expended. It is also true that we have not yet overtaken the speed of light, although we have so far exceeded our calculated necessary fuel consumption to get as far as we have that we are pretty disgusted with Mr. Newton and his laws by now. But all the same, our ship is still inching forward. As for other things, we have noticed no changes whatever. We look the same to each other, seeing no evidence of this ridiculous "contraction" of our physical dimensions that Dr. Einstein had predicted ought to be occurring. The table in the ship's galley was 2½ yards long at the beginning of our trip, and according to our own yardstick it is still 2½ yards long even though we are traveling with a velocity of 185,500 miles per second. No one on board has grown taller or longer or wider or thinner as far as *we* can observe.

What is more, our clocks and watches continue ticking at the normal rate of speed. We still seem to get hungry at the usual mealtimes, and we still seem to need eight hours of sleep out of twenty-four (all measured by our clock). True, the trip so far has taken months or even years longer than we expected it to take, but no one on the ship seems to be growing perceptibly older or younger. In fact, the only change that we really can observe at all is that we seem to be receiving poorer and poorer quality fuel all the time, and that we are in fact *not* approaching the velocity of light very fast, close as we have come. Indeed, the closer we approach it, the more slowly we seem to accelerate through that last 500 miles per second. And now, to add insult to injury, we receive this ridiculous message that the entire universe is running out of fuel just to meet the insatiable demands of our ship's engines!

What is wrong? What is happening? To find a clear answer to these questions, we must step aside from the ship and look at this whole imaginary voyage from the position of an outside observer who is *not* traveling along with the ship.

## A VOYAGE TO THE LIGHT LIMIT: VARIATION II

Suppose that from the very start there were a group of observers interested in watching our ship's voyage "from the outside looking in" so to speak, and suppose (since this is an imaginary voyage anyway) that they were able to watch our progress, make measurements of our ship and things aboard it, even observe the dials on our watches, all from some kind of imaginary fixed platform out in space—in other words, from a vantage point that was not moving at all. Of course we recall that Einstein contended that no such "fixed immovable reference point" exists anywhere in the universe, but there is no reason we cannot imagine such a thing just to go along with the game. Granted this, what would the observers watching from that imaginary space platform have observed during our ship's remarkable voyage?

They would have observed a number of things that were quite different indeed from what we aboard the ship observed at the same time.

At first, near the beginning of the voyage, these observers would have noticed very little that was different from what we noticed. While the ship was still traveling at relatively low velocities, they would have come up with much the same measurements of dimension, time, and mass that we came up with, and if any discrepancies were noted at all between their observations and ours, we would be tempted to chalk them up to a very minor degree of instrument error.

But by the time our ship had expended its second tank of fuel and had reached a velocity approaching one-fifth the speed of light, these outside observers would have begun noticing changes that we aboard the ship did not notice at all, and by the time the third tank of fuel had been consumed these changes would have become quite dramatic. First, these outside observers would have noticed that our ship seemed to be getting measurably shorter from bow to stern than it was at the beginning of the voyage, and that everything aboard it (including the crewmen themselves and the yardstick they were using to measure things aboard the ship) was also beginning to contract or shrink in the fore-and-aft dimension.

Furthermore, about the time that our ship's captain was beginning to complain seriously about the quality of the fuel or the question of sabotage because the ship was not increasing in velocity as rapidly as it ought to be, these outside observers discovered something that our captain had failed to notice at all. "The man's crazy," they would say. "There's nothing wrong with his fuel, nor any evidence of sabotage that *we* can see. But look what's happening to the ship's mass! It's taking on ballast like crazy from somewhere; our measurements show that the ship is far more massive than it was at the beginning. Of course a tank of the same grade of fuel can't move that ship forward as fast as the last tank did when the ship's mass has increased so much! Doesn't he know what's happening to his own ship? Something must be wrong with his measuring instruments. Why, look! *Even his clock is running slow!*"

And indeed, so it is, as observed from this fixed observation post out in space. Furthermore, as more and more fuel is poured in and the ship's velocity continues to increase, these changes in the linear dimensions of the ship and its crewmen, this increase in the mass of the ship, and this slowing down of the ship's clock all become increasingly more noticeable to these outside observers. By the time the ship has reached the velocity of one-half the speed of light (approximately 93,000 miles per second) their measurements show that the total mass of the ship has increased by almost 14 per cent, though where the ship is taking on all this ballast remains a mystery. From that point on, every extra thousand miles per second increase in the ship's velocity is found to be accompanied by a progressively greater increase in its mass, so that by the time it has achieved a little over eight-tenths the speed of light (approximately 150,000 miles per second) our observers find that the ship's mass has become fully *twice* what it was at the beginning of the trip.

At the same time, other measurements these observers have been making have become grotesque indeed—it seems that either their eyes are lying or their instruments are. Observed with radar, the ship by now appears very markedly squashed from front to back, and while the men aboard remain quite as tall as ever in the radar, they are beginning to look more and more like cardboard figures, very thin from front to back. But when observed by direct vision, neither ship nor men show any front-to-back contraction at all; rather, they seem to be twisted around some 45 degrees from their normal orientation even though both ship and men seem physically undistorted. Most puzzling of all, the ship's clock is seen to be running progressively slower and slower until at this point it seems to be losing over two hours for every three that the clock on the observation platform ticks off.

"Can't they see what's happening?" one of the men on the observation platform cries. "Look at them! They are going around measuring things with that ridiculously twisted yardstick and they don't even seem to notice any difference at all! And don't they know that their clock is running slow? See how slowly they're moving! The whole ship's activity is like watching a slow-motion movie. And according to our clocks, they're sleeping more than twenty-four hours at a stretch and then standing forty-eight-hour watches! How can anybody keep going that way?"

No one on the observation platform has any answers to these questions, but one of the observers comes up with an even more baffling question—the key question in the whole mystery: "*What's happening to all the fuel that they're burning?* That ship is gulping down fuel like a bottomless pit, and all it does is increase in *mass*. Its *velocity* isn't increasing worth a curse any more."

Although he might not realize it, that acute observer has hit the nail on the head. The truth is that that accelerating ship has become nothing less than a bottomless pit. As long as this insane adventure is continued, as long as that ship is fed all the fuel that it demands to produce more and more energy to increase its velocity further, it will continue to gulp down more and more fuel and continue to increase its velocity by less and less until, quite literally, *the universe runs out of fuel.* And even at that incredible extreme, the ship will *still* not have achieved the speed of light.

## THE MYSTERY OF THE VANISHING FUEL

What has been going on here in this imaginary space flight? Two things are obvious: first, that neither the crewmen on the ship nor the outside observers see the whole picture; and second, that the key to the mystery is somehow related to the energy provided by the fuel and the increasing velocity of the ship.

The captain's main problem was that in calculating the ship's fuel needs for this journey he had assumed Newton's classical laws of motion to be universally valid, and then used the Galilean transformation equations to determine how much fuel he would need to increase the velocity to and beyond the velocity of light. At first his calculations fitted the situation splendidly, up to and even somewhat beyond one-tenth the speed of light. But from that point on the results of his calculations began to vary more and more widely from what he actually found happening.

According to Galileo's equations and Newton's classical laws of motion, so much fuel producing so much thrust over such and such a period of time should have increased the ship's velocity by a specific calculable amount. And it did just that, so long as the ship's velocity was relatively slow—up to 20,000 or 25,000 miles per second. From that point on, Galileo's equations did not work so splendidly. Having moved past and through the "low-velocity" area in which the Galilean transformation equations are so close to accurate that any inaccuracy is negligible, the ship then entered into higher and higher velocity areas in which the Galilean equations simply went farther and farther askew. The captain and the crew began to notice only that the fuel no longer produced the increase in velocity it was expected to. The outside observers began to notice other changes, and the faster the ship moved, the more widely the measurements of these two sets of observers—those inside the ship and those observing from outside—began to differ.

In fact, if we could have polled both sets of observers at the same time, we would have found that there was *one and only one thing* that both the ship's crew and the outside observers were measuring exactly the same: the velocity of the ship with respect to the velocity of light. Everything else they saw differently. The ship's crew were *not* aware of any increase in the ship's mass, while the outside observers noticed this more and more. The ship's crew were *not* aware of any change in their linear dimensions; yet these changes became more and more noticeable to the outside observers. The crew on the ship were *not* aware of any change in the accuracy of the ship's clock, while the outside observers were more puzzled by the progressive slowing down of the ship's clock that *they* observed than perhaps by anything else. But both sets of observers were clearly aware that the ship was increasing its velocity more and more slowly all the time at the cost of an increasingly enormous amount of energy in the form of expensive fuel sent to the engines.

*Where was all that fuel going?*

The answer lay in Albert Einstein's special theory of relativity. Both the crewmen on the ship and the outside observers could have figured out the answers in advance had they used the Lorentz transformation equations instead of the transformation equations of Galileo. Certainly they would have discovered that the amount of fuel planned for the trip was going to be grossly inadequate. They could also have predicted that the outside observers would observe a "contraction" or shrinking in linear dimensions of the ship (which the crewmen could not observe) as well as an increase in the ship's mass (which the crewmen could not observe, either).

What about this change in mass observed by the outsiders but not by the insiders? Did the ship increase in mass or didn't it? Indeed it did, just as everything inside the ship increased in mass proportionately. In fact, the ship and its contents began increasing in mass from the very beginning of its voyage, although the increase at first was imperceptible and unmeasurable by anyone. This increase in mass continued on a progressive scale as the ship and its contents moved faster and faster—but everything on board including all the ship's measuring instruments, scales, and yardsticks also increased in mass so that there was no way that those aboard could discover that this change was taking place. *Only an outside observer* could have noticed or measured this change. The only clue that the captain of the ship had to indicate that his ship's mass was increasing more and more rapidly as its velocity approached closer and closer to the speed of light was the one thing that he *did* observe: that he was dumping more and more fuel into his engines and getting less and less power in return.

But this still does not solve the mystery of the disappearing energy. It was clear to all observers, inside the ship or outside it, that the increasing amount of energy produced by the increasing amount of fuel burned was *not* increasing the velocity of the ship proportionately. Then where was all that energy going? Again, Albert Einstein had the answer. More and more of that increasing amount of energy supplied by that increasing consumption of fuel *was being transformed into mass instead of providing additional thrust.* And with this happening, the ship was caught up in a vicious circle: the more its mass increased, the more inertia the ship had, and thus the more energy had to be applied to increase the ship's velocity further, and yet as still more energy was applied, a greater and greater proportion of it was transformed into mass to start the circle all over again.

If the ship's captain had known of Einstein's special relativity theory and had known enough to use the Lorentz transformation equations in figuring his fuel consumption on the trip, he would have realized that his expedition was doomed from the start. Even so, of course, he might have been stubborn and insistent. He might have decided to go to the outside extreme in quest of his goal, and commandeered all the fuel in the universe for his ship. Would he not, then, finally have reached light speed?

The answer, of course, is no—and now we can see why. "All the fuel in the universe," taken literally, would have to mean all existing energy in the universe in any form whatever, together with all the solid matter in the universe transformed completely into energy, with all this applied to the goal of increasing his ship's velocity a little bit more and a little bit more. This would mean that every star, every planet, every single tidbit of solid matter in every galaxy of the universe, combined with all the bits and pieces of matter scattered hither and yon in the form of dust and gas throughout the empty reaches of space would somehow have had to be transformed into pure energy to supply this one ship. And even at this extreme, with all this supply of energy, the ship would at best merely have approached *close* to the speed of light. The sacrifice of nine-tenths of the total mass of solid matter throughout the entire universe, all transformed into energy for the ship, would still have left the ship running short of its goal of light speed, because the vast proportion of all that energy would have been retransformed into the ever-increasing mass of the ship itself.

In order for that ship's velocity merely to *equal* the velocity of light, even the totality of all the mass-energy in a finite universe, including the energy derived from the mass of the ship itself, would not have been sufficient. An *infinite* volume of energy would have been required. Thus, if our entire universe, unthinkably huge as it is, is nevertheless finite in dimension, the sum total of all the mass-energy it contains could not have increased the ship's velocity quite to the speed of light. If, on the other hand, our universe is infinite in dimension, then we must suppose that the goal could barely have been reached—the ship's velocity could have *just equaled* the velocity of light—only upon the sacrifice of all the infinite amount of mass-energy in that entire infinite universe. And by that time all of the mass of the ship and its crewmen would also have been transformed into energy as well, in the form of light or other electromagnetic waves, to race endlessly at the speed of light through a universe empty of anything else. From a practical point of view, it would seem to be a rather poor bargain.

## RELATIVITY THEORY AND THE REAL WORLD

In practical terms, of course, our imaginary space flight must be considered as nothing more than a wild-eyed fantasy. We knew from the beginning that there was no conceivable way that such an experiment could be conducted, any more than there is any way to fix an observation platform absolutely motionless in space from which to observe what is going on in the space ship. We have simply used an imaginative situation to show us in broad terms the effect of Einstein's special relativity theory and the meaning of the Lorentz transformation equations as adapted by Einstein.

But to come back to earth, does this mean that we must rely upon imaginative experiments alone to demonstrate the validity of Einstein's work? What proof can there be that Einstein's predictions are correct if there is no experimental way to demonstrate them? Is this just another case of a strangely fascinating but utterly unprovable theory? Not quite. If it were, relativity theory would certainly be suspect, if not completely discredited. True enough, at the time Einstein first outlined his theory of special relativity there were no experimental techniques available to test it. But bit by bit experimental evidence began to appear suggesting very strongly that Einstein's predictions about the behavior of material objects moving at very high velocities were indeed accurate.

Some of this evidence has appeared only within the last few decades, with the construction of machines designed to accelerate elementary particles to higher and higher velocities, machines such as the cyclotron, the bevatrons, and other types of "atom smashers" used in modern physics laboratories. Of particular interest is the linear accelerator currently being enlarged at Stanford University, a machine designed to accelerate electrons faster and faster in a straight line by means of a succession of "pushes" supplied by electromagnets stationed at intervals along the electrons' course. The acceleration of an electron in such a machine is very much analogous to the voyage of our imaginary spaceship: an electron starting from a measured minimum velocity is accelerated and further accelerated, its velocity increased push by push (we might say fuel tank by fuel tank) until it increases closer and closer to the speed of light. But as the electron's velocity increases, the pushes must become greater and greater, and they must come closer and closer together in order to keep this tiny particle accelerating faster and faster.

The results of such linear acceleration of an electron have been observed and measured by experiment, not imagination. Of course we might argue that trying to accelerate a single electron up to the speed of light is quite a different matter from trying to accelerate an entire spaceship up to the speed of light, but in point of fact it actually matters far less than we might think. The results, in terms of energy required and in terms of what would happen in the long run would be virtually identical, regardless of the size or the initial mass of the object or particle we were accelerating. In either case, whether we were accelerating a single tiny electron or a whole spaceship, we would find that at low velocities the increase in acceleration would occur in accordance with the classical laws of motion in direct proportion to the amount of energy expended. But as higher velocities were attained—one-tenth the speed of light, for example—we would find that the rate of increase in the electron's acceleration would begin to fall off more and more sharply, and we would have to push it ahead with more and more energy, either by increasing the strength of the electromagnetic field in which the electron is moving or by placing the magnets closer and closer together, in order to keep increasing the electron's velocity. Simultaneously, we would find that the mass of the electron, a measurable quantity, was beginning to increase, first very slowly, then more and more sharply.

This is not to say that the electron would appear to be getting bigger or more bulky. A single electron is a trifle too tiny to observe directly anyway, even with the most powerful magnifying devices. It would be impossible to observe any change in the electron's linear dimensions for the same reason; if the electron was indeed shrinking or "foreshortening" in the direction it was moving, we would be at a loss to prove it. But it *would* be possible to detect and measure an increase in the electron's mass as its velocity increased. The nonscientist would be inclined to say that the electron seemed to be getting heavier or increasing in density as it accelerated, whether there was any change in its dimension or not. Physicists would prefer to say that the increase in mass in the electron was demonstrated by an increase in its inertia—merely a fancy way of saying that it is getting noticeably more difficult to push forward because of its increase in mass. In any case, when this electron-accelerating experiment was actually performed in the laboratory, the increase in the electron's mass and the concomitant increase in its inertia—in the difficulty of making it move forward faster—began to reach ridiculous proportions as the speed of light was approached. Where at low velocities a small increase in the amount of energy had increased the electron's acceleration greatly, now an enormous additional amount of energy barely increased its acceleration at all.

And where was all this additional energy going to? The answer is simple if we remember Einstein's contention that matter and energy are but two different forms of the same thing. We know that some energy, however small the amount, was necessary to start the electron accelerating even at low speeeds; the electron had a certain tiny amount of mass to begin with. We can then see that by the time the electron has reached a very high velocity, say 150,000 miles per second, most of the energy that we are adding in an effort to increase the electron's acceleration even a small amount is not exactly *vanishing,* any more than the great increase in the electron's mass that we can measure has come about as a result of magic. If we think of the moving electron and the energy that has been applied to it from the start of its accelerating journey as a "mass-energy system," we can see that at the beginning of the acceleration, at low velocities, the energy part of the mass-energy total was very small, with almost all of that mass-energy total composed of the mass of the electron. Now, when we actually can measure this strange increase in mass of the electron at high velocities, it is perfectly plain that the additional mass the electron has been accumulating must be coming from the energy being applied to it.

In other words, the constantly increasing amount of energy that is continually being added to this moving mass-energy system is, at this very high velocity, being transformed almost totally into its *mass* form just as fast as the energy is supplied. Thus, with each new increment of energy that is added at this high velocity in an attempt to increase the particle's velocity a little, more and more of the energy becomes *mass* that is associated with the particle, thus increasing the particle's inertia or resistance to being further accelerated and leaving less and less energy left over to provide additional thrust.

Of course, just as in our imaginary space flight, we *do* see the electron's velocity continuing to increase, even if more and more slowly. Surely then if we had an utterly unlimited supply of energy on hand, we should ultimately be able to accelerate this electron up to the speed of light regardless of what Dr. Einstein said!—but the key words here are "unlimited supply of energy." And again, as in the case of our spaceship, if we followed this laboratory experiment out to its ultimate end we would find that at the point at which the electron had reached a velocity just a bare whisker short of the velocity of light, we would have expended virtually all of the energy available in the entire universe, and that virtually all of it would have been transformed into the mass associated with this one single electron. Actually to reach the light speed limit we would need not merely all the energy in the universe but an *infinite* amount of energy, and the particle would then have attained not just all of the mass in the universe but an *infinite mass* and would thus have *infinite inertia.*

We would get no sympathy from the mathematicians if we tried this experiment and found ourselves beaten. They would merely look smug and say, "We told you so," because this is precisely what the Lorentz-Einstein transformation equations and Einstein's mass-energy equivalence equations predicted would happen. This was the reason that Einstein contended that the velocity of light was the one absolute limiting constant in the universe, for he reasoned that no material object, not even a single tiny electron, could ever actually achieve this velocity under any circumstances. In imaginary experiments, perhaps; in the real world, never.

Of course Einstein did not pick this limiting velocity of 186,000 miles per second out of a hat, nor seize upon it as a handy arbitrary figure to use for lack of any other. In thinking out the implications of his special relativity theory and his adaptation of Lorentz's mathematics, he found that he had no arbitrary choice at all. He was forced to the conclusion that the velocity of light *had to be* the uttermost limiting velocity at which any material object could travel precisely because he saw that an infinite quantity of energy would be required to accelerate any material particle, no matter how tiny it might be, up to that velocity; and that by the time such a velocity had been reached, all that energy would have been transformed totally into an infinite mass associated with the traveling particle. He saw that at such a point there could not possibly be further energy available to push the particle faster, since an infinite amount of energy had been required to get it that far—and even if there were somehow some energy left at that point, the particle would have been 100 per cent infinitely unbudgeable because it already had acquired 100 per cent of all the mass that existed anywhere.

As we have seen, the Einstein-Lorentz transformation equations predicted that in the case of particles or objects moving at low velocities, the increase in mass, the contraction of linear dimensions, and the change of time scale would all be so negligible that the old Galilean equations and the classical laws of motion would be perfectly valid and usable. At higher velocities, however, all of these apparently absolute and unchanging measurable characteristics of the particle or object would indeed begin to change, as measured by an outside observer in a different frame of reference. In fact, *none* of these characteristics was actually fixed, rigid, and unchanging, and never had been. They had been thought to be so only because such scientists as Galileo and Newton had never encountered objects or particles moving at extremely high velocities, nor even imagined that such high velocities were possible.

Granted that much of this discussion of Einstein's "new laws" and their implications seemed awkward and difficult to comprehend when Einstein first set them forth. Even today, when many of Einstein's then unproven predictions have been verified by actual experiment again and again, and when physicists all over the world have come to recognize that Einstein's new notion of relativity is indeed closer to the true natural law of the universe than Newton's comfortable laws of motion were, we may still wonder what difference it makes to us in our everyday lives. After all, if the old laws remain valid for us here on earth where the velocities we ordinarily encounter are very slow, in a part of the universe that is familiar to us, who needs any new laws or cares about them? So these new laws do apply to particles or objects that might be thrashing around through the universe at enormous velocities somewhere else—so what? Are there any areas at all in which these new laws of Einstein's actually affect *us*?

Fortunately or unfortunately, for better or for worse, our lives are indeed affected by Einstein's new laws in a whole succession of vital areas.

Our lives are affected every moment of the day by the behavior of light and other electromagnetic waves—and the new laws certainly apply to them.

The very shape of our everyday lives, the way in which we live, the societies that we build, our powers to create or destroy—all depend upon the sources of power that are available to us. And the greatest of all sources of power known to man, the power locked up in the interior of atoms, could never have been released either in the violence of thermonuclear bombs or under the controlling leash and harness of modern technology without knowledge and application of the new laws.

In another area, we know today that we are constantly bombarded by cosmic rays from outer space, the only known material particles that actually *do* travel from place to place at anything approaching light velocity, and our lives are affected in a variety of ways by the energy and the end results of this bombardment. It is quite possible that this cosmic ray bombardment of our upper atmosphere provided the initial key to the appearance of life itself on our planet—the radiant energy that initially changed simple molecules into more complex molecules and further transformed them into the living, metabolizing, and reproducing molecules within the nuclei of earth's earliest one-celled living creatures. We know also that the high-energy bombardment of atomic nuclei in our physics laboratories, requiring the artificial acceleration of material particles to very high velocities, has already had irrevocable effect upon our lives and will affect us and the way we live more and more in the future. The behavior of subatomic particles in transistors and semiconductors affects us. All of these things have been discovered, detected, understood, and brought under useful control only through application of the new laws.

In fact, we could never have developed the world of technology in which we are living today without the help of those new laws, any more than the farmer of today can well survive operating his farm with a horse-drawn plow, without radio contact with the world, without telephone contact with his markets, or without advance warning of storms, floods, and droughts. Already in our country the farmer who tries to live and maintain himself that way is quite literally a dead farmer. Within a few decades he will be a dead farmer anywhere in the world.

So we make peace with the strange ideas of Einstein and his successors, however uncomfortable they may be. As we go along we will see more and more how important these ideas were to the further development of physicists' understanding of the universe. We will see that at first the predictions of relativity theory had to be accepted on faith, supported only by mathematical proofs, but were later actually demonstrated through laboratory experiment or observation to hold true in the real world as well. When we actually observe and measure the transformation of energy into mass or of matter into energy right before our eyes in accordance with Einstein's mathematical predictions, it is difficult to dispute the theory or the mathematics that predicted these effects, no matter how improbable they might have sounded or how difficult they might be to understand. When we actually observe a material object twisting around so we can see its opposite side when it is traveling at high velocities, it becomes difficult to dispute what we see with our eyes simply because we can't understand how this could possibly be. When we have ruled out the impossible, what remains must be the truth, however improbable—and Einstein's work was what stood towering over the world of physics after the Michelson-Morley experiment had failed.

So peace was made with the special relativity theory as scientists and nonscientists came to understand the reasons and the reasoning behind it. But it was an uneasy peace. One could, perhaps, accept the idea that there is no fixed dimensional reference point in the universe. One could accept the idea that there is no such thing as absolute motion of anything in the universe. One could even concede that perhaps the mass of a given object was *not* always a fixed quantity under all circumstances, or that under certain circumstances various forms of energy might indeed behave more like a physical entity with mass than like plain energy, or that perhaps the linear dimensions of a material object might not be a fixed value but rather might change to some degree under special circumstances.

But of all Einstein's contentions, there was one that was perhaps more difficult to comprehend or accept than all the others put together. If there was one single characteristic of the universe that human beings throughout all history had always depended on as being absolute and unchanging, a single measurement that men had always considered fixed, rigid, and totally dependable under any imaginable circumstances, it was our measurement of *time*. Yet Einstein insisted that in a universe of relative values the measurement of time, too, would have to vary and change according to the relative speed of the observers. He contended that time was *not* fixed, rigid, unchanging, and steady in its flow from past to present to future. He predicted that the measurement of time by an observer within a moving system would have to be different from the same time measurement taken by an outside observer—indeed, that the very time clock controlling the life processes or metabolism of a living organism on a moving system would have to differ from the same biological time clock of the observer outside the system. Einstein went further: he even predicted that two events that were observed to occur simultaneously by one observer might quite as accurately be observed to take place *at different times* by another observer, and that *both observers could be entirely correct about when they observed the same two events to occur.*

To many scientists as well as many laymen, this apparent absurdity was by far the hardest to accept of all the paradoxes of the special theory of relativity. Einstein was not merely saying that time passes and things change. Anyone could agree to this, insofar as they could agree that time did indeed pass at a steady and reliable rate, the same for all people. The famous *March of Time* newsreels emphasized the inexorable passage of time through the affairs of men, a relentless unchanging and unbroken rhythm of time moving from past to present to future, the same here as anywhere in the universe. "Time marches on!" the newsreel insisted. Perhaps a more realistic view was expressed in the equally famous paraphrase: "Time is; Man marches on." From our everyday viewpoint, we could argue that neither expression is entirely accurate. It might be better to say that, "Man marches on through time, which passes him inexorably as he marches."

But if so, how fast does man march on through time? How fast does time travel past him? And in what direction? Strangely enough, really *meaningful* answers to such questions as these can be found only in the relativistic concept of the universe advanced and followed out to its logical conclusions by Albert Einstein and the other relativity theorists. To understand how this can be so, we must pause for a moment and consider precisely what we *do* consider time to be, how we observe time to behave, and in what manner modern physicists regard it.

# CHAPTER 16

## *The Puzzle of Time*

Throughout the entire history of Western civilization men have always been preoccupied with time, and never more so than in our present day. Always the month and the year in which events occurred have been the substance of recorded history; today we identify the time of occurrences in hours, minutes, or seconds. Centuries ago men reckoned the length of journeys in days, weeks, or months, used the turn of the seasons and the phases of the moon to determine planting and harvesting times, or clocked off earth-shaking events in terms of years or scores of years that had elapsed since earlier earth-shaking events. Today we use alarm clocks to awaken us at the proper time each morning, set our wristwatches by common time standards, arrive at work at fixed hours, count on the clock to tell us when to eat lunch, when to go home from the office, or when to expect dinner.

Of course we know that in certain Oriental religions and in the teachings of certain famous mystics and philosophers the passage of time is regarded as far less important than we consider it, but in our rigidly time-oriented societies we tend to view such an attitude as mysterious and slightly impractical. Even when we are engaged in activities in which time simply doesn't matter—in the midst of a vacation holiday, for example, or on a prolonged ocean cruise or wilderness camping trip—we still become acutely uncomfortable, perhaps even severely distressed, at the loss of a wristwatch because then we can't tell what time it is from one moment or hour to the next. Our preoccupation with time has invaded our very language with a multitude of such casual expressions as "take your time," "it's high time that they were arriving," or "time out until we decide what to do!"

But for all of this exaggerated time-consciousness, what exactly *is* time and how do we regard it? Obviously it has no color, no taste, no smell—it is not to be detected by our ordinary senses. We cannot see it; yet we are aware that time is constantly passing us by. Or, inversely, that we are passing through time. But what is it? Is time, like the world-ether that we discussed in earlier chapters also an undetectable, unmeasurable, and invisible entity surrounding the entire universe like some kind of vast, placid, and motionless fluid through which everything in the universe moves inexorably in one direction only?

If time *is* such a motionless medium through which we are moving, how do we know it is there? Why do we not create a "time wind" when we pass through it, or when it passes us by? How can we be aware of the passage of time, and why must time move inexorably in one direction only? Why can't we move backward in time as well as forward, to live again in ages past? Why can't we move forward in time a bit faster in order to gain a clear view of what is going to happen next? For that matter, why can't we find some way to halt our movement with respect to time in order to perpetuate some ecstatic moment indefinitely at will?

The greatest of the ancient Greek philosophers were unable to come up with any plausible answers to questions such as these, and generations of modern scientists, it seems, have not had much better luck.

Yet if we cannot describe or define the precise *nature* of time, there is still no question in our minds that in the world around us there is indeed some entity, or phenomenon, or condition, or universal property, or part of the orderly nature of things which we speak of as "time" and of which human beings have always been aware. Clearly, time is an abstract concept rather than some physical entity which can in itself be measured, tested, or observed. We are aware of time at all only in its passage. Yet in many ways, time seems to resemble a physical entity, and certainly it seems to be inextricably bound up with virtually all natural phenomena in the universe. Every motion of every object clearly involves the passage of an interval of time, even if that motion is nothing more than the blinking of an eyelid or the tiniest vibration of a single atom. And men have known for millenniums that if time itself cannot be observed or measured as an entity, the *passage* of time can be measured and divided into large or small intervals.

In early days only the most crude means were available to measure the passage of time: the movement of the sun across the sky from dawn until dark, for example, or the cycle of the seasons or the waxing and waning of the moon. Later, more sophisticated methods were devised. The passage of equal intervals of time could be measured indirectly by means of the seeping of sand through an hourglass, the steady dripping of water from the spout of a water clock, or the rhythmic sweep of a pendulum to and fro. In modern times clocks and watches measure the passage of time, again indirectly, by means of the rocking of a balanced flywheel or even by the vibration of a tuning fork. In each case, you will notice, the passage of time intervals was accomplished indirectly through comparison with the regular ryhthmic motion of some object or another. We measure the passage of time by "clocking" it against some tangible and observable form of rhythmic motion.

Thus it makes perfect sense to us to say that "time passes" as a flywheel rocks back and forth or a pendulum swings. To the classical physicists it seemed self-evident that time passed at a steady, constant, utterly reliable, and unchanging rate everywhere in the universe, no matter what form of clock was used to measure its passage. It also seemed self-evident that time should pass at the same steady rate and in the same direction regardless of who was measuring its passage, and regardless of whether the observer was in motion, standing still, or whirling like a dervish. The passage of time, it seemed, was a comfortably invariable constant throughout the universe.

Even so, the classical physicists were aware of certain other characteristics of time which could be observed and measured indirectly from the effect of time upon other natural phenomena. The passage of time, for example, seems to be a strictly one-way affair. We can turn our clocks back an hour, if we wish, but there is no way that we can move backward in time by so much as a single second. Many events that occur during a given interval of time can never be recalled or undone. If we throw a dry maple log onto the fire and wait for a while we find the log soon reduced to ashes, smoke, and heat during that time interval. This is an irrevocable and irreversible change; there is no way to recapture the smoke and the heat and add them to the ashes to recreate a dry maple log again. Even the basic life processes of the human body are inevitably linked to the one-way passage of time: men grow perceptibly older, never younger; body tissues form, age, and die; but the order is never reversed, nor is there any scientific evidence that anyone has ever been able to move so much as a second into the future, although there are those who have believed that they could. Of all kinds of movement in the universe, the passage of time seems to stand alone, unchanging and utterly irreversible in direction, with all events in the universe irrevocably bound to occur at the precise stately rate of the passage of time, no faster and no slower.

Or so the classical physicists assumed. Then Einstein, in his special theory of relativity, introduced a startling paradox with regard to the passage of time. Incredible as it seemed, the relativity theory contended that time might move faster for one observer than for another observer taking the same measurement. Relativity theory even contended that two events which seemed to occur simultaneously to one observer could be seen by another observer to be separated by a discrete, measurable interval of time. But how could this be? How could it be possible that the very biological time clock determining the aging of cells and tissues and determining the rate of chemical reactions in the cells of living creatures would actually slow down under certain special circumstances, so that the crew aboard an extremely high-velocity spaceship, for example, actually *lived more slowly* than outside observers stationed on a fixed observation platform? Did time pass at a fixed and unchanging rate throughout the universe regardless of the observer, or did it really change along with an object's linear dimensions and mass when the object attained an extremely high velocity?

## THE BIOLOGICAL TIME CLOCK

The notion that the time-scale of a moving object slowed down progressively more and more as the object approached the velocity of light was indeed one of the predictions set forth by Lorentz, Einstein, and other relativity theorists in the early 1900s. But even then this was not really an entirely new idea. On the contrary, thousands of years before Einstein was born individual human beings had undergone subjective experiences that seemed very much akin to relativistic slowing down or speeding up of the passage of time—experiences which had always seemed puzzling, and often were regarded as mystical or even supernatural manifestations of some sort.

Today we know that such experiences have nothing whatever to do with relativity theory, but they continue to occur. In fact, strange as it may seem, *our own perception* of the passage of time—not as measured on clocks, but as we perceive it subjectively in our own mind—may change radically from day to day or from circumstance to circumstance.

Consider, for example, the experience reported by an army engineer stationed in northern Labrador for several winter months during World War II. The place was a frigid refueling depot, and the landing strip was continually buried under drifted snow. To keep the runway clear, the engineer and his partner scooped off the snow with a bulldozer, wearing standard arctic gear for protection against the intense cold. During one such operation with temperatures as low as 50 degrees below zero, the engineer was surprised to find his partner turning the bulldozer back toward shelter after only half an hour of work instead of the usual four-hour run. When he asked his partner why they were quitting so soon, the other seemed surprised. "Soon?" he said. "We've already been out four hours, and if we stay out any longer we're likely to freeze!" Back in shelter, the engineer checked his watch and found that his partner was perfectly right. They had been out for four full hours, yet his own perception of time seemed somehow to have gone awry: In his mind, that four hours had been telescoped into a mere half hour, so that the long time interval as measured by the clock had seemed far shorter to him that it really was.

Another service man reported a different and apparently contradictory experience. While swimming in the surf off a South Pacific island, this man found himself caught in a swift undertow. Swimming for his life, the man struggled for hours trying to get back to shore; when he finally made it, he was so exhausted that he lay on the beach for another hour or two, barely able to breathe. Finally staggering to his feet, he lurched down the beach toward his friends who were still searching for him in the surf. Only then did he learn that the "hours and hours" that had passed from the moment the undertow caught him had really been less than thirty minutes, little as he could believe it. Throughout the rest of that afternoon and evening the man reported that things still seemed to be happening in slow motion; he simply could not believe that what had seemed such a prolonged period of time to him had really been so brief. Oddly enough, his body could not believe it either, for on the following morning the man was as ravenously hungry as if he had not eaten for a week.

These were by no means isolated incidents; similar experiences are noted every day. Every one of us has experienced some such distortion of time perception at one time or another in our lives, instances in which a brief span of time has seemed far longer to us, or in which a long span of time has somehow seemed to be telescoped or shortened. Such experiences of course do not mean that time is actually *passing* more slowly or more swiftly than normally. They simply mean that our *individual perception* of the passage of time can be influenced by certain biological factors. We perceive time passing according to the rate of metabolic activity in our body's cells, according to the rate of speed that biochemical reactions are taking place within us. Thus, when an individual's metabolism is slowed down (as, for example, when his body temperature has fallen several degrees below normal because of external cold) or when an individual's body is very inactive (as, for example, during an afternoon nap) the passage of time as perceived by that individual will seem to be squeezed down or telescoped so that he doesn't notice time passing. On the other hand, during periods of intense activity (swimming to escape an undertow, for example) an individual will perceive time as being slowed down or prolonged, often to a very marked degree.

We can find even more commonplace examples of such differences in individual perception of the passage of time. To a child, for example, time seems to be drawn out indefinitely; a single day seems an extremely long period of time, crammed with activity though it may be. We say that children have "an extremely poor time sense." To an adult, on the other hand, an hour seems like a minute, a day seems over before it has begun and even the years seem to be racing by. When we are waiting for an overdue airplane minutes seem like hours and the second hand barely creeps around the clock; yet when we are concentrating intensely on some fascinating problem a whole afternoon may pass without our knowing it.

Understand that what we are speaking of here really has nothing to do with physics. We are not attempting to define what time is or how it behaves; we are merely discussing how we, as individuals, perceive the passing of time under different conditions. In other words, we are talking about the individual biological time clocks built into our bodies and minds which tell us how much time we *think* has passed. These biological time clocks have no influence whatever upon the actual passage of time; they merely influence the way that we individually interpret what we experience. Even so, the biological time clock clearly illustrates that in our own human experience the passage of time has never really been a constant, steady, unchanging flow which is always the same for all persons under all circumstances—certainly not as we perceive it. Indeed, our subjective awareness of the passage of time under any given circumstances is in a sense *relative to the rate of metabolic activity within our tissues and cells.* And although this is quite a different kind of "relativity" from that which concerned Einstein, we can at least begin to see that time and its measurement may indeed have to be considered relative to other things, rather than as a fixed, unchanging, and absolute entity.

But even if we agree that our individual perception of the passage of time may vary greatly according to the operation of built-in biological time clocks, we recognize that this says nothing at all about time itself, and the measured passage of time seems to remain constant regardless of how we happen to perceive it at a given moment. Thus we are still grappling with a perplexing abstract concept without finding any way of defining it. Is there no way whatever that we can somehow find a physical definition or description of time, of what time is and how it behaves?

Clearly what we need is some way of visualizing time (or the passage of time) so that we can pin it down and examine it. And, in fact, there is such a way of regarding time as a physical property or characteristic of the universe—an odd way of regarding time, perhaps, but a way that will help us to see the important part that time plays in Einstein's relativity theory. This way of regarding time is sometimes called "the statics of time," for it is an imaginary attempt to make time "stand still" long enough for us to examine "what it is." In its simplest form this involves nothing more than visualizing time as a *new physical dimension* very much akin to the three ordinary spatial or linear dimensions of length, width, and height so familiar to us all.

## THE ELUSIVE SALT SHAKER

Most of us by now have encountered the flat assertion (usually ascribed to Einstein) that "time is the fourth dimension." In fact, in these days of extensive popular science reporting the statement is tossed off with such free-handed authority that we are discouraged even from trying to figure out what it means, much less from challenging it. Like many other such "authoritative" statements, this one builds a mountain of confusion upon a grain of truth. Besides being slightly inaccurate, the statement does not really mean what it seems to mean at all.

The fact is that numerous physicists and mathematicians at the beginning of the 1900s, Einstein included, did indeed hypothesize that time is a dimension that exists in addition to the three familiar spatial dimensions (although none of them would have ventured to call it *the only* "fourth dimension"). What was more, these scientists hypothesized that time was a *linear* dimension, a dimension following a straight-line direction which could be used together with the spatial dimensions of length, width, and height to describe the shape of an object and its location in space and time. But these men were not playing word games or seeking to confuse anyone by claiming that time was a dimension. To them that statement had a very concrete and down-to-earth meaning which anyone could readily understand and visualize.

How could time be a dimension? First we must be sure we know what "dimension" means in the scientific sense of the word. Definitions of this term tend to be more confusing than helpful; derived from a Latin word meaning "to measure out," the term "dimension" could best be defined as an imaginary extension or line in a given direction used to determine the position of a given object either in space or in time or both. To a scientist this simply means that a dimension is a device that can be used to pinpoint the exact and specific location of any object or particle or even an imaginary point anywhere in the universe as measured from some fixed reference point.

We are all familiar, of course, with the three commonplace linear dimensions. "Go one block south," we tell the policeman, "then turn two blocks west to a red brick building, and on the second story up you will find a man beating his wife." With such simple dimensional instructions we have directed the policeman from the place where we are standing along three successive straight lines or *coordinates,* each one perpendicular to the other two and each one in a different but specified direction from the other two, and led him to the precise location of a family brawl, hopefully in time to prevent a murder. Ships at sea locate or "fix" their positions on two-dimensional (length and width) navigational charts by means of two intersecting dimensional lines or coordinates, the lines of latitude and longitude. For the pilot of an airplane to identify its position, the third dimension of height must be added to the dimensional lines of length and width in order to "fix" its position in space at any given moment. Unlike the ship, the airplane might be anywhere from one foot to forty thousand feet above the ground, and without that third or "height" coordinate running up and down perpendicular to the "length" and "width" coordinates on the navigational chart there would be no way to tell how high in the air the airplane might be.

But are these three spatial dimensions—length, width, and height—not sufficient to locate any object anywhere in the universe? Not quite. We can see why not with a simple example. Suppose we are entertaining a friend for lunch and he thinks the soup needs more salt. We know that a salt shaker is sitting on a coffee table in the living room. How can we go about telling our friend where to locate it?

In all likelihood we would simply say, "There is a salt shaker sitting on the coffee table in the living room," and expect him to find it with no difficulty. But if the friend were a scientist, he might find those simple directions extremely crude and imprecise. And in all fairness we would have told him nothing whatever that would be really useful to a total stranger groping around in search of a salt shaker. "That won't do," he might say. "I need to know *precisely and exactly* where the salt shaker is located. Imagine that I am totally blind and have no knowledge of the location of any of your living room furniture; all I possess is a perfect sense of direction and a yardstick. Then how would you direct me to the salt shaker?"

Obviously we would then have to give him different directions, this time more scientifically meticulous. "Very well," we reply, "the salt shaker is located at a point in space eighteen inches above the floor, three yards in-

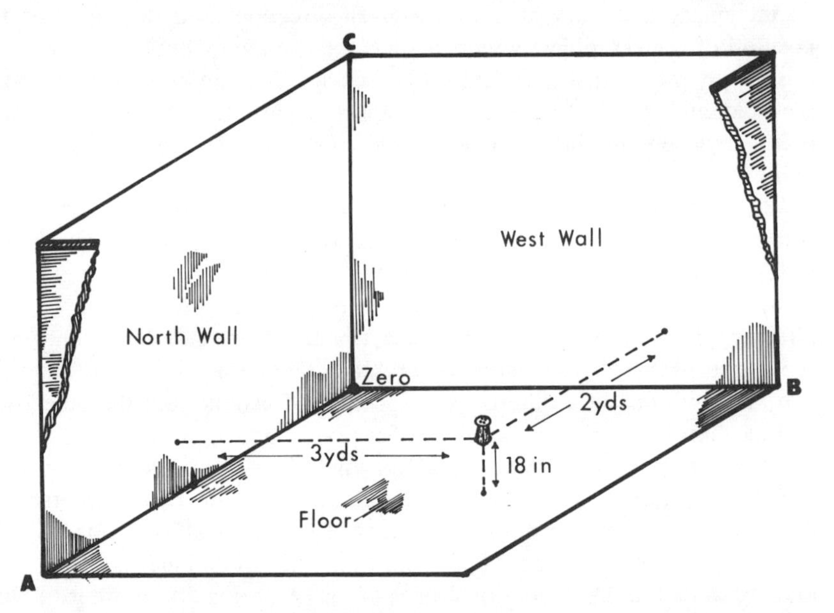


*Fig. 25* The location of an object by means of measurement of three linear dimensions or coordinates: length, width and height.

side the north wall and two yards inside the west wall." Surely, we think, with those directions even a blind scientist would be able to find the salt shaker.

What are we actually doing when we give these directions for locating that salt shaker? Essentially, we are "fixing" the salt shaker at a point where three linear dimensional coordinates—one for height, one for length, and one for breadth—intersect in space. We can think of it as lines drawn from the salt shaker in three different directions, each line perpendicular to the others like three sticks nailed together to form the corner of a box, and then measured the length of each of these lines from the salt shaker to the floor or the walls of the room.

To put it another way, as illustrated in Figure 25, we might say that the salt shaker was located somewhere on an imaginary horizontal plane eighteen inches above the floor; that at the same time, it was located somewhere on a vertical plane three yards away from the north wall; and still at the same time that it was located somewhere on another vertical plane two yards away from the west wall. We can see that the two vertical planes must be at right angles to each other (since west is 90 degrees away from north) and that both vertical planes cut through the horizontal plane perpendicularly. We can also see that if the salt shaker is located on all three of these planes (representing the three spatial dimensions or dimensional coordinates) at once, then it has to be located at the single point where all three coordinates intersect. Thus the salt shaker can be located by anyone simply by measuring to determine where that "magic intersection point" is.

It may well seem by now that we are going to extremes to make a perfectly obvious point. All we have been talking about is the simple measurement of three commonplace linear dimensions or coordinates, length, width, and height. Any schoolboy knows that these three dimensions are needed to describe accurately the location of an object. Indeed, we are a little annoyed at our scientific friend for making us go to all this work for no reason, until he calmly points out that we really have not fixed the true location of that salt shaker at all.

"But of course we have!"

"Not so," says the scientist. "All the time that you have been giving me those three coordinates, that salt shaker has constantly been changing its true location within the room."

"But it hasn't moved at all," we insist.

"Not in *space,* perhaps," the scientist replies, "but it has moved a distance of some five minutes from the position it occupied *in time* at the moment you started describing its location."

At first this might seem a ridiculous way to speak of time. Whoever heard of an object moving a *distance of five minutes?* But then we pause and reconsider. After all, we have always known that the passage of time was an inescapable natural condition of the universe. Time passes inexorably, one instant following the next. We could say just as accurately that every person or object in the universe is constantly and inexorably moving through time, instant by instant, even though it may not be moving *at all* through the linear dimensions of space. We could imagine, for example, the motion of an object through time as if it were being carried at an absolutely constant velocity along an endless railroad track. The track goes in one direction only: straight ahead. With the passage of each successive instant, the object would be found located farther and farther down the time track; each instant it would have moved ahead a certain distance, always in the same direction. If we then used a swinging pendulum to divide up the passage of time into equal segments each one second long, and then by measurement found that the object we were observing moved five feet forward on the track for every swing of the pendulum, we might very reasonably reach the conclusion that that object was moving through time a *linear distance of five feet per second along its time-track.*

When we consider time in this way, we can see that it begins to make sense to think of it as an additional linear dimension somehow inextricably tied up with the three more familiar linear dimensions of length, width, and height. If the movement of an object through time is constant and inescapable, then we can see that that object must inevitably move a certain fixed linear distance along its time-track (or along the time dimension, or along the time coordinate) with the passage of each successive unit of time, whether it is moving in any of the three spatial dimensions or not. We can also see that there is no possible way of completely separating the spatial dimensions of our universe from the passage-of-time dimension. We can see that our universe is not merely a universe of space alone with objects scattered about in it, but rather a *continuum of space and time* in which objects may or may not be moving about through spatial dimensions that are always continuously moving along a time-track in a single direction.

If this is so, then we obviously cannot fix the true location of our salt shaker in *space-time* just by using the three spatial dimensions of length, width, and height. Those three dimensions are not sufficient to locate the salt shaker in time as well as in space. If we wish to fix its location *exactly* in our universe of space-time at a given instant, we must realize that it is constantly moving along in time even if it is completely at rest in space, so that *we have no choice* but to use a time coordinate as well as a length-width-height coordinate to locate its precise position in our universe.

Now this is all very well when our salt shaker is sitting motionless on the coffee table, unmoving in any of its spatial dimensions. We can nevertheless visualize it as moving along some unspecified distance per second in some unspecified direction along a time-track or time dimension. But what happens to its motion in the time dimension if we deliberately move the salt shaker from one end of the coffee table to the other? We can see as we move it that it is constantly changing its location in space-time along at least one spatial coordinate and perhaps even two if we happen to slide it diagonally. Does this then mean that its motion in the time dimension is also changing?

Not at all. The salt shaker's progress through time is no faster nor slower than when it was sitting motionless in space, and it proceeds in precisely the same direction—forward in time—however that "time direction" may be oriented to the three spatial coordinates. But still, if the salt shaker has moved in space, something must have happened to keep the salt shaker and its time-track together. After all, it would not do to have the salt shaker wandering aimlessly around in space at the far left-hand side of the coffee table while its time-track continues to extend along the right-hand side of the table. What must happen, then, is that the entire time-track, salt shaker and all, must move through space as the salt shaker is moved to a different location in space, with the time coordinate always maintaining precisely the same orientation with the three spatial dimensions at all times, as illustrated in Figure 26.

Exactly the same thing must happen if the salt shaker were tossed in an arc through the air so that all three of its spatial coordinates were constantly changing in the course of its movement. Here again the salt shaker's motion and direction along the time dimension would remain unchanged, but the

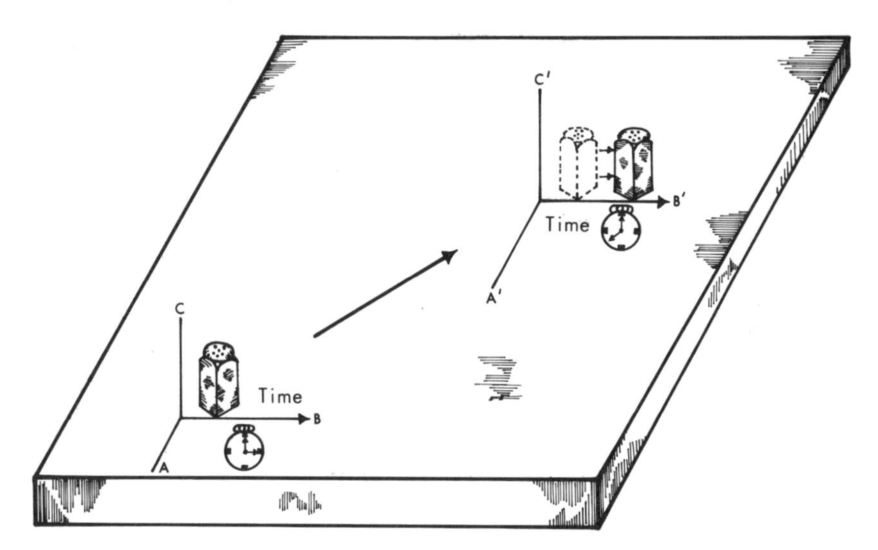


*Fig. 26* A projection of an object's time coordinate, remaining in the same orientation with the three spatial dimensions at all times. Salt shaker is moved along spatial coordinates from C-A-B to $C^1$-$A^1$-$B^1$; but it also moves along its time coordinate, so ends up in a different position in *time* as well as in *space*. Unfortunately, we have available no fourth linear coordinate we can use to indicate the time coordinate or time dimension in a drawing.

time coordinate itself with the salt shaker on it would have to be "carried along," so to speak, by the salt shaker as it arched through the air and struck the floor.

## THE DIRECTION OF TIME

So far we have been able to imagine time fairly well as a linear dimension analogous to any of the three spatial linear dimensions. We have not, however, been able to visualize the *direction* the time coordinate must take with respect to the three spatial dimensions of length, width, and height.

As we have seen, each of these three spatial dimensions is perpendicular to the other two; so it might seem logical that the time dimension must lie perpendicular to each of the three spatial dimensions, but if we try to visualize this, or draw a picture of it, we find we have painted ourselves into a corner. There can be no possible way we can visualize the direction of the time dimension in terms of space, since space in the universe as we know it is constructed in only three linear dimensions and we cannot comprehend any possible direction for a fourth linear dimension perpendicular to each of the other three. We could, of course, "visualize" the direction and orientation of the time dimension as illustrated in Figure 26, in which we see an object which is moving along its time coordinate depicted as if it were sweeping through an area of space—but here we are really only visualizing a crude *projection* of a four-dimensional object drawn in three-dimensional perspective on a two-dimensional sheet of paper.

In fact, we find that there is no way whatever accurately to illustrate or "visualize" the movement of an object along its time dimension. We can see our salt shaker moving in its spatial dimensions perfectly well when we toss it in an arc through the air, and we can *imagine* it moving simultaneously through time, just as it was moving through time while it was sitting motionless on a coffee table, but we cannot actually *see* the salt shaker moving through time, nor can we even imagine a direction perpendicular to length, width, and height as the direction of the time dimension.

In fact, we can be sure that the salt shaker *has* to be moving through time only because we know that time is passing as we sit and watch it. A series of time-lapse photographs of the salt shaker could be taken at the rate of a frame a minute, but they would reveal no apparent motion or change in location of the salt shaker; all the films would show would be a progressive darkening of the room in which the salt shaker was located, frame by frame, as the sun set and night fell, then a progressive lightening of the room around the salt shaker, frame by frame, some time later as dawn broke and the sun moved high in the sky again.

Thus, while we can visualize the three spatial dimensions without difficulty, this curious "time dimension" of ours differs from the others in the respect that it cannot be visualized in space. Nor is this the only problem that we encounter when we try to compare a linear dimension of "distance per second through time" with the three spatial dimensions. Perhaps the most awkward and glaring difference of all lies in the fact that an object can be moved quite readily *in either direction* along any of the three spatial coordinates, yet can move in one direction only along its time coordinate. It is quite impossible for an object to be moved *backward* in time, or to be moved forward faster in time than at a fixed, constant, predetermined velocity. In fact, it seems quite impossible for any outside force to cause any change whatever in the movement of an object through time, no matter how much energy might be expended in the effort!

Now this is very curious indeed. If time is really a fourth linear dimension, or behaves like one, then why can't an object be moved back and forth in time at will, just as our salt shaker could be moved back and forth in space from one side of the coffee table to the other at will? What prevents us from reversing the motion of an object through time? What velocity do objects ordinarily have through time, and what determines what that velocity must be?

These questions are not quite as unanswerable as they may seem, although the logical answers mean bad news for all those who enjoy daydreams about time travel. There is a very sound reason why an object cannot move backward in time, and why no one will ever find any way of forcing an object to do so. And curiously enough, that reason is very closely related to the reason an object cannot be accelerated in space beyond the velocity of light. In fact, as we shall see, the very same universal constant that limits the velocity of any material object traveling through space to the velocity of light or less also freezes the motion of objects to a fixed constant velocity of 186,000 miles per second *through time,* never under any circumstances either more or less.

To understand this, consider what we already know about the movement of objects in space, setting the question of time aside for the moment. We know that an object at rest or in a state of uniform motion in some direction will remain at rest or in its state of uniform motion until acted upon by some outside force. When an outside force is applied to an object in the direction of any spatial linear dimension, the object's state of rest or motion will be changed by the force so that the object will move with a new velocity in the direction (or along the spatial dimension) that the force has pushed it. In other words, before the outside force acts on an object, the object has a certain inertia of rest or inertia of motion as a result of its mass, and will tend to continue doing what it is already doing—whether resting or moving at a constant velocity—until an outside force acts upon it. Thus an apple might remain at rest on a table forever unless some force comes along and starts it rolling in a given direction.

We are by now fully familiar with this classical law of motion, but now we notice something that we have overlooked before. Where the motion of an object through space is concerned, the classical laws of motion place no restriction on the direction of motion that is possible. A force applied to the apple on the table could start it rolling due north at a given velocity if the force acted upon the apple in a south-to-north direction; but the same force could just as readily start the same apple rolling due south if the force happened to be applied in a north-to-south direction.

The same thing is true regardless of which spatial dimension happens to be the direction through space that a force acts on an object. Motion of any object in any of the three spatial linear dimensions of length, width, or height is completely and readily reversible, the actual direction depending solely upon the direction in which the force acts. There is no natural law that insists that an object can move up but not down, to the right but not to the left, forward but not backward. Of course, *how fast* the object moves in any spatial dimension depends upon a number of factors: the mass of the object (and thus its inertia or resistance to being moved), the interference of other outside forces such as frictional forces or gravitational forces tending to limit the object's freedom to move, and the magnitude of the force acting upon it. But whatever the magnitude of the force, the object is as free to move in one direction as in another along any of the three spatial dimensions without any restriction and without requiring any greater force to move it in one direction than to move it in another direction.

Now consider that when an object is moving in a given direction at a given velocity through space (and thus within the framework of three spatial

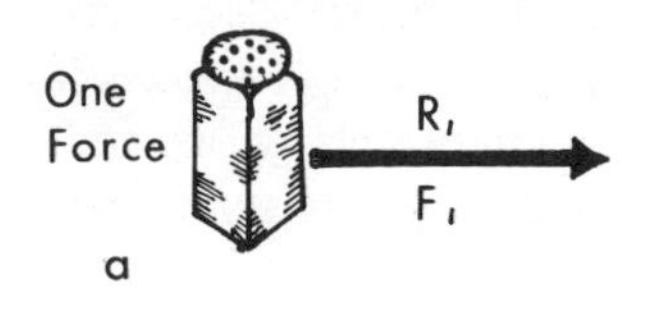


*Fig. 27* Motion of an object as a result of vector forces. The heavy arrow $R^1$ in each case represents the *resultant* course followed by the salt shaker in response to action of vector forces $F_1$, $F_2$ and $F_3$ respectively.

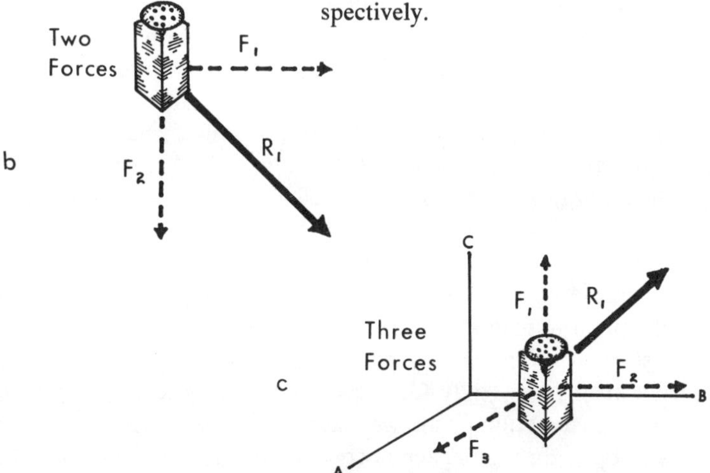


dimensions) as a result of the action of a force upon it, that object behaves like a vector quantity—that is, its motion has some measurable speed per unit time and it moves in a given direction so that we could represent its motion as in Figure 27a, with an arrow pointing in the direction of the object's motion and with the length of the arrow representing the object's speed per second. If another force acts upon that object perpendicular to its path of motion, that force could be regarded as a vector quantity also—a force of such and such a magnitude in a direction at right angles to the object's motion. As we see in Figure 27b such a "crosscurrent" vector can also be represented by an arrow, with the direction of the force shown in the direction the arrow is pointing and the length of the arrow representing the magnitude of the crosscurrent force.

When such a force acts, the moving object will shift its direction and speed until its motion perfectly balances the two vector quantities acting upon it, one seeking to keep it going straight ahead, the other seeking to make it go 90 degrees to the right as in Figure 27b. Thus we see that the object's new course can be represented by a new vector arrow which is the *resultant* of the combined but still competing forces of the other two.

Now suppose a third force acts on this object in the third spatial dimension, perpendicular to each of the other two forces. If the object were a salt shaker being tossed straight ahead (force 1) but also pushed off course by a crosswind from the left (force 2) the third force might be the force

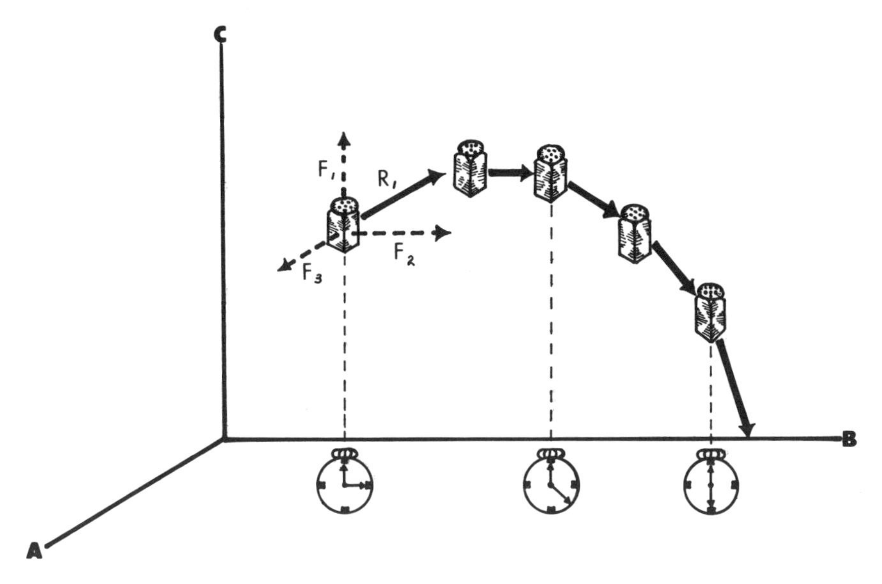


*Fig. 28* Continuous path of motion of the tossed salt shaker is shown as a succession of resultant motions, each different from the previous, as a result of changing forces acting on the object from instant to instant.

of gravity tending to pull the salt shaker down to the ground. This force also could be represented as a vector, and again the actual path of the object follows the course of a vector (or more accurately a succession of changing vectors) that is the resultant of the three combined yet competing forces acting upon the object, as we see in Figure 27c. Thus when we tossed our salt shaker across the room, it moved in an arc through all three spatial linear dimensions, but its actual path of motion at any given instant was the sum of the three vectors related to the three spatial dimensions: the height vector, the length vector, and the width vector. In Figure 28 we represent the path of the salt shaker as an arcing path made up of a succession of small arrows, each representing the resultant vector sum of the three dimensional vectors at uniform instants in time throughout the course of the salt shaker's journey. Actually, we know the motion of the salt shaker is not discontinuous, made up of a number of jerking changes of motion, but rather a smooth, continuously changing path—but if we could show a vector arrow for each of an infinity of instants during the salt shaker's flight, the path we have shown would be a fair representation of the salt shaker's flight.

But again, we see that the salt shaker could just as readily move in exactly the opposite course it takes in this three-dimensional flight through space if the forces were acting in the opposite direction. We would have to imagine a "force of gravitational repulsion" in order to see this, and from a practical point of view such a motion through space is unlikely, but the point is that there is no built-in restriction against its occurrence. The salt shaker could move either way along any of the three spatial dimensions with equal ease if the forces causing the motion so determined the salt shaker's flight.

Once again, this all seems pretty obvious, but by this time we have learned to be leery of these trains of thought that seem obvious on the surface; more often than not we find them to be based on a shaky assumption, or hedged with unexpected limitations. In this case we are fine as far as we have gone: objects at rest or in motion do indeed respond to the action of outside forces by moving quite readily in either direction in any of the three spatial dimensions, as long as they are moving at the ordinary low velocities we encounter in everyday life, and thus Newton's first law of motion describes their behavior splendidly. Within certain limits, that is. But now with a closer look we see that those laws of motion were all based upon a rather shaky assumption that we have not even noticed before. Although they do not say so, they refer only to the motion of objects *in space*—that is, within the framework of the three commonplace spatial dimensions. They say nothing about an object's motion *in time*. The reason Newton and the classical physicists neglected this important aspect of an object's motion was simply that the classical laws of motion were based on the assumption that the passage of time was an absolute—a fixed, unchanging constant which remained the same for all objects anywhere in the universe at all times. And, time being constant for all objects, and always the same for any given object, passage of an object through time was not even considered as a possibly changing aspect of its motion, and the laws of motion simply ignored it.

But now we see that that assumption may have been incorrect. As we have seen, time appears to be analogous to a fourth linear dimension—a dimension which is clearly *not* oriented in the same direction as any of the three spatial dimensions. Nevertheless, if time is a linear dimension, we see that an object's motion through time must occur in *some* direction. Even if we cannot quite visualize it because of the three-dimensional limitation upon our thinking, we see that an object moving along its time-track can also be regarded as a vector quantity going in a specific direction—the direction in this case being *forward in time.* But whereas we saw that our salt shaker could readily move in either direction in any of the three familiar spatial dimensions, that salt shaker's motion along the time dimension is *not reversible in any way.* In fact, it appears that the salt shaker, or for that matter any other object in the universe, is irrevocably *fixed in motion* at all times in a given direction in the time dimension, and that the time vector, that is, the velocity of an object moving in the forward-in-time direction, is quite impossible to reverse.

Why should this be? Why is it impossible to reverse an object's motion along its time coordinate when its motion can so easily be reversed along its length coordinate, its width coordinate, or its height coordinate? Is it possible that the difference lies in the fact that the object is *always in motion at all times* along its time-track, regardless of whether an outside force is acting upon it or not?

By applying what we already know about the characteristics of objects at rest and of objects in motion, we can see that this may indeed be the answer provided that certain things are true about the motion of an object along its time-track.

We know, for example, that an object such as our salt shaker moving through the air possesses *inertia of motion* related to its mass. If that salt shaker were moving in a straight line in one spatial dimension in a given direction, the only way its motion in that direction (a vector quantity) could be reversed would be to apply sufficient energy (in the form of some force acting upon it) in the opposite direction to overcome the salt shaker's inertia of motion, bring it to a stop, and then accelerate it again in the opposite direction. We also know that in the case of an object traveling at a relatively low velocity, the amount of energy necessary to reverse its direction of motion, or in any other way alter its pattern of motion, will always be proportional to the object's inertia, whether that inertia is expressed by the object's velocity or by the amount of kinetic energy the moving object possesses. If it were barely moving at all, only a slight amount of energy or a small force would be required to reverse its direction in space. If it were moving faster, it would take proportionately more energy, or a proportionately larger force to alter its course at all.

But suppose that the object were moving at a very high velocity through space. In that case, we have seen that some of the energy used to set it in motion at such a high velocity remains associated with the moving object in its "alter ego" of increased mass. Thus the object moving at a high velocity has a greater mass than it had when it was at rest or moving at a low velocity, and thus greater inertia. Consequently, a proportionately greater force would be required to slow that object down and reverse its direction of motion, with part of that force acting to overcome the inertia of the object's mass when it was at rest, and the remainder of the force required to overcome the *additional* inertia of the *additional* mass the object had acquired as a result of its high velocity.

Remembering our imaginary spaceship voyage, we can see what happens if we carry this to the extreme. If an object were moving through space in a straight line (that is, along one spatial dimension) with a velocity close to the velocity of light, its mass would have increased enormously, so that a comparatively enormous amount of energy would be needed to oppose it in order to reverse its direction of motion in that spatial dimension. If the object were moving at just a whisker less than the velocity of light, it would have taken on virtually all of the mass-energy in the universe, mostly in the form of mass, so that almost all the energy in the universe would be required to reverse its direction of motion. And if the object were moving with the full velocity of light, it would have infinite inertia so that *an infinite amount of energy,* from somewhere, would be required to oppose it and reverse its direction. But if an infinite amount of energy was needed to accelerate the object to light-speed, there could be none available anywhere to oppose its motion. The result would be another paradox: Under such conditions our object would be moving at a velocity of 186,000 miles per second along a spatial dimension *and its motion in that dimension would have become irreversible!*

What, then, about the motion of an object along the time dimension? If we are to consider time as a linear dimension analogous to the three spatial dimensions, then as we have seen, our salt shaker is always constantly in motion in a straight line in one direction along the time dimension even when it is completely at rest in all three of its spatial dimensions. We can even translate its motion along the time dimension into linear terms by saying that it is moving a certain distance along its time-track every second —the same sort of terms that we use to describe the velocity of an object moving along a spatial dimension. If the object is moving through space at a low velocity we can reverse its direction easily just by applying enough energy in the opposite direction to overcome the object's inertia and accelerate it backward. By analogy, then, we also ought to be able to apply energy to the object in the opposite direction of its motion along its time dimension in order to overcome its inertia of motion, bring it to a halt in its time dimension, and then send it accelerating backward through time.

Fine, but how much energy would this require? The amount of energy needed would have to be related to the object's velocity through time —the distance per second that it was moving forward along its time coordinate. Thus if the object's velocity through time were comparatively low, there should be no problem in reversing its direction in that dimension. If its velocity through time were quite high, a great deal of energy might be needed to reverse its direction. But we have seen that it has *never* been possible to bring sufficient energy to bear to alter the motion of any object through time in the slightest degree. And this fact would suggest that the velocity with which any object in the universe, however large or small, normally moves forward along the time dimension must be disproportionately huge compared to the same object's velocity (if any) along any of the spatial dimensions.

Then how huge *is* "disproportionately huge"? At first we might think that regardless of what an object's velocity through time might be we ought somehow to be able to apply enough energy to reverse its direction—but this has never occurred. Is there *any* possible velocity forward in time that would be so huge that no amount of energy could ever possibly bring an object moving at that velocity through time to a halt and send it moving backward? Of course there is such a velocity: the velocity of light, 186,000 miles per second. In fact, this is the *only* possible velocity with which an object might be moving forward through time that would render its motion through time totally unalterable and irreversible even if we had all the energy in the universe at our disposal to oppose that object's motion.

And now we see that the solution to the problem is obvious: The reason that no object has ever been reversed in its motion forward through time and sent moving backward in time is simply that everything in the universe is already racing forward in the time dimension at the maximum velocity possible. With every second of time that passes, everything in the universe has moved forward in time a distance of 186,000 miles!

No wonder, then, that no object's motion through time has ever been reversed! Any object, even the tiniest grain of sand, is already moving forward in time 186,000 miles each second; it must then already have infinite inertia of motion in the direction it is traveling in time forward, and its direction of travel can never be reversed simply because there is not enough energy in all the universe to reverse it. Furthermore, since everything in the universe is already moving in the forward-in-time direction at the ultimate limiting velocity that any material object can move, there is no conceivable possibility that any object's velocity through time can ever be increased. We could no more make an object move ahead into the future—that is, accelerate it to a velocity faster than light velocity—than we could muster sufficient energy to overcome its infinite forward-moving inertia and push it backward in time. Indeed, the very fact that nothing has ever been observed to travel from the present into either the future or the past only serves to confirm Einstein's contention that the velocity of light is indeed the maximum limiting velocity at which any object or signal can move in the universe, whether through the three linear dimensions of space *or* through time.

## TIME AND RELATIVITY

From the above, we see that the concept of time as a fourth linear dimension holds up splendidly, if we accept the necessary conclusion that objects must move forward in time with a velocity equivalent to the velocity of light. This concept of time confirms what has actually been observed in nature with regard to the behavior of objects moving through time. Of course, we may be going too far in assuming that the velocity of objects along the time dimension is equal to light velocity; all we can really be sure of is that that velocity must be very great, and since we have no way actually to bring huge quantities of energy to bear on an object in an attempt to alter the direction of its passage through time, there is no way that we can experimentally verify that objects move along the time dimension at 186,000 miles per second. The only experimental support for this idea is purely negative: no scientist has ever objectively observed an object moving either forward or backward in time.

On the other hand, this assumption does have the positive support of mathematics behind it. Mathematicians discovered that in any calculations involving velocities of moving objects (including calculations using the Lorentz transformation equations) the velocity for anything moving through time has to be 186,000 miles per second in order for the observed and measurable behavior of the object in motion in the three spatial dimensions to balance out. Furthermore, it is only by regarding time as a fourth linear dimension analogous to the three linear dimensions of space that the behavior of times makes sense in relativity problems, and it is in precisely this way that Einstein visualized time. Thus, we have seen that changes occur in the linear spatial dimensions of objects traveling at very high velocities: These spatial dimensions contract or shrink in the direction of the object's motion. But this shrinkage of the spatial dimensions is accompanied by an analogous contraction of the passage of time. If we think of such a high-velocity object moving through time as a fourth linear dimension as well as through three spatial dimensions, we see that the "slowing down of time" that Einstein predicted fits in just as logically as the shrinking of objects in their spatial dimensions.

From the first it seemed impossible to demonstrate experimentally that objects moved through time at the velocity of light; and for years, moreover, physicists had little hope of finding any experimental confirmation that the time scale of a moving object actually does slow down when the object is moving at a very high velocity. There was obviously no way to plant an observer on a fixed platform out in space so that he could watch the behavior of the clock aboard a spaceship racing by at the velocity of light.

Recently, however, quite unexpected experimental observations have been made which strongly confirm that such a "slowing down of time" does indeed occur in the real world, provided that an object's velocity is great enough. For many years physicists have been aware that a variety of tiny subatomic particles or high-energy electromagnetic waves commonly known as "cosmic rays" are constantly bombarding our atmosphere from outer space. These cosmic rays are known to reach the earth moving at velocities greater than 99.5% the speed of light and to collide with molecules of gas in our upper atmosphere with extremely high energy. Furthermore, these high-energy collisions between cosmic rays and atoms or molecules of gas high up in our atmosphere result in the manufacture of a peculiar kind of subatomic particle which had, until recently, never been observed on earth simply because most of these peculiar particles, known as "mu-mesons," immediately collide and react in turn with other molecules of atmospheric gas almost as fast as they are formed. A few, however, do reach the surface of the earth and in recent decades have been detected and studied by nuclear physicists.

These mu-mesons were found to be very unstable; once formed in a cosmic ray collision, they exist for only about two microseconds on the average before spontaneously changing into a different type of particle (assuming that they don't collide and interact with an atom or molecule sometime sooner). Yet physicists calculated that the mu-mesons that do reach earth's surface after they are created high in the atmosphere must take some *twenty* microseconds to travel from the place where they are formed down to the surface. But how can a particle with a known average lifetime of only *two* microseconds survive long enough to reach the earth's surface if the trip requires *twenty* microseconds, even with the particle traveling at enormously high velocities?

The answer is obvious: According to our measurements, the average lifetimes of mu-mesons that are moving at these high velocities are about ten times as long as they ought to be; the internal time-scale of these high-velocity particles has slowed down by a factor of 10. From the point of view of a mu-meson, however, (assuming that a mu-meson could have a point of view) its lifetime still seems to be the normal two microseconds, but it reaches earth's surface sooner than it should because the thickness of earth's atmosphere seems to the mu-meson to be contracted or shrunken to one-tenth the thickness that our measurements on earth estimate it to be.

Here again we have a relativistic paradox: Two sharply differing measurements of the same event made simultaneously by two observers moving at different velocities. Nor is there any way in the world for any third unrelated observer to tell which of the two observations is correct. By our measurements, the mu-meson's time-scale slows down; according to the mu-meson's measurements, our atmosphere is much shallower than we think it is. All that we can be sure of is that the mu-meson traveling at such a very high velocity does in fact survive a journey so long that it should need ten average mu-meson lifetimes to complete it, yet—impossible as the observed fact may be—it does indeed make that trip in one mu-meson lifetime and still sticks around long enough upon arrival for its existence to be confirmed before it turns into a different particle altogether!

Here, then, is at least one tidbit of laboratory evidence which can only be explained on the basis of Einstein's prediction that the time-scale of objects moving at velocities close to the velocity of light do indeed slow down, according to our measurements. Far out as this may seem, it is merely one of a multitude of experimental observations gathered over the past fifty years that tend to confirm that Einstein's relativistic concept of the universe is closer to the true picture than the classical concept that time and space are separate and distinct entities, and that objects have absolute dimensional characteristics under all circumstances. And soon we may have more direct experimental evidence of relativistic "time dilation." At this writing there are plans to launch an extremely accurate atomic clock into orbit—a clock so very accurate that even the tiny time dilation resulting from the satellite's orbital speed can be measured to two decimal places. If successful, this will be the first such time dilation measurement that has ever been made.

## THE DYNAMICS OF TIME

So far our discussion has been focusing on the *statics* of time—the concept of the movement of physical objects in the universe through time as a static, fixed linear dimension. This view of time is necessary and valid to help us understand that the special relativity theory predictions of slowing time-scales for moving objects is not really as absurd as it seems, and to understand why modern physicists regard our universe as a four-dimensional structure with three spatial dimensions and one time dimension all permanently interconnected with each other—in other words, to see why physicists regard the universe as a *space-time continuum*.

But all this still does not give us any clue as to the exact *nature* of time, nor is it necessarily the only fruitful way of thinking about time. Time may well be a static dimension through which all objects in the universe are constantly moving, but we are nevertheless totally unable actually to perceive this movement along the time dimension or measure it in any way. The best we can do is observe indirect evidences of it and say, "Well! Such and so has happened, and since such and so could never have happened unless our view of time as a static dimension were valid, then that view must be valid."

It is odd, therefore, that in our everyday lives we are much more accustomed to thinking of time as a *dynamic, moving entity* and of ourselves and the physical universe as the static entity past which time flows. Certainly we have the perception, or at least the illusion, of time passing. But how can we be sure that this elusive entity is not simply a product of our imaginations? What evidence do we have that time is passing? How can we even be sure that there is any such thing as time, or rather that there would be any such thing as time in the universe if there were no intelligent minds around to perceive it? Is time truly a built-in aspect of our universe that would exist whether we perceived it or not? Or is it another of these things like the world-ether which we have dreamed up simply because its existence seems to be necessary or convenient?

These are really not such vague philosophic questions as they may seem. If time is a built-in aspect of our universe, and if the passage of time has dynamic effects on the universe (as well as upon ourselves) whether we can perceive them or not, then changes ought to be occurring in the universe as a result of these dynamic effects—perhaps changes that are vitally important for us to know about. We wonder when and where the universe first came into existence and where it is ultimately going, and this speculation leads us to wonder if everything occurring in the universe may not be occurring as a result of the effect of time upon it. If the velocity of light is one absolute and unchanging constant in the universe, then perhaps the effects of time upon the universe are also absolute and unchanging. But if time does indeed exist as a built-in aspect of the universe and not something created from our own imaginations for our own comfort, what clues do we have that time is in fact passing?

There are, in fact, a number of very convincing clues that time is indeed passing. One of the most convincing of these clues is closely related to our old friend entropy. As we saw in an earlier chapter, energy exists in the universe in a variety of forms, and can be converted from one form to another, but of all its various forms energy in the form of heat tends to accumulate the most readily. We have seen that in almost any conversion of energy from one form to another at least some of the energy is converted to heat, whereas it is very hard to convert heat energy back into any other form of energy. As a consequence, the total amount of energy in any closed system is gradually being converted bit by bit into heat, with less and less all the time remaining in the form of kinetic energy, potential energy, chemical energy, electrical energy, or whatnot.

But we also know that the conversion of energy from one form to another requires more than just the interaction of energy-bearing forces. Any conversion of energy from one form to another, regardless of what those forms are, requires the *passage of time*. Since any conversion of energy requires some interval of time in which to occur, and since it is known that throughout the universe energy interactions of all sorts do occur and lead to a steady increase in entropy, we can see that this demonstrable increase in entropy through the years, months, and centuries is actually a *direct function of the passing of time*. If time were not passing, entropy in the universe could not be increasing, and since we know it is, we therefore have a surprisingly solid bit of evidence that something which we call "time" does indeed exist and is indeed passing.

A second clue, of course, lies in our own subjective perception of the passage of time (as opposed to some mechanical measurement of the passage of time) determined by our individual biological time clocks. The fact that we encounter discrepancies between our individual biological perception of time and the passage of time as measured by some kind of mechanical system suggests that one or the other must be an erroneous measurement, and we concede that our subjective perception of time may indeed be demonstrably erroneous. But at the same time these very discrepancies suggest that there *is* such a thing as time passing which can be measured, whether erroneously or not.

Finally, there is an even more valid reason to believe that time does indeed exist in the universe whether intelligent minds perceive it or not: the simple fact that intelligent creatures do not tolerate sensory deprivation very well at all. In recent experiments well-trained healthy human subjects have been exposed to conditions in which all sensory contacts with the outside world were deliberately obliterated. For example, human beings have been experimentally suspended in tanks of water, in total darkness and in complete silence, with even their fingers so thoroughly bandaged up that these human guinea pigs could neither feel with them nor move them. The purpose of these experiments was simple: The experimenters sought to find out what would happen to human beings when they were as completely deprived of all sensory information about their environment as possible.

It was expected that presently, after a prolonged period of sensory deprivation such as this, any individual's mental processes would begin breaking down. But the results of these experiments far exceeded expectations. The investigators were amazed to discover that even the best-trained and best-prepared individuals could not tolerate such total sensory deprivation even as long as 72 hours, and usually began suffering mental disturbances (fortunately reversible) in as little as 12 to 24 hours. In recounting their experiences, these individuals reported a pleasant, even comfortable sense of isolation for a while, but this pleasurable feeling soon gave way to irrational panic. These individuals seemed to be "losing track of time" and soon found their minds creating purely hallucinatory colors, sights, and sounds and irrational emotional reactions in an effort to substitute for normal sensory input. One individual reported an overwhelming paranoid conviction that the experimenters were going to leave him there to go mad, although he knew he could signal them to release him at any moment. He even imagined that he had pushed the signal button repeatedly, with no response, and was fixedly convinced afterward that he had been "down in the tank" for weeks when the actual time had been only 17 hours!

So far no one is quite sure how to interpret such early appearance of mental confusion under these conditions of sensory deprivation, but it is quite possible that the thing that upset these individuals more than anything else was their utter inability to find any clues whatever as to the passage of time. In other words, it is quite possible that an intelligent human being's continuing moment-by-moment awareness of his own movement through time, or of time passing him by, may be utterly vital to the maintenance of mental stability and healthy mental function. And if so, there must be some true time "entity," of the existence of which all human beings are continuously aware.

Today most physicists agree that the passage of time is indeed a true and "built-in" attribute of the universe, existing whether intelligent minds perceive it or not, existing before any intelligent minds had evolved in the universe and destined to continue to exist long after intelligent minds have vanished from the universe. This is not to say that time necessarily exists or passes in term of hours, minutes, seconds or any other particular units we might choose to measure its passage. These are merely arbitrary units that have been devised for measuring time, and we could just as well use any other units we chose. Thus, while it is agreed that time exists, the units we use for clocking the passage of time are not built-in characteristics but only arbitrary scales men have set up to divide the passage of time into convenient and comprehensible segments.

But if there is agreement that time exists, there is considerable disagreement with regard to some of its characteristics. We saw earlier in this chapter that it is impossible for any person or object to move either backward or forward in time by even the tiniest amount; everything in the universe moves forward in time at a constant, prescribed velocity. This means, of course, that every person or object at any given instant in time exists *only* in the present, unable to move back into the past or ahead into the future. In fact, we could define "in the past" and "in the future" indirectly by stating that any event which we can find out about through observation has necessarily occurred "in the past," while any event which we are still, at least in principle, able to alter in some way must lie "in the future." But if we exist at each instant only in the present, what exactly is "the present"? How long is "an instant"? Is the passage of time a steady flow or is it divided up into a succession of some kind of "smallest possible units"? In other words, is time *continuous* or *discontinuous?*

## THE MYSTERIOUS "CHRONON"

It might seem that there is a pointlessly fine line of distinction between time moving in a steady stream and time broken up into immeasurably tiny units, until we consider more closely what these terms "continuous" and "discontinuous" really mean. We normally think of time flowing continuously without any jerks or breaks in its motion—but do we really mean that time is *continuous*? If we do, this means that each instant of time must be infinitely small. We would then have to think of "the present instant" as a moving, dimensionless interface between all the time in the past and all the time there is to come in the future: an interface such as we might find between oil and water. But such an interface could have no thickness or any other measurable characteristic other than the fact that it is there. Such an interface could have no dimension whatever either in space or in time. But if "the present instant" is such an interface between past time and future time—one which has no dimension at all—then we are saying that "the present instant" has no duration in time whatever and therefore cannot exist at all except as an infinitely thin (i.e., dimensionless) plane of cleavage between past and future.

Thus if we agree that time is indeed a continuous flow, as it appears to us to be, we are automatically trapping ourselves into saying that *there is no such thing as the present instant* as far as duration in time is concerned. Yet if we cannot exist in the past nor move ahead into the future, and if there is no "present instant" in which we exist either, then where *do* we exist in relation to time? If, on the other hand, there is some "smallest possible unit of time" that has any finite duration at all, and if we call that smallest possible unit of time "an instant," this would mean that time must flow in a *discontinuous* fashion as a homogeneous but interrupted stream or succession of these tiny time intervals. If this is the case, why does the flow of time *appear* continuous?

An analogy will help clarify things. We are accustomed to seeing the apparently "continuous" motion of images on the motion picture screen; yet we know that this is merely an illusion of motion created by moving a succession of individual still pictures past the lens of the projector so rapidly that our eyes and minds cannot separate or distinguish the individual pictures. Incapable of registering each frame of the motion picture frame individually, our minds simply *interpret* the succession of images we see as moving continuously across the screen, even when we know this is not what is really happening at all. Physicists recognized that "the present instant" *had* to have some duration in time, and thus concluded that time does flow in a discontinuous fashion, instant by instant, but that the intervals are so very tiny and moving in such rapid succession that the flow of these successive instants of time *appears* continuous to us.

The next question, of course, is obvious: What is the duration in time of that "shortest possible instant of time"? How long does it last, and how can the pause from one instant of time to the next be measured? Is all motion in the whole universe occurring like a strip of motion picture film moving through a projector, one frozen frame followed by another frozen frame in such rapid succession as to create the illusion of smooth homogeneous motion? If a given object is moving at a velocity of 186,000 miles per second along the time dimension, how far does it move during this briefest possible instant?*

* You will notice that this question is very similar to questions we have already raised twice before regarding other natural phenomena. In the early 1900s physicists were questioning whether the motion of light through space was continuous or discontinuous; did light flow in a continuous wavelike motion, or as a succession of

Physicists could agree that if the flow of time was divided into "shortest possible intervals" each with a tiny but finite duration, then the duration of those intervals should be subject to investigation. It may have been a harried physicist who first proposed that the shortest possible interval of time might be defined as the period of time that elapsed between the instant the stop light changed and the moment the car behind you began honking its horn. Certainly any "smallest interval of time" would have to be of very short duration indeed, and many physicists became more and more convinced that the flow of time was indeed fragmented into a rapid-fire succession of such tiny individual units following each other in a steady stream. But the problem of defining these mysterious hypothetical "tiniest possible units" of time has so far proved insurmountable. Champions of the hypothesis spoke of these "tiniest units" of time as "chronons" (from the Greek *chronos* meaning "time") but so far no known event in nature has been successfully matched with the name. And there are physicists today who maintain that no such thing as a chronon can exist. Certainly any event in nature with a duration taken as the "tiniest possible unit" of elapsed time will have to face the physical paradox of relativistic time dilation with high velocities, in which time is not a constant but a relative measurement, before it can ever be accepted as anything but vague speculation.

Thus we see that with the development of the special theory of relativity by Einstein and others, a totally abstract concept—the concept of the passage of time in the universe—was given concrete and visualizable dimensional structure. Physicists recognized that nothing could happen in the universe independent of the passage of time. On the contrary, time was so closely interlocked and interconnected with all events in our universe of three-dimensional space that it became necessary to think of the universe not structured merely in three spatial dimensions with time passing by like the "ether wind," but as an intimately connected continuum of

---

tiny individual pellets? At the same time, physicists were also pondering whether the flow of electricity through wire was continuous or discontinuous—whether electricity was a wave phenomenon, so to speak, or a succession of tiny individual units of electricity moving along one after the other so swiftly as to create the illusion of a continuous flow of current. After much study, Maxwell hypothesized that electricity was discontinuous, composed of a succession of tiny charged particles which he called "electrons," and all subsequent investigations of electricity supported this view. As for the problem of light, physicists before Einstein had found a great deal of evidence that light travels as a continuous flow of waves; yet as we shall see, Einstein himself reported one natural phenomenon—the photoelectric effect—which could be explained only by assuming that light, in spite of its wavelike properties, also traveled through space in discrete "chunks" or "packets" which became known as "photons." Thus it is not quite as strange as it might seem to question whether the flow of time is continuous or discontinuous.

space and time in which the time dimension, just like the spatial dimensions, is subject to relativistic change according to the speed of one observer relative to another.

Almost as soon as it was first expounded, Einstein's theory of special relativity began reshaping physicists' concepts of the way things behaved in the universe and forcing modification of some of the most tried and true natural laws that had ever been worked out. But the theory of special relativity did not cover enough ground. It could be applied only to objects moving in straight lines at constant velocities—circumstances which were rarely indeed found to apply to real objects in the real universe. From the very first Einstein realized that he would ultimately have to extend this "theory of relativity for the special case of constant velocities" to cover *all* possible kinds of motion occurring at *all* kinds of varying velocities. By 1936 Einstein had extended his relativity theory to apply to general and realistic situations of increasing and decreasing acceleration of objects, and again caused physicists all over the world further to alter their concept of the way things behaved. In particular, Einstein's theory of general relativity forced scientists to take a new look at Newton's law of universal gravitation, and at the same time to revise completely their deeply entrenched ideas of the geometry in which the continuum of space and time is constructed. But most important of all, it led physicists to ask questions about a universe which lay far beyond the universe that men had so far come to know through observation and experiment—a universe so incredibly large and so incredibly strange that human imagination could hardly encompass it.

# CHAPTER 17

## *The House That Einstein Built: General Relativity*

There is a very common impression among nonscientists that the theory of special relativity arose in some divinely inspired fashion, cut out of whole cloth through the efforts of a single man. According to this contention, if Albert Einstein had never been born, then this whole confusing and perplexing new way of regarding the universe might never have been dreamed up at all.

It is not surprising that such an idea prevails, however erroneously. Einstein did indeed develop his special relativity theory virtually single-handedly. He was somehow capable of seeing further than other physicists of his day could see, and led the first spearhead thrust against the old, inadequate classical view of the universe quite by himself.

Even so, however, the chances are very good that if Einstein had not come up with the basic principle of special relativity when he did, someone else would have come up with it in virtually the same form and from virtually the same approach within a matter of a few years. Daring and revolutionary as Einstein's ideas were, they received the excited reception that they did because *the climate was right* for this theory throughout the world of physics. Physicists had come to an intolerable dead end, and a complete reappraisal of underlying assumptions had become vitally necessary. Other men like George Fitzgerald and Lorentz were thinking along the same lines; Einstein was merely the first with enough confidence in his own logic to be able to take the few necessary further steps that special relativity demanded.

In doing so he performed a feat that was destined to become increasingly familiar in the decades to come. He developed a profoundly revolutionary hypothesis on purely mathematical and theoretical grounds. Einstein was not an experimental physicist; he left the experimental proof of the predictions of his special theory of relativity up to others; yet as the years passed, even some of the most bizarre of these predictions were subsequently confirmed again and again by actual experimental observation in the laboratory. When the special theory of relativity was first announced in 1905, its impact on the world of physics was mainly theoretical, mathematical, and philosophical; it seemed to have little relationship to the "real" world that we live in. But in the ensuing years, as physicists struggled to understand both the structure and function of atoms in the microworld and the overall structure of the enormous macro-universe beyond our immediate solar system and galaxy, special relativity came to be recognized as one of the most practically useful and best-substantiated theories ever to appear in the history of physics.

But the theory of special relativity was limited in its application. Einstein knew from the first that it could make reliable predictions in regard to just one special kind of motion: the motion of objects or electromagnetic waves traveling in straight lines at constant velocities. Even though it was a basic assumption of the theory that light waves and other electromagnetic waves always did travel through space at a very high constant velocity, Einstein recognized that this type of motion was by no means the most common pattern of motion in the universe. In fact, except for the rather special case of high-velocity electromagnetic waves traveling through space, completely straight-line motion at constant velocities was virtually nonexistent in the observed universe. Rather, objects or particles nearly always were found to be moving with either increasing or decreasing velocities—or, as physicists preferred to describe it, moving with *positive acceleration* or *negative acceleration.*

Einstein picked the special case of motion at constant velocities as a starting place largely because it was the simplest form of motion upon which to try to build theoretical predictions. Once special relativity was worked out to his satisfaction, he then turned his attention to the relativity characteristics of accelerating objects, hoping to extend his "relativity in the special case of constant velocities" into a broad general description of the relative motion of *all* moving objects or particles, whether they were moving with constant velocity, undergoing positive acceleration (forward acceleration or speeding up), or negative acceleration (deceleration or slowing down). In approaching such a "general theory of relativity," it was logical to begin by examining the nature of a force long known to cause moving objects to accelerate or decelerate: the force of universal gravitation which Sir Isaac Newton had so clearly defined as a force of mutual attraction existing between any two objects in the universe that possessed mass.

And once again Einstein demonstrated that another of the most elegant and firmly established of all natural laws, Newton's law of universal gravitation, was not really universal in its application at all but had to be modified if it were to describe accurately all kinds of accelerated motion of objects anywhere in the universe. Indeed, he forced the world of physics to stop and reconsider the entire classical picture of the geometrical structure of the universe, and to take a completely new look at the question of what the force of gravity was and how it really did affect moving bodies.

## IN WHICH GRAVITY IS FOUND PERHAPS NOT TO BE GRAVITY AFTER ALL, BUT RATHER, ACCELERATION, AND VICE VERSA

After the dismal failure of the Michelson-Morley experiment to demonstrate absolute motion through a fixed "world-ether," and after the publication of the theory of special relativity, physicists recognized that uniform and unchanging motion of any object had to be considered relative and not absolute. In other words, an object moving at a constant velocity (neither increasing nor decreasing in velocity as it moved) might be shown to be in motion relative to any number of other moving objects in the universe—relative to the surface of the earth, for example, or to the movement of the earth around the sun, or even to the movement of the galaxy in rotation around its own axis—but it could never be shown to have absolute motion relative to any fixed reference point, no matter what kind of mechanical or optical experiment were used to try to demonstrate such absolute motion.

Obviously the next step was to try to extend or generalize this theory of special relativity to include *nonuniform* motion as well—the motion, for example, of an object which is speeding up or slowing down. If such a generalization could be made, then all kinds of motion in the universe, whether uniform or nonuniform, would have to be considered relative rather than absolute.

Unfortunately, in studying what happened to an object when its motion was accelerating or decelerating, physicists soon realized that they had to deal with an additional observable effect that was not present at all when an object was moving at a uniform velocity in a straight line: the effect of inertia.

There was no getting around the fact that in any accelerating system (just another way of saying "in any case of an object undergoing positive or negative acceleration") the effects of inertia play an important part in what happens. We all encounter the effects of inertia in our everyday lives any time we become involved in speeding-up or slowing-down types of motion. When a driver accelerates his automobile from a standstill to 60 miles per hour in a quarter of a minute, he feels a powerful force thrusting his whole body back against the seat of the car. The more rapidly he accelerates, the greater the force he will feel squashing him back into the driver's seat. Then when he reaches 60 miles per hour and takes his lead foot off the accelerator, his *accelerating* motion suddenly changes into *uniform* motion at a constant velocity of 60 miles per hour, and this backward force of inertia is suddenly gone. If the driver then accelerates negatively (i.e., decelerates) by breaking hard, he will feel a very similar inertial force thrusting him forward against his seat belt (assuming he is wearing one) just as vigorously as it was previously shoving him back against the seat; and this forward-thrusting force will continue as long as the moving car continues to decelerate. The moment he takes his foot off the brake, however, the negatively accelerating or decelerating motion of the car again is suddenly changed into uniform motion at some constant velocity—say 30 miles per hour—and the forward-thrusting inertial force again suddenly vanishes.

We are so accustomed to the effects of inertia in our everyday experience that we rarely even notice it except when it happens to catch us unaware and unprepared. It is only the novice unused to the city who lurches forward off balance when the subway car or bus he is riding slows down for a stop, and then is hurled back the other way when the train or bus starts up again. Regular subway riders have long since learned to brace themselves against both these effects of inertia. This inertial force is precisely the same thing that was described in Newton's first law of motion as the tendency of an object to remain either at rest or in uniform motion and to resist any force acting upon it to change its state of rest or uniform motion. Thus inertial effects appear only *when a change occurs* in an object's state of rest or motion, and then under such circumstances the inertial effect always seems to be *opposing* the change in motion, no matter in what direction that change in motion is taking place.

We have also seen that this inertial effect is a direct function of the mass of an object. When the mass of an accelerating object is very small, the inertial effect is minimal, whereas it is proportionally greater when the mass of the accelerating object is greater. In fact, we have seen that measurement of the inertial effect on an accelerating object is a way of accurately measuring that object's mass, even somewhere deep in space beyond any gravitational field where weighing of the object can be of no help in determining its mass.

It is interesting to note that Newton's first Law of Motion—the law of inertia—did a splendid job of describing what *happened* to an object when some force made it accelerate or decelerate, but it did *not* explain *what this inertial force was*. Nevertheless, the law of inertia seemed to imply that this was one way to prove that at least one kind of motion—accelerating motion—was absolute and not relative. When a driver feels himself being pushed back in the seat by an inertial force while he is accelerating his car, is this not proof positive that he is indeed moving forward? Wouldn't the inertial force tell him this even if he were blindfolded and unable to pick up any other clue to show him that he was moving forward? If accelerated motion were relative in the same way that uniform motion is relative according to the special relativity theory, we could just as well say that the man in the accelerating car was at rest while the earth and the entire cosmos around him were moving backwards—but if that were true, how could we explain this mysterious force pressing the man back against the driver's seat? Indeed, the very fact that he can actually *feel* the inertial force pushing him back would seem to prove beyond any possible question that it is the car and the driver that are moving forward and not just the rest of the universe moving backward.

Another example will help clarify this point. We are all familiar with the action of a common laboratory centrifuge, in which two balanced test tubes are whirled around a central axis at high speed. While the tubes are whirling around, centrifugal force (which is nothing more than the inertial force accompanying rotational or angular acceleration) will force heavy substances carried in the tubes out to the periphery of the centrifuge circle. Thus, if muddy river water is placed in the centrifuge tubes and the machine turned on, the muddy sediment will be forced by inertia to the bottom of the test tubes. If the river water happens to contain three or four substances with different masses, these substances will be separated out into layers by the centrifugal force, with the most massive substance thrown down to the bottom of the centrifuge tube, the next massive substance forming a layer above it and so on.

We would certainly think that this effect of inertial forces, which we can easily demonstrate with any laboratory centrifuge, ought to constitute positive proof that the angular motion of the centrifuge is absolute! We could, of course, argue that perhaps the arms of the centrifuge are in fact standing still, absolutely at rest in space, while all the rest of the universe is rotating around them in the opposite direction—but in that case, why would the muddy sediment be driven out to the ends of the test tubes?

To Isaac Newton, this line of reasoning seemed to constitute undeniable proof that the angular motion of a centrifuge *had* to be absolute. He could see no other way to explain the observed effects of inertia. This was also a conundrum which puzzled Albert Einstein for ten whole years after his special theory of relativity had been published. While scientists could agree that *uniform* motion might be relative, as the special theory asserted, the presence of observed inertial effects in cases of *accelerated* motion seemed to force the conclusion that accelerated motion was absolute.

To Einstein this was a disturbing notion. He could not believe that there could be one group of natural laws governing one special form of motion in the universe—uniform motion at constant velocities—and a completely different group of natural laws governing accelerated motion. He was convinced that all kinds of motion anywhere in the universe had to be relative, and searched doggedly for some way to extend his special relativity rules to account for accelerated motion or any other kind of motion, not just uniform motion at a constant velocity. The search took him ten years, but the solution he finally came up with was both so simple and so shocking that it made young physicists shout with excitement and old physicists weep.

How did Einstein approach the enigma of accelerated motion? It seemed certain that the answer had something to do with inertia or inertial effects. He therefore began by carefully reexamining everything that classical physics had to say about inertia and the effects of inertia on accelerating systems. There seemed to be no loophole in Newton's reasoning—but one thing about inertial effects struck Einstein as very odd. There seemed to be a strange similarity between the effects of inertial forces on moving objects and the effects of yet another familiar and powerful physical force: the effects of gravity. It had been known for centuries that if an object suspended above the ground were released, the pull of gravity exerted on the object would cause it to fall toward the earth with a constant acceleration of 32 feet per second every second. Galileo had proved beyond question that this rate of acceleration for any falling body was always the same, regardless of the mass of the object that was falling. But although he could demonstrate that this was true, Galileo could not explain why a light wooden ball and an iron cannonball ninety times as massive would fall side by side; it certainly *seemed* that the pull of gravity on the cannonball ought to have been ninety times greater than the pull of gravity on the wooden ball.

Newton, of course, explained this apparently strange behavior of gravity by pointing out that the rate of acceleration of any object in a gravitational field was opposed by that object's inertia—that the tendency of the object to remain at rest suspended in the air and to resist the downward pull of gravity when it was released held the object back from accelerating downward (just as we might say that the driver's inertia, related to his mass and thrusting him back against the driver's seat as he accelerates his car, is really acting to "hold the car back" from accelerating forward). Since the inertia of an object resisting the acceleration of gravity would be greater the more massive the object, Newton explained, the iron cannonball would have ninety times the inertial resistance to accelerating downward under the force of gravity that the wooden ball would have, so that the two balls would have to fall side by side.

This is merely another way of saying that the force of gravity acting on an object is always proportional to the object's inertia. If object A is four times as massive as object B, its inertia is also four times as great, so that four times as much force would be needed to accelerate it to a certain speed as would be necessary to accelerate object B to the same speed.

If we now look again at some of our earlier examples of the effect of inertia in cases of accelerated motion, we begin to see some very odd similarities between the gravitational force tending to pull an object down toward the center of the earth on the one hand, and the effect of that object's inertia tending to resist or fight against any acceleration of that object from its previous state of rest or uniform motion. For example, we can see that the force of inertia pressing the driver back into the driver's seat while the car is accelerating seems very similar to the force of gravity which might pull the same man down into the cushions of a reclining chair. In fact, if the driver were accelerating his car at the rate of 32 feet per second each second—a rate of acceleration precisely the same as the rate of acceleration of an object downward due to the pull of gravity if the object is allowed to fall freely—the inertial force the driver of the car feels would be *of exactly the same magnitude* as the force he would feel gravity exerting upon him if he were resting in a reclining chair.*

In the case of the centrifuge, the particles of mud in the murky river water "fell" to the ends of the centrifuge tubes with a certain rate of acceleration during rapid rotation, and this acceleration of the mud particles due to inertial force driving the particles out to the periphery of the whirling centrifuge circle appears very similar to the acceleration of the mud particles we would observe if they were being pulled to the bottom of an upright test tube by the force of gravity.

Indeed, we can see that the angular motion of the centrifuge has the effect of creating a local and artificial "gravitational force" that pulls the mud particles out to the periphery. And if we strain our imaginations a bit and look at the whole business from the point of view of one of those mud particles, there might be no way that a given mud particle could tell *which* force was acting to pull it to the bottom of the test tube—the force of gravity or the force of inertia.

The same line of reasoning has been followed by modern space engineers trying to find some way for astronauts on some future manned space station in orbit around the earth to avoid the awkward effects of weightlessness or "free fall" as they go about their daily chores. One possible solution to this problem would be to make the whole space station spin like a wagon wheel around an imaginary axis as it turns in orbit around the earth, so that the centrifugal force of the space station's angular motion would cause all of the men and other objects inside the space station to be pressed against the outer edge. If such a station could be rotated at precisely the right speed so that the inertial force acting upon astronauts and other objects inside at the rim of the wheel exactly equaled normal gravitational pull, the astronauts could then walk and move comfortably around the inside of the space station's rim. It would seem to them that they were walking on a curved floor, and they might get dizzy watching the sun, the moon, the earth, and all the stars apparently whirling madly by the windows, but otherwise it would *appear to them* that the space station was in a gravitational field just as strong as that upon the surface of the earth. When they dropped things, the objects would "fall" to the "floor," and if

* It was precisely because of the close similarity between inertial force and gravitational force that the inertial force encountered by aviators and later by astronauts during acceleration came to be measured in "gravities" or "G's"—multiples of normal gravitational acceleration. Today we know that astronauts are subjected to inertial forces equivalent to about seven "gravities" throughout the time that an Apollo rocket is accelerating to escape-velocity during blast-off.

they weighed themselves on the scales they would seem to weigh the same as on earth. In fact, *all* the effects of a normal earthly gravitational field would seem to be present.

## THE MYSTERIOUS "ELEVATOR EFFECT"

Thus we can see that in an accelerating system the effects of inertia can literally "mimic" the effect of gravity. But under certain circumstances, it can even *counteract* gravity. The classical example of this is the case of the unfortunate man standing in an elevator at the top of a tall building when the cable breaks and the elevator begins falling uncontrolled toward the bottom of the shaft. If we can ignore for a moment the poor fellow's plight when the elevator hits bottom, we can see that while the elevator is accelerating downward, the apparent effect of gravity on the man inside the elevator cab would be temporarily eliminated. From his point of view, he would be quite weightless all the time that the elevator was falling. Instead of having his feet firmly planted on the floor of the elevator by the force of gravity, both he and the elevator would suddenly be accelerating downward at the same rate, so that he would suddenly find himself floating free and topsy-turvy in the air. He would be in exactly the same situation as a mountain climber who falls off a cliff, or a sky diver before he pulls the ripcord of his parachute. The apparent "elimination of gravity" experienced by the man in the falling elevator is precisely the same as the "weightlessness" experienced by astronauts in orbit around the earth; the orbiting space capsule is constantly falling toward the earth under the accelerating pull of gravity, just as a cannonball is constantly being pulled toward the earth throughout the course of its flight, except that the space capsule is moving at such a high velocity horizontal to the surface of the earth that the centrifugal force tending to hurl the capsule out into space in a straight line exactly balances the earth's gravitational pull upon it. When a space capsule is in such a state of "free fall," gravity seems to have been eliminated within the space capsule and the astronauts have no weight.

Of course in the case of the man in the falling elevator, as in the case of the astronauts, we know that nothing has changed the gravitational field tugging both vehicle and passengers toward the center of the earth. But in each of these cases the *effect* of earth's gravitational force has disappeared, at least as far as the passengers can determine. As far as the man in the elevator is concerned, the moment the cable breaks and the elevator begins falling, the gravitational force holding him to the floor of the elevator suddenly vanishes—as far as he can tell, suddenly *there is no gravity*. We recall that the inertial force pressing the driver back into the driver's seat while he is accelerating also suddenly vanishes, in exactly the same fashion, the moment he stops accelerating the car or increasing its velocity and allows it to assume a constant uniform velocity forward.

This, of course, raises a rather curious question. When the cable breaks and the elevator with its passenger starts plunging downward at a rate of 32 feet per second every second as a result of earth's gravitational attraction, *does gravity then exist for the man in the elevator or doesn't it?* Someone observing this disaster from outside the elevator would say, "Of course gravity exists; you can see it pulling both the elevator and the man inside it downward with constantly increasing acceleration, and heaven help them both when they hit bottom." But if we ask the man inside the elevator the same question, he might well say, "Gravity? What gravity? Gravity has suddenly vanished. See—I can float!" as he does a double somersault in the air and ricochets from one side of the elevator to the other.

Then who is right—the outside observer or the man inside the elevator?

We can sit and whittle away at this conundrum indefinitely, but we will not come up with any answer. Each observer is 100 per cent right, *from his own point of view.* This was the conclusion that Einstein was finally forced to accept when he had worked his way through this problem. *No one* could say with absolute certainty *which* was right; it all depended upon the viewpoint.

But Einstein carried the rationale a step further in a famous "thought-experiment." He said, in effect, "Let's turn the situation around, and see where it takes us. Suppose that this elevator, instead of falling, is being towed upward through empty space by a cable suspended from some kind of celestial sky hook, so that the elevator is accelerating upward with its velocity increasing 32 feet per second every second. Suppose further that this imaginary elevator is accelerating in some area of empty space far from any measurable gravitational field whatever. But there are no windows in the elevator, and no one has remembered to mention to the passenger that his elevator has left the confines of earth. With the elevator accelerating through space at this rate, what would the passenger be able to observe then?"

Einstein pointed out that the answer would have to be that the passenger might just as well be standing in a closed closet at home on earth, for all that he could tell. Far from being weightless under these circumstances, his feet would be firmly planted on the floor of the elevator. His inertia resisting the acceleration of the elevator would press him down to the floor with precisely the same force that gravity would press him down to the floor if he were on earth. If he should happen to slip in this accelerating elevator, he would fall on his face just as he would on earth. If he happened to drop an apple, or a ball, or a toothpick, or anything else, it would fall to the floor precisely as if it were acting under the influence of earthly gravity.

In fact, if no one told him that he was really aboard a crazy elevator being towed upward through space at a constant acceleration, and if there were no windows through which he could see the stars and planets moving by in the distance, you might have considerable trouble convincing this man that he was not merely standing in an elevator that had stalled between the seventieth and seventy-first floors of the Empire State Building. Still more confusing, if you were there with him trying to convince him, there would seem to be *no experiment of any kind* that you could perform inside that accelerating elevator that would demonstrate that he was wrong! (Actually, there *is* one, but in a room so small the results would not be very convincing, as we will see later.)

It was just such a thought-experiment as this which led Einstein to the simple yet shocking basic idea of his general theory of relativity. It was clear from this simple and logical thought-experiment that the effects of inertia could *not* be relied upon to prove that accelerated motion was absolute. There was *no possible way* that the man in the accelerating elevator could tell that he was moving at all on the basis of forces he could feel, because under these circumstances the inertial forces that kept him planted to the floor of the elevator would be, to him, virtually indistinguishable in any way from the commonplace pull of gravity to which he was accustomed all his life.

But suppose something happened to *increase* the acceleration of this celestial elevator. Surely then the man inside would know that something strange was afoot. An outside observer watching from a distance would say, "They must have stepped up the power; that elevator is accelerating faster than it was before." But what about the passenger inside? All he would notice would be that he suddenly seemed to weigh more and would find it more difficult to lift his feet and move around. "That's odd," he would say. "Something must have happened here on earth to increase the pull of gravity! Somehow our planet must have moved into some new and stronger gravitational field!"

On the other hand, if the towing cable suddenly broke so that the elevator ceased to accelerate and merely continued moving upward at a constant velocity, the outside observer would say, "The elevator has stopped accelerating. Therefore, the inertial force pressing the passenger against the floor of the elevator has disappeared, and look! He's floating weightless, since there isn't any gravitational field anywhere near." As for the passenger in the elevator, spinning head over heels and bouncing from one wall to the other, suddenly weightless with no explanation, he might say, "Help! The cable's broken and the elevator is falling down the shaft! Somebody stop it before we crash!"

So who would be right—the distant observer or the passenger in the elevator? Obviously, each is just as right as the other from his own point of observation. Isaac Newton himself could not have answered the question, but he was a brilliant enough man that if this thought-experiment had ever occurred to him he would have become very uneasy indeed trying to distinguish what was inertial force as a result of acceleration and what was gravitational force as a result of the gravitational attraction between two massive bodies. Albert Einstein's answer to the question was a quiet blockbuster: *Both* observers were equally right, for the simple reason that *both were describing the self-same and identical effect.* They were merely using different terms and expressions to describe that effect.

Indeed, it was no wonder that gravity and inertia seemed to have similar effects, Einstein contended. They *had* to have similar effects, because they were precisely and identically equivalent, literally one and the same thing. This was not a matter of two different forces having similar effects under certain special circumstances. The reason that the effect of inertia and the effect of gravity could never be positively distinguished one from the other with certainty was simply that gravity and inertia are *two different words for exactly the same thing.*

## THE IMPLICATIONS OF EQUIVALENCE

What does this so-called principle of equivalence—the notion that gravity and inertia are identical and equivalent in every way—imply? For one thing, it implies that *any kind of motion, including accelerated motion, must be relative*; there can be no such thing as "absolute motion"—motion of an object that will always and invariably be observed to be the same by any possible observer. Thus even the *source* of the force acting on our elevator passenger cannot be identified for certain. A little way back we imagined that some kind of celestial tow rope was causing the elevator to accelerate through space, thus generating an inertial force that kept the passenger's feet pressed to the floor. But we could just as well imagine that the elevator was sitting dead still somewhere out in space, and that the entire universe around it was moving past it in the opposite direction with accelerating speed. The man inside the elevator would observe the same thing as before; he would be held to the floor of the elevator exactly as if he were in a gravitational field.

But in this case we would have to agree that it was the accelerating motion of the universe past the motionless elevator that generated a gravitational force inside the elevator. In such a case, there can be no point whatever in asking ourselves which is really happening. This would not even be a proper or "acceptable" question, since there is no "real" absolute motion of any sort, only motion of the elevator relative to the universe or vice versa. The "gravitational" or "inertial" force felt by the man inside the elevator, whatever the nature of that force, must therefore be the result of some kind of force field which is generated by the motion of any accelerating object relative to the universe, or vice versa. In some situations we see, feel, or interpret this force as an "inertial force." At other times we see, feel, or interpret the force as a "gravitational force." Which it *really* is depends entirely upon whether we want to think of the elevator as fixed in space with the universe accelerating past it, or of the universe as being motionless with the elevator accelerating through it. Since it is generally more comfortable for us to think of the universe as being motionless, with various objects moving around in it, we would generally describe the force field generated by the relative motion of the elevator and the universe as an inertial force field, rather than a gravitational field —but this would be purely an arbitrary choice, a mere matter of convenience or comfort to us. There is simply no way to tell but what the reverse is the true state of affairs!

Now granted, if our passenger in the accelerating elevator happened to be a clever scientist with exquisitely sensitive instruments, he might well be able to detect certain very minor differences between the force holding him down to the elevator floor and the effect of a gravitational field surrounding a planet such as the earth. If two boys are standing ten feet apart on the surface of the earth and simultaneously drop rocks to the ground, the rocks will not fall along *precisely* parallel paths. Each rock will fall in a straight line directed toward the center of the earth, so that the two boys would in effect be standing at two corners of a triangle whose apex is at the center of the earth, and the rocks would angle very, very slightly toward each other as they fall to the ground. If the man in the elevator were to drop two rocks ten feet apart at the same time, however, those rocks would fall along *precisely* parallel paths. If the man were to measure the paths that his two rocks followed very accurately, as might be possible inside the elevator, he could indeed conclude that the force field pressing him to the elevator floor must have a slightly different mathematical structure from the gravitational force on earth pulling rocks from the boys' hands.

On the basis of this, he might well conclude that, wherever he was, he was *not* standing in *earth's* gravitational field. But these measurements would still not permit him to distinguish between inertia and gravity. All they could tell him would be that the force field in the elevator, which he has chosen to regard as a gravitational field, has a slightly different mathematical structure from another gravitational field familiar to him— the one affecting objects on the surface of the earth. He could still no more distinguish whether the force field present in the elevator was "really" inertial or gravitational, from such an experiment, than he (or anyone else) could tell whether the elevator was accelerating in one direction through a fixed universe or whether the universe was accelerating past a motionless elevator!

But an even more serious implication of Einstein's contention that gravity and inertia are identical and equivalent, and that accelerated motion can therefore be no more absolute than uniform motion in a straight line, has to do with the very geometrical structure of the universe. Since the time of the ancient Greeks, Euclid's concept of geometry—the "plane geometry" and "solid geometry" we have all encountered in high school—had been accepted as a completely valid description of the structure of the entire universe in three linear dimensions. But now the principle of equivalence of inertia and gravity implied very strongly that Euclidean geometry was not, in fact, a valid tool for describing the universe at all—or at least that it could be considered valid only in the description of very limited areas of the universe at once.

This was perhaps an even more bitter pill for the scientists of the world to swallow than any other that Dr. Einstein had prescribed. We recall that when the theory of special relativity was first proposed in 1905, physicists and other scientists of the time were able to accept it and cope with it to a great extent because it seemed to be largely a purely theoretical and essentially unreal abstract notion that seemed to bear very little relationship with the world of here and now. After all, special relativity dealt essentially with imaginary forces and circumstances. It dealt with objects moving at uniform velocities, whereas uniform motion of objects was rarely encountered in the universe. It dealt with objects moving at velocities so high that they could hardly even be imagined. Thus the areas in which the effects of special relativity might become evident and measurable were essentially "unreal" areas as far as human experience was concerned.

Furthermore, special relativity applied only to the motion of objects moving in uncomplicated straight lines (as classical Euclidean geometry defined straight lines) and through vast reaches of totally empty space, even though scientists knew that space was really filled with a vast multitude of such massive conglomerations of matter as stars, solar systems, galaxies, or what have you. Finally, special relativity necessarily ignored the presence of any gravitational fields for the simple reason that gravitational fields were known invariably to influence the motion of objects passing through them by accelerating the motion of these objects, and thus taking them out of the realm of special relativity which, by definition, dealt *only* with uniformly moving (i.e., nonaccelerating) objects.*

* Here we must recall once again that in physics "acceleration" is defined as a speeding up, a slowing down, or a *change in the direction of motion* of a moving object. Thus, by definition, an object such as the earth traveling in its orbit around the sun as a result of the sun's gravitational attraction is always regarded as constantly "accelerating" around the sun, since its direction of motion is constantly changing, even though its velocity in orbit—the distance it travels per unit of time, approximately 18,000 miles per second—remains uniform. Similarly, if we swing a ball around our heads at the end of a string so that it is moving at a constant angular velocity and thus is making a constant number of revolutions per unit of time, we would nevertheless have to regard the motion of that ball as "accelerated motion," since the direction in which the ball is moving constantly changes instant by instant throughout any given revolution.

In other words, special relativity seemed to apply to such *very* special circumstances, such totally unreal and impractical circumstances, as far as human experience was concerned, that physicists could accept it as a lovely theoretical analysis of things that might happen in a part of the universe beyond our conception, but as no real threat to their concept of the world of here and now. Today we know of many everyday applications of special relativity, but when it was first proposed by Einstein it seemed no more than an interesting intellectual exercise; there seemed little likelihood, at first, that it would ever find any practical application in the "real" universe.

Of course special relativity provided a strange new way of looking at space and time as an interlocking four-dimensional continuum, a new way of regarding all objects in the universe as moving through this four-dimensional space-time continuum along a so-called world line. It was fascinating to consider that even an object that was completely at rest in the framework of the three spatial dimensions of height, length, and width, was nevertheless constantly moving along the time dimension, and therefore could not be said to remain in the exact location in space-time from one instant to the next. But even here, this concept was comfortable enough because the three spatial dimensions under consideration were straight-line dimensions in a universe described by classical Euclidean geometry, and time as a fourth linear but nonspatial dimension was also regarded as a straight-line Euclidean dimension.

Even the contention of special relativity that objects moving at constant velocities in straight lines underwent changes in their spatial and time dimensions was not too hard to swallow. These dimensional changes were still seen to be mere physical changes in the object within the limits of classical Euclidean geometry. "Fine," scientists could be heard to say: "Let's agree that uniform motion must be relative, and accept the idea that *if* the special conditions of the theory ever were fulfilled or could ever exist in the universe, then the effects mathematically predicted by Einstein would indeed be observed. We still have the effects of inertial forces to prove that accelerated motion—the most commonplace motion in our experience—is indeed absolute. Certainly an observer can determine, from the effect of these forces, whether or not he is really moving, and thus these rules of absolute motion are the only ones that have any serious application in the real universe we live in."

Thus physicists could make peace with special relativity by concluding that certain natural laws applied to one kind of motion (unreal and seldom occurring in nature) while another set applied to another kind of motion (the accelerated motion most common to our experience).

But this was, of course, precisely the concept that Albert Einstein could not accept and would not believe. This was what led him to reconsider the whole question of differentiating between inertial force and gravitational force. Einstein believed passionately that the universe was orderly, and that any natural laws describing the behavior of things in the universe simply *had* to be universal laws governing all things in all situations in precisely the same fashion. Those natural laws might well be exceedingly complex. We might only be able to see fragments of them, or inaccurate approximations of them. Indeed, there might be some natural laws which had so far been totally obscured from human comprehension—never yet discovered, nor even suspected, in centuries of observation of nature. But whatever the true laws of nature really were, Einstein was certain that they were orderly and uniform throughout all existence. He could not abide the intellectual complacency among his scientific colleagues that could make peace with the idea that uniform motion might have to be relative but that accelerated motion could at the same time be absolute.

His principle of equivalence—the contention that gravity and inertia were identical—shattered his colleagues' intellectual complacency once and for all. *All* motion in the universe is relative, not just uniform motion in a straight line. The effects of gravitational forces and inertial forces are nothing more than clues tipping us off to certain hitherto unsuspected characteristics of the universe of space-time in which we live. These clues only come to the attention of our senses in cases of accelerated motion—but since gravity and inertia are the same thing, there is no possible way that anyone can distinguish one from the other. To express it differently, Einstein was contending that no one could ever determine for certain which term, the term "gravitational force" or the term "inertial force," was the proper one to describe a single great force which we may experience from time to time. Which term is the proper one to use depends upon our viewpoint in the given circumstances, and in any event we are always describing the same thing.

One immediate implication of the principle of equivalence was that the great classical distinction between inertia and gravity—a distinction carefully analyzed and described by Newton in his laws of motion—had to be thrown out altogether. Gravity had classically been regarded as a force of attraction existing between two massive objects (such as the earth and an apple) tending to draw those objects toward each other. Newton's laws of universal gravitation provided a formula for calculating the magnitude of this force in any given case: The more massive the objects, the greater the force called "gravity" attracting them to each other; the closer together the two objects, the greater that force; the farther away the two objects, the weaker that force became. The *motion* of the objects did not even enter into Newton's thinking, except in terms of how rapidly the objects would accelerate toward each other under the influence of their mutual gravitational attraction, and this was represented in Newton's formula by the letter $G$, representing the acceleration constant of gravity.

On the other hand, Newton regarded inertia not so much as a force but as an inherent characteristic or quality of any object possessing mass. As long as a massive object was at rest or as long as it remained in uniform motion in a straight line, Newton considered that no inertial force was present at all; an observer would have no clue that the massive object possessed the characteristic of inertia. It was only when some outside force accelerated that object, either by altering its velocity in a straight line or by changing its direction, that the object's inherent inertia was suddenly revealed by the appearance of an inertial force acting to resist the acceleration.

Thus Newton considered that an object moving freely in empty space in a straight line at a constant velocity would continue to do so *because of the inertia it possessed as an inherent quality,* until such time as some force acted upon it to change (i.e., to accelerate) its motion. But a cannonball fired in a gravitational field such as that on the surface of the earth would continuously be pulled downward toward the center of the earth by a constant powerful force, gravity, with a constant downward acceleration, the acceleration of gravity, until it struck the ground. Thus the cannonball would trace a curved line or arc as it traveled from the mouth of the cannon through the air and ultimately down to earth.

Now Einstein was insisting that an inherent quality of matter, inertia, and a force that was exerted in the neighborhood of massive objects, gravity, were identical. But this was tantamount to contending that *gravity was not a force at all.* Einstein insisted, rather, that "gravity" and "inertia" were merely two names for the same inherent quality of massive objects moving through the space-time continuum of the universe, and that this inherent quality of matter revealed itself by certain measurable effects only when objects possessing this inherent quality moved with accelerated motion. In other words, he was saying that the measurable and perceptible effect of "inertial force" on an object, or the effect of "gravitational force" on an object was in either case nothing more than a sort of reflection or shadow picture of some local disturbance of that object's motion through the universe of space-time. In the one case, this disturbance is *interpreted* as an inertial force resisting acceleration. In the other case, the disturbance is *interpreted* as a gravitational force trying to tug the object off-course from its normal motion through space-time.

In either case, the disturbance could be considered as arising because the object had entered into some *area of distortion* in the normal four-dimensional structure of space-time. This area in which space-time was distorted could be regarded, say, as a sort of trap or deadfall into which objects moving through space-time would tend to fall, thus altering their paths temporarily, if they approached too closely. A simple way of visualizing this is diagramed in Figure 29. We can imagine the universe as a huge and tightly stretched rubber sheet on which a number of very massive bodies (representing planets or stars, for example) are supported. As we see in the diagram, areas between these massive objects (that is, areas of empty space) are represented by perfectly flat surfaces on the rubber sheet, but the area of the sheet around each of these massive bodies is stretched and distorted, with this effect more pronounced the more massive the body is. An object traveling on a normal course through space-time might be represented as a tiny marble rolling along the vast flat surfaces of the rubber sheet. But any time such an object approaches

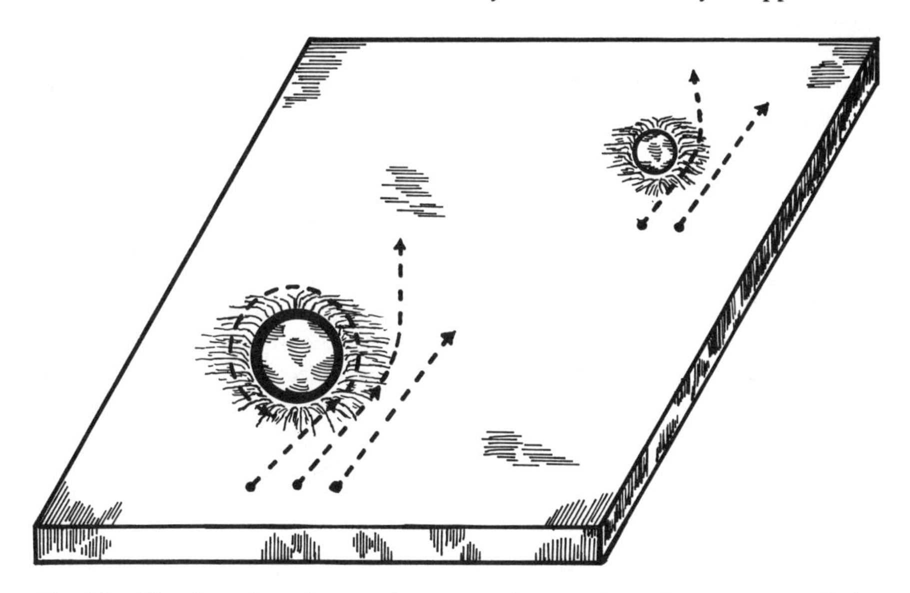

*Fig. 29* The distortion of space-time caused by massive objects in space. Paths of motion of objects passing nearby are altered by such areas of distortion as the objects follow the "shortest distance" of geodesic curves. If they pass too close, moving objects may be "trapped in orbit," perpetually following the geodesic curves of the distorted area.

near a massive body, its path of motion would be distorted from the "normal" by the abnormal stretching of the rubber sheet in the vicinity of such a massive object (the distortion in space-time that exists in the vicinity of a massive object) and thus the "normal" path of the object through space-time would be temporarily altered.

Einstein spoke of such areas of distortion in space-time as "fields" or "force-fields" in which the degree or magnitude of the distortion could be described by specific field equations. Since these field equations treated gravitational effects and inertial effects as identical reflections or manifestations of areas of distortion in the space-time continuum, his equations describing the magnitude of the distortion in these areas under various circumstances came to be known as "gravitational field equations."

## DISTORTION AND GEOMETRY

It is important for us to understand clearly at this point that the distortion of space-time that Einstein was imagining was a *geometric* distortion—an actual bending or twisting of the geometric structure of the universe. As he saw it, such a geometric distortion existed in various places and to varying degrees throughout the universe as an inherent, inescapable characteristic of the structure of space-time. Furthermore, these areas of geometric distortion were invariably found in the areas where massive objects existed in space-time.

In fact, Einstein maintained that *every* object with any mass at all invariably causes *some* degree of distortion of the geometric structure of space-time around it as a direct result of its mass, and huge massive bodies such as the earth or the sun caused a marked degree of distortion of space-time in a wide area around them—an area of distortion which extended into space beyond the physical limits of the earth or the sun according to the inverse square rule. In other words, the distortion caused by a given massive object would affect the path of an object moving through space-time in its vicinity in inverse relationship to the square of the massive object's mass and the distance of the moving object from it. In any case, the moving object's path through space-time would be altered or bent in such a way that the object would continue to follow a curving path representing the shortest distance traveled per unit of time in the area of the distortion.

Regarded in this way, we can see gravity as something quite different from a mysterious force reaching out into space from a massive object such as the earth, and as something quite different from a mysterious field of force existing between two massive bodies such as the earth and the sun. Regarded this way, we see that the earth is not falling toward the sun in response to some mysterious force tugging at it; rather, it is traveling through space-time in an area in which the geometry of space-time has been severely distorted by the presence of the massive sun. Earth's elliptical orbit around the sun, observed in our three spatial dimensions of classical astronomy, is not (as Newton supposed) the result of a constant tug of war between the sun's gravitational pull on the one hand and the earth's inertia constantly tending to keep the earth moving on a straight-line course moving away from the sun at a tangent. Rather, earth's orbit around the sun, as we observe it, is nothing more than a three-dimensional projection of the earth's four-dimensional route through space-time, following the shortest path it can follow through an area of geometrical distortion. Indeed, according to Einstein there is really no force whatever acting upon the earth's motion, either gravitational or inertial. The earth is merely trapped in an area of distorted geometrical space-time structure from which it cannot escape—trapped in a force-field created in space by the sun's presence—and its elliptical orbit represents nothing more than the shortest path it can take from one instant to the next along its constantly changing path through space-time.

This elliptical path is not a straight line or an elliptical line in the sense of Euclidean geometry, but rather a curving path through four dimensions described as a "geodesic line" or simply a "geodesic." Einstein hypothesized that all matter in the universe moves not in the straight lines or the curved lines of Euclidean geometry at all, but always along curving geodesic lines through space-time, their paths of motion merely being bent or altered more or less according to whether they are moving in an area of greater or lesser geometric distortion.

On the surface it might seem that Einstein's concept of gravity as a geometric distortion in the structure of space-time was simply a matter of using new and fancy terms to describe precisely what Newton was describing in his concept of gravity as a mysterious force of attraction between massive objects. Whatever they chose to call it, it was soon apparent to physicists that the observed effects on objects moving in gravitational fields, whether on earth or elsewhere in the solar system, seemed to be almost precisely the same whether they used Newton's equations to calculate the behavior of objects influenced by universal gravitation or whether they used Einstein's unified field equations. Just as the Lorentz-Einstein transformation equations boiled down to almost precisely the same results as Newton's classical laws of motion in the case of objects moving at relatively low velocities, so Einstein's field equations produced virtually the same results as Newton's equations in predicting the behavior of accelerating objects within our own solar system. The discrepancies between Einstein's field equations and Newton's classical equations only began to appear when those equations were applied to the behavior of heavenly bodies of enormous mass located at almost inconceivable distances from earth.

Thus at first it was very difficult to find any kind of actual observation that could be used either to prove or disprove Einstein's theory of general relativity. Once again, it appeared to be a theory which could never really be subjected to the kind of experimental confirmation that science demanded, no matter how fascinating and logical it appeared.

Nevertheless, one or two observations had been made which seemed to offer at least *some* support to Einstein's contention that his gravitational field equations provided more valid descriptions of the behavior of gravitational forces than Newton's equations did. There was, for example, a long-standing mystery surrounding the curious behavior of the planet Mercury in its orbit. This small planet, the closest of all to our sun, has a comparatively high orbital velocity and at the same time has a very eccentric, elongated elliptical orbit, so that it moves in remarkably close to the sun at one point in its orbital cycle and out a great distance away from the sun at the opposite point in its orbit.

Such an elongated orbital path is not in itself particularly remarkable; many comets, by comparison, follow a far more lopsided and elongated orbit than this, moving progressively faster in orbit as they swing in close to the sun, then slowing down more and more as they reach the opposite point in their orbit far away from the sun at the outer edges of the solar system. But in the case of Mercury, astronomers had been puzzled since 1850 because the planet's elliptical orbit had been observed to be shifting slightly in a continuous fashion with each circuit of the planet around the sun. In other words, Mercury's orbital path had been observed to be "walking forward" or preceding itself by a tiny bit each time the planet revolved around the sun, instead of remaining in the same identical orientation to the sun each Mercurial year. Astronomers had tried to explain away this so-called perturbation in the orbit of Mercury as the end result of the gravitational pull that was exerted on the tiny planet by the huge and massive gas-giant planets far out in the solar system—Jupiter, Saturn, Neptune and Uranus. Unfortunately, the calculated gravitational attraction of these gas-giant planets upon Mercury had never seemed to be quite enough to account for the amount of "walking forward" that Mercury's orbit exhibited.

Einstein's gravitational field equations seemed to solve the mystery splendidly. Using these field equations to predict the apparent effects of gravity, Einstein demonstrated that any planet's path through the distorted area of space-time that existed around the sun would have to follow a geodesic curve representing the "shortest distance" through the area of distortion. Thus the orbit of any planet observed by astronomers with their telescopes and charted in three-dimensional space would have to "walk forward" a small amount with each completed orbit, and this observable "walking forward" of a planet's orbit ought to be far more noticeable and measurable in the case of a planet with an elongated elliptical orbit such as Mercury's than in the case of a planet with an orbit that was almost circular, like the orbit of Venus, for example. By using his field equations actually to calculate the amount that Mercury's orbit should be expected to walk forward each cycle if his theory of general relativity were correct, and by adding this amount of walking forward onto the amount known to be caused by the influence of the outer planets' gravitational fields, Einstein showed that the predictions of his field equations coincided very closely with the observed precession of Mercury's orbit.

Another possible way of confirming or disproving Einstein's theory of general relativity seemed to lie in the observation of the path of light from various distant stars. Since light, as a form of radiant energy, ought to have a certain tiny amount of mass and even exert a certain minute amount of pressure on any surface that it strikes, Einstein reasoned that the path of light ought to be bent slightly any time that light passed through an area of strong space-time distortion—that is, any time it passed in the near vicinity of a massive body anywhere in space. All attempts to measure such a bending of light rays traveling from the sun to the earth had met with failure; if there was any bending at all, it was so very slight that there was no way it could be measured. But if Einstein's predictions were correct, it seemed possible that light traveling from a very bright star located at a very great distance away from the earth might well be bent to a measurable degree if it passed close enough to the earth's sun on its way

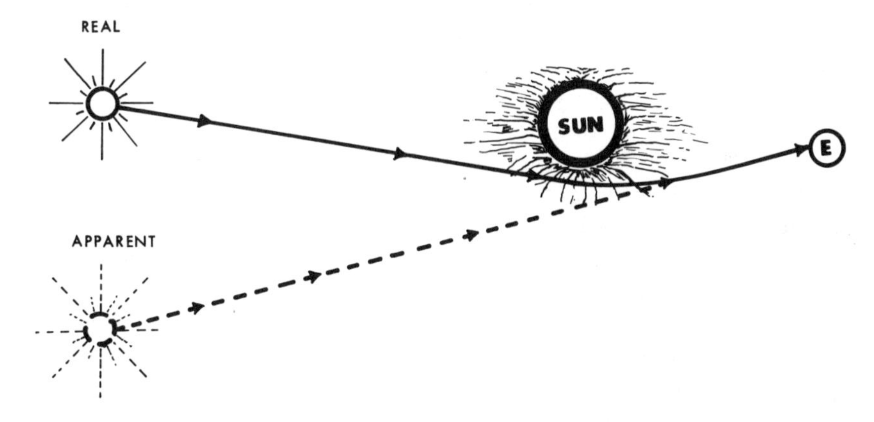


*Fig. 30* The bending of light rays passing near the sun on their way to earth was one of the famous early confirmations of Einstein's General Theory of Relativity.

to an earthly observatory. The trouble with actually observing and measuring such a phenomenon, if it actually existed, was simply that the glaring brightness of the sun ordinarily made it quite impossible to identify the light from any distant star shining through the sun's corona.

Clearly, if any such light-bending phenomenon were ever to be observed, it would have to be observed under very special conditions. In 1917 just the right conditions arose; an expedition of physicists and astronomers traveled to an area of the earth that lay in the path of a total eclipse of the sun that was due to occur, in order to try to observe the path of light past the sun from a very bright distant star, Sirius. With the sun's brightness temporarily blotted out by the intervening shadow of the moon, the necessary observations were made, and the results of the experiment were a very satisfactory confirmation of Einstein's predictions (see Fig. 30).

Light rays emanating from Sirius were actually being bent inward toward the earth as they passed the vicinity of the massive sun itself.*

Finally, a third type of experiment provided an even more sweeping confirmation of the theory of general relativity. These experiments were particularly convincing, for they confirmed Einstein's contention that his theory of special relativity was merely a part of a much more universal theory of general relativity. Thus general relativity not only contended that the path of motion of objects is altered as they pass through areas of distortion in space-time in the vicinity of massive bodies, but that time itself, as a built-in part of the space-time continuum, was also altered in the vicinity and under the effects of massive bodies.

Einstein reasoned that all time processes would have to be considerably slowed down in the neighborhood of massive heavenly bodies, and he calculated that a clock on the surface of the sun (if one could place one there and then observe it in comparison to a clock on earth) would run one second slower in every six-day interval. By the same token, he reasoned, the vibration time or frequency of radiations arising on the sun ought to be slower due to the sun's greater mass than the frequency of the same radiation on earth. Thus he contended that the light emitted by the sun should (in its various color components) be shifted toward the red end of the spectrum of wavelengths in comparison to the spectrum of white light from some earthly source. A few years ago, laboratory measurements of the slowing of the frequency of radiation due to the effect of the sun's mass confirmed Einstein's formula to an accuracy of 10 per cent. At first, laboratory attempts to measure a slowing down of *time* from this cause failed. If any such slowing down occurred, it seemed too slight to measure. Unperturbed by this failure, astronomers made observations of the light emanating from a very distant star which was not particularly bright, but which was known to be extremely dense, a white dwarf. By studying the faint light emitted by this star, an American astronomer named W. S. Adams proved that the frequency of light waves on that distant star was sharply shifted to the red end of the spectrum—in other words, that time on the surface of that star indeed moves much more slowly than it does on earth. In 1954 similar experiments were done with greater accuracy by another American, D. M. Popper, and once again actual observation of a phenomenon in nature served to confirm one of the basic principles of general relativity: the principle that the structure of space-time is altered or distorted by the presence in it of massive material objects.

* The first measurements were actually made not of the light from the star Sirius, but from a number of unidentified stars. The light from Sirius was measured in a third repetition of the experiment—but the results of all three observations, and a great number which have been made since, all have confirmed Einstein's predictions.

## BOLYAI, RIEMANN, AND THE STRANGE GEOMETRIES

Thus, as confirmation after confirmation began to be found for the theory of general relativity, physicists began to realize that they had to grapple with the theory very realistically if they were to hope to extend further their understanding of the nature of the universe. And one of the most obvious and difficult implications of the theory was that the whole concept of a four-dimensional continuum of space-time constructed in conformity with the principles of Euclid's geometry had to be discarded. It was not that Euclidean geometry was suddenly found to be invalid—it was perfectly valid within certain severe limits. The trouble was that Euclid's geometry simply could not be validly applied to describe the structure of the entire universe in space and time. And the further afield from the limited local structure of things on earth and in our own solar system you went, the more invalid Euclid's geometry became.

Consider for a moment just what this means. Before any relativity theory was devised, when the behavior of electromagnetic waves was being considered without any knowledge of distortion of space-time caused by massive bodies such as planets and stars, everyone had assumed that light and other electromagnetic waves always traveled in a straight line. But electromagnetic radiation is known to be traveling in the vicinity of stars, planets, and galaxies as well as through totally empty space, and now the general relativity theory contended that the direction of motion of these high-velocity waves is altered whenever they approach such massive bodies in space. Similarly, before the general relativity theory was discovered, it made scientific sense to think of space and time existing primarily as a vast matterless emptiness in which various chunks of matter were scattered more or less at random, and thus to regard the space-time continuum described by the theory of special relativity as conforming perfectly to Euclid's geometry, with light and other electromagnetic waves always traveling in a straight line. But the general relativity theory pointed out that the presence of massive bodies in the space-time continuum causes marked distortion of that continuum in huge areas surrounding those bodies, so that the path of anything moving in space, light included, is constantly being altered as it travels and has to be curvilinear, not straight.

Now it would be possible to argue that there is no way to decide whether the observations of nature confirming this general relativity theory—the precession of Mercury's orbit, the bending of light rays from distant stars, or the slowing down of the frequency of radiations in the vicinity of massive bodies—were in fact due to the effects of a mysterious force called gravity, or whether they were due to a distortion of the space-time continuum in the vicinity of any massive celestial body. But as Einstein pointed out, it really didn't matter in the least which of these two alternatives was blamed for the phenomena that had been observed; he merely reiterated that the force we normally interpret as "gravitational force" could just as well be the effect produced by distortion of the four-dimensional space-time continuum that occurs in the vicinity of massive bodies like the sun or the earth. And clearly, if such distortion of the geometric structure of space-time occurred, then there must be something very strange indeed about the geometry—the four dimensional *shape*—of space-time. Whatever its geometry, it was certainly *not* the geometry of Euclid!

At first glance, such a contention seems like nonsense. All of us are acquainted with the propositions and theorems of Euclidean geometry, and we know perfectly well from lifelong experience that they apply splendidly in the world around us. For over a thousand years Euclid's geometry had been recognized and accepted as the *only possible* accurate geometric description of the structure of our three-dimensional universe. Starting from a mere handful of self-evident axioms which were not themselves subject to proof because their validity was so obvious, Euclid had built up a splendid and flawless, logical mathematical and mechanical structure, first devising proofs that certain simple geometric relationships had to be true, based upon the original axioms, then using those proofs as further evidence to help prove that more complex relationships were true.

Euclid's geometry described a geometric structure for space involving all possible linear dimensions up to and including three. For example, Euclid described a *point* as an imaginary dot in space which had no dimension at all. If such a point were moved in a perfectly straight line in any direction for a certain distance, the point would have described a one-dimensional figure in space: a single line bounded by two points. If that line were then moved a certain distance in a direction perpendicular to the direction the point had moved in tracing the line, a second linear dimension was outlined: a two-dimensional figure composed of four lines bounded by four points described on a plane two-dimensional surface. If the resulting two-dimensional square were then moved a certain distance perpendicular to the plane upon which it was inscribed, a third linear dimension was depicted—height—and a three-dimensional figure or cube would have been created, composed of twelve lines and bounded by eight points, and encompassing a three-dimensional area of space bounded by six two-dimensional squares. We can see this familiar geometric exercise diagramed in Figure 31.

Euclid's geometry dealt carefully and progressively with all varieties of geometric forms described on a plane surface: squares, rectangles, triangles, parallelograms, circles, ellipses, etc. It also dealt with three-dimensional "solid" objects such as spheres, cubes, spheroids, pyramids, cylinders, hexagons, etc. And the structure of Euclid's plane and solid geometry was so ruthlessly logical and true to its original axioms that it had been accepted not merely as "a geometry" but as "the only possible geometry" that could accurately describe our three-dimensional universe. It was also assumed that if we could somehow become capable of visualizing higher dimensions —a fourth linear dimension, a fifth, a sixth, etc.—then Euclid's geometry would also present the only possible geometry that could be used to describe a universe constructed in those dimensions also.

How, then, could it be that relativity theory should force scientists to look to some other type of geometry to describe the structure of space-time? A simple imaginary example will help us see why Euclid's geometry was found to be so useless in describing the four-dimensional space-time continuum of relativity. In his system of geometry, Euclid had made a careful study of the relationship of the circumference of a circle, any circle, to its diameter. One of the earliest mathematical constants ever to be discovered was the constant known as pi, a number equal to approximately

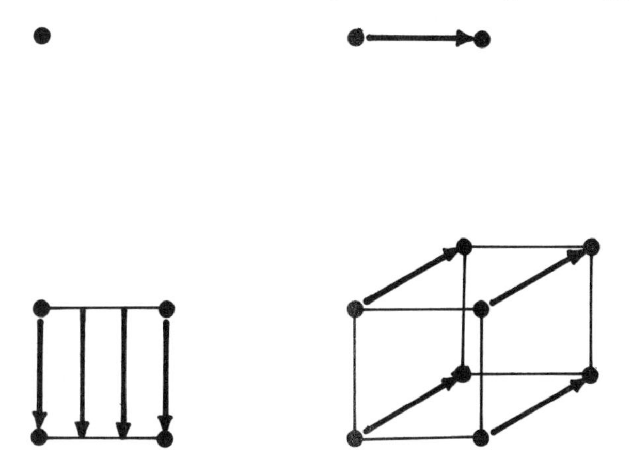

*Fig. 31* The development of three linear dimensions.

3.1416. According to Euclidean geometry, this number would always be the result obtained if one divided the circumference of any circle by its diameter. The size of the circle would make no difference in the value of pi, nor would the time at which the measurement and calculation were made.

But now suppose that we had a huge centrifuge with a cabin at either end of the centrifuge arms and a pivot point at the center. With the centrifuge at rest, an observer standing at the pivot point and another observer in a cabin at the end of one centrifuge arm could both measure the diameter of the circle (twice the length of one centrifuge arm) and could both measure the circumference of the circle simultaneously. Both observers could then divide the circumference they had measured with the diameter they had measured and both would come up with a value for pi that was precisely the same, even if their measurements had been accurate to fifteen decimal places and they had carried their calculation of pi out to ten or twelve decimal places.

But then suppose that the centrifuge was started in motion, with the cabins at the ends of the centrifuge arms moving at a high angular velocity around the pivot point. To the observer standing at the pivot point, the ratio of the diameter of the circle to its circumference would remain completely unchanged, with pi continuing to equal 3.1416 carried out to ten or twelve decimal places. But as the velocity of the centrifuge cabins climbed higher and higher, the observer taking measurements from one of the cabins would discover that according to *his* measurements the *diameter* of the circle remained the same but the *circumference* of the circle was *less* than it was when the centrifuge was at rest, since the yardstick he was using to measure the circumference had shrunk slightly. If the centrifuge velocity began approaching the velocity of light, the circumference of the circle as measured by the moving observer would become quite significantly shortened.

What is more, this observer inside the centrifuge cabin would notice that a new "force" had appeared which seemed to press him against the outside wall of the cabin. That observer then might conceivably ascribe the change in the circumference of the circle to the effects of gravity, whereas the observer at the pivot point would ascribe the "force" the cabin observer felt to centrifugal force, and would observe no shrinkage of the circumference of the circle at all.

Whichever interpretation of this "force" is accepted as correct, its effects would be precisely the same, and if the observer in the centrifuge cabin tried to calculate pi by dividing the new circumference of the circle as he had just measured it by the diameter of the circle that he had also just measured, he would come up with a value for pi that was not 3.1416 but some smaller value.

If his velocity were relatively slow, his new measurement of the circle's circumference would be almost the same as when the centrifuge was at rest, so that his calculation of pi would differ only negligibly from that of the observer at the pivot point, perhaps no more than a difference in the eighth or ninth decimal point. But if this observer were traveling at a progressively higher and higher velocity, he would find his calculated value for pi to be growing smaller and smaller, perhaps 2.1416, then 1.1416, then 0.3416, then increasingly still smaller the closer his velocity approached light velocity.

Einstein saw that the geometric structure of space-time which was distorted by the presence of massive bodies scattered about in it could not possibly be the same geometry to which most people were accustomed. Euclid's classical plane geometry simply would not work; a new geometric form had to be sought for describing the shape of space-time in which points, lines, planes, and solid bodies were regarded as events in time continuously moving along curved or geodesic lines, more and more sharply distorted the closer they moved into the vicinity of massive bodies. But could some other geometry exist—some kind of four-dimensional geometry to which Euclid's postulates did not apply at all? It was hard to challenge the self-evident truth of most of those postulates; from observation of nature, it seemed obvious beyond necessity of proof that one could draw a straight line to connect any point to any other point, or that all right angles were equal to one another. In fact, there was really only a single one of Euclid's postulates which could not in any way be proved or derived from the others: the postulate which states (in the form familiar to most of us) that "through a given point only one straight line can be drawn parallel to another given line that is not passing through that point."

For centuries mathematicians had been toying with this postulate, convinced that it *ought* to be provable, but never succeeding in finding any mathematical proof for it. One of the first mathematicians to begin to glimpse just *why* this postulate had never been proved was Karl Friedrich Gauss (1777–1855), a brilliant German mathematician who had earlier concerned himself with the use of so-called imaginary numbers in algebra —unreal quantities involving the square roots of negative numbers which could not be derived by any mathematical means at all.

Gauss began to speculate, very plausibly, that perhaps the reason that the "parallel line postulate" of Euclid had never been proved was simply that it was not capable of proof. This speculation was then carried a step farther by a young Hungarian mathematician named Junos Bolyai (1802–1860) who declared in the mid-1800s that the reason the parallel line postulate could not be proved was simply because it was *not true,* at least, not under all possible circumstances. The parallel line postulate might apply splendidly in the three-dimensional space of Euclid's plane geometry, but Bolyai insisted that it was perfectly possible to construct other systems of geometry describing other, perhaps imaginary, constructions of space in which *no straight line* could be drawn parallel to another given line through a given point, or in which an infinity of straight lines could be drawn, each parallel to another given line and all passing through a given point.

In short, it was Bolyai's idea that any number of so-called non-Euclidean geometries could be created without violating any part of the structure of Euclidean geometry. They could simply be created *outside the range* of Euclidean geometry, so to speak. This startling idea stimulated the German mathematician Bernhard Riemann (1826–1866) to go to work and actually construct two such new and non-Euclidean geometries from scratch. And Riemann found that Bolyai was indeed right; it was perfectly possible to construct such new geometries while by-passing Euclid's postulates altogether.

One significant difference between the Riemann geometry and Euclidean geometry, for example, was that Euclid's plane geometry involved two-dimensional geometric figures drawn on a flat plane surface which was assumed to extend through all infinity in all directions and always to remain perfectly flat. Riemann drew his two-dimensional figures on the surface of a sphere, first on the outer convex surface and then on the inner concave surface, and then in each case carried his constructions on to higher-dimensional levels. In three dimensions, Riemann's geometry derived from two-dimensional figures drawn on the convex outside surface of a sphere described a form of three-dimensional space which had *positive curvature,* while his constructions deriving from the concave inside surface of the sphere described a type of three-dimensional space which came to be known as having *negative curvature.* Riemann chose first to concentrate his attention on a new geometry to describe a space possessing positive curvature, while Bolyai proceeded to work out a strange new geometry related to an imaginary space with negative curvature.

In each case the geometry that was evolved was a queer geometry indeed in terms of "flat" or "normal" Euclidean geometry of three-dimensional space. Suppose that instead of using a flat sheet of paper, we use the inside (negative) curvature of a large hemisphere on which to draw triangles, circles or parallel lines. We would soon discover that we could not draw *straight* lines on such a curved surface—only *geodesics* (lines of shortest distance from one point to another). We would further discover that while we could draw one such geodesic on such a negatively curved surface, we could *not* draw a *second* geodesic that remained parallel to the first. The second geodesic might appear parallel to the first if we considered only a short segment of it, but if we extended it in either direction for any distance at all we would find that the second line inevitably began to diverge from the first in either direction. We would find that we could indeed draw a triangle on this negatively curved surface using three perfectly straight lines to form it, but if we then used a protractor to measure the three angles of any such triangle we might construct, we would discover the sum of the angles of that triangle to add up to *less* than 180 degrees! Still worse, if we drew two triangles similar in shape but different in size, making one proportionately larger than the other, we would find that the corresponding angles of the two triangles were not identical at all. (In Euclidean plane geometry an important proposition states that the corresponding angles of two similar triangles are *always* identical.) The sum of the angles in the larger triangle would be markedly less than the sum of the angles in the smaller one! Finally, if we carefully measured first the circumference and then the diameter of any circle drawn on this negatively curved surface, we would find that we inevitably came up with a wrong value for pi, and that the area of the circle would *not* be equal to pi times the square of the radius, unless you used the funny value for pi calculated from the measured circumference and diameter for each individual different-sized circle. Indeed, we would find that the areas of a series of enlarging concentric circles drawn on such a negatively curved surface would increase faster and faster in proportion to the square of the radii of the circles!

We would see a similar but opposite situation in the relationships of Riemann's strange geometry constructed on a positively curved or convex surface. In fact, we could easily test these strange relationships simply by drawing figures on the surface of any globe or large ball. In this case all the lines of any triangle that we might draw represent arcs of great circles around the ball, and since any two such lines that we might draw will ultimately intersect, no parallel lines can possibly be drawn. No matter what the size of a triangle we might draw on the outside of a ball's surface, the sum of the angles will always add up to *more* than 180 degrees. In fact, if we were to trace a triangle on a globe with its apex at the North Pole and with the apex angle measuring 90 degrees, and then extend the sides of the triangle to the equator and draw a third line around the equator to close the triangle, we would discover that *each* angle of the triangle measures 90 degrees, a total of 270 degrees in all. Again, as in the case of circles drawn on a negatively curved surface, the circumference of any circle drawn on such a positively curved surface will not be constantly proportional to its diameter, but in this case we would find that the areas of increasingly large concentric circles would increase progressively more slowly than the squares of the radii of those circles as the circles became larger.

## THE CURIOUS SHAPE OF SPACE-TIME

When he worked out his relativity theories, Einstein did not have any particular geometric construction for space-time in mind, except that he recognized that Euclidean geometry simply would not do. Ultimately he picked Riemann's non-Euclidean geometry of positively curved space as the four-dimensional geometric construction that seemed to do the best job of defining the four-dimensional space-time continuum of his relativity theory. But there was no possible way that Einstein or anyone else could actually *visualize* the true "shape" of space-time's geometric construction, or observe the true "path of motion" of any object moving through that space-time. The best visualization anyone could manage would have to be a three-dimensional projection or "shadow" of the structure of that four-dimensional space-time, or of the path of objects moving through it. Furthermore, any given shadow, or projection, that we happened to see would depend upon the angle from which we were observing—except that in the case of four-dimensional space-time, that "angle" would not be a spatial three-dimensional relationship, but an "angle" observed from deep within an area of distortion (that is, observed from the near vicinity of a massive body like the earth). In other words, even the shadow, or projection, of an object's motion that we happened to observe would have to be distorted as long as we were observing it from an earthly laboratory.

Why is this so? Why is it not possible for us to visualize the "shape" of space-time's geometric construction in any way? The reason is that we human beings, born and raised in an area of distortion of space-time, and so small in size that three-dimensional Euclidean geometry has always provided enough dimensional room, so to speak, for our everyday needs, simply have no means of visualizing any other dimensional structure than one made up of straight lines in three linear dimensions. We can, at least, observe, feel, and examine objects in all three of these dimensions —but *only* in three dimensions. For example, we can hold a sheet of paper in our hands and see that it is 9½ inches wide, 11 inches long, and a fraction of an inch thick. We cannot see or examine the shape or movement of that sheet of paper in the time dimension; we can *imagine* it, but we cannot *visualize* it. We would find it just as impossible to visualize a *fourth* linear dimension in our accustomed area of space-time as an imaginary two-dimensional being living on a flat plane surface like the floor would be able to visualize a *third* linear dimension. Such a two-dimensional being would have length and width, but would have no thickness whatever. He would see others of his kind around him as nothing more than faintly discernible lines, although we looking down on him might see that he had a square, circular, or triangular shape. That two-dimensional being not only can have no means of visualizing a third or height dimension, he would have no means even of imagining it. All that he could "see" or perceive of a three-dimensional object like our sheet of paper would be the shadow the paper cast on the two-dimensional plane that makes *his* total universe.

Thus, if we held the sheet of paper in a certain way, it might cast a 9-inch by 12-inch shadow on the floor, and our two-dimensional being might become aware that there was an area of shadow in his universe which he could measure by, say, finding out how long it takes for him to cross from one side of the shadow area to the other. Suppose, then, that we began rotating the sheet of paper until it was rotated 90 degrees and oriented at right angles with the floor. As we rotated it, the width of the shadow on the floor would grow narrower and narrower until when the paper was at a right angle to the floor it would cast only the thinnest hairline of a shadow.

Of course, since we are able to see the sheet of paper in all three directions, we realize what is happening. A sheet of paper is being turned in three dimensions in such a way that the two-dimensional projection or "shadow" of the paper falling on the floor seems to grow narrower and narrower. But all that our two-dimensional being could see, being incapable even of conceiving a third or height dimension, would be a patch of shadow for some mysterious reason changing in shape and growing narrower and narrower until it had become merely a thin hairline. What is more, if we moved the paper in other three-dimensional orientations, the shadow on the floor our two-dimensional being could detect would constantly be changing in shape, apparently for no reason at all and quite arbitrarily, shifting all the way from a patch of shadow that was exceedingly thin to a large rectangle, then to a triangle, etc. And without any concept of what a three-dimensional object was or what kind of shadow it would cast on the floor, our two-dimensional observer would be hard put indeed to explain the phenomenon that he was observing: the apparently arbitrary shifting and changing in shape of a patch of shadow in his universe.

Just as the two-dimensional being could not visualize a universe of three dimensions, we as three-dimensional beings cannot visualize a space-time continuum of four dimensions. It was for this reason that Einstein's description and definition of four-dimensional space-time structure, and of the motion of objects through it, had to be expressed entirely in mathematical terms. He could not, for example, have built a meaningful model of a four-dimensional object, using only the three-spatial dimensions he had readily at hand to work with. And it is just because of this built-in impossibility of ever visualizing or constructing a model of the space-time continuum of general relativity theory which has made that theory so difficult for nonscientists without mathematical training or experience to understand or comprehend. After all, most of us have spent our lives learning to understand those things that we *can* visualize or lay our hands on, and the general relativity theory eludes us in both cases.

It is also true that in the everyday universe as we see it around us, in our everyday world of slow velocities, Newton's laws of motion, his law of universal gravitation, and Euclid's three-dimensional straight-line geometry all fill the bill perfectly. The mirrorlike surface of an undisturbed lake *is*, to all intents and purposes, a perfectly flat plane surface, just as a tabletop is, and if we were to draw triangles on such a surface those triangles would have angles that added up to a total of 180 degrees—or so very close to 180 degrees that the discrepancy could not be measured. But if we can imagine that that lake stretched all the way from the equator to the North Pole, and that the triangles we were trying to measure were so large as to cover most of the surface of that huge lake, we would find that Euclidean plane geometry would fail us: the angles of such a huge triangle, when measured and added together, would total more than 180 degrees (see Fig. 32).

Similarly, Einstein's general relativity theory and the curvature or distortion of space-time geometry he described became relevant only over enormous areas of space. These things were not *necessary* to understand the behavior of moving objects in earth's gravitational field, but it became very necessary to take such strange geometry into account in trying to learn anything about the world of the inconceivably large—the universe in which stars and galaxies are born and die, for example. Consequently, if any clue was to be found as to the true shape and structure of the entire universe in all its enormous expanse, if there was to be hope ever of learning anything about the origins of that huge universe and its future, general relativity theory and the strange new geometrical concepts of space-time had to be applied.

Granted that for centuries physicists had been content to "stay at home," to study, observe, and try to learn about the everyday universe of human experience. Einstein's work had expanded the horizons of physics and

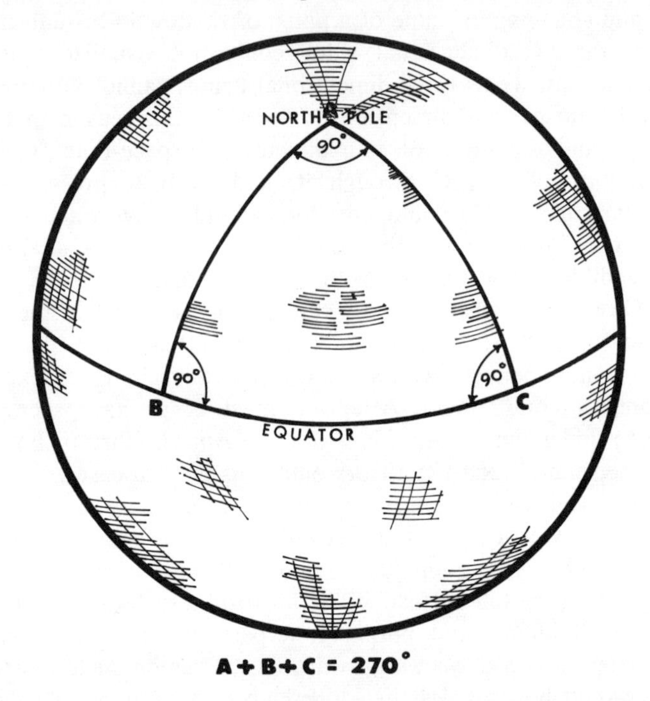


*Fig. 32* In this huge triangle drawn on the globe, if the angle at A (North Pole) is 90° the angles at the equator will also be 90° each, a total of 270°.

shown scientists a glimpse of the vast universe still waiting to be explored. And faced with these new frontiers, physicists could no longer be satisfied merely to peer myopically at the world around them. In their centuries-long quest for universal natural laws governing all kinds of phenomena everywhere in the universe, they had reached a point beyond which they could not go. From that point on it was necessary that they look at the universe through relativity-colored glasses and begin to ask questions about the universe as a whole—not even limiting themselves to that portion of the universe which could actually be observed and measured with telescopes. They had to ask questions about the staggering hugenesses of the universe beyond reach of the human eye or of human instruments of observation—areas of the universe no man could ever hope to observe or measure directly.

In short, with the impetus of Einstein's work behind them, physicists set out to grapple with the mysteries of a universe beyond human experience, a universe so incredibly huge that the part that we can visualize and measure directly is only a tiny fragment. In such a universe our own earth is, by comparison, as tiny and dimensionless as the "ideal point" of Euclid's geometry.

The branch of physics, astronomy, and mathematics which is concerned with exploring this world of the inconceivably large is called "cosmogony," or more commonly in this country, "cosmology." What is the over-all shape and nature of the universe? How large is it in all? Does it have finite limits somewhere, or does it extend to all infinity in all directions beyond us? Does the part of the universe that we can observe resemble all other parts, or are things somehow different in the part which we inhabit from other parts? In the part of the universe that we are able to observe we know that objects, entities, and forces exist; but are there in addition objects, entities, and forces which we are unable for some reason to perceive at all? Of that which we do see, how much is truly direct observation of what is actually happening, and how much is some distorted shadow that we see of events happening in an unvisualizable four-dimensional space-time continuum? Did the universe ever have a beginning? If it did, then when did it begin and how? Will it ever cease to exist, or will it simply continue existence throughout infinity? And if the physical universe will indeed sometime come to an end, will time also come to an end? Is there any possibility of some future rebirth?

It is questions of this kind, often as much philosophical as scientific, that the cosmologists have been asking for decades and continue asking today. Most of these questions have not found satisfactory answers. Many of them have found a multitude of plausible or possible answers, all conflicting with all the rest. Some of these questions may in fact prove to be unanswerable. Nobody knows for sure—yet. In the chapters which follow, we will take a brief look at some of the possible answers that have been proposed, and at some of the physical clues and the disciplined reasoning that have led scientists to such answers as have been found to date.

# Part IV

*The Universe of the Inconceivably Large*

# CHAPTER 18

## *Macro-Universe: The Problems of Observing*

In an earlier section of this book we found it helpful to divide or separate our consideration of the entire broad area of investigation of the physical nature of the universe into discussions of four individual "worlds of physics," in order to understand the history of discovery in physics more readily. As we have seen, one of these worlds of physics is the world of our everyday experience, the world of natural phenomena which can be directly observed and measured by the human senses, or at least at second-hand or indirectly by means of instruments which extend or amplify the human senses—microscopes, galvanometers, telescopes, etc. Much of the book so far has been devoted to a consideration of the great classical laws of physics that arose from studying this "universe of everyday experience," a world of physics which we might describe by the word *macro-universe* or "universe of large things that we can get our hands on," so to speak.

We have seen that the study of the macro-universe led finally to the discovery of quite a different world of physics: an odd and unsettling "relativistic" view of the universe proposed at the beginning of this century by Albert Einstein and others.

But relativity theory did not stop with the macro-universe of everyday experience. It led physicists to revise their views of another world of physics about which we have said very little so far: the universe of the inconceivably tiny, of molecules, atoms, electrons, and other elementary particles making up the minute building blocks of matter—a world of physics we will discuss later as the *micro-universe*. Relativity theory also led men to reassess their earlier views of the outer limits of the macro-universe—our sun, the planets of our solar system, the nature of nearby stars, even the great galactic swarm of stars of which our sun is only a small and insignificant part. And finally, relativity theory led men's minds even farther out, into an investigation of an inconceivably large universe beyond our immediate galactic neighborhood—a *mega-universe* so huge that men could not even comprehend its vastness, much less make meaningful direct observations of it.

Essentially both the micro-universe at one end of the scale of size and the mega-universe at the other end lie forever beyond the range of direct sensory observation. Physicists have never yet been able to observe the tiny elementary "particles" of the micro-universe directly, and probably never will. Neither can physicists, mathematicians, astronomers, or cosmologists travel even to the nearest stars in our galaxy, much less cross the enormous gulfs of space necessary to study at close hand the details of other galaxies. Thus, although phenomena in our everyday macro-universe are directly observable, the discoveries of physics have inexorably led physicists to explore farther and farther beyond the limits of sensory observation. It is in the realm of the micro-universe and the mega-universe that the modern work in physics is being done and it is in such realms beyond the range of observation that the next new discoveries and revolutions in physics will occur.

In this and the following few chapters we will look more closely at the nature of the incredibly huge mega-universe which has been explored in a multitude of fascinating ways and with increasing enthusiasm and wonder in just the past few decades of history. The search is on for a coherent and comprehensible view of the nature and structure of the universe at large, but the first steps were taken long before the Einstein revolution in physics, and the first chapter in this story of exploration and discovery is still concerned with the outer limits of the macro-universe, the outer edges of the world of our everyday experience. For it was only when astronomers and physicists had achieved some grasp of the nature of our own sun, of neighboring stars, and the great galaxy of stars of which we are a part, that they were then able to look and think beyond.

Actually, that exploration had its beginnings thousands upon thousands of years ago. Throughout all history men have been deeply preoccupied with the nature of our sun, of the stars in the heavens, and of their relationship to our own earth. From earliest times men have observed the movement of the sun across the sky in its daily rounds, the waxing and waning of the moon, and the shapes of the great constellations of stars. For thousands of years the odd behavior of certain "wandering stars" in the sky was one of the most inexplicable physical phenomena observed by men, and hence the subject of uneasy speculation. It is hardly surprising that even the most primitive scientists were drawn to a study of the heavens, seeking relentlessly to discover the size, distance, and physical nature of the heavenly bodies they observed, to pry into their origins, seek to explain their movements, and ultimately to try to determine the shape, extent, and possible future of the universe.

Nor is it surprising that each time an answer was found to some astronomical puzzle, a dozen new questions seemed to arise, for early answers came from careful observation, and early scientists found that the more they observed in search of answers, the more strange and inexplicable the things they discovered which required still closer observation.

Thus over the centuries scientists found their concept of the universe growing from that of the comfortably small and manageable cosmos described by the ancients to that of the unimaginably huge cosmos of the modern physicists, astronomers, and cosmologists. When the ancient Greeks really stretched their imaginations to the limit in order to try to comprehend the over-all size and shape of the universe, they were still thinking of that universe as a discrete, finite expanse of flat earth that extended so many days' journey in one direction and so many days' journey in another and then dropped off, with the whole of it covered by a hemispherical dome of sky on the inside of which the stars were inscribed. The modern physicist today discusses the extent and shape of the universe in terms of an expanse or continuum of space-time so incredibly huge that not even the most scientifically trained and disciplined human mind can hope to visualize its true four-dimensional shape. Yet now it is upon this huge expanse of universe, with its stars and galaxies scattered like dust through the vastness of space, that the modern cosmologist has turned his most sophisticated instruments and his highest mathematical skills and powers of reason, in hopes of finding some orderly answers to the questions that have plagued men for millenniums.

So far the answers found have proved to be somewhat less than satisfactory. To many nonscientists, much that has been "discovered" about the size, shape, nature, and behavior of this incredibly huge mega-universe seems paradoxical or arbitrary, if not just plain incomprehensible. But for all the confusing and conflicting views held by the experts, scientists today have a far clearer idea of the wide variety of things which are *probably* true about this mega-universe than at any time in history.

What is more, some of the things that are true—or probably true—about the stars and galaxies have very down-to-earth applicability to our everyday lives. Thus it will be worth while to digress slightly at this point and turn our attention for a while to an area of discovery that is perhaps more in the field of the astronomer than the physicist, to see the picture of the vast mega-universe around us that the astronomers and cosmologists have slowly been piecing together, before we turn our attention to the world of the inconceivably small which occupies the attention of so many physicists today.

## THE EARLY OBSERVERS

In order to comprehend the theories, discoveries, and speculations of modern cosmology, we must look at them in perspective. First, we need some notion of the background of astronomical history from which these modern ideas arose. Second, we must have some idea of the powers and the limitations of the instruments that have made any kind of study of the mega-universe possible at all.

In an earlier chapter we discussed in some detail the step-by-step development of a visualization or "model" of our solar system that was built up bit by bit from the observation and discoveries of early astronomers. Ptolemy's earth-centered solar system gave way to the sun-centered system of Copernicus, and later Tycho Brahe and Kepler further modified that sun-centered system. But since the beginning of time, man's understanding of what he could observe and understand of all the things happening in the heavens depended sharply upon the instruments he had available for observation, and those instruments were terribly limited. In the earliest civilizations, men gazed at the stars to see what they could see with the naked eye. Of course they could observe the movement of the sun, the moon in its various phases, and a broad band of hazy light stretching across the heavens on dark nights—the band of starlight we know today as the Milky Way. In addition, they could distinguish perhaps some two thousand individual stars, including five "wanderers" which appeared to be pinpoints of light just like the other stars, but which moved in regularly predictable, sometimes even zigzagging patterns against a backdrop of "fixed stars."

Those early astronomers made amazingly accurate observations of the movement of these "wandering stars"—now known to be the five planets of our solar system that are most readily observed: Mercury, Venus, Mars, Jupiter, and Saturn. But it was not surprising that those early astronomers assumed without question that our planet Earth was the center of the universe, with the sun, the moon, and all the stars and planets revolving around it. This, after all, was what they *saw*, and there was no outside perspective to help them.

With the development of the earliest telescopes, a great step forward was taken in the observation and understanding of this outer fringe of the macro-universe. The telescope that Galileo invented—probably after hearing about earlier spy-glasses invented in Holland—was not particularly good. But even that crude telescope did allow a great scientist to look farther and see more than anyone had seen before. Galileo's telescope still showed the fixed stars only as pinpoints of light—just as modern-day telescopes do—but even so it revealed far more than the naked eye could see. With this crude instrument an astronomer could at last be certain that the Milky Way was really an enormous aggregation of individual stars, and not merely a hazy glowing halo of light across the sky. Galileo's telescope revealed the planets as disks rather than as pinpoints of light, and made it possible for the first time to see the four largest of Jupiter's moons, always invisible to the naked eye. (The eight remaining moons of Jupiter, far smaller, were not discovered until more powerful telescopes were invented.) Galileo's telescope also revealed the rings around the planet Saturn, and with it Galileo discovered that a number of stars which had always appeared to be single pinpoints of light to the naked eye were really double stars or "binaries."*

Above all, Galileo's telescope made truly accurate observation and measurement of the movements of the planets possible. Thus it stimulated great interest in the study of these heavenly bodies. In fact, most of the great astronomical work from Galileo's time clear down to the end of the nineteenth century was concentrated on the study of the planets in our own solar system, to the exclusion of anything else. Many stars beyond our solar system could be seen, of course, but relatively little could be learned about them except that they were there. And even though astronomers from the time of Copernicus on agreed that the sun was the true center of our solar system, the idea persisted that it must also be the "natural center" of the entire universe. It seemed obvious to those early men that the earth was the most significant of the planets, with the sun existing for the purpose of providing warmth and light. The slow annual movement of the other stars across the sky served as a guide for the planting of crops, harvesting, etc. What could be more natural than to assume that all the rest of the universe existed to serve the convenience of men on earth?

## THE GALAXY AND ITS COMPONENTS

This rather limited view of the nature of the universe beyond the solar system could not last for long, of course. Bit by bit evidence to the contrary began to accumulate, until the notion of our solar system as the center of the universe began to look ridiculous. With the invention of more powerful telescopes, and the persistent work of a few determined men, the riddle of the movement of the planets in our own solar system was finally solved. Tycho Brahe recognized that the careful measurements of planetary motions that he made could not be possible if the orbits of the various planets were perfect circles around the sun. Brahe's student Johannes Kepler studied Brahe's observations in detail—a lifetime's work in itself—and finally found for the first time a geometric pattern for the orbits of the planets that consistently agreed with observations; the orbits of all the planets around the sun were not circles but ellipses, with the motion of the planets in their orbits speeding up or slowing down according to whether they were comparatively close to the sun or com-

* The star located at the outer lip of the Big Dipper, for example, is such a star. To the unaided eye it looks like a fairly bright single star, but with a good pair of binoculars you can see that it is really two stars set very closely together, not just one.

paratively far away. On the heels of Kepler's discovery came Newton's work in outlining the natural laws of motion, and the law of universal gravitation which explained why the earth and the other planets traveled in their endless orbits around the sun.

It was natural, then, for astronomers in the early 1800s to begin to look beyond our solar system and to study, as best they could, other stars in the universe beyond our own sun. These men soon came to realize that these other stars, even the closest, lay at enormous distances across space from our own solar system, and were separated from each other by equally enormous distances. Nevertheless some parts of the sky, such as the broad hazy band of the Milky Way, seemed virtually peppered with stars while other parts revealed relatively few. Why this uneven distribution? Gradually, astronomers began to realize that our sun was one of a great coherent aggregation of stars, a so-called galaxy, which numbered millions upon millions of stars, and that this huge galactic star cluster seemed to be floating quite by itself in an immense void of empty space.

What was the shape and nature of this galaxy? In over-all shape it resembled a great cartwheel or disk of stars, comparatively thin at the edge, with a cental "bulge" at the pivot point or "hub" of the wheel—the part of the galaxy which contained by far the greatest concentration of stars. This disk, together with all the stars composing it, revolved slowly around this central hub like a giant fiery pinwheel; it is now believed that stars at the rim of this galactic wheel require in excess of 400,000,000 years (as measured on an earthly calendar) to make a single revolution, and travel 40,000,000,000,000,000,000 ($4 \times 10^{19}$) miles in the process, some 1,000,000,000 times the total distance of earth's orbit around the sun.

As we might guess, astronomers at first took it for granted that our own sun was at the center of this galaxy, so that everything else in the galaxy revolved around it. But careful observation of the apparent motion of stars in the heavens over periods of decades and centuries proved this idea wrong. Far from being at the center of the galaxy, our sun was found to be located far out toward the rim of this great rotating wheel. Nor is our sun the magnificant blazing attention-getter that astronomers once imagined it might be to the eyes of alien observers. We now know that it is a rather ordinary and pedestrian star, commonplace in size, brightness, and color, and lost among hundreds and hundreds of thousands like it, obscure and quite remote from the center of the galactic concentration.

Gradually study of our Milky Way galaxy began to reveal a number of interesting things about it. First, the galaxy seemed to be populated by an amazing variety of different kinds of stars of various sizes, colors, and temperatures. At first astronomers were hard put to judge just how far away other stars in the galaxy were. Presently the *relative brightness* of the various stars came to be accepted as the most reliable yardstick for measuring a star's distance from the earth, and the perfectly enormous size of the galaxy began to strike home to the astronomers. Some stars were found to be very large and intensely hot, blazing with a fierce blue light. Others, somewhat smaller and cooler, were blue-white, white, or even yellow like our own sun. Still others, much cooler, glowed pale orange or red.

It was also found that the relative brightness of a star is closely related to its temperature and color. A cool red-glowing star comparatively near to our own sun might appear much fainter in the telescope than a hot blue-white star that was many times farther away. But to say "comparatively near to our sun" is not to imply that other stars are exactly cheek-by-jowl neighbors. The closest of them all, a small bright star named Alpha Centauri, is some 25 trillion miles away—a distance so great that light waves leaving that "nearest neighbor" and traveling 186,000 miles per second through space require over four years to reach our telescopes. And of all the other stars in the galaxy, only a mere dozen or so are closer than 300 trillion miles away from our own sun.

By studying and classifying stars according to their colors, relative brightness, and estimated distance from our sun, astronomers in the 1850s had found that a great many stars in our galaxy could be catalogued in an orderly sequence, ranging from large, hot, blue stars through the smaller white or yellow stars to the cool orange or red stars. But certain stars seemed to fall completely outside this so-called *main sequence* of stars. For instance, a number of pale orange and orange-red stars were discovered to be many hundreds of times brighter than other stars with the same color, and came to be known as "red giants." A few remarkable stars were found to be shining as much as a hundred times more brightly than even these giants, with all varieties of color from blue-white to orange —the "supergiants." At the other end of the scale, a few stars glowing with white light were so very faint that they could hardly be seen at all, and thus came to be called "white dwarfs."

Still another kind of curiosity was noted as well. Stars which had previously been very faint were occasionally seen to flare up with surprising brightness which might last anywhere from a few hours to several days before these stars subsided back to their former undistinguished light. It was as if these stars had suddenly undergone some kind of cosmic explosion for some reason or another. A star showing such surprising "new brightness" was called a "nova." Ordinary novae were found to occur quite frequently, but on very rare occasions a novalike explosion of a star would occur with such extraordinary brilliance and last for so long—months or even years—that for a while it would appear brighter than any other object in the heavens. No one knows for certain whether the Star of Bethlehem, for example, was merely a symbolic embellishment of the Nativity story, or whether it actually existed in nature as a rare "supernova" display lighting up the sky for a period of weeks some two thousand years ago; no reliable record of such a display exists, and modern astronomers speculate that it may have been nothing more than a rare conjunction of three of the planets in an area of sky about as big as the moon. But history does record that Chinese astronomers observed a supernova in the heavens on the evening of July 4 in the year A.D. 1054, and even today astronomers can see the remains of that ancient explosion in the still-glowing and expanding cloud of gas that we know as the Crab Nebula. Subsequently Tycho Brahe and Johannes Kepler each was lucky enough to observe a supernova during his lifetime.

Still other characteristics and differences among the stars in the galaxy were noted by classical astronomers. With the use of highly refined and powerful telescopes, observers were startled to discover that many stars—perhaps as many as half the stars in the entire galaxy—were not single stars like our own sun at all, but rather were *multiple stars*: combinations of two, three, four, or even more stars in tightly grouped clusters, revolving perpetually in intricate orbital patterns around each other. Furthermore, a number of stars were found to vary greatly in the intensity of their light at regular intervals. Unlike our own sun, which glows with a steady and reliable intensity without significant change from one day to the next or even from one century to the next, these *variable stars* would change in their intensity of brightness in a rhythmic pattern, flaring up in gaudy display for a period of time, then subsiding to faintness for a period, then becoming bright again, then faint again in a recurring cycle. There seemed to be no rhyme nor reason to the way such variable stars behaved; some would flare and subside regularly every four or five hours, while others flared at intervals of weeks, months, or even years. When these pulsating stars, now known as "Cepheid variables" or "Cepheids" were first discovered, no one had any idea why they behaved as they did, but it became clear that all stars were not by any means as stable in their light-producing qualities as our own sun.

Naturally, as astronomers explored the immense variety of stars in our galaxy, they began to wonder if our own sun was in any way unique at all. Certainly the sun has one very striking characteristic: It has a family of at least nine planets traveling in orbit around it, and several of these planets even have satellites of their own. Was this a truly unique characteristic of the sun, or was it possible that other stars might also have solar systems, perhaps even life-bearing planets? The more that was learned about other stars the more this began to appear not only possible but probable. More and more physicists were coming to realize that for all its apparent disorder and complexity, there was a certain underlying order in the structure of the universe. Even such apparently unique events as the appearance of novae or supernovae appeared "unique" only because men had not been around watching for them very long. Such "unique" events, physicists realized, might actually be fairly commonplace occurrences in the galaxy when viewed in the perspective of geological ages of millions or even billions of years.

Thus, it seemed unlikely indeed that the formation of a solar system of planets around a given star could be a truly unique event. At first some astronomers insisted that the formation of solar systems, if not exactly unique, might at least occur only rarely under very unusual circumstances of star formation, so that if there *were* other solar systems in the galaxy they would be few and far between. But even this hope has begun to look forlorn in recent years. True, no planets have yet actually been observed in attendance around any star. But among the relatively few stars close enough to our sun to have been observed and studied in careful detail, at least two—the binary star systems 61 Cygni and 70 Ophiuchi—are believed almost certainly to have at least one huge planet apiece. In the case of each, *something* causes slight deviations in the expected orbits of these two pairs of stars moving around each other, and deviations of this sort are most probably the result of the gravitational effects of huge planets which are just too dim to be seen with our telescopes.

If true, the implications are startling. If *two* such star systems as close to us as these in our own galaxy have planetary satellites revolving around them, it is hard indeed to believe that solar systems can be anything but commonplace in the universe. Most astronomers today believe that in our own galaxy alone there are probably millions of stars with planetary systems of one sort or another, and most concede that among so many planetary systems there must be multitudes of planets—tens of thousands, perhaps even millions—which have the conditions necessary to support some form of life similar to the kind of life that has developed on earth.

## THE WORKING TOOLS OF ASTRONOMY

Obviously a great amount of study and work over a period of centuries was needed to accumulate this mass of information about such a widely varied population of stars in the galaxy. And much of this information was gathered under extreme difficulty. For one thing, astronomers were limited by the very instruments they were using to study the heavens.

The usefulness of a telescope depends upon three things: the light-gathering power of its lenses, the efficiency of those lenses in focusing images of heavenly bodies and the ability of other lenses to magnify those focused images. Galileo's telescope was little more than an overgrown spyglass. It permitted him to look farther out than he could see with the naked eye, but not a great deal farther out in terms of astronomical distances. Later on, larger telescopes were built, with larger lenses to provide better light-gathering power and sharper focusing, but with these instruments another problem intruded. When a glass lens gathered light from distant stars and then bent that light to a focal point, the bending of the light rays took place according to the refractive power of the glass, and light of varying wavelengths was refracted at varying angles, broken up into its various color components much as light was broken up into a rainbow spectrum in passing through Newton's prism. Thus the star image reached the eye of the observer as a sort of a chromatic blur, especially marked at the outer edges of the telescope's field (see Fig. 33a). Special "achromatic" lenses had to be built into the telescopes to correct this aberration, but inevitably these lenses produced other aberrations of their own, so that the image the astronomer actually observed came to him "third hand," so to speak.

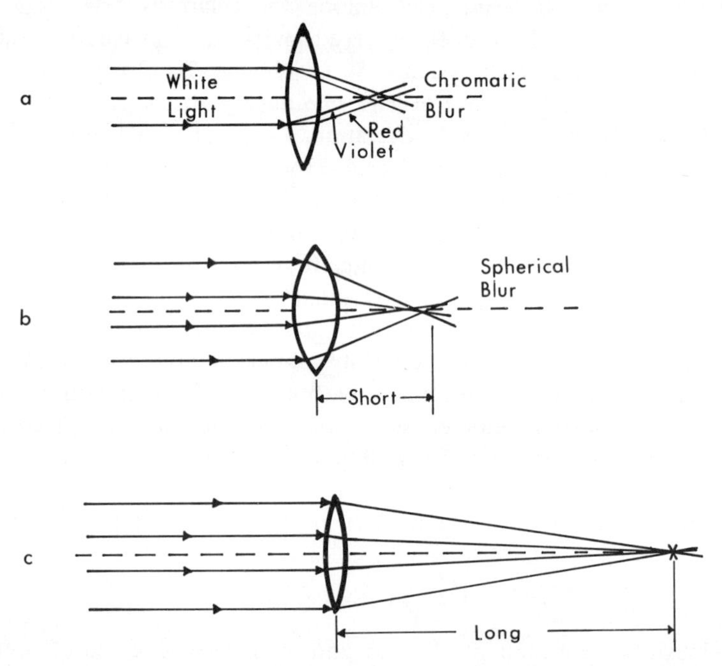


*Fig. 33* The trouble with refractor telescopes: chromatic and spherical aberration were chronic nuisances. One solution: to lengthen the focal length—the distance between the lens and the eyepiece—by special grinding of the lens. But this led to telescopes that were too long to be either practical or maneuverable.

Furthermore, these "refractor" telescopes (so called because starlight was passed through a glass lens and *refracted* to bring it to a focus) were rather sharply limited in size, since any imperfection in the glass, however minor, introduced additional distortion. Yet the only way to increase the light-gathering power and the resolving power of a refractor telescope was to increase the diameter of the refracting lens.

In addition, early astronomers were plagued with still another distortion of light known as "spherical aberration." The lenses of those early telescopes were normally ground in a spherical contour, and in large telescopes such spherical lenses could not bring a distant star image into absolutely clear focus (see Fig. 33b). Attempts were made to overcome this spherical aberration by grinding lenses with longer and longer focal lengths, so that the light-gathering lens at the front end of the telescope had to be mounted farther and farther away from the eyepiece (see Fig. 33c).

In the late 1700s and early 1800s such long-focus refracting telescopes reached an extreme of clumsiness, with the light-gathering lenses, suspended 60 feet, 120 feet or even as far as 210 feet away from the eyepiece, held in position by an immense framework. Refracting lenses up to 50 inches in diameter were built for these long-focus telescopes, but unfortunately the problems involved in using such awkward devices increased in direct proportion to their size. By the year 1847 these ugly ducklings of astronomy had been abandoned, and the largest functioning refractor telescope was one with a lens 15 inches in diameter located at Harvard College Observatory and a similar one in Russia. Today the 40-inch Yerkes telescope, housed in a Wisconsin observatory eighty miles northwest of Chicago, is the largest refracting telescope in the world. It is a splendid instrument of its type, but it requires highly skillful manipulation of the light passing through its lenses in order to obtain images without aberration, and can be used effectively to observe only a limited range of celestial objects.

Isaac Newton sought to overcome the problem of chromatic and spherical aberration in refracting telescopes by building a telescope that operated on a different principle entirely. Instead of placing a refracting lens in his telescope to collect light, he built a long empty tube with a *reflecting mirror* at the bottom end (see Fig. 34). This mirror was ground until it had a highly polished concave reflecting surface. Starlight traveling down the tube was reflected by this concave mirror to a focal point somewhere along the length of the tube where a tiny flat mirror set at an angle could reflect the focused image into an eyepiece lens.

The basic structure of such a "reflecting telescope" is diagramed in Figure 34. Reflecting telescopes had the enormous advantage that the light from the stars never actually had to pass *through* glass at all until it reached the thin magnifying lens of the eyepiece. Consequently it was not broken up into chromatic bands because of the refraction of light of varying wavelengths. Furthermore, imperfections deep in the glass of the reflecting telescope's mirror could be tolerated as long as the surface of the mirror itself was entirely free of blemishes, so it became technically possible to cast extremely large mirrors and thus build telescopes with vastly increased light-gathering power. Absolute perfection of focus could be obtained by grinding the mirrors in a concave *parabolic* curve rather than in a spherical curve, so that spherical aberration was avoided and perfectly focused pinpoint images of extremely distant objects could be observed through the eyepiece.

Today refractor telescopes still remain the most suitable instruments for the study of the surface of the moon, the planets, or other nearby celestial objects, but the job of searching the vast distances of the heavens has fallen almost exclusively to reflecting telescopes. Gradually the art of making these instruments improved, reaching a high point in the casting and grinding of the 200-inch mirror that was mounted in the Hale telescope on Mt. Palomar, the largest reflecting telescope ever made, and even today the single most powerful and useful astronomical instrument ever designed.

But while the techniques for making telescopes steadily improved, astronomers faced still other problems in observing. Even with the best of telescopes, everything they studied in the heavens had to be observed

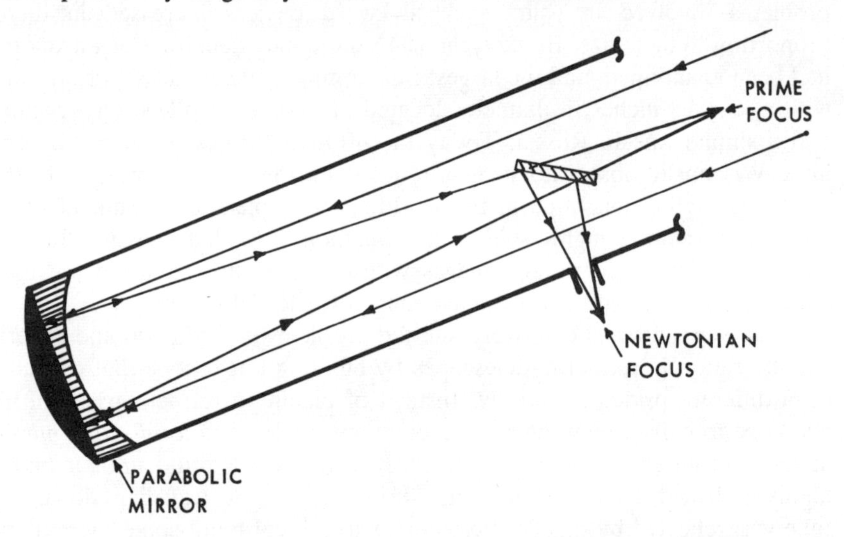


*Fig. 34* A simplified diagram of a reflector telescope. Starlight is collected and reflected without aberration by a parabolic mirror at the bottom of a long tube, and focused at some point along the tube's length. A small mirror suspended within the tube reflects focused image through an eyepiece at the side of the tube. Small suspended mirror does not interfere significantly with light-gathering by the large parabolic mirror.

through a thick layer of the earth's atmosphere. Often, cloud cover made observation impossible. Even when the sky was clear, atmospheric turbulence could distort the light coming from the stars, so that astronomers were much like sea captains trying to see a lighthouse beacon through rain and fog. As our cities grew in size, increasing clouds of smoke and dust around them made astronomical observation more difficult and sky glow at night became a constant annoyance. Finally, astronomers took their instruments and fled to the country in order to escape these man-made disturbances, and began building observatories on high mountain tops in order to get above the lowest and thickest mile or so of the earth's atmospheric blanket.

Even so, atmospheric turbulence remains a major headache to astronomers today, and there are few who do not dream of that utopian day when a reflecting instrument of the size and power of Mt. Palomar telescope can be mounted in an observatory on the surface of the moon where no atmosphere can interfere with observation—or better yet, on an orbiting satellite far beyond earth's atmosphere, so that observation of the heavens in all directions will be possible at any time.*

## THE PROBLEM OF ASTRONOMICAL DISTANCES

Another difficulty in observing the stars and learning much about them arose from the extremely great distances that separated them from the observer on earth. In observing the planets in our solar system, early astronomers were able to make fairly accurate estimates of their distance by measurement of "parallax"—the apparent movement of the planets in the sky against the background of more distant stars from one time of observation to another. By measuring the angle of parallax of a planet and then calculating the distance by simple trigonometry, surprisingly accurate values were found.

Even within our own solar system these distances were almost unimaginably great: the sun, for example, is an average of 93 million miles from the earth, the moon is over 200,000 miles away, and the planet Mars may be anywhere from 30 million to 60 million miles away even when it is in conjunction with the earth—that is, lined up on the same side of the sun as the earth is and thus as close to the earth as it ever comes. Such huge numbers were clumsy to work with, so astronomers agreed to measure and describe distances within the solar system in terms of *astronomical units,* with one astronomical unit equal to the average distance from the sun to the earth, 93 million miles.

The measurement of parallax was an extremely useful technique for estimating distances of other objects within the solar system. We use the same technique quite unconsciously every day in judging the comparative distance of objects around us, except that we speak of it as "depth perception." For example, if we look at a drinking glass on a table across the room from us first with one eye and then with the other, we see that it *appears to move* in relation to some fixed object on the wall beyond it. If we try to judge the distance of the glass using a single eye, we run into difficulty; we cannot tell whether it is a large glass clear across the room or a

* To date there is no really large and technically perfect telescope *anywhere at all* in the southern hemisphere, so that even now virtually all of our knowledge of the heavens is gathered from observation of northern skies. An orbiting telescope would solve that problem; even a comparatively small instrument on a manned space station would reveal more than the best existing telescopes located on the earth's surface in the southern hemisphere.

small glass much closer to us. But by measuring the apparent motion of the glass against the background wall when we look at it first with one eye and then with the other, we can find the angle of parallax (a triangle from one eye to the glass and back to the other eye) and then determine the distance of the glass quite accurately by trigonometry.*

Thus the measurement of parallax was very useful in judging distances within the solar system. But by comparison, the distances even to the nearest stars are so very much greater than "local solar system" distances that parallax measurement became totally useless even for rough guesses. In 1838 an astronomer named Friedrich Vessel attempted to measure the parallax of one of the nearest stars to earth, the 61 Cygni binary star we mentioned earlier. To do this he sighted the star against the background of more distant stars at six-month intervals, but after much labor found the angle of parallax to be only about three-tenths of one second of an arc. Considering the possible margin for error just in measuring such a tiny angle of parallax, Vessel estimated that 61 Cygni had to be in the neighborhood of 65 trillion miles away from the earth—65 *trillion* miles! Soon thereafter Thomas Henderson in South Africa found that a closer star, Alpha Centauri, was only about 25 trillion miles away—and Alpha Centauri is our sun's nearest neighboring star. It became clear that astronomical units would not do in measuring such incredibly great interstellar distances. Instead, astronomers began describing these distances in terms of "light-years," the number of miles that a beam of light will travel in the course of a whole year, moving at a velocity of 186,000 miles per second. We can calculate that a single light-year is a distance of approximately 6 trillion miles.

Thus in the case of a few "nearby" stars the trigonometric methods of parallax had some limited usefulness in estimating distances, but when stars were some 300 light-years or so away, any observable angle of parallax became so immeasurably tiny that the method was hopelessly inadequate for estimating distances. Of course, this reminds us forcefully of the perfectly staggering immensity of our galaxy, but it also illustrates one of the odd curiosities connected with observing any of the distant stars that lie within it. If we measure astronomical distances in terms of light-years, we see that our own sun is a mere 8 "light-minutes" away from the earth. That means that light waves from our sun require 8 minutes in order to reach the earth, so that anything out of the ordinary that we observe happening on the surface of the sun—the appearance of a solar flare during an

* Fortunately we don't ordinarily have to go to all that trouble; when we look at the glass with both eyes open, our mind does the judging automatically. We see the room and the glass "in depth" in three-dimensional perspective, and then make a quite accurate instinctive guess of the distance to the glass, without even the aid of a tape measure.

eclipse of the sun, for example—has actually occurred *eight minutes before we can see it.* When we say that Alpha Centauri is 4.3 light-years away, or 61 Cygni is just less than 11 light-years away, we are really saying that the *image we observe* of these stars at any given moment as a result of light traveling from them has taken 4.3 years or almost 11 years, respectively, to reach us. Thus the images we see in telescopes are really only shadows of what those stars looked like 4.3 years ago and almost 11 years ago respectively. If 61 Cygni were to undergo a novalike explosion today, the light from that explosion would take almost 11 years to reach the earth, so that we would have no way to know this cosmic catastrophe had happened at all until more than a decade after it was all over!

This "observational time lag" obviously becomes even more marked in the case of stars that are far more distant than these nearby neighbors. Astronomers today have studied stars estimated to be thousands or even millions of light-years away—but what we observe of such distant stars in our earthly telescopes is actually not what is happening *now* on those stars, but rather what *did* happen on those stars thousands or even millions of years ago.

In fact, astronomers peering at these distant stars are actually peering *back into time,* observing images carried by light waves that were emitted by those stars in the very remote past, in some cases, long before man had even appeared on the earth, in other cases long before our own sun had even formed! How, then, is it possible to tell what is happening on those stars *now*?

The answer, odd as it may seem, is that it is *not* possible to know anything about what is happening now in those distant areas of the universe. There is no way even to guess, for instance, that some cosmic catastrophe might be occurring today on such a distant star unless we are prepared to wait the necessary thousands or millions of years required for the light being created by such a catastrophe today finally to reach the earth. Since we know that no signal of any sort can travel faster than the speed of light, there is no possible way we can *ever* learn what is happening on distant stars and in distant galaxies at the present time. The best that we can ever observe is what happened at some time in the distant past, perhaps millenniums ago.

But if measurement of parallax is inefficient or useless for measuring or even estimating the vast distances of other stars, then what other method can be used? For centuries astronomers puzzled over alternative techniques. One method depended upon the relative brightness of stars. While it was true that there seemed to be a wide variety of different kinds of stars in our galaxy—stars of different colors, different sizes or stars traveling in pairs, trios, or clusters—there was still some limit to the variety. Astronomers found that most stars fell into very clearly defined and easily recognized categories, and that stars in each such class or category were roughly about the same size, had the same color and temperatures, and shone with the same brightness no matter whether they were comparatively close to our sun or immensely distant. But the *observed* brightness of such stars depended greatly upon how far away they were; the more distant a star of a given class, the dimmer it appeared.

Thus when stars were found to be outside the range of measurement by angle of parallax, astronomers could still estimate their distance away by the apparent decrease in their brightness. And as new techniques were discovered in the mid-1800s for classifying stars, predicting their movements, and estimating their masses, it became possible to make more and more accurate estimates of the distance of any star under observation.

In addition, astronomers fortunately did not have to depend entirely upon direct vision by means of telescopes to gather information about the stars. In the early half of the nineteenth century, another instrument was invented capable of providing a completely different type of information which no telescope could hope to supply. This instrument was, of course, the spectroscope. Very soon after Newton had discovered that white light from the sun could be broken up into a rainbow spectrum of light of differing wavelengths, certain clever astronomers tried passing light from the stars through prisms. They found that the light from any star could also be spread out into a rainbow band. As might be expected, however, stars which appeared red or orange in the telescope showed a predominantly red or orange spectrum when passed through a prism, while light from a blue-white star fell primarily in the green, blue, and violet wavelengths of the spectrum.

Then in 1814 a German optician named Joseph von Fraunhofer discovered a curious thing: A beam of sunlight passing through a prism did not, in fact, form quite as perfect a rainbow spectrum as Newton had thought. Inspecting such a spectrum closely, he found that the colored band of light was actually interrupted by hundreds of very slender, vertical, dark lines. Fraunhofer carefully described the location of as many of these lines in the sun's spectrum as he could find (today we know that there are thousands of them) even though he had no idea what caused them or why they were there.

These so-called Fraunhofer lines in the sun's spectrum remained a mystery until 1859, when Gustaf Kirchhoff demonstrated that the light thrown off by a number of common chemical elements, heated to incandescence in the laboratory, showed extra-bright lines of color in their spectra which coincided perfectly with the lines of darkness that Fraunhofer had observed. Indeed, it was found that each different chemical substance, when heated, gave off such an individually characteristic pattern of spectral lines that the pattern alone could be used to identify the substance accurately, just as fingerprints alone can be used to identify an ax murderer once his prints have been catalogued and filed.

Today we know the reason for these curious findings. A continuous rainbow spectrum is produced only when some *solid* or *liquid* material is heated glowing hot, or when a dense gas is heated vigorously under very high pressure. Thus, the light thrown from a tungsten filament in a light globe, heated to incandescence, will produce a smooth rainbow spectrum when passed through a prism. But hot gases of low density (such as vaporizing iron or calcium, for example) produce characteristic and unique spectra broken into narrow bright lines of color such as Kirchhoff observed. On the other hand, when a hot solid or hot dense gas which might ordinarily be expected to produce a continuous rainbow spectrum is surrounded by a mantle of cooler gases, those cooler gases absorb or *blot out* the light in exactly those parts of the spectrum where bright lines would appear if those surrounding gases were hot and compressed.

But this, astronomers realized, was exactly what was happening in the case of stars emitting light: An intensely hot and dense central core of gas was glowing fiercely and emitting light waves which had to pass through cooler outer layers of gas before they reached outer space. By matching the black lines seen in Fraunhofer's "absorption spectrum" of the sun with the bright lines of the "emission spectra" of various known elements studied in the laboratory, astronomers were able to identify exactly which chemical elements were actually present in the outer layer of gases of the sun! The same was found true of other stars when their light was analyzed by the spectroscope, and soon astronomers had found that various stars could be classified in characteristic "spectral groups"—another highly useful means of classifying stars into categories or families, and thus of judging their distances from the earth.

This in itself was a giant step forward in the study of stars, but astronomers found the spectroscope useful in other ways as well. This instrument was useful, for example, in helping to determine the surface temperature of stars. A blue-white star such as Rigel with a surface temperature of some 20,000 degrees Fahrenheit or more, shows relatively few lines in its absorption spectrum. Physicists today know that comparatively few elements can exist in their natural state at such high temperatures. In the case of cooler stars such as Aldebaran, an orange star with a surface temperature in the neighborhood of 7,000 degrees Fahrenheit, far more lines are present in the absorption spectrum, indicating that many more elements are present there in their natural state. Thus the presence or absence of absorption lines in a star's spectrum provided astronomers with a rough guide to the surface temperatures of a wide variety of stars.

The spectroscope also provided an important means of judging the speed with which a given star might be moving relative to the earth. When a star is in motion toward the earth its spectral lines are found to be shifted slightly toward the blue or left-hand end of the spectrum, as though light waves traveling from it are being squeezed together slightly. But if the star is moving away from the earth, the shift of its spectral lines is toward the right or red end of the spectrum, as though the wavelength of light traveling from it is being stretched out or lengthened. It was found that the faster a star was moving in either direction, the more its light wavelengths were shifted. Thus, by calibrating this shift of the spectral lines of a given star in comparison to laboratory standards, it was possible not only to judge in which direction a given star was moving relative to the earth but to make a good estimate of the speed with which it was moving!

Important as the spectroscope became in astronomy, a far more simple device became even more important: the ordinary photographic plate. The silver emulsion on a photographic plate had a great advantage over the human eye in making astronomical observations: It could "collect" accumulated light from very, very dim stars over a period of hours, and thus reveal the presence of stars that human eyes could never hope to see. A photographic plate placed at the focal point of a reflecting telescope and exposed for a minute would reveal many stars in the heavens on a particular night. Exposed for an hour instead of a minute, it would reveal thousands more stars that are too dim to register in the briefer exposure. An all-night exposure would demonstrate multitudes of still more dim and distant stars which could never have been detected or observed in any other way.

Furthermore, unlike the transient observation possible with the human eye, a photographic plate of a given section of the heavens could be kept as a permanent record. Future photographs of the same segment of the heavens taken by the same instrument with the same exposure at various future times could then be carefully matched and compared with earlier plates to reveal any change that might have taken place in the heavens too slowly for direct vision to detect it.

This use of photographic plates produced some startling and historical discoveries. After Neptune, the eighth planet, had been discovered astronomers had begun to suspect strongly that yet a ninth planet existed in our solar system still farther out from the sun, but no one had been able to find it. The planet Pluto was finally revealed in the examination of a series of photographic plates taken of a given section of the heavens on successive nights. The fixed stars on those plates remained serenely in their place from one night to the next, but one "star" was found to appear at a slightly different place on each successive photograph, moving, as it were, against a backdrop of fixed stars. Even though it appeared as nothing more than a very dim star on any individual plate, there was little doubt that this "wandering star" had to be the elusive distant planet so many astronomers had been searching for.

Considering the limitations of their crude telescopes, early astronomers had accumulated a surprising amount of information, but they had actually not scratched the surface. Really good telescopes, spectroscopes, and photographic plates all became available to astronomers within a fairly brief period of time in the middle and late 1800s, and with these refined tools to aid in observation, an intensive wave of study of the skies began. Stars were identified and classified according to their size, brightness, location in the heavens, variable characteristics of light emission, etc. Astronomers for the first time began to grasp the truly enormous size of our galaxy, to learn about its shape, about the manner in which it moved, and about the individual stars that moved within it. The exact physical characteristics of the galaxy were hard to determine at first because earth-bound observers were in such an awkward place from which to observe. It was like trying to judge the size and shape of a house by standing inside and peeking out through a crack. But bit by bit knowledge accumulated. Groups of stars within the galaxy were identified and studied as astronomers came to realize that probably only one star out of every four in the galaxy is a "singleton" such as our own sun. Three out of four stars are teamed up in pairs, trios, quartets, or closely associated groups made up of still larger numbers of stars moving as a unified star complex.

And as astronomers worked, striving to classify the stars in the heavens and groping for explanations of how stars were formed, why there were so many different kinds, how this island universe happened to be formed, and what might happen to it as time went on, they made a disturbing discovery. It became increasingly clear that the great reaches of "empty" space between the stars in our galaxy were not really empty at all, but were filled with great swirling clouds of gas and dust, more tenuous and rarefied by far than any gas known on earth yet dense enough, when gathered into huge clouds, to obscure or blot out completely the starlight shining from many parts of the galaxy.

These huge clouds of interstellar dust and gas at first were ignored as a nuisance—they interfered with astronomical observation. But soon their very existence forced astronomers to begin reassessing another age-old assumption about the nature of the universe. Men had always assumed that the solid material substance of the universe was concentrated primarily in the stars and planets which moved about the galaxy in groups and clusters through virtually empty space. But as more and more immense clouds of dust and gas were found between the stars and calculations were made of the actual mass, volume, and density of matter scattered throughout those clouds, investigators began to wonder if this concept might not be completely backward. More and more it began to appear that these clouds of gas and dust, in aggregate, must actually contain *far more* of the total amount of matter in the universe than all of the stars and planets put together!

What was more, astronomers began to realize that these gas clouds were, in fact, the very stuff from which the stars and the planets themselves were made in the first place, and that the prime mover in the formation of individual stars—perhaps even in the formation of the whole galaxy—was none other than our old friend the force of universal gravitation. And slowly, haltingly, as more and more observations were made, astronomers and physicists for the first time began to develop a plausible concept of how these islands of matter in the universe had originally come to be—a picture, however fuzzy, of the possible natural history of stars, solar systems, and galaxies.

# CHAPTER 19

## *Macro-Universe: The Birth of Stars and Planets*

If we try to choose the single oldest and most persistent of all the questions that have ever been tackled by scientists throughout history, we would unquestionably find the following among the strongest contenders:

*Where did the stars and the planets come from?*
*How, and from what, were they formed, and when?*

Oddly enough, for all the growth of science since Galileo's time, there were no answers to these questions even remotely plausible until the early years of our own twentieth century.

All of us at one time or another have observed smoke rising from a cigarette or pipe and spreading out in whirls and eddies through the air of the room. The motion of smoke is endlessly fascinating to watch; it rises, sinks, concentrates, diffuses, moves in great swirling circles or settles back upon itself, subject in its motion to the action of even the tiniest imperceptible air current that may be present. We know that tobacco smoke is not a single substance, but is composed of a variety of gases—carbon dioxide, nitrogen, and water vapor, for example—as well as minute particles of dust (carbon particles and tiny bits of ash) and droplets of vaporized tars and resins. Whatever its exact composition, however, we know that the tobacco smoke will constantly be in motion, no matter how still the air in the room, until it is completely diffused and dissipated. Even if the smoke were released into a vacuum tube, where no air currents could affect its motion, we would still observe that the smoke was never still. Molecules of gases racing about the container would constantly be bumping and joggling the larger dust particles and droplets; the motion of the smoke cloud might not be as dramatic as if the smoke were released in an air-filled room, but motion would still be present.

Compared to the thick concentration of gas and dust in a cloud of tobacco smoke, the great clouds of dust and gas in the empty reaches of space between the stars would seem so tenuous and rarefied as to be practically nonexistent. We know today that the most perfect vacuum ever achieved on earth still contains *more* molecules of gas per cubic centimeter than would be found in the most dense concentration of gas in the most dense of all celestial gas clouds ever observed. Yet what these enormous billowing clouds of dust and gas in outer space may lack in concentration and density, they make up in sheer volume. These clouds are immense, filling up areas of space hundreds and thousands of light-years across.

## THE COSMIC WHIRLPOOL

But if the clouds of gas and dust filling such huge areas of our galaxy are so rarefied, how was it possible for astronomers to discover their presence at all? These celestial dust and gas clouds were first detected in one of two ways. In some areas of the galaxy they were so vast and so deep that they totally obscured the stars that lay beyond them. Thus astronomers of a century ago, studying some portion of the heavens that seemed literally peppered with stars, would come across a huge area immediately adjacent in which no stars could be seen at all. Since it did not make sense that there should be such star-free "holes" in the heavens extending as far as telescopes could probe, astronomers were forced to conclude that the stars in these "empty" areas were simply obscured from their sight by great clouds of gas or dust that lay between them and their earthly observers. Such clouds (which themselves seemed to contain no stars) came to be called "nebulae," a sort of wastebasket term used to describe any vague, ill-defined cloud of apparently insubstantial stuff discovered in the heavens.

As we mentioned, these star-obscuring nebulae were at first considered an astronomical nuisance. But many of them were of such great size or of such characteristic shape that they became familiar landmarks in the heavens, and were identified by their own individual names. One such is a great dark cloud of cosmic dust shaped like the head of a horse which has been known for centuries as the Horsehead Nebula.

But others of these nebulae, far from being dark, were found to glow with such a brilliant luminosity that they were themselves at first mistaken for stars. Over the years astronomers discovered that many such immense glowing dust clouds had intensely bright stars in their midst, so that the glow of the gas in the nebulae seemed to be a sort of fluorescence not unlike the fluorescence of the neon Grandma's Cookies sign which shatters the nighttime beauty near Seattle's university district with its repellent glare.

Still other such glowing nebulae also contained much fainter stars, but seemed to be mostly composed of the debris of past celestial explosions of enormous magnitude. As we mentioned earlier, at least one such nebula has an extremely well-documented history. Almost a thousand years ago astronomers in the Orient observed and recorded what appeared to be an immensely brilliant explosion in the sky—a faint star which suddenly became incredibly bright, outshining even the nighttime brilliance of our huge planetary neighbor, Jupiter. This display of star fire continued for over two years before it gradually subsided; but astronomers of today studying the area of sky in which this cosmic accident occurred find not a star but a huge irregular glowing cloud of dust and gas which, as we noted before, is shaped for all the world like a crab and hence is known as the Crab Nebula.

Modern astronomers are certain that prior to the year A.D. 1054 astronomers would have seen nothing in that area of the heavens but an ordinary star of ordinary brightness, quite indistinguishable from multitudes of others. The sudden change those ancient observers witnessed was, in fact, a rare astronomical event, the explosion of a supernova; and what can be seen today is nothing more than the still-expanding cloud of gas, dust, and rubble thrown off in that mighty blowup. Of course we know that what we see in our telescopes today (and what those early astronomers saw almost a thousand years ago) in the Crab Nebula actually occurred thousands of years earlier. The Crab Nebula is estimated to be approximately three thousand light-years away from earth, so that light traveling from the supernova explosion long in the past took three thousand years to reach the earth and be seen by the early astronomers. Similarly, what we are witnessing today is actually what *did* happen to that exploded star three thousand years ago.

Whatever the source of these great clouds of gas and dust, whether they were dark enough to obscure the stars beyond them, or were fluorescing nebulae that were themselves emitting light, there was no question that they were indeed massive, and were constantly in motion. It is hard for us to understand how any collection of gas and dust as rarefied as this could be considered to have any significant amount of mass at all, until we begin to think of the enormous *size* of these clouds. If we found a total of only one lone hydrogen atom in every cubic centimeter of otherwise empty space, we would have to admit that the tiny mass of that lone hydrogen atom might hardly deserve acknowledgment. But in an area of space encompassing 5 million billion cubic *miles,* with each cubic centimeter containing one hydrogen atom, we can begin to see that such a "cloud" of gas existing in that unthinkably huge expanse of space would indeed add up to a significant amount of matter with a significant mass if it were all taken together.

What is more, those atoms of gas would have to be in constant motion. Any heat at all above absolute zero would excite them into *some* degree of motion or other, so that in this vast expanse of space there would be atoms of gas constantly moving and colliding with each other, even if one atom had to travel an enormous distance to find another. We also know that such clouds of gas are being "pressed in" constantly on all sides by the pressure of light from multitudes of stars, since we know that light has some mass of its own and therefore exerts a small but significant pressure on solid matter of any sort anywhere in space.

Astronomers reasoned, therefore, that any small portion of such a celestial cloud of dust and gas might indeed have little mass and thus exert very little gravitational attraction on anything else, but the *cloud as a whole* could have a perfectly huge aggregate mass and thus be governed by the law of universal gravitation. However tiny the gravitational pull such a cloud might exert, the central part of the cloud, in aggregate, would exert *some* gravitational pull upon the even more rarefied outer portions of the cloud. Thus the matter in such a dust cloud would constantly be kept in motion by the influence of its own gravity acting upon its component molecules. And from this point it began to seem possible that these enormous dust clouds in space might, in fact, be the raw material from which all stars and solar systems were formed.

## THE BIRTH OF A STAR

Over the centuries many theories have been proposed to explain the formation of stars and solar systems. All such theories took into account the enormous hugeness of the areas of space in which stars are formed, areas thousands of light-years across. And according to the most plausible and widely accepted of these theories, stars and solar systems do indeed form from great swirls of moving gas and dust that lie in such empty spaces between the stars.

As we have seen, the gas and dust particles in such a cloud or nebula are constantly in motion. They are also constantly subjected to an infinitesimally small but very real tug of gravity arising from the center of the cloud. Over millions or billions of years in which a given gas cloud is swirling and twisting like a cloud of cigarette smoke in a still room, more and more of the mass of that cloud is drawn bit by bit toward its center. The more concentrated the gas at the center becomes, the greater the aggregation of mass in that area and the stronger the gravitational pull it must exert on outlying atoms of gas, particles of dust, or frozen droplets of vapor.

But the motion of such gas atoms and dust particles is not determined entirely by gravitational pull. Collisions with other atoms or particles send the collision fragments hurtling off into space in all directions. Some such particles might continue traveling outward for light-years before ever colliding with another particle, but others which are moving away from the center of the gas cloud must fight the tiny inward gravitational pull exerted upon them, and instead of flying away free are "caught" by the gravitational field of the concentrating cloud nucleus and forced into a circular orbit around it. Over eons of time more and more such particles that are too far out to be pulled into the center are at least held captive; given sufficient time the entire cloud of gas begins to take on a disk-shaped configuration, thick and dense in the center, far more tenuous at the edges, with the whole disk beginning to rotate first very slowly and then faster and faster around a central hub or axis, like a huge celestial pinwheel.

Thus an observer watching from a great distance might see a formerly disorganized cloud of gas and dust begin to take a peculiar shape and configuration, bit by bit. But if he could observe what was happening *within* this great rotating cloud in greater detail, he would find that whirls and eddies of concentrated gas were beginning to form outside the central area. Out toward the periphery of the disk there would be special areas where particles of dust and molecules of gas happened to be a bit more dense than in surrounding areas, and had begun to form central concentrations of their own, each exerting its own gravitational pull. If the observer came back a few million years later, he would find that the larger of these eddies had themselves become foci of mass within the larger disk, not nearly as large as the huge central aggregation, perhaps, but significantly large enough to be gathering orbiting streams of gas and particles around themselves even as they themselves continued to rotate around the center of the great pinwheel. Presently these eddies or foci become concentrated into "protoplanets" while the huge main nucleus of the entire gas cloud has become denser and denser due to the mass it had attracted inward with its growing gravitational force, as shown in diagram form in Figure 35a.

But this central nucleus does not merely become denser; it also becomes hotter and hotter. The pressure of the gases concentrated here becomes greater and greater, and when gas molecules are concentrated under pressure, more and more of their kinetic energy is converted into heat. Presently the main nucleus of the gas cloud—which we might now call the "protostar"—has become so hot that it begins to glow like any other aggregation of hot solid matter, and bit by bit, spread out upon an incredibly huge canvas and moving in incredibly slow motion, we see the first unmistakable evidence that a new star is being born (see Fig. 35b).

Dramatic as this sequence of events has been so far, the major critical event that changes our protostar into a true living star has not yet occurred. The protostar may have begun to glow and radiate heat and light from its condensing center, but this heat and light so far result merely from the increasing concentration and pressure of gas and dust as its center. It glows much as the point of a needle begins to glow when it is held in a candle flame. The heat that produces that glow in the protostar arises from precisely the same processes that are commonplace occurrences on earth, for the gases concentrating there are in the same form and follow the same laws of behavior as gases that we know on earth. Nothing has happened yet to ignite the kind of star fire that could sustain this protostar as a glowing beacon in the sky for billions of years.

But the force of compression at the center is still continuing, a self-perpetuating and self-enlarging process. As more and more of the mass of the original nebula begins to concentrate at the center of a huge rotating disk, the gravitational field of that concentration of mass increases, so that

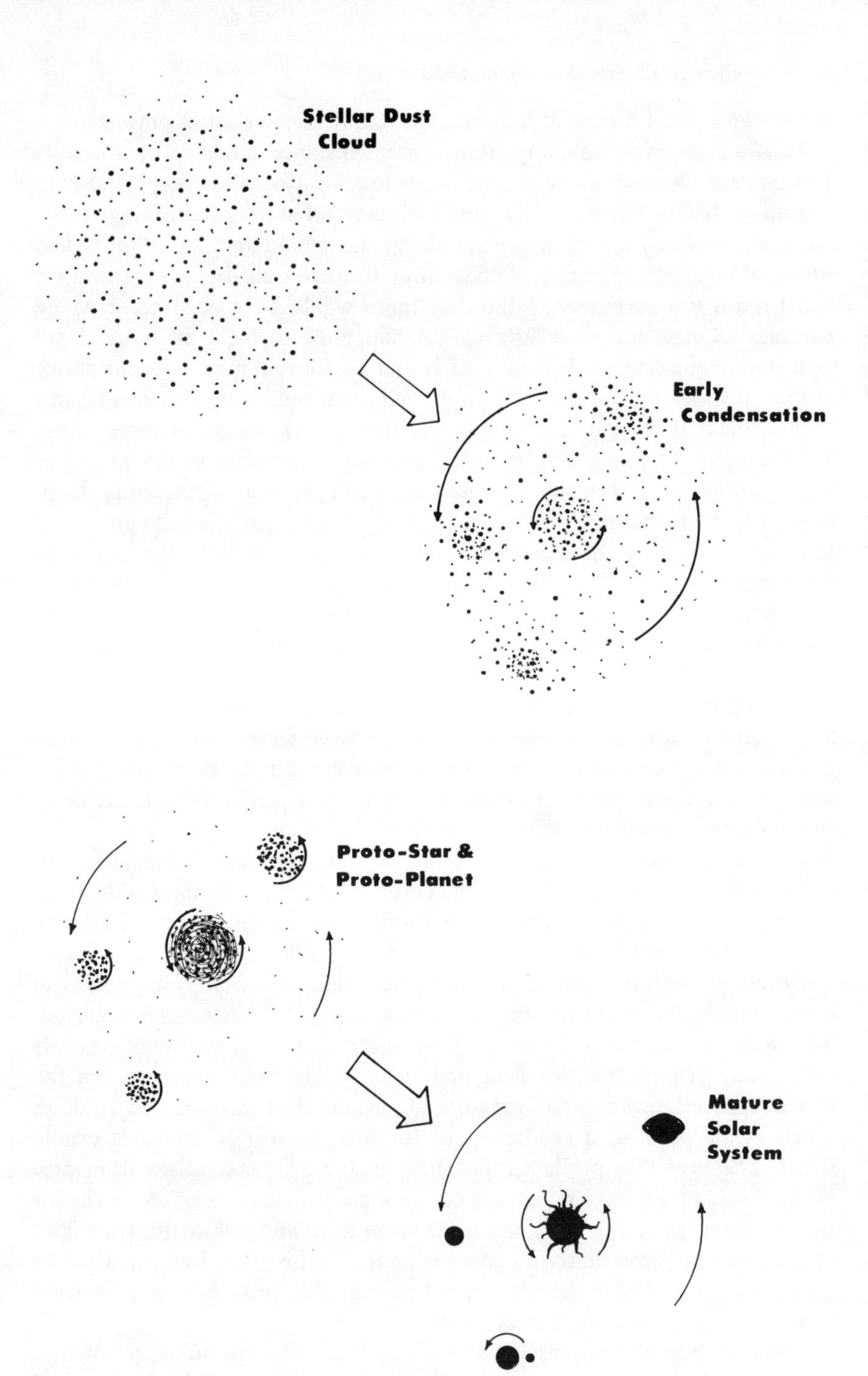


*Fig. 35a and 35b* The evolution of a star system, from a homogeneous collection of interstellar dust and gas to the formation of a mature solar system.

it pulls more and more of the outlying gas and dust in toward the center from farther and farther away, faster and faster. Gases that are already near the center are pulled deeper into the core of the central protostar mass with ever-increasing speed and pressure. Thus, as this central concentration grows ever more massive, it also grows ever hotter and hotter.

If this is happening, we might ask, why isn't *all* the gas in the nebula eventually drawn to the center and packed tighter and tighter, under ever-increasing gravitational force, until virtually *all* the mass of the original nebula is crammed into an exceedingly dense, exceedingly hot ball in the center? Part of the answer, of course, is that the disk of gas is rotating with ever-increasing speed, so that centrifugal force tending to pull peripheral particles of dust and molecules of gas *outward* helps counteract the gravitational force tending to pull them *inward*. But even this would not be sufficient to counteract the growing gravity if some other opposing force did not appear *within* the growing protostar and start acting to push mass and energy out and away from the center at the same time that gravity is pulling it in.

And, in fact, something does indeed happen which has precisely this effect. As the concentration of gas and dust in the center of this huge cosmic whirlpool reaches temperatures first of hundreds of thousands of degrees and then millions of degrees Fahrenheit, a certain critical degree of temperature and pressure is reached—a point we might speak of as the "kindling temperature" or "ignition point" of our protostar. And when that ignition point is reached, something most remarkable begins to happen. Under such extremes of temperature and pressure, hydrogen molecules in the center of the protostar are first torn apart into individual hydrogen atoms, and then those atoms themselves are stripped of their electrons. Since virtually all of the gas in the nebula was hydrogen gas to start with, practically the entire central core of the condensing star mass is composed of intensely hot compressed hydrogen, and when electrons are stripped away, the core of the protostar becomes a dense aggregate of exceedingly high-energy hydrogen nuclei—a form of matter known as a "plasma" in the modern physics laboratory. As we shall see later, hydrogen as a plasma is a very strange substance, far different from hydrogen in the normal gaseous state with which we are acquainted. And under these extreme conditions, hydrogen nuclei begin fusing together, first in pairs and then in fours, to form the nuclei of a totally different element roughly four times as heavy as hydrogen and far more stable and more chemically inert: the element helium.

This fusing together of hydrogen nuclei into helium nuclei is not a reversible reaction; helium does not "unfuse" or break down again into its four hydrogen nuclei components. But strangely enough, the final mass of the helium atom formed in such a fusion is never quite as great as the combined masses of the four hydrogen nuclei that formed it. In each such reaction a tiny portion of the mass of the four hydrogen atoms—approximately seven-tenths of 1 per cent of their total mass—disappears completely in the process of fusion. Disappears as *matter,* that is; but the tiny bit of matter that "disappears" in each reaction is converted or transformed into a huge quantity of energy which is thrown out and away with incredible violence.

Much of this newly formed energy is in the form of heat, but a good deal of it is thrown off as radiant energy: gamma rays, light waves, and other electromagnetic waves that are hurled away from the center of the stellar mass toward the more rarefied areas of gas near its surface, and then on into comparatively empty space beyond. In the interior of a forming star huge quantities of hydrogen begin fusing into helium, so that the outpouring of radiant energy, once started, is almost unbelievably great. In addition, if any carbon atoms happen to be present, still another kind of fusion reaction may occur, far more involved and complex than hydrogen fusion, but with the same end result: the production of helium atoms and the disappearance of small amounts of matter which are simultaneously transformed into disproportionately huge amounts of radiant energy and heat.

Once such matter-consuming reactions begin to occur, both the heat and the pressure within the core of the protostar increase sharply, and an endless chain of such reactions is triggered. These continue at a fantastic rate of speed, consuming enormous quantities of hydrogen, as an amount of radiant energy is poured out which is incomparably greater in quantity than the dull "heat glow" that had been present before. Between the hydrogen fusion reactions and the so-called carbon cycle reactions that are occurring within our own sun, it is estimated that some 615 million *tons* of hydrogen gas are converted into helium every *second,* with the simultaneous disappearance of some 4.5 billion tons of matter and the transformation of that matter into radiant energy.

When such reactions begin occurring at the center of a forming star, "ignition point" is a good descriptive term, for a radical change has taken place within the core of the protostar. If we compare the heat glow of the star-mass just before ignition point is reached to the red-hot glow of a needle point held in a candle, then this sudden violent outburst of energy would be more comparable to the ignition of a Fourth of July sparkler—incredibly more hot, and flaring into a fierce, self-perpetuating fire which hurls radiant energy at great speed (and with great pressure) in all directions away from the center.

What would happen if all that outpouring of energy were somehow to be trapped and unable to escape to the surface of a newly forming sun so that it could pour forth into surrounding space? Best we not think about it too much, for the result would soon be a cosmic disaster. The interior temperature of the newly forming star would very soon increase to such a point that the star itself would explode and hurl virtually all of its concentrated matter outward. Fortunately, however, the newly formed energy is not trapped inside the star. Instead, it pushes its way out to the surface layers of the starlike nucleus and escapes into space beyond, helping in its outward thrust to counteract the gravitational force acting to draw more gas in toward the center.

Thus a newly "ignited" star flares suddenly into beaconlike brilliance, pouring light and other radiant energy out in all directions into space, and then continuing to "burn" with its nuclear fire in a steady, reliable fashion until all the concentrated mass of hydrogen within the star has been consumed. Once ignited, such a nuclear fire may last anywhere from 10 billion to over 100 billion years, depending upon how huge and massive the central star core is, how great its internal temperature may be, and how fast its hydrogen is converted into helium and other nuclear ash.

## BINARIES, MULTIPLE SUNS, AND ABORTIVE STARS

Of course, as we have seen, all stars that form are not single stars like our own sun. In fact, at least in our own galaxy, it seems more common for a star-forming nebula of gas and dust to develop two, three, or even a cluster of central foci, sometimes located very close together, sometimes separated from each other within the huge dust cloud by dozens of light-years. In such cases each focus condenses and forms its own protostar, reaches its own ignition point, and flares up to form its own beacon in the sky with its own peculiar characteristics of size, heat, and color. Being at least relatively close to other such protostars or newly formed stars in the same nebula, each star holds the others in the grasp of its own developing gravitational field. Each tries to pull the others toward itself, and is pulled toward the others in turn. Ultimately these multiple foci can enter into stable orbital patterns around each other, and continue to move in those patterns throughout their existence.

But just because two or three or even multiple stars may form in the same nebula does not necessarily mean that these stars will be equal or even similar to each other in size, mass, or any other characteristics. More often, in fact, the stars in binary and multiple systems are quite different from each other. Many binary star systems, for example, are composed of huge, cool, red supergiants teamed up with small and intensely hot blue-white stars. In some cases a cool supergiant may be associated with a burnt-out white dwarf star, and it was from such odd bedfellows that astronomers obtained a vital clue to the true history and evolution of stars. For no matter how different from each other the two stars of such a binary system might be, it seemed likely that both were formed from the *same cloud of gas and dust at approximately the same time*. How, then, could they end up so different? In terms of years, such pairs of stars had to be equally old, yet in terms of health and vigor one might seem still to be enjoying an exuberant youth while the other had reached senility.

Today most astronomers agree that the life span of any given star depends upon how massive it is, how hot it is, and how fast the nuclear fires within it are burning. Our own sun, for example, is enjoying a staid, unremarkable, vigorous, but not very violent middle age. Our sun is probably very much the same in size and brightness today as it was some 5 billion years ago when it first reached ignition point, and it will probably continue to burn its hydrogen in the same stable and reliable fashion for another 5 billion years. But other stars, more massive because they formed from larger nebulae, reaching ignition point sooner and at higher temperatures because of the increased mass of the forming stellar mass, and expending their hydrogen supply far faster than our sun, will glow with an intense blue-white light instead of the sun's yellow light, radiating a hundred or a thousand times more heat and light than our sun does every second, but burning out within a billion years or so instead of surviving for 10 billion.

Still other stars, never so hot nor so massive as our sun, may proceed upon their self-consuming course far more slowly, never reaching the yellow brilliance of our sun but instead shining with a dark orange or even red color, fusing their hydrogen nuclei at a lower temperature and at a far slower rate, and going on to radiate youthfully for perhaps 90 or 100 billion years before the inevitable end comes.

But because the condensation of nebular gases so commonly involves two, three, or more focal points or major aggregations of matter does not mean that all of these focal points necessarily evolve into protostars and then reach ignition point. Most authorities today agree that the planets of our solar system probably arose as concentrated areas of gas within the swirling nebula from which our sun was also forming. Ultimately these localized concentrations of gas, not massive enough to become stars themselves, condensed and cooled into solid matter, collecting gas from their immediate surrounding areas and from the supply of gas driven from the sun's surface by the sudden burst of radiant energy when it reached its ignition point. Some of these developing protoplanets were massive enough and had sufficiently strong gravitational fields to hold a surrounding envelope of gas as an atmosphere. Those with too little mass to have a significantly strong gravitational field allowed such gases as they captured to leak away a bit at a time into outer space. All of them, of necessity, entered into orbit around the much more massive central protostar of the nebula, our newly forming sun.

It is even quite possible that one or more of the planets in our sun's solar system may, in fact, actually have been an abortive second star which never quite accumulated sufficient mass or internal heat and pressure to reach ignition point. Perhaps at one time one or more of these planets had become massive enough and hot enough, short of reaching ignition point, to radiate a great deal of heat energy. But once condensation was completed these planets cooled down so that even their radiation of heat diminished greatly. Probably not even the largest planet in our solar system ever got hot enough to shine with its own heat glow, much less to ignite any nuclear reaction—yet we know that even today Jupiter is still emitting some three times as much heat energy as it receives from the sun, which means that Jupiter is still contracting and giving out energy of its own.

Even our own earth, far less massive than Jupiter, still holds at its center a huge mass of red-hot molten iron and rock, but the cool crust of the earth contains and holds the interior heat, so that the earth appears as a cold solid planetary body illuminated only by reflection of the sun's light. Our earth could never have been massive enough to have had any remote chance of becoming a "partner sun" or binary to Sol—but Jupiter might have, if things had been slightly different than they were. It is entirely possible, at the time that our solar system was forming, that if Jupiter had captured and condensed a bit more nebular gas and dust than it actually did, it might have developed into a second star of a binary system, with Sol as a moderately massive yellow star and Jupiter as a far smaller red-glowing partner. But alas, Jupiter never did gain quite enough mass nor get quite hot enough to approach ignition point and make the grade as a self-perpetuating star. As it was, Jupiter's maximum temperature probably never exceeded 100,000 degrees F., with never quite enough heat or pressure to initiate a nuclear fusion reaction.

This, then, is the probable pattern of events in which stars were formed, and are still being formed, within our galaxy over unimaginable eons of time. Even today astronomers are witnessing what is believed to be the formation of a new triple or quadruple star in one of the great gas nebulae visible in the heavens. The process of star formation, then, has not ended even in our own "local area" in the universe.

But we also know that all stars in our galaxy are not the stable beacons of light that our sun is, pouring out radiant energy in a steady flow millennium after millenium without varying by so much as 1 per cent from one millennium to the next. Some stars, for example, flare up in brightness and then subside to comparative dimness, their luminosity waxing and waning in steady, endless cyclic patterns. These so-called variable stars (also known as "Cepheids" or "Cepheid variables" after the star Delta Cephei, one of the first to be studied) flare up and die down in regular periods of variation from a matter of a few hours for very small ones to as much as 100 days for very massive ones—the greater the star's mass the longer the period. Today astronomers believe that the pulsation of these variable stars is the result of cyclic gales of heating and cooling of the hydrogen and helium gas at the outer surface of these stars. As these gases heat up they become more transparent, allowing more radiant energy to escape and pour forth from the star's interior. Then, as the surface gases cool down somewhat, they become more opaque and allow less light to escape from the interior, until the energy that is steadily being generated at the star's core again heats up the surface gases, giving rise to a steady pulsation.

Other stars make an even more spectacular display of themselves. Whereas Cepheids pulsate in a steady, controlled fashion, growing regularly brighter and dimmer, one of these other stars will remain quiet for a long period and then seem, quite suddenly, to explode, pouring out a great surge of radiant energy that may continue for days or weeks before it gradually subsides to become a quietly glowing star again. Such a star is called a "nova," from the Latin word for "new," since early astronomers thought that such nova explosions represented new stars being born. We know today that this is not the case; the nova explosion is merely a temporary, cataclysmic change in a pre-existing star. But even modern astronomers cannot satisfactorily explain what may be causing such novalike flareups. They do know, however, that a small portion (perhaps 1/100,000) of the mass of the star is hurled outward in the course of such an explosion, to diffuse into space as gas and cosmic dust. Nor is such a nova explosion necessarily a one-time event; many novae have been observed to undergo repeated explosions at intervals varying from 10 to 120 years. Whatever forces cause these stars to explode in this fashion apparently persist, causing explosions to recur again and again.

Much more rarely, a far more violent kind of cosmic catastrophe can occur as a part of the death agony of certain extremely massive and extremely hot stars. These are the "supernova" explosions, of which only three (and possibly a fourth) have been observed in our galaxy in the last thousand years. The explosion of a supernova is an awesome event indeed, a catastrophic blowup in which up to nine-tenths of the star's mass is hurled out into space, leaving behind nothing in its place but a tiny, exceedingly dense, white dwarf star.

What causes such an incredible explosion to occur? Two possible mechanisms have been proposed to explain such a massive catastrophe, each of which may be correct, depending upon the type of star. In either case, the supernova explosion is believed to be a sort of terminal event in the life history of a very hot, massive star, occurring at a time when most or all of the nuclear fuel in the interior of the star has been exhausted. With various types of energy-producing nuclear reactions flickering out, the star begins to shrink or collapse upon itself under the incessant force of its own gravity. According to one theory, the core of the star heats up so much at the expense of gravitational energy that the atoms of iron at the star's core begin breaking down again into helium fragments, a process that absorbs energy rather than yields it. As gravitational pressure at the center increases ever more rapidly, this iron-to-helium breakdown accelerates in a vicious cycle until there is a sudden, catastrophic collapse of the star. When this occurs, all the nuclear fuel—hydrogen—remaining in the outer surface of the star is ignited in one huge burst of fusion and the star explodes, hurling most of its enormous mass outward in an expanding cloud and leaving behind only a small, dense, dying chunk of nuclear ash.

Another explanation for such a supernova explosion was proposed in 1961 by the Chinese-American astronomer Hong-Yee Chiu. As we have seen, the radiant energy of stars arises from nuclear reactions—primarily the fusion of hydrogen atoms into helium—occurring under conditions of great heat and compression in the hearts of stars. Ordinarily most of this radiant energy makes its way slowly to the surface of the star before being radiated away, and the outward pressure of radiation balances the inward press of the star's gravitational field to keep the star stable for eons. But part of the radiant energy continually leaks away in the form of "neutrinos" —tiny particles which, as we will see later, travel at the speed of light and can, because of their peculiar properties, pass through the entire diameters of stars without ever striking or interacting with other particles.

In a small, relatively cool star like our own sun, neutrino production does not cause a significant "leak" of energy. But in a very massive, very hot star, this energy leak can be very great. Furthermore, as nuclear fuels are exhausted in the interior of such stars, and as the star begins shrinking under the force of its own gravity, temperatures as high as 6 billion degrees centigrade may occur. At this critical point, according to Chiu's calculations, so much energy has leaked from the star's core through the escape of neutrinos that the nuclear reactions still going on there cannot produce enough energy to keep the star from collapsing suddenly with a resulting massive explosion.

## SUPERNOVAE AND THE HEAVY ELEMENTS

As we said, no one knows for certain which of these mechanisms account for supernova explosions. Very possibly both may be valid under slightly differing circumstances. But the fact is that not a great deal is known at first hand about supernova. They have occurred only rarely in our galaxy, the last one observed by Johannes Kepler in the year 1604. *None* have appeared for study with modern astronomical equipment (although astronomers feel that one is due to appear any time now!). Even so, it is quite possible that these cosmic catastrophes may have contributed far more to the processes of star formation than would seem likely, considering their rarity.

Many astronomers now believe that supernovae occur only in stars in which a whole sequence of different kinds of fusion reactions are taking place layer upon layer from the inside core of the star outward, each type of reaction forming atoms of heavier and heavier elements as their end product or ash. According to this view, when such a star explodes as a supernova, the gaseous nebula of debris that inevitably expands from that cosmic explosion contains a wide variety of heavier elements as well as hydrogen, and these clouds of gas and dust mix and intersperse with other clouds that are becoming involved in the birth of new stars. Thus we have a picture of one generation of stars building up aggregations of heavy elements which are then seeded into a second generation of stars, and then from this second generation still more heavy elements may be formed in some stars and be further seeded, through supernova explosions, into yet another generation of stars. Astronomers who theorize in this fashion believe that the original primeval matter in the universe was in the form of hydrogen and helium atoms and nothing else. They contend that all the heavier elements in existence anywhere in the universe were formed within the interiors of stars through a process of fusion which began with a fusion of hydrogen into helium, and then the fusion of helium into heavier elements. These elements were subsequently hurled out into space in supernova explosions to join with more hydrogen to form new gas clouds from which new stars arose.

Not all astronomers agree that this is a true picture of what has happened and is continuing to happen in the universe, but this theory does offer a plausible explanation to account for the fact that the planets in the solar system of our own sun have such a surprisingly rich distribution of heavy elements, not only of oxygen and nitrogen in our atmosphere but of silicon, iron, gold, even radium and uranium. It could also explain why our sun contains in its outer layers so many heavy elements which could not ever have been formed in the particular kind of nuclear fusion reactions that we know keeps our sun alive. Indeed, proponents of this theory suspect that our own sun may be a fourth- or fifth-generation star in our galaxy. Of course our sun itself may not be heavy enough or hot enough ever to develop the peculiar conditions that would make it become a supernova, but multitudes of other stars in our galaxy may well fulfill the necessary criteria and thus eventually contribute to still another generation of stars virtually identical to our own.

In any event, we can be sure that the star-forming process continues in our galaxy, and more is being learned every day about the nature of this cosmic process of generation and regeneration. We also learn more every day about the over-all characteristics of our galaxy as a huge but isolated aggregation of stars and solar systems. But is our galaxy the only one that exists, containing all the stars and planets in the universe? As little as a hundred years ago many astronomers and physicists thought that it was, until a group of men began making some discoveries that indicated more and more that huge as it was, our galaxy might be no more than the merest fly speck in a vastly more enormous universe, just one of millions, perhaps billions of galaxies scattered through the universe at distances so staggering that the human mind could not begin to comprehend them.

# CHAPTER 20

## *Mega-Universe: The Puzzle of Distant Galaxies*

To the astronomers and physicists of the late 1800s and early 1900s, it was a great step forward to have developed some sort of plausible explanation of how stars and solar systems formed—a theory which also suggested reasons for the wide variety of stars that had been discovered and studied. These notions were only theories, to be sure, but a growing body of evidence seemed to support them. Most important of all, these theories made it possible for scientists to begin to develop an orderly mental picture or "model" of the universe around us. Instead of the stars in the heavens being regarded as a disorderly random scattering of matter thrown about willy-nilly, it was becoming clear that the stars were ordered in a very distinct pattern, and that the galaxy of which our sun was a member had an identifiable shape with certain identifiable characteristics which seemed to corroborate once again the ancient thesis that everything in the universe was orderly, whether men could understand *why* it was put together as it was or not.

But as we have already seen, it is a characteristic of scientific advancement that the discovery of answers or even of possible answers to puzzling questions about nature invariably opens up even more puzzling and complex questions which must then be answered in their turn. So it was that the development of a plausible theory for the birth of suns and solar systems, while answering questions that had dogged men's minds for centuries, brought physicists and astronomers face to face with an even more difficult and basic cosmological question: How and when did this entire galaxy of stars have its origin in space and time? How old was it, why did it have the shape and structure that it had, and what might be going to happen to it in the future?

There were no obvious or self-evident guidelines to follow in searching out answers to these questions; astronomers had learned that lesson in the past as it had become clear that their "self-evident" assumptions about our sun's place in the galaxy had proven consistently wrong. Even though scientists had evolved a fairly clear picture of the shape of the galaxy, it still presented perplexing problems, for a number of things about it had been observed which seemed to have no explanation. It was clear that our sun was nowhere near the center of the galaxy, but was located approximately two-thirds of the way out from the center, one of millions of stars lying near the periphery of the galaxy in a great spiral arm. It was also clear that the region of the galaxy in which the sun lay was a region of great star-forming activity. Stars in this region of the galaxy exhibited a great many widely varying characteristics. New stars were forming, old ones were dying, and some were exploding in brilliant nova or supernova displays. This outer edge of the galaxy was also the home of many of the apparently unstable Cepheids or variable stars that flared and died down and flared again in rhythmic pulsations.

Yet there were other regions of the galaxy visible to the telescope which seemed to exhibit no such evidence of star-forming activity at all. In the great central hub of the galaxy, and in the halo of stars found above it and below it, there seemed to be a completely different population of stars from those in the more outlying regions. Here all the stars appeared very similar one to the other, with no evidence of new star formation and no evidence of gaseous nebulae from which stars could form. Neither was there any evidence to be seen that stars were dying in this region.

In fact, despite the star-forming activity in some of the more remote parts of the galaxy, the greatest part seemed inactive and homogeneous, as though the great bulk of the stars in the galaxy had formed at approximately the same time in the distant past and had reached the same age together in the same way. Could it be that the activity in the "active" areas was no more than a delayed reflection of activity which once had taken place throughout the entire galaxy but had now slowed down to stable and undistinguished old age? Or was the truth quite the opposite of this? Was it possible that the activity in the "active" parts actually represented the beginning of some degenerative process that had not yet spread to the great central hub of the galaxy, so that these "active" regions were once quite as stable and inactive as the rest of the galaxy? In short, did our galaxy form as a unit, all at once at approximately the same time, or had it been "built up gradually" a bit at a time throughout the span of billions of years, and was it destined to continue building and reforming and regrouping endlessly?

There seemed to be no satisfactory answer to this question. Nor was this the only perplexing question about the galaxy that remained unanswered. What was the nature of the "primeval matter" that had originally entered into the formation of the galaxy? Was it, as some believed, originally all hydrogen and nothing but hydrogen, or were all the heavier elements also present in the great primeval dust cloud that must have existed before the first star was formed? In either case, where did this primeval matter come from? When did it first gather here? And where, after repeated eras of shifting and regrouping as new stars formed and old ones died, would all the material substance of the stars ultimately go? Most perplexing of all was the one basic question that dwarfed all the others into insignificance: Why was it that all the matter in the universe had focused and condensed into one great conglomeration of stars making up our galaxy, and what lay in the empty space beyond it? Where did that empty space ultimately end—or *did* it ultimately end?

This was a perplexing question, with its roots in the earliest philosophical beginnings of scientific inquiry. For a long time there seemed no possible way that an answer could ever be found. And then, in the early decades of our twentieth century, astronomers startled themselves by coming up with a breathtaking answer: that for all its enormous size and for all its billions of stars, our own galaxy was not by any means the only galaxy there was. In the space of ten brief years in the early 1900s, astronomers and cosmologists who had already become used to "thinking big" in their work found that they had to think absolutely huge. For they discovered beyond question that the space extending beyond our galaxy was not "empty space" at all but contained innumerable other whole galaxies of stars, each lying at staggering distances from any other, and each containing billions of stars just as ours does. It was not a matter of discovering a dozen or a hundred such galaxies; slowly it became clear that there were *billions upon billions* of such galaxies scattered through space as far as telescopes could reach, extending to the uttermost limit that human beings could ever hope to observe, and extending probably even far beyond that—perhaps on to infinity.

## NEBULAE OR GALAXIES?

Of course, the idea that the universe might contain galaxies other than our own was not an entirely new consideration in the early decades of the 1900s. Many astronomers had speculated that this might be the case. There appeared to be definite outer limits to the extent of our galaxy and the number of stars within it, so that the galaxy was commonly believed to be a huge but sharply defined cluster of suns moving like a vast "island universe" through space, and many astronomers did not think it reasonable to assume categorically that this island universe had to be the only one that existed. But if other such galaxies existed similar to our own, why had no one ever seen them? Surely *some* ought to have been observed, and scientists were puzzled and frustrated that no such observation had been made.

Today we know that other galaxies had indeed been seen—literally thousands of them—and recorded on astronomers' photographic plates. They had simply not been recognized as distant galaxies. Many of them were so incredibly remote from our own galaxy that they had been interpreted on the plates as nothing more than very faint stars; accustomed as they were to thinking in terms of vast distances, astronomers had never dreamed that a *whole galaxy of stars* might be so remote as to appear as nothing more than the faintest speck of light on the most sensitive star photo.

Many other distant galaxies were not as remote as all that, but had been overlooked or missed for a different reason. There are at least three galaxies separate from our own which are close enough that they can easily be seen with the naked eye. Indeed, one of them, the galaxy now known as Andromeda, had even been observed and catalogued as one of the "fixed stars" by Persian stargazers as early as the tenth century A.D. But when early astronomers focused their crude telescopes on this "star" they observed something different from the pinpoint of light they expected to see. In the telescope this "star" appeared like a hazy gas cloud; it was misinterpreted as just another of the many "nebulae" of gas and dust floating in the spiral arms of our own galaxy.

The same misinterpretation was made in the case of two other "nebulae" that appear quite prominently in the skies in the southern hemisphere, and have been known as celestial reference points since the days of the Portuguese sea explorers of the fifteenth century. These two nebulae appeared just like all the other fixed stars in our own galactic family, until telescopes revealed that they also had no pinpoint definition but looked like glowing dust clouds—so much so that they came to be known as the "clouds of Magellan" or the Magellanic Clouds, one of them significantly larger than the other.

By the middle of the eighteenth century, astronomers such as Charles Messier had found over a hundred such "nebulae" in the heavens. Messier, who lived between 1730 and 1817, regarded these "nebulae" as nuisances, and actually catalogued them by number in order to remind himself to ignore them as he proceeded with his search of the heavens for new comets. Among these "nebulae" identified by Messier and others were a number of large galaxies comparatively near to our own Milky Way galaxy, including Andromeda and the Magellanic Clouds. Ironically, when some of these nebulae were finally identified as true galaxies beyond the Milky Way, they became known among astronomers by Messier's own catalogue numbers and are even today identified by "Messier numbers" or "M numbers."

Even as early as Messier's time some astronomers speculated that these "nebulae" might be whole systems of stars like the Milky Way, but located so far outside it that the individual stars making them up could not be "resolved" or separated one from another, even with the best telescopes available. By 1850 the Earl of Rosse had discovered that a number of these "nebulae" seemed to have a curious spiral shape to them, even though they were still generally regarded by astronomers as strictly local spiral-shaped gas clouds within our own galaxy. But when a new general catalogue of stars was published by Jele Dryer in 1888, listing some *eight thousand* such "nebulae," over half of which were spiral in shape, more and more astronomers began to suspect that there was more to these celestial "nuisances" than was immediately apparent.

In spite of this suspicion, it was not until 1908 that *some* sort of definite evidence appeared that these "nebulae" were not nebulae at all but true distant galaxies, and even then the evidence was not very convincing. In that year an American astronomer named J. W. Ritchie, using a 60-inch reflector telescope that had recently been built on Mt. Wilson in California, *thought* that he had been able to distinguish individual star images in a nearby galaxy, although the images were so fuzzy that he could not be entirely certain that they were not merely pockets of dense gas in the midst of a nearby gas cloud. Tentative as it was, Ritchie's report caused a flurry of excitement among astronomers, for they recognized clearly that if a nebula ever *could* be resolved into separate star images it would prove beyond doubt that the nebula under observation actually was a complete star system in itself, similar to the Milky Way but far outside of it.

Another bit of suggestive evidence turned up in 1912 when a new method of gauging the distance of stars was devised by American astronomer Henrietta Swan Leavitt, using variable stars or Cepheids as a yardstick. Measurements taken by this new method suggested strongly that the smaller or Lesser Magellanic Cloud might be much farther away from the earth than the diameter of our whole galaxy. Between 1916 and 1918 this lead was followed up by Harlow Shapley, an American astronomer, and Ejnar Hertzprung, a Dane, who were able between them to produce a great deal of new information about our own galaxy. They were able to prove, from their studies, that the Milky Way was unquestionably a single system of stars; in addition, they came up with a canny estimate of its over-all size, of the direction of its hub from our sun, and of our sun's distance from its center. All of these estimates were well supported by observation; but if they were true, it meant that the Magellanic Clouds, whatever they were, *had* to be completely separate systems of stars located far outside the main bulk of the Milky Way galaxy.

Finally, in 1917, Ritchie and another astronomer, H. D. Curtis, discovered a number of exploding stars or novae in the midst of several of these mysterious "galactic nebulae." These novalike explosions looked just the same as many others which had been observed except that they were far fainter than they ought to have been. One such star explosion had an apparent brightness some twenty thousand times less than it should have had, suggesting that it had occurred in some area enormously remote from the earth, far beyond the limits of our own galaxy.

But for all these disturbing findings, nothing yet had been found to constitute inarguable *proof* that these celestial events were not events happening within our own galaxy. Ritchie and Curtis were convinced by then that our galaxy was only one of many galaxies rotating in a universe immensely more huge than had ever before been suspected. But the majority of astronomers were not yet ready to accept this concept. They wanted proof to go on, not supposition and speculation. Particularly in the case of such comparatively "nearby" nebulae as Andromeda or the Greater and Lesser Magellanic Clouds, they argued, at least *some* clear-cut and inarguable star images ought to have been identified if these "gas clouds" were actually true separate star systems.

In large measure, it was a situation in which both scientists and the progress of science had to wait for the development of better instruments for observation and more reliable means of measuring distances. Thus, it was not until early in the year 1925 that a California astronomer, Edwin Hubble, working with a brand-new 100-inch reflecting telescope at Mt. Wilson, was able to report to his colleagues that he had succeeded in resolving three so-called galactic nebulae into clear-cut collections of individual stars, and was convinced by the behavior and brightness of certain Cepheid variable stars within those "nebulae" that they were really complete galaxies enormously far removed from the galaxy that contained our own sun.

## THE FAMILY OF GALAXIES

Edwin Hubble's discovery in 1925 opened the floodgates. Data began piling in from all sides establishing beyond doubt or debate that the universe, far from encompassing only a few billion stars in our own galaxy, was actually enormously more huge than that, containing hundreds, perhaps thousands, or even millions of separate and distinct galaxies—huge island universes separated from each other by staggering expanses of apparently empty space. None of these galaxies could reasonably be called "close" to ours in space. Hubble and an assistant, Milton Humason, calculated that a sphere of space 6 million light-years in diameter with our own galaxy located at the center actually encompassed no more than some twenty or so other galaxies. A sphere of space 60 million light-years in diameter still contained only perhaps two hundred more galaxies—but going even farther out there were more galaxies, literally millions and perhaps billions more galaxies, each one containing a billion stars on the average, although some were smaller than that while others were much larger.

Today we know that the faintest and most remote of these clearly identified galaxies that are visible with modern telescopes is over 5 billion light-years away. The light from that galaxy received by our telescopes today must have left its source on its way to us at approximately the same time that our own sun and earth were in the process of condensing from a swirl of interstellar gas! Many of these galaxies are so remote that their approximate distances must be gauged not by the brightness of individual stars within them but by the brilliance of the whole galaxy.

One result of Hubble's discovery and the subsequent flood of new observations that poured in was that in the course of a few short years the horizons of astronomy were rolled back far beyond the previous limit. Hubble calculated that there were probably as many whole galaxies outside the Milky Way as there were stars in it; today, astronomers would regard this as an extremely timid and conservative guess. These island universes are scattered through space outward and outward to the very limits our modern telescopes can reach.

But with such a discovery old questions were raised anew. Where did all these galaxies come from? When were they formed? Could we learn anything about the shape and magnitude of the universe from observing them? Other questions were more practical and down to earth: What could be learned about *our* planet and *our* sun and *our* galaxy from the others beyond us? If these "island universes" were really so remote from us that it took thousands of years for astronomers just to discover that they actually *were* other galaxies and not single stars, was it possible to learn *anything* useful from them other than the fact that they, like Mt. Everest, were there?

As it happened, the answer to this last question, at least, was affirmative. A very great deal was learned about the nature and magnitude of our universe, and about our own galaxy from the careful study that was undertaken of these others. In fact, astronomers and cosmologists were quick to realize that a perfectly staggering opportunity lay before them, one hitherto unheard of in science: the opportunity to look back in time and observe our earth's remotest past and to gaze forward in time at her most distant future, all at the same time. Since 1925, much of the work of modern astronomy, astrophysics, and cosmology has been concerned with doing just that.

This apparent paradox arises from two key facts, both of which were apparent to astronomers from the very moment that Hubble proved beyond question that distant galaxies really did exist. First, the galaxies stretching across the visible universe differed from one another in very marked degrees and in certain definite patterns. Second, since light was known to require a finite period of time to travel from one place to another, astronomers recognized from the first that whatever they *were* able to observe of far-distant galaxies had to represent conditions that had been prevailing and events that had been occurring in the distant past. With these two facts to guide them, astronomers began to develop some very clear-cut and often sharply conflicting ideas about just when and how the universe was formed, how our own galaxy evolved, how it compared with the multitudes of other galaxies in the universe, and what its ultimate future might be.

We might well imagine that these scientists would have been appalled at the apparent task before them when they began to realize the sheer numbers of galaxies that existed remote from our own, just as early microscopists must have been appalled at the seemingly endless numbers and varieties of previously unsuspected bacteria and other single-celled organisms that their microscopes revealed were thriving on the face of the earth. But appalled or not, astronomers dug into the task.

One of the first jobs, clearly, was to devise some kind of meaningful classification system for these newly discovered galaxies, since they differed so greatly from one another. Some were unimaginably huge, many times the size of our own Milky Way. Others were, by comparison, relatively small and insignificant. Some poured out light like blinding beacons even though they were enormously remote from us, while others that were far closer seemed to glow very feebly by comparison. In some of these galaxies the individual stars could be "resolved" by our telescopes into multitudinous individual pinpoints of light, but in the case of others even the most meticulous efforts of astronomers failed to separate out individual stars. Certain galaxies or parts of galaxies seemed to remain fuzzy blurs of light even though they were now known almost certainly to be galaxies.

Other distinctions were discovered as well. Some galaxies appeared to contain enormous clouds of dust and gas that hid large sections from view, while others seemed to be totally free from such cosmic gas and dust clouds. Some, such as the beautiful Andromeda galaxy, seemed to have very sharply defined spiral shapes, appearing in the telescope as flattened disks with dense concentrations of stars at the center and sprawling spiral arms spreading out from the hub so that they looked for all the world like glorious celestial pinwheels. Others seemed to be perfectly homogeneous fuzzy masses of stars, brighter at the center than at the periphery, but with stars diffusing out evenly in all directions to form nothing more distinctive than an amorphous blob of light on a photographic plate. Some galaxies seemed to travel together in groups; others seemed to have "satellite galaxies" closely associated with them. Some galaxies appeared to have enormous gaps of space between them bridged by immense columns of gas and dust, while others actually appeared to be *colliding* with each other! The variety of shapes and sizes seemed endless.

But for all this apparent riot of disorderly variation among the galaxies, astronomers gradually came to recognize a certain kind of order in the chaos. Hubble, who continued working on the front lines of research with the 100-inch telescope at Mt. Wilson after his first discovery of an unmistakable galaxy apart from our own, soon noticed that *most* distant galaxies seemed to fall into three great families which could be distinguished according to their shapes or configurations. He further found that each of these three families could be subdivided into a relatively few distinct types of galaxy within the family.

For example, Hubble found that by far the greatest proportion of all the galaxies identified as such—something like 80 per cent—were definitely spiral in shape, with a dense central hub of stars surrounded above and below by a halo of other stars, and with great spiral arms of stars spreading out in a flat disk around the hub. Characteristically, these spiral or "S-type" galaxies presented the appearance of great cartwheels in the sky, sometimes seen full face so that the bright hub and the circle of spiral arms were clearly visible, sometimes seen edge-on, so that they appeared like flattened disks with a bright bulge in the middle (see Fig. 36). Among these spiral galaxies there was still individual variation to be found: Some seemed to be tightly packed spirals with very fat hubs and smooth homogeneous spiral arms held in closely—the so-called Sa galaxies. In others, the hubs

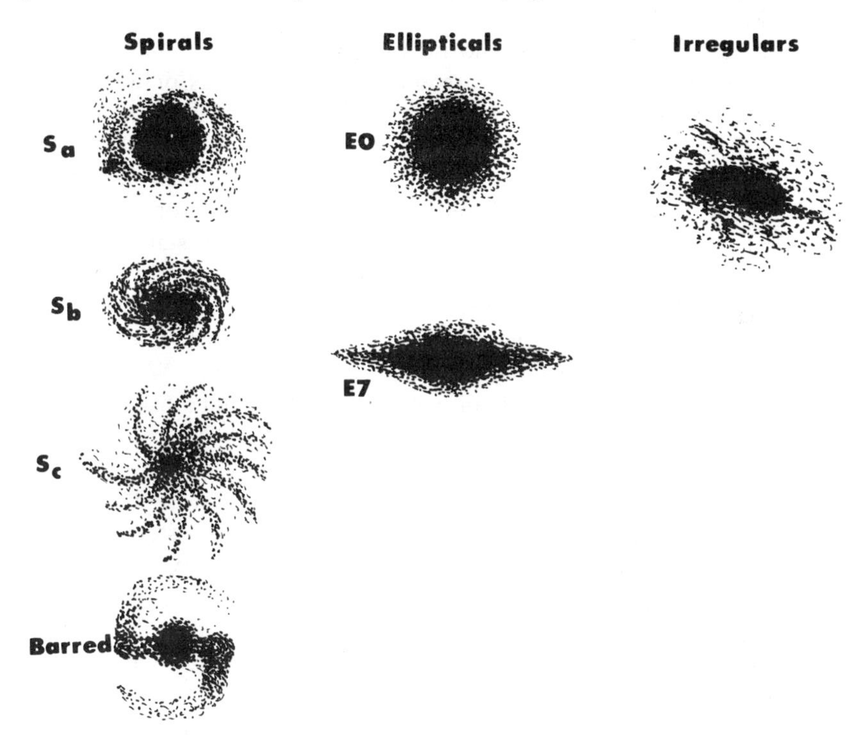


*Fig. 36* Edwin Hubbel's classification of galaxies. Astronomers today speculate that all galaxies tend to progress through an "evolutionary chain" from Irregular to Spiral to Elliptical.

were smaller and flatter, and the spirals spread wider into distinctly definable arms (the Sb and Sc types), while still others seemed to have elongated hubs with the spirals streaming from either end of an almost cigar-shaped center—the so-called barred spirals.

The second family of galaxies Hubble described, comprising perhaps 16 or 17 per cent of all known galaxies, appeared quite different from the spiral galaxies. These so-called elliptical, or E, galaxies were far smoother and more homogeneous in appearance than the spirals, with their shapes varying from elongated ellipses to spheres, with dense bright hubs gradually fading out to a fainter halo of stars around them. Hubbell subclassified these

E-type galaxies from type E–0 and E–1 (tightly packed spherical galaxies) to type E–6 and E–7 (flattened ellipses verging almost into the spiral types).

Yet a third family of galaxies, composing fewer than 3 per cent of the total number, were quite irregular in shape, neither smoothly homogeneous like the ellipticals nor sprawled out in distinct cartwheels like the spirals. These "irregulars" or "IRR" galaxies seemed to defy classification altogether. Most of them appeared to be composed of great roiling clouds of dust and gas with many large and intensely hot blue and white stars distinguishable, but a few seemed to contain little or no gas or dust and to contain mostly dim, cool, red stars.

These three major families of galaxies, the spirals, the ellipticals, and the irregulars, each had other characteristics than merely their shape that seemed to distinguish them. Very early in the game, astronomers studying the spiral galaxies found great difficulty in getting all the stars in a given galaxy to show up clearly as pinpoints of light. They frequently found that while the stars in the spiral arms were clearly enough defined, they seemed to dissolve into fuzzy blurs in the central hub of the spirals.

Then in 1941 one of Hubble's colleagues, a man named Walter Baade, decided to make a detailed study of one of the nearest and most spectacular of all the spiral galaxies known, the great spiral of Andromeda. This galaxy was of special interest to astronomers not only because it seemed to offer a better opportunity for "closeup" study than the others but more particularly because it was thought to be a sort of "twin sister" galaxy to our own, generally similar to the Milky Way in size, shape, and distribution of stars. Our own galaxy, of course, was awkward to study well simply because we were looking at it from within it; Walter Baade reasoned that a great deal more might be learned about the nature of our own galaxy by a careful study of this giant "twin."

Baade used the 100-inch telescope on Mt. Wilson to begin his study, and that year, 1941, was a singularly fortuitous time for good observation, with the United States just plunged into World War II, with Los Angeles under blackout, and with gasoline rationed so the area was not steeped in smog. But even under such "ideal conditions" Baade was plagued by a puzzling difficulty from the very first. Using the normal blue-sensitive photographic plates that were commonly used with telescopes at the time, he found that he could resolve the stars in the sweeping spiral arms of Andromeda into individual pinpoints of light quite well, but that individual star images in the vicinity of the galaxy's central hub simply could not be resolved at all; they appeared always as a hazy blur. This did not seem seem right; even assuming that the density of the star population at the galaxy's center was much greater than in the spiral arms, Baade still felt that the telescope he was using *should* have been powerful enough to resolve the central haze into individual star images. Then, in a moment of inspiration, Baade substituted special red-sensitive plates in place of the blue-sensitive ones in the telescope, and found a dramatically different picture of the Andromeda spiral appearing on these new plates. The stars in the spiral arms now appeared far less dramatic than before, but the fuzzy haze of stars in Andromeda's hub was sharply resolved into countless myriads of distinct red pinpoints of light.

The meaning of this observation was startlingly clear. This great spiral galaxy obviously contained two distinctly different kinds or "populations" of stars, with one type predominating in the hub and the other type most prevalent in the spiral arms. The stars in the hub were for the most part cool, red giant stars—generally considered to be aging or dying stars—while the predominating stars in the spiral arms were younger, hotter white- or blue-shining stars, including many blue supergiants.

What was more, these two populations of stars in the Andromeda galaxy included two distinctly different types of Cepheid-variable stars. The Cepheids that Harlow Shapley had first studied in our own galaxy had clearly been aging stars of only moderate brilliance; but the Cepheid variables found in the spiral arms of Andromeda were far younger and far more brilliant. When Hubble and other astronomers had first begun using these variable stars as a yardstick for measuring the distance of such galaxies as Andromeda, they had assumed that *all* variables were of approximately the same brilliance, so that these younger and brighter "super-Cepheids" in Andromeda had been equated with our own galaxy's older and duller ones.

Thus, as a totally unexpected side effect, Walter Baade's discovery of two quite different populations of stars in Andromeda forced astronomers to conclude that the distance of that galaxy (and of all other galaxies measured on the basis of their variable stars) had been grossly underestimated. According to Baade's new calculations, the "nearby" Andromeda galaxy was actually *twice* as remote from our galaxy as Hubble had estimated, and the same had to be true of more distant galaxies as well. With one single awkward discovery, Baade had effectively doubled astronomy's best estimate of the magnitude of the known visible universe overnight!

## A YARDSTICK FOR MEASURING

In order to understand the impact of Baade's discovery, we need to pause a moment and see clearly just *how* these Cepheid variables were used as gauges of distance. We can see, of course, that in the comparatively limited spans of distance within our own galaxy two stars of identical size and color might appear either dim or brilliant depending upon whether they were far away or nearby; when one such star appeared far dimmer than the other and we were certain that they were of the same size and real brilliance, we could reasonably conclude that the reason one was dimmer than the other was because it was located at a much greater distance from us than the other. But in attempting to gauge intergalactic distances, immensely greater than any distance from one point of our galaxy to another, no such comparison of the apparent brilliance of ordinary stars is of the least help. Astronomers had to turn to the Cepheid variables and rely upon them as gauges of comparative distances between the galaxies.

How could this be done? As we saw earlier, these variable stars appear to be highly unstable. Unlike the majority of stars, the Cepheids are constantly "out of balance," varying in brilliance in a regular rhythm from one phase to the next.

But as we have seen, cycles of the Cepheid variables are by by no means the same; they seem to vary widely. By 1912 Cepheids had been identified both in our own Milky Way and in such nebulae (later found to be different galaxies) as the Lesser Magellanic Cloud. Some flared up and died down at intervals no longer than every twelve or fifteen hours; others became successively brighter and dimmer at more dignified intervals of months or longer. But in 1912 Henrietta Leavitt, working as an assistant at Harvard Observatory and studying a number of Cepheid variables, noticed something about them which previously had either been ignored or overlooked. She discovered that there was a distinct correlation between the brightness of variable stars and the speed of their fluctuations. Dim variable stars fluctuated rapidly while those that were brighter showed a much slower and more stately cycle of fluctuation.

Almost immediately the Danish astronomer Ejnar Hertzprung recognized that this relationship between the brightness of a variable star and its period of fluctuation might provide a splendid means of measuring the enormous distances between galaxies. He reasoned that even a "nearby" galaxy such as the Lesser Magellanic Cloud was so very far away from earth that *all* of its stars could be considered to be roughly about the same distance from earth. Of course, they might vary in distance by the diameter of the galaxy, but this distance was so small compared to the staggering distance from one galaxy to the next that it could be considered insignificant.

Therefore, Hertzprung proposed, any differences in the apparent brightness of various Cepheids in the Lesser Magellanic Cloud had to reflect real differences in the actual brightness of those stars and not merely differences in distance. Furthermore, he reasoned that the relationship between the rate of pulsation of a variable and its apparent brightness must also apply to its real brightness as well. If this were so, it should be possible to calculate the real brightness of any variable star anywhere, no matter how distant it might be, simply by watching it with a telescope and clocking its period of pulsation. Then by comparing the calculated real brightness with the apparent brightness observed in the telescope, one could always find out how far away that star really was.

To put it in a different way, any one characteristic of variable stars alone would be of no help in distinguishing how far away they (and other stars in their vicinity) might be, but if *two* characteristics, such as brightness and period of fluctuation, were *always related one to the other in simple proportion,* then these two characteristics taken together could serve as a useful measuring rod for distance.

Obviously any star, no matter how bright or how dim it actually was, would *appear* brighter in our telescopes if it was close to us than it would appear if it were far away. Two ordinary stars might appear identical in brightness in our telescopes when their real brightness was quite different; one of the two could be a very bright star extremely far away while the other could be a dim star quite near by. But with only one characteristic to go on—apparent brightness—we would have no clue to this fact. If both these stars happened to be variable stars, the fact that the brilliance of each of them was fluctuating in a cycle would still be of no help in telling us which one was close and which one was distant—unless we knew, as Miss Leavitt discovered, that a dim variable star fluctuates far more rapidly than a bright one, with the period of fluctuation directly related to the brightness.

But knowing that relationship, if we then observed our two apparently "identical" stars, both variables, to be fluctuating at different rates of speed, one with a cycle of six days, for example, and the other with a cycle of two months, we could then deduce quite accurately that the one with a cycle of two months must have a real brilliance ten times greater than that of the one with the six-day cycle of fluctuation, and thus, in order to appear "identical" to it in our telescope, must actually be ten times as far away from us as the short-cycle star.

Following out his reasoning, Hertzprung calculated that any variable star that had a period of fluctuation of 6.6 days had a real brightness nearly seven hundred times that of our sun, and (assuming that all variable stars were of the same class or population) he reasoned that such variables could be used as accurate measuring rods to gauge distances any place in the sky that they were visible. Thus, gradually, astronomers all over the world had come to accept this method of distance measurement as universally accurate. Harlow Shapley, for example, had used Cepheids as a basic measuring rod in studying the distance of stars in our own Milky Way, and when Cepheids were discovered in other galaxies as well, astronomers proceeded to use them in the same way to calculate the distance of the galaxies.

Walter Baade's discovery showed that Hertzprung's basic assumption—that all variable stars everywhere in the universe were of the same class or population—was in error. He showed that there were, in fact, two distinct classes of variable stars, the dim ones and the bright ones. Shapley had used the dim ones in studying our own Milky Way, and subsequent estimates of the distance of other galaxies based on the same measuring rod therefore suggested that those galaxies were much closer to our own than they really were. By using the brilliant variables in the spiral arms of the Andromeda galaxy, rather than the dim ones, as a new measuring rod Walter Baade had simply demonstrated that the galaxy (with a multitude of others whose distances had been calculated) was much further away than it had ever been thought to be.

But Baade's discovery of two distinct classes or populations of stars in the spiral galaxies had even more far-reaching implications. For one thing, this discovery suggested that just as individual stars formed out of huge turbulent clouds of hydrogen atoms and dust, reached an "ignition point," and began burning with nuclear fire, then ultimately aged and died, so it was also possible that whole galaxies had formed from dust and gas clouds, evolved through stars of various sizes, and then ultimately aged and finally died.

This notion that whole galaxies might have evolved over a period of time seemed to be supported by other observations. For example, the fact that all the galaxies that had been observed fell into three major families—spiral, elliptical, and irregular—seemed to support the idea that all galaxies everywhere went through a characteristic *pattern of evolution* over unthinkable ages of time. The spiral galaxies which comprised the majority of all galaxies seemed to be made up of two classes or populations of stars, one population concentrated in the galactic hubs and the other in the spiral arms. Elliptical galaxies, on the other hand, seemed to be made up uniformly of the same kind of star—for the most part, aging red giant stars. Furthermore, the elliptical galaxies showed very little evidence of clouds of unused hydrogen or interstellar dust, if any evidence at all, and all of the stars in these elliptical galaxies appeared to be of approximately the same age, with no evidence of new stars forming. By way of sharp contrast, the irregular galaxies seemed to be filled with gigantic clouds of hydrogen and dust, and to be liberally populated with young, hot, and recently formed stars. It even seemed quite possible that in some irregular galaxies the total mass of interstellar gas and dust in the galaxy greatly outweighed the total mass of all the lighted stars taken together.

Thus, bit by bit, an awesome picture began to emerge, a pattern of evolution of magnificent orderliness extending farther into the past than men could comprehend and destined to extend farther into the future than we could imagine. It was a pattern of evolution in which whole galaxies formed, aged, changed, and died as new ones began to enter into the same endless pattern. According to this picture of evolution, in the beginning great clouds of gas and dust first began to develop localized whirls and eddies which one by one evolved into new stars. As these new stars were formed, they were drawn together by gravitational or other forces to form dense clusters at the centers of unimaginably huge gas clouds, the earliest precursors of shapeless and irregular galaxies filled with young, hot, vigorous new stars.

Then, as time passed, all the components of such irregular galaxies would begin slowly to rotate, flattening out to form huge galactic disks with the older stars clustered at the center and bright blue-and-white-shining newly formed stars and the remaining clouds of dust stringing out around the center in sprawling spiral arms as the irregular galaxies over the span of billions of years gradually evolved into spiral galaxies. As more stars formed and ever-greater gravitational mass was drawn to the centers of these galaxies, the remaining free gas and dust would be drawn into the central hub to be consumed there, leaving behind in their orbits around the hub stars that had already been formed.

Still more billions of years later, these spiral galaxies would have become flattened and elliptical in shape, now empty of gas and dust clouds and filled more and more with aging and dying red giant stars. If a given spiral galaxy had been a fast rotating spiral, it might evolve into a very flat elliptical galaxy, while those spirals rotating more slowly formed into more spherical elliptical galaxies. In any event, as all the gaseous "raw material" within a galaxy was finally consumed through the formation of new stars, the galaxy would enter into an extended period of old age as the stars within it winked out one by one and no new stars were formed to take their places. Finally, at long long last, an end point in the evolutionary chain would be reached in which all the stars in a dying elliptical galaxy were old or aging, expending their last nuclear fuels bit by bit and gradually dying, one by one, until all were finally gone.

## THE QUESTION OF EVOLUTION

The picture we have just described—the idea that galaxies have followed and are still following an evolutionary pattern of this sort throughout countless eons of time, yet always in the same inexorable pattern—was a very attractive theory to astronomers and physicists in the 1920s. It was a very neat picture, and it seemed to tie things up into a tidy bundle. Unfortunately, even today it remains no more than a very questionable hypothesis. Indeed, many leading astronomers today flatly reject the whole idea of evolutionary changes in galaxies, contending rather that galaxies are born precisely the way they look now, that is, that each type of galaxy represents a steady configuration that has been present since the galaxy first formed. In this view an irregular galaxy always remains as an irregular; a spiral was born the way it is now, spiral arms and all; an elliptical galaxy has been elliptical since shortly after its creation; and so forth.

One reason that modern astronomers are not comfortable with the "evolution-of-galaxies" hypothesis is because it leaves one insuperable question unanswered: the question of the *length of time* that such an evolutionary chain of events would require. Critical astronomers have pointed out that billions upon billions of years would be needed for a formless mass of intergalactic gas and dust to form an irregular galaxy of stars, and then to move through a series of evolutionary changes to become a spiral galaxy and then an elliptical galaxy or a spherical galaxy. In fact, these critics insisted, so much time would be required for this evolutionary process to be fulfilled that it would be quite impossible for billions upon billions of stars to form and ultimately organize themselves into an elliptical system before most of these stars had in fact burnt themselves out. From this point of view, an evolutionary progress of galaxies would require unchanging epochs of time, just as the Darwinian evolutionary theory of life on earth required its proportionately briefer epochs of time to occur—except that the galactic epochs of time would be far too great to permit individual stars to survive.

There is still no good answer to this criticism, but at least part of the answer seems to lie in the question of the total mass of each galaxy. Curiously enough, most of the elliptical galaxies that have been observed are extremely massive and distant, while most of the known irregular galaxies—far younger according to the evolutionary theory—are relatively tiny by comparison, containing far less mass than the supposedly older elliptical galaxies. One school of astronomical thought maintains that the huge and massive galaxies have in fact evolved much more rapidly than the tiny lightweights, just as a star formed from a huge mass of interstellar material appears to form much faster, to become a much bigger star, and to burn hotter and more rapidly than another star with barely enough material to reach ignition point at all.

If this viewpoint were correct, then elliptical galaxies observed today might well represent huge galactic aggregations that went through the evolutionary stages from irregular galaxy to spiral galaxy to elliptical galaxy at such a fantastically accelerated rate of speed that most of its stars are still alive and shining, while the small irregular galaxies that we observe today are evolving much, much more slowly—or, to be more accurate, are remaining irregular far longer than we might expect because they do not contain enough star material within them to organize into spirals. And if this viewpoint were correct, it might be quite possible that all galaxies, whether extremely massive and highly organized ellipticals or small and disorganized irregulars, might indeed have been formed at approximately the same time, unimaginable ages ago, but that some because of their great mass have gone through the evolutionary process far more rapidly than others.

For the time being, the question of evolution of galaxies must remain unresolved; although many astronomers reject the evolutionary hypothesis outright, more and better observation of galaxies must still be undertaken before a final answer can be found. But in searching for an answer to this question, we come face to face with the great underlying problem that has tormented astronomers, physicists, and cosmologists for over half a century and is still tormenting them today: the question of how all of these massive galaxies in the known universe came about in the first place, the question of how large the universe actually is, and the question of what is going to happen to it ultimately over geological and astronomical ages of time that may lie ahead. Of all these scientists, the cosmologists perhaps bring the widest spectrum of scientific knowledge to their study; these are men using the tools of mathematics, physics, and astronomy all together in an effort to breach the gulf of the unknown that surrounds us, and these are the men who for the past forty or fifty years have been asking the most awkward questions imaginable about the over-all nature of our universe and the over-all way that it works.

If we were lazy, or indifferent, we might well question whether the work of these men actually has any relation to physics at all; certainly their work has carried them far afield from the classical laws of Newton or even the revolutionary relativity theory of Einstein. And yet the cosmologists basically are doing no more than Galileo sought to do, no more than Einstein sought to do, no more than the physicist studying the baffling interactions of the tiniest subatomic particles seeks to do. They seek to learn something more of the over-all shape and nature of the universe. And because of this, their work is pertinent to physics and to anyone interested in the questions and discoveries of modern physics.

Cosmologists began asking questions about the origin of the galaxies, about the size of the universe and its possible future. But aside from these questions, they asked other questions that were perhaps even more penetrating and more difficult to answer. Was the cosmos infinite, containing an infinite number of galaxies with no outer limits whatever? Or was it finite, containing a certain limited number of galaxies (no matter how great the number) and no more? If the cosmos was finite, then what lay outside its outermost boundaries? Could it be that the cosmos that we know, although finite, was only one of an infinite number of similar cosmoses? In any event, what was the over-all shape and pattern of its construction? What would one see if one were somehow miraculously able to stand aside and observe it from afar? How and when did it begin, and how and when might it end?

These were questions that demanded answers. The cosmologists set about earnestly to find answers. And then, in the 1920s, in the midst of their galactic research, a discovery was made that was so amazing, and so staggering in its implications, that it completely revolutionized our entire picture of the universe and how it works. It was found beyond question that the universe of stars and galaxies, far from existing in a static, unchanging form in which neat answers to these awkward questions could be found, was in fact in the midst of a great expansion, in which galaxies were flying apart from each other in space with enormous velocities—velocities up to and even beyond the velocity of light! Just how this expansion was found to be taking place, and just what this discovery came to mean in our world of here and now, remains today one of the most exciting and important chapters in the history of physics.

# CHAPTER 21

## *The Puzzle of the Expanding Universe*

The questions that we raised at the end of the last chapter can hardly be called new or startling questions in any way. They had been raised, in one form or another, long before Galileo had a telescope, and astronomers had debated them seriously long before Einstein ever thought of his relativity theories. As early as 1826 a German astronomer named Heinrich Olbers had taken sharp issue with the classical viewpoint espoused by the ancient Greeks, which was that the universe was infinitely large and contained an infinity of stars spreading out forever in all directions without end. If such a thing were really true, Olbers argued, then it followed that any straight line drawn from the earth and going in any direction would sooner or later *have* to intersect a star. It would not matter in what direction the line went; at *some* point or another such a line would inevitably encounter a star. But this would mean that the entire sky would have to appear filled with blazing brilliant starlight both day and night. So how could it be that the night sky was *dark*?

It was a crafty and compelling question. Faced with such a paradox, many astronomers of Olbers' day felt compelled to conclude that the universe was *not* infinite in size; instead, they contended, it must contain a fixed and finite number of stars located in a cluster around our sun, with nothing at all beyond. But a hundred years later this contention was exploded, in its turn, with the discovery of distant galaxies in great profusion. Other attempts to explain Olbers' paradox proved no less unsatisfactory; the paradox remained even if you were talking about galaxies instead of stars. In other words, if the universe were populated by an infinity of galaxies stretching out in all directions, then the night sky *still* ought to be a blanket of blazing light—and quite obviously it was not.

When Albert Einstein propounded his relativity theories early in the 1900s, he provided a possible answer to the question of why we are not blinded by starlight in the night sky. Space in the universe, Einstein pointed out, is not flat like a Euclidean plane, but is curved in its over-all configuration, and in addition to that is distorted by innumerable bumps and dents caused by the presence of massive bodies such as stars or galaxies. Light, traveling through vast expanses of such a distorted universe, could not possibly follow the Euclidean straight line but rather would follow a succession of changing *geodesic curves*—lines following the shortest distance between successions of points on a curved surface or medium. Any given beam of light would further be distorted in its path any time it approached a gravitating mass, such as the sun.

Precisely what *form* the over-all curvature of the universe might take in its four dimensions Einstein was unable to say, but he speculated that it ought to take one of two forms. One possibility was space-time constructed with positive or "spherical" curvature, so that light rays would ultimately follow "great circle" geodesics and return to the place from which they started (although much later in time), much as a modern

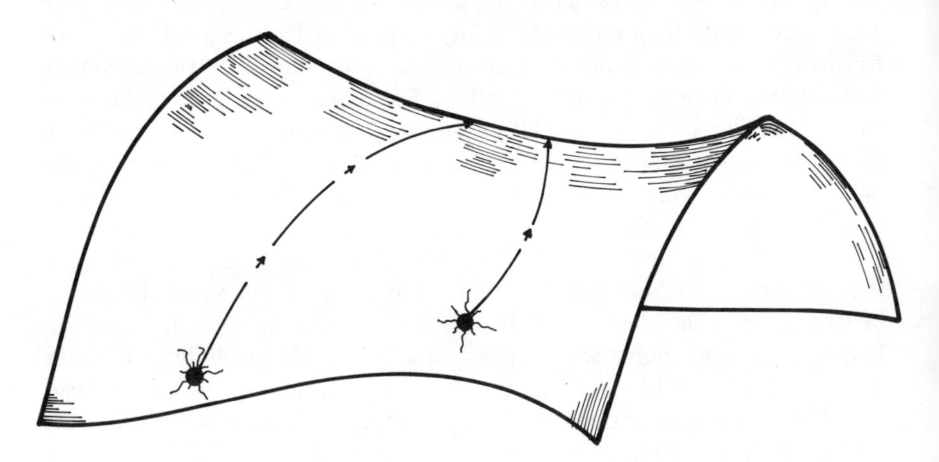

*Fig. 37* The "western saddle" configuration of negatively curved space. Light rays following geodesic curves through negatively curved space will follow courses shaped like hyperbolas. The diagram, of course, is an imperfect representation, since four-dimensional space-time cannot be represented in a two-dimensional drawing, even in perspective, without distortion.

astronaut follows a great circle in orbit around the earth and ultimately returns ninety minutes later to the area from which he started. The other alternative was space-time with a negative curvature, constructed so that its geodesics curved the opposite way in the other direction, rather in the form of a western saddle, as seen in Figure 37. If space-time had such a configuration, Einstein reasoned, then light rays would follow geodesic curves shaped like hyperbolas, and thus would always follow curves which invariably took them farther and farther away from their source in an infinite and nonreturning process.

These two possible configurations or curvatures of space seemed equally plausible, but they had quite different implications. If space were positively curved, then all the galaxies and other matter in space would make up a finite quantity of matter which had no specific boundary, totally encompassed, as it were, within a spherical area however huge it might be, and with that entire spherical area moving forward in the time dimension at a velocity of 186,000 miles per second. Thus, the night sky would be dark because there was a finite number of stars and galaxies in it and no more.

On the other hand, if the space-time of the universe were negatively curved, then the observable galaxies lying at greater and greater distances from the earth would automatically have to be farther and farther apart, so that a straight line drawn in any given direction would become increasingly less and less likely to encounter a star or galaxy the farther away from earth it got. Thus this type of space-time configuration also explains Olbers' paradox, but in addition implies the existence of a completely unbounded universe with the possibility of an infinite and endless number of galaxies within it.

Which of these "models" of the universe could possibly be true? No one could say fifty years ago, and there is still no one who can say today. Even modern astronomers with the finest telescopes and the most sophisticated equipment have not yet actually been able to observe *enough* of the universe to be able to draw any general conclusions to support one premise or the other. But if Einstein's question about the configuration of the universe was awkward, he also came up with an even more uncomfortable idea, an idea which he himself found so awkward and difficult to believe that he finally was compelled to abandon it. This idea, which he derived from the mathematics of his relativity theory, was simply that the universe and the matter within it could not possibly be static or fixed in space. Einstein's calculations indicated that with all of the gravities and masses of the planets, stars, galaxies, and interstellar dust clouds that existed in the universe, gravitational forces would have to be drawing all of the matter in the universe together toward some kind of universal center of gravity, wherever that might be. In other words, Einstein's calculations suggested that the universe actually had to be *collapsing in upon itself*.

This idea proved so uncomfortable (and, as we shall see, so completely at variance with what was actually observed by astronomers) that Einstein felt compelled to introduce another hitherto unheard-of and unobserved force acting in the opposite direction: a "cosmic repulsive force" which acted exactly the opposite of universal gravitational forces to thrust matter of the universe *apart* to exactly the same degree that universal gravitational forces were drawing it *together*. Such a "cosmic repulsive force" would thus perfectly balance gravitational forces, and would result in a *static* universe in which all the massive collections of matter—planets, stars, galaxies—remained in an unchanging relationship to each other, neither moving apart nor collapsing in toward some universal center of gravity.

Even Einstein freely admitted that his idea of a cosmic repulsive force was cut out of whole cloth to serve what he regarded as a need in explaining how things worked. But by introducing such a factor into his general relativity equations, Einstein was at least able to come up with a solution indicating the sort of model of the universe that he imagined—a stable, static universe of positive or spherical curvature, in which the three spatial dimensions were curved like the latitude and longitude coordinates on the surface of the earth, but in which the time coordinate ran straight. An alternative solution to Einstein's equations, also using the factor of a cosmic repulsive force, was worked out by a Dutch mathematicians named Willem de Sitter. This solution also indicated a static universe with spherical curvature, but with the time coordinate also curved.

Unfortunately, both these models of the universe depended upon the assumption that the mysterious cosmic repulsive force existed to keep the universe static and unchanging, and neither model remained unchallenged for long. In 1922 a Russian mathematician, Alexander Friedmann, found that Einstein himself had made a basic mathematical error in his cosmological calculations. Proceeding with his own solution of Einstein's equations, Friedmann by-passed the questionable cosmic repulsion factor altogether and came up with two different possible cosmological models for the universe. In contrast to the Einstein and de Sitter models, both of Friedmann's solutions indicated not a static universe, but a dynamic, constantly changing universe. According to one solution, the universe was steadily contracting with the passage of time; according to the other, equally possible, the universe was *steadily expanding*. Einstein himself was one of the first to recognize the importance of Friedmann's discovery—in fact, he later admitted that he considered his idea of a cosmic repulsive force to be the greatest single blunder he ever made in his life.

It appeared, then, that the stars and galaxies of the universe could not be in a fixed and static relationship one to another if Einstein's general relativity theory were correct. But the real impact of Friedmann's mathematical discovery became apparent almost at once as a result of actual observation of the heavens. Nearly as soon as galaxies were discovered to exist at all, Edwin Hubble and Milton Humason came up with apparently indisputable astronomical evidence that all galaxies were indeed *actually constantly moving* in relation to our own Milky Way and to all other galaxies. In the case of the "local group" of galaxies lying comparatively near to our own Milky Way, some were found to be moving toward us and others away from us—but this was strictly a "local" phenomenon. Beyond this small cluster of nearby galaxies, gathering evidence began to indicate that *all* galaxies were in fact moving *away* from our own, and the farther away from us a galaxy might be, the faster it was moving away from us. In short, Hubble and Humason found that the entire observable universe was actually *expanding in all directions!*

What possible basis could there be for such an idea? How could the velocity of some distant galaxy possibly be measured? Obviously, any measurement had to be made on the basis of what we could learn of that galaxy from the light from it which reached our telescopes. Up until a few decades ago we had no way to tell *anything* about a given galaxy excepting from what we could observe of its light in the sky. But light travels in the form of waves, and it is a well-known characteristic of any form of waves, including sound waves and light waves, that the wavelength measured by a stationary observer on earth might vary according to whether the waves are *approaching* that observer or *receding* from him.

Consider a simple example. When an ambulance with siren blaring is approaching us we hear a high-pitched siren whine. As it reaches us and passes us, we suddenly hear a distinct change in the pitch of the siren: as the ambulance swings by, the pitch of its siren drops sharply. This familiar "Doppler effect," named after Christian Johann Doppler, the Austrian physicist who first described it in the mid-nineteenth century, is not hard to explain. As the ambulance approaches us, its speed toward us has the effect of pressing the sound waves together so that they reach our ears at a faster rate (that is, at a higher frequency) than if the siren were sounding from some stationary point in the distance. But as soon as the ambulance swings past us, its speed moving away from us has the effect of stretching the sound waves farther apart so that they reach our ears at a slower rate or lower frequency than if the siren were sounding from some fixed point. We hear and interpret the sudden shift from higher-frequency sound waves to lower-frequency sound waves as a sudden drop in the pitch of the sound.

This same Doppler effect can also be observed in the case of light waves traveling from a moving source toward a stationary receiver. When the light source is moving toward the observer, the wavelength of the light reaching him is slightly shortened, so that that light appears slightly more blue than if the source of the light were stationary—a so-called blue shift of the light's spectrum. If, on the other hand, the source of light is moving away from the observer, the wavelengths of the light are slightly stretched out and appear longer to the observer than if the source of the light were stationary; and the spectrum of that light appears to shift slightly to the longer wavelength side or the red side. Physicists have come to speak of this lengthening of light wavelengths transmitted from an object moving rapidly away from the observer as a "red shift."*

It was this "Doppler shift" in the wavelength of light reaching the earth from distant galaxies which created a wave of excitement in astronomy

* This "red shift," of course, has nothing to do with the *speed* of the light waves. As we have seen, the speed or velocity of light waves remains completely unchanged at all times regardless of any motion of the source toward or away from the observer. The Doppler effect, and the consequent blue shift or red shift of the spectrum of the light in such cases, has to do only with the light's wavelength, not with its speed or velocity.

during the first decades of this century and forced astronomers to take a completely different view of the shape and nature of the entire universe than they had held before. As early as 1912 workers had noticed that many of the mysterious "nebulae" (which were only later identified as distant galaxies) seemed to emit light that was shifted toward the red end of the spectrum, as though those nebulae were moving swiftly away from the earth. Then in 1925 Hubble and Humason, in the course of their studies of their newly discovered galaxies, were amazed to find that virtually *all* galaxies that were more remote from us than those few clustered in the "nearby" vicinity of our Milky Way showed a characteristic red shift in their spectra, and the farther distant a given galaxy was, the greater red shift it seemed to demonstrate.

The conclusion appeared inescapable: All of the galaxies in the universe were *rushing away* from our own galaxy and from each other as if the entire universe was expanding like a balloon at a uniform rate of speed. Nearby galaxies were moving away from our own Milky Way at comparatively leisurely and sedate velocities, but remote galaxies were racing away from us at perfectly staggering velocities, perhaps even approaching the velocity of light *or even faster!*

To the nonscientist today such an idea may well seem ridiculous and unbelievable. Many astronomers had the same impression in 1925. There was no arguing with the observation; any trained astronomer with a good telescope and spectroscope could measure for himself the red shift of the light from distant galaxies. But Hubble's interpretation of his discovery seemed to defy reason, and a multitude of attempts were made to explain away this apparent "expansion" of the universe. In this case, however, every alternative explanation failed. As more and more observations of galaxies were made, it became increasingly clear that Hubble's conclusion had to be correct—that the universe was indeed expanding, with each galaxy drawing away from every other galaxy at a uniform rate of speed. Furthermore, this expansion was not the result of any distortion or misinterpretation of observations; the expansion was real, and had to be taken into account and explained by any cosmologist seeking to describe the origin, shape, nature, and future of the universe.

How can we best go about visualizing this continuous expansion of our universe? Certainly we observe no evidence of it with the unaided eye as we gaze at the night sky. We see the same stars in the same location and relationship in the sky one night that we saw the night before or the night before that. If the universe with all its galaxies is indeed expanding, from what central point is it expanding? And how could some distant galaxy be moving away from our own at a velocity exceeding the velocity of light, if the velocity of light is truly a limiting constant in the universe?

## THE INCREDIBLE EXPANSION

In order to answer these questions, we must exercise our imaginations a bit. The facts are fairly simple. Although comparatively nearby galaxies seem to be moving more or less at random with respect to each other in a sort of local motion pattern, galaxies a bit farther away from our own "local" group were found to be moving slowly away from us in all directions. This was indicated by the fact that light reaching us from those galaxies was shifted slightly to the red end of the spectrum. Galaxies still more distant were found to be moving away from us more rapidly yet, showing more dramatic red shifts in the spectroscope. And *really* distant galaxies were racing away at perfectly incredible velocities.

At first we might conclude from this observation that our own galaxy was standing absolutely still in space while the rest of the universe was expanding away from us, as if we were planted squarely at the center of things. But this is merely an illusion. The truth is that everywhere in the universe *every galaxy is moving away from every other galaxy at precisely the same velocity*. It seems to us that we are at the center of this expansion, but observers in a distant galaxy might be equally certain from their observations that *they* were at the center of expansion. It would merely be a matter of an illusory viewpoint. But the over-all uniform expansion of the universe is decidedly no illusion.

There are several ways we can visualize this expansion. Suppose we have a child's balloon covered with tiny ink spots set at uniform distances from each other. If we were to blow up the balloon, we would observe each of the spots moving away from each of the other spots at a uniform rate of speed. From the viewpoint of an observer sitting on one of these spots, it would appear that all of the spots were moving away from him while his spot remained stationary. He would see the spots closest to him moving away slowly, while spots further away were moving much faster. Precisely the same thing would be observed by an observer sitting on any other spot. Of course this picture is not quite accurate, since we would see that each of the ink spots on the balloon would itself grow larger as the balloon expanded, and there is no evidence that the galaxies themselves are expanding in size. It is merely the *space between the galaxies* that is expanding.

We might see this more clearly if we could imagine the entire cosmos as a bubble of gas suddenly released into a vacuum so that it could begin expanding in all directions. We might imagine that each molecule of the gas represents one galaxy. As the expansion of the gas bubble continues, each molecule of gas would retain its original size, but the distance between each molecule and all the rest would increase constantly at a uniform rate of speed.

Both of these imaginary pictures of the expanding universe are easiest to visualize if we think of the universe as *limited* or *finite,* containing just so many galaxies within it and no more. There are, after all, just so many ink spots on the balloon—we can count them if we wish—and a certain finite (though very large) number of gas molecules in a bubble of gas. Therefore these visualizations would suggest to us that somewhere there must be an outer boundary, a sort of division point between galaxy-containing universe on the one hand and empty space or nonuniverse on the other.

But this is merely a limitation of the model of the universe we have been visualizing—of necessity, a three-dimensional model—and not necessarily of the four-dimensional space-time continuum we believe the actual universe to be. As we have seen, there is no observed evidence to suggest that any such limit to the size of the universe exists. Indeed, many astronomers and cosmologists today are convinced that the space-time continuum of our universe has negative curvature, so that there can be no physical limit to the extent to which this expanding universe can expand. So far, no one knows. Nor has anyone yet been able to prove whether the universe is finite—for example, a universe of positively curved space in which space turns back upon itself in a four-dimensional spherical fashion—or whether it is infinite, with negative curvature and no outer limits, so that everything in it is endlessly moving farther and farther away from everything else.

But oddly enough, the fact that the universe *is* expanding in uniform fashion places a very definite and finite outer limit upon the *amount* of the universe that we on earth will ever be able to observe. This is true simply because, from our viewpoint here on earth, the expansion of the universe around us seems to increase in velocity the farther out we look. While nearby galaxies seem to be retreating from us very slowly, possibly only a few dozen or a few hundred miles per second, galaxies farther away are receding from us at velocities of tens of thousands of miles per second. In fact the velocity at which a given galaxy is observed to be moving away from our own (as revealed by the red shift in its light reaching our spectroscopes) is directly proportional to the distance the galaxy is away from the earth.

Thus, on the scale of staggering distances we are talking about, a reasonably "nearby" cluster of galaxies measured at perhaps 50 million light-years distant is moving away from our Milky Way at 750 miles each second. Another cluster estimated to be 1,000 million light-years away is rushing away from us at over 13,000 miles per second. One of the most distant galaxies that has ever been observed and measured with the 200-inch Hale telescope, one of a group of very faint galaxies in the Hydra constellation estimated to be 2.5 billion light-years away, is racing away from us with a velocity of 38,000 miles every second if the red shift of its light reaching us is to be believed—approximately one-fifth of the speed of light. A still more distant galaxy studied by other measuring instruments is receding from us at a velocity of more than one-third the speed of light, or rather, it *was* at the time that the light from it that is now reaching us was emitted, over 4 billion years ago. By now that galaxy must have traveled so much farther away that its velocity of recession is even greater. And although accurate measurements are difficult at such enormous distances, there are certain exceedingly faint galaxies that can barely be seen at all with the 200-inch telescope which are thought to be receding from us with velocities as much as two-thirds of the speed of light.

## THE CURTAIN OF DARKNESS

Obviously, in observing such galaxies at all, we are approaching the limit of capacity of our finest present-day optical instruments and other astronomical measuring devices. There is obviously more out there to be seen than the 200-inch telescope or the most modern radio telescopes are capable of detecting. Yet we face the odd fact that we could not see a great deal further out than this even with vastly larger and more powerful telescopes.

We can see the reason why if we carry this train of thought to its logical conclusion. Supposing we had a fine new telescope with much greater light-gathering power than the 200-inch Hale telescope that is being used today. We might reasonably expect to see galaxies located still further out than the faintest ones detected by the 200-inch telescope. But we would also logically expect those galaxies to be receding from us at still greater velocities than the most distant ones we can detect with present-day instruments. We would find one receding from us at 3/4 the velocity of light; another, still more distant, receding from us at 4/5 the velocity of light; perhaps another receding at 9/10 or even 99/100 the velocity of light. But if this were true, there would also eventually have to be a galaxy somewhere out there which is retreating from us at a velocity *exactly equal to the velocity of the light that is leaving it.* What would *that* galaxy look like in our superpowerful telescope? We could never possibly know. The light leaving such a galaxy, or any galaxy still farther away, could never possibly reach us, and thus we could never see such a light source, no matter how powerful our telescopes might be. We might carefully search and probe those immensely distant outer reaches and find many galaxies receding at *almost but not quite* the velocity of light, but beyond them we would encounter a blank dark wall of lightlessness.

Furthermore, if we were observing a galaxy receding from us at a velocity just a whisker short of the velocity of light, and thus theoretically visible to us, we would find that the further away it moved the faster it moved until its velocity of recession reached the "light barrier." What would we see if our most modern and powerful telescope happened to be trained on that galaxy at that critical instant when its velocity of recession *just exceeded* the speed of light? We would see the light reaching us from that galaxy suddenly blink out, never again to be seen, as though someone had thrown a cosmic light switch.

At first glance this seems to violate the rule that "no material object or signal can travel faster than the velocity of light." How then can a whole galaxy conceivably be moving away from us at the speed of light or even faster? Actually this "contradiction" is not a contradiction of scientific fact at all but only a matter of the terms we use. That distant galaxy we mentioned is "moving faster than the speed of light" only in the sense that the expanse of space between it and our own galaxy is increasing at a very rapid rate. The velocity of recession of that galaxy is greater than light velocity *only relative to the earth.* An observer standing on a platform *located halfway between our Milky Way and that fast-receding galaxy* would see that both galaxies were moving away from each other—and from him—but that each was receding at only half the speed of light. And, of course, if the observer were stationed on a planet in that distant galaxy, it would seem to him that his galaxy was not moving anywhere at all; it would be our Milky Way galaxy that seemed to be moving away from his own faster than the speed of light!*

This confusion about "faster-than-light" velocities of distant galaxies can be avoided altogether merely by changing the wording. Rather than saying that the galaxy is receding from us at a velocity greater than the velocity of light, so that light waves from it can't reach the earth, we could just as accurately say that such a galaxy is receding from us so swiftly that its light waves arriving on earth are red-shifted completely out of the visible spectrum and thus cannot be seen. However we prefer to express the phenomenon, the end result is the same: There is a point beyond which galaxies (if any exist that far away—and there is no reason to believe that they don't) are receding from our own so fast that we can no longer detect any light traveling from them. Therefore, regardless of whether our uni-

* See Milton A. Rothman, "Things That Go Faster Than Light," *Scientific American,* July 1960. As Rothman and others have pointed out, we can find many examples of things which *seem* to be moving at a velocity in excess of the velocity of light. For instance, if we consider the earth as a nonrotating, nonmoving frame of reference, we can see that many distant stars in our own galaxy seem to be sweeping through space at a speed relative to earth that is far beyond the speed of light in the course of a single night's observation. For that matter, if a child sets a top spinning, he can "create" a rotational speed for the moon relative to some fixed coordinate system on the surface of the spinning top which far exceeds 186,000 miles per second. But no such "created" relative speed in excess of the speed of light actually violates the light speed limitation. That limitation merely comes into effect when a *signal* is sent from one material object to another, or when a material object is moving from one fixed location to another.

verse is finite or infinite in size, the amount of the universe that is *observable* by us is and always will be a fixed and definite size.

But what is that size? We can only estimate this by estimating how long the universe must have been expanding. After a great deal of study of many galaxies and their respective rates of recession from us, the best guess of astronomers today is that the expansion of the universe which we are currently observing must have begun somewhere between 10 and 13 billion years ago and the universe has continued uniformly expanding ever since. But this implies that the universe must have started expanding from some different kind of state than it was in 10 billion to 13 billion years ago. What was that state of the universe then? What is our universe now expanding *from,* and how long will it continue to go on expanding? It was precisely questions like these that led cosmologists into a knockdown, drag-out battle several decades ago, a conflict that still has not been completely resolved.

Granted that these are tricky questions to approach without immediately finding ourselves ensnarled in paradoxes and conflicting definitions. In fact, they are loaded questions, simple as they may seem, for they imply that the universe must be expanding from some prior unexpanded state, from some sort of dense primordial conglomeration of material. They further imply that that dense collection of material must have been gathered at some particular place in space-time; that is to say, that all the matter in the universe must at some time have been concentrated in some densely packed central focus located *somewhere* before it started expanding. Consequently, these questions also imply that there must *still* be a center of the expansion located *somewhere.*

Oddly enough, in spite of the questions, modern cosmologists agree that *none* of these implications are necessarily true (although they *may* be) and that if any of them *are* true, precious little solid supporting evidence has been found to prove it so far. And indeed, until very recently, one group of cosmologists doggedly maintained that none of these implications were true at all, that the expanding universe never did have any "center," nor any dense primordial concentration of matter that existed at some time in the past.

To understand these two conflicting views we must consider certain uncomfortable implications that follow inescapably from each. If there ever actually *was* a dense concentration of all the matter in the universe at some localized focus at some time in the remote past, it follows that that concentration must have existed somewhere. But where? From what we can observe with our telescopes, all other galaxies are expanding away from us, faster and faster the farther away they are. We might accept this as an implication that our galaxy is the center of expansion, until we realize that observers on any other galaxy, no matter how distant, might come to precisely the same conclusion on the basis of what *they* could observe. There is no answer to this conundrum other than a purely philosophical and rational one: we know that throughout the history of science no one has found any kind of scientific evidence to suggest that our planet, our sun, or our galaxy is in any way unique or different from multitudes of others scattered throughout the cosmos, so that to assume that we are located at the center of expansion of the whole universe is as intellectually offensive as the ancient Greek idea that the sun and all the rest of the universe revolved around a unique and stationary earth.

Obviously our location in the expanding universe, if not at its center, must depend to some degree on whether the universe is finite in volume and turned back upon itself in spherical curvature, or infinite in volume, stretching away forever in all directions. Here again we come up short with our own reasoning: If the universe is finite, and if we are not located at its center, it ought to follow that we would be able to see more starlight on one side of our galaxy than on the other. But no such observation has ever been made. From what we can see, there is no more nor less total amount of starlight reaching us from one direction than from another. On the contrary, astronomers probing and cataloguing the heavens find that the multitudes of galaxies seem to be scattered in remarkably uniform fashion on all sides of our own Milky Way. Many have interpreted this simple observation as convincing proof that our universe is in fact infinite in size, stretching out in endless successions of galaxies and clusters of galaxies forever in all directions. Yet even this interpretation would not hold true if, as we have seen, we are in the midst of a bubble or sphere of "observable" universe that is surrounded by an unknown and totally unobservable and forever invisible quantity of universe that may be stretching out beyond the curtain of darkness imposed by the limitation of the speed of light.

Indeed, there might be whole segments of universe which we can never even hope to observe, either directly, or indirectly, at any time, because everything beyond our "observable" segment of the universe is receding from us faster than the speed of light. Even that "observable" part of the universe is fantastically huge—but is it not possible that somewhere out beyond the "curtain of darkness" there still exists a finite boundary to the universe that we simply can't detect?

It is questions like this that make mathematicians fumble learnedly with their slide rules and cause physicists and astronomers to glance anxiously at the sky as if they expect a meteor to hit them at any moment. So far such questions have no answers, and although some cosmologists are optimistic that answers will be found, there is no particularly convincing reason to feel that their optimism is justified.

## THE THEORY OF COSMIC EXPLOSION

The best that we can do, at least so far, is to imagine "models" of a universe which we cannot yet (and perhaps never can) fully observe, and then try to fit what we *can* observe—an increasing amount each year—into one of these theoretical and speculative "models." Physicists have always considered the use of imaginary experiments and imaginary models to be a fruitful approach to understanding things which cannot be ideally observed, and astronomers and cosmologists have felt that in many ways such models offer the *only* possible approach to the answers of the great cosmological questions facing them. Even though our capabilities for observation may be limited by the speed at which distant parts of the universe are rushing away from us, it still seems reasonable to consider that the observable part is at least a respectable fragment of the whole, and to assume that this part reveals characteristics that are representative of the universe as a whole. As we will see later, some recent and startling discoveries tend to throw even this contention into doubt, but at least it is a plausible place to begin. And in recent decades two strikingly different "models" of the universe have been proposed, each with some fascinating characteristics, each raising some new questions of its own, and each worthy of attention.

According to the first of these models, the expansion of our universe that we observe today is exactly what it appears to be: the late aftermath of some kind of ancient cosmic explosion that took place somewhere in the universe at some remote time in the past, probably some time between 10 billion and 13 billion years ago. This model of the universe was first proposed as a theoretical, mathematical possibility by Alexander Friedmann, and was further developed and applied to the question of how the universe might have begun expanding by Belgian astronomer Georges Lemaître. Later popularized by the Russian-born American physicist George Gamow, this model of the universe is today supported (at least in its basic premise) by the vast majority of astronomers and cosmologists, and is commonly referred to as the "big-bang" theory of the creation of the expanding universe that we now observe around us.

According to this theory, there was a time in the enormously remote past—nobody knows exactly when or where—when all of the matter and energy currently in existence in the universe was concentrated into a single densely packed and exceedingly hot mass. Although we can only guess at the the size of that glob of matter, we can imagine it existing not as a solid chunk of matter like a lump of coal, but rather as a dense magma of primordial nuclear fluid, more raw energy than physical matter, crammed down into the most extreme physical density that either energy or matter or any combination of the two could possibly achieve.

In this blob of nuclear fluid (which Gamow rather romantically called

"Ylem," from a Greek word meaning "primordial matter") there could have been none of the atomic or even the nuclear structure of matter that we know today. Perhaps at that critical point there was no physical matter whatever present. But since we are assuming that all of the *matter and energy* of the universe was concentrated here, then there must have been absolute nothingness everywhere else. Certainly in comparison with the magnitude of the matter-and-energy-filled universe we know today, the total volume of this primordial mass must have been very small; mathematicians have speculated that it might have occupied a volume no greater than a sphere which could have been encompassed within the present orbit of our planet Neptune, but there is no real reason to assume that it might not have been packed down into a sphere the volume of a tennis ball, or a marble, or a grain of sand or even a pinpoint.

Nor do we have any clear-cut idea where this dense mass of Ylem came from. Some contend that it might have arisen in that form as the end result of an infinite period of *contraction* of matter previously existing in more expanded form at some even more remote era of time past. Perhaps such a period of contraction had been going on infinitely long in order for all that matter and energy in the universe to have collapsed into this dense core of concentrated energy and matter. But whatever its shadowy origin, the "big-bang" theorists contend that there came an instant in time when something catastrophic happened to this mass of inchoate primordial material, more intensely concentrated and more intensely hot than anything we can imagine, so that it suddenly began to expand with explosive violence, splattering itself out in all directions from its center.

During the first few microseconds, minutes, or hours following this cosmic explosive instant, according to this theory, a series of extraordinary things must have happened. As we have seen, at the instant of the explosion this mass of "nuclear fluid" must have been composed almost entirely of raw energy with very little actual matter existing in any form, and with virtually none in the atomic and subatomic form we know today. But when the explosion had occurred and expansion had begun, the temperature of the exploding mass must have begun to drop sharply, and radiant energy must have begun transforming into physical matter in one form or another very swiftly. According to some calculations, some "matter" must have condensed from the radiant energy present within a few thousandths of a second after the explosion, although not in any form we would ordinarily recognize. This "matter" was probably in the form of primordial "strange" particles and antiparticles such as we will discuss in a later chapter. By the end of the first *second* after explosion more familiar particles were forming —protons, neutrons and electrons, for example, together with their corresponding antiparticles—only to annihilate into radiation again a second or two later as the still incredibly high temperature dropped a bit further.

As the expansion continued, more stable physical matter began accumulating, so that within thirty *minutes* of the beginning of this unthinkable explosion thermonuclear reactions must have transformed a vast amount of radiant energy into atomic nuclei in their familiar form: nearly 80 per cent hydrogen, 20 per cent helium, and perhaps 1 per cent made up of heavier nuclei. But in spite of this swift sequence of events immediately following the explosion, huge quantities of radiant energy must have continued to dominate this scene of furious expansion for as long as 100 million years more before enough hydrogen, helium, and heavier elements had been formed for physical matter at last to predominate over radiation and become available for the formation of stars and galaxies.

Of course once the expansion of radiant energy and newly formed matter had begun, it continued and continued in all directions. Material particles—atoms, electrons, neutrons, etc.—were scattered uniformly out into space in the form of gas and dust. After hundreds of millions, perhaps even billions, of years, this gas and dust which had billowed out in ever-expanding fashion must have begun to form whirlpools and eddies which then presently began concentrating clusters of matter at their centers to form into stars, which subsequently gathered together in the huge stellar galaxies we observe today. Stars flared to brilliance, contracted, exploded, formed newer and heavier elements, and then scattered them in gaseous form far and wide. Other stars formed, grew old and died, with or without solar systems. Galaxies formed and flared up in beaconlike brilliance as matter continued to expand and scatter farther and farther from its primordial center. And this expansion continues to this very day as we stare in wonder at the star-studded night sky.

The proponents of the big-bang theory do not pretend to be able to explain where the original primordial matter came from or how it reached such a stage of extreme concentration and heat that it had to explode, but they do contend that this indeed must have been what happened. As Dr. Allen R. Sandage, of California's Palomar Observatory, recently expressed it in an interview: "Something happened, independent of any theory, about ten billion years ago, which changed the world into its present state from an unknown previous state. And you might call that the creation of the universe in its present form."*

According to this viewpoint, the universe continues to expand today and will continue to expand tomorrow and next year and over the next century and the next millennium. But will this expansion continue forever, or will it sometime somehow come to an end? In answering this question the big-bang theorists differ very sharply in their opinions and predictions,

* See C. P. Gilmore, "The Birth and Life of the Universe," *New York Times Magazine*, June 12, 1966.

depending upon whether they regard the universe as a finite, closed system with positive curvature of space-time, or as an infinite open system with negative curvature of space-time.

If the universe is finite and closed, then it must have a certain fixed total of mass and energy within it. Astronomers argue that if that is the case, gravitational forces acting between the galaxies, however minuscule they might be over such great distances, will inevitably work to counteract the outward force of expansion, so that the galaxies will gradually slow down in their expansion, and finally at some time in the future will come to a stop. When this happens those same gravitational forces will continue to act, so that ultimately, ever so slowly, the universe will begin contracting again, at first imperceptibly, then ever and ever more swiftly, until once again it has contracted down to a superdense, exceedingly hot, and massively concentrated nuclear core, with all its matter and energy once again broken down into a gas of elemental protons and neutrons and finally reconverted into an energy-rich primordial nuclear fluid. According to this theory, such an era of contraction would then ultimately once again reach a critical point at which further contraction was no longer possible, and another "big-bang" would be triggered to start the universe expanding again in another stage in an endless cycle of expansion and contraction.

Such an idea is not based entirely upon speculation; there is a certain small amount of evidence to support it. For example, certain recent measurements of the speed of recession of extremely distant galaxies do seem to reveal some slight slowing down of the velocity of expansion. Unfortunately, such observations can only be made at the very extreme edge of the range of our present-day telescopes and other measuring instruments, and the error in such measurements can be so great that such tentative results as these may prove totally erroneous. In this case the truly convincing evidence—if any such evidence is actually found—must wait upon larger and more powerful telescopes or technological refinements of other astronomical instruments.

Other astronomers take an opposing view. They argue that if the space-time of the universe is negatively curved, or even if it were not curved at all, galaxies would be expanding away from each other far too swiftly for gravity to overcome their expansion and cause it to slow down. These men therefore contend that the universe will go on expanding forever until all galaxies have burnt out and nothing is left but an endless and still-expanding scattering of dust and cinders. Those who support the notion of a pulsating universe rather than a "single-trip" expansion argue that the theory of pulsation at least would explain where the nucleus of the present stage of expansion came from. But there is no known mechanism that might cause a closed universe to oscillate between expansion and contraction in this fashion, and even proponents of the theory have no way to explain where the matter of the universe might have come from in the first place. To contend that it had always been there through an infinity of time is really no more satisfactory scientifically than to contend that at some point in time it was created en masse by a flick of the finger of God.

## THE CONCEPT OF A STEADY-STATE UNIVERSE

If the questions still unanswered by such a "big-bang" model of a universe are perplexing, the questions raised by an alternative and quite different model of the universe are perhaps even more so. This alternative theory of the cosmological nature of the universe was originally proposed and supported by British astronomers Fred Hoyle and Herman Bondi, and Hoyle still considers it a conceivable alternative to the big-bang theory (although with certain significant modifications). According to this theory, the expansion of the universe that we observe going on around us is the result of a totally different process than the grand initial explosion proposed by the big-bang theorists. According to Hoyle's theory, there is no need to speculate about some extraordinarily dense and hot concentration of matter as the source of the universe's expansion at some time in the remote past. In fact, according to this theory, there never was any such concentration of matter, nor any explosion setting off the current expansion of the universe.

Instead, this theory contends that throughout all time the concentration of stars and galaxies throughout the universe has been precisely the same as it is now—never more dense, never more diffuse. It was untenable, these men insisted, to imagine that all matter of the universe might have come into being at once at some time in the remote past. Rather, they contended, new matter is constantly being created, bit by bit, in the form of single, solitary hydrogen atoms, appearing out of nowhere, one by one, throughout the vast reaches of space between the fleeing galaxies. This piecemeal creation of new matter, they argue, has been occurring in a quiet and steady fashion throughout all eternity in the past and will continue in precisely the same fashion throughout all eternity in the future. Thus the "steady-state universe" proposed by this theory had no beginning and will have no end. It is utterly infinite in either direction through time, and utterly infinite in its expanse through space-time.

But what, then, makes the universe expand as we know that it is doing today, with galaxies flying apart from one another at tremendous velocities? According to Hoyle's calculations, new hydrogen atoms must appear in space at a rate of approximately one hydrogen atom in each square meter of space every 10 billion years. As these solitary atoms form or appear, from nowhere in the empty space between the receding galaxies, they gradually gather into huge tenuous gas clouds which presently become more dense, collect fragments of dust left over from nova and supernova explosions, and presently enter into the evolutionary chain that creates new stars and galaxies to fill in the empty spaces left by the expansion of the old galaxies. Thus, the universe is not only uniform in its composition of galaxies everywhere in all directions; it has been essentially uniform at every time in the past and will continue in this steady state of matter creation and expansion forever into the future. All the heavier elements making up the stars and the planets are considered to be second-, third-, fourth-, or tenth-generation products of stars that have collapsed and gone nova, undergoing thermonuclear reactions in the process that built up these heavier elements and then scattering them in the form of dust and gas throughout the surrounding areas of space, later to be incorporated into new stars which then undergo the same sort of process.

As we might imagine, this steady-state theory was attacked from all sides when it was first proposed. There was no scientific evidence whatever, critics complained, that matter or energy had ever been created out of nothing or ever could be, not even a single lowly hydrogen atom. Indeed, one of the most basic and thoroughly substantiated universal natural laws indicated exactly the opposite: that matter-energy could *not* be created out of nothing, any more than it could be destroyed and done away with in some fashion. Even so, the steady-state theory provided a fascinating alternative explanation to the big-bang theory for those seeking to explain the things that had actually been observed of the expansion of our universe. According to the steady-state theory, the universe is constantly being fed by the appearance of new matter, never formed very rapidly nor in great quantity in any given area of space, but actually being formed in enormous total quantities if we consider the immensity of space that exists throughout the universe. The newly created hydrogen is formed in sufficient quantities to lead to the evolution and constant formation of new stars and galaxies, creating a dynamic balance with the consumption of existing hydrogen within existing stars.

But if this occurred and if there was no corresponding obliteration or annihilation of matter-energy, then one of two things would have to happen. Either all the matter in the universe, in ever-increasing amounts, would have to become more and more condensed as time went on, or else the already existing matter in the universe would have to move out of the way in some fashion to make room for the constantly created new matter.

The first alternative, of course, would imply a finite end point some time in the future when all the matter in the universe would be packed into a supercondensed mass beyond which no further packing was possible. But astronomers have actually observed another alternative: the constant and uniform expansion of the galaxies away from each other, creating empty space in between them. Thus the steady-state theory maintains that the expansion of the universe is not the end result of some kind of cosmic explosion of a supercondensed mass of matter some time in the remote past, but simply a means of getting existing matter moved out of the way fast enough to make room for the new matter being formed.

Both the big-bang theory and the steady-state theory had certain attractive features, and both were obviously vulnerable to criticism. At least the steady-state theory seemed to offer scientists the possibility that it might somehow, sometime, be confirmed or refuted by means of scientific observation, and this was a very strong advantage. In contrast, the big-bang theory rather unfairly put the critical period of universe formation—a period before the supposed cosmic explosion occurred—forever beyond the reach of either observation or proof. For this reason many physicists and astronomers, who like above all to believe that things in nature are demonstrable and provable, felt a strong emotional attachment to the steady-state theory, even when the preponderance of evidence seemed to suggest more and more strongly that some form of a big-bang model of the universe was most likely the correct one.

But curiously enough, in the heat of debate over these two theories, astronomers tend to overlook the simple fact that both theories have one remarkable basic premise in common. Both imply that at some point in time, at some place in our universe, *something must have been created out of nothing.* It really doesn't make sense to argue that the creation of a single hydrogen atom out of nothing is scientifically insupportable, but that the idea of the creation of all the matter and energy in the entire universe in the form of a supercondensed mass of Ylem out of nothing is scientifically plausible. In either case, we are talking about the same basic thing; we are merely arguing about the order of magnitude. Whether matter is being created a tidbit at a time or all at once in a magnificent display, we are still talking about the creation of something from nothing—the "finger of God" that we mentioned before. Nor does the idea of a pulsating universe undergoing a regular cycle of expansion and contraction of the matter and energy within it allow us to escape the dilemma. The matter and energy that is expanding and contracting (if it is) *ultimately had to come from somewhere.*

Until very recently there had been little conclusive or even suggestive evidence to favor either of these two major theories over the other. The big-bang theory and the steady-state theory seemed equally supportable or insupportable, depending upon how you looked at it. But in the past few years bits of evidence began to appear which seemed to offer solid indication that some form or another of the big-bang theory was indeed the correct one. The most interesting—and enigmatic—evidence was the discovery in 1965 of a mysterious "background radiation" permeating the universe with no apparent contemporary source to account for it. Years back such big-bang-theory proponents as George Gamow, and later C. Hayashi and others, had predicted that the primeval material thrown out by a very hot initial explosion at the beginning of the universe as we know it should have left behind traces of very weak so-called "black-body" radiation which should still be detectable throughout the universe with a temperature ranging somewhere between 3 degrees and 20 degrees Kelvin. This prediction, unconfirmed at the time, had been thoroughly forgotten until 1965, when A. A. Penzias and R. W. Wilson, of the Bell Telephone Laboratories, quite accidentally detected a mysterious background radiation at 3 degrees Kelvin, and R. H. Dicke, of Princeton University, connected this radiation with the black-body radiation predicted earlier. Today many cosmologists are extremely sanguine about this 3-degree Kelvin radiation, contending that it does indeed represent the continuing remnants of the violent radiation storm of a big-bang explosion and provides strong evidence that that theory is true.

But before such hopeful (if still inconclusive) evidence had appeared, both the big-bang theory and the steady-state theory raised apparently irresolvable questions—questions which appeared to be beyond the realm of science altogether. Each was philosophically untidy. The best that could be hoped was that possibly more observation and improvement of observational techniques might add the necessary bits of demonstrable scientific evidence to tip the balance one way or the other.

But was there really any hope of ever gathering such evidence? Indeed there was, and the past decade has proved to be one of the richest and most exciting periods in the long history of astronomy. One of the most fascinating things in any study of the history of science is the frequency with which the whole direction of study and understanding can be changed abruptly by the unexpected appearance of some totally unpredictable variable. Time and again scientists have found themselves facing stalemates, their work halted by apparently insuperable roadblocks, only to find a whole new horizon of investigation opening up at just the critical moment.

Sometimes the unpredictable variable is the development of a new instrument for observation; sometimes it is the discovery of some hitherto completely unsuspected fact. In the case of modern astronomy, both these things happened on the heels of each other, and as a result astronomers began searching the skies again with renewed excitement and vigor. Even cosmologists found sudden new hope that some sort of scientifically supportable answers might after all be found regarding the birth, life, and future of the universe.

# CHAPTER 22

## *The Riddle of the Quasars*

It is perfectly understandable that most of us naturally and instinctively think of astronomy as a *visual* science. Throughout history, mankind's knowledge of the skies arose primarily from what he could see with telescopes. Even in the past two hundred years, when such instruments as the spectroscope and the astronomical photographic plate have come to his aid, the basic tool of astronomy remained the reflecting telescope such as the 200-inch "eye in the sky" mounted atop Mt. Palomar.

It is an odd paradox, therefore, that one of the most startling breakthroughs in all of the history of astronomy arose not from visual observation at all, but from one scientist's irritation at the stream of garbled noises that he found pouring into his ears from the heavens.

At first this curious phenomenon was considered more of a nuisance or scientific curiosity than a discovery of any major proportions. As early as 1931 Karl Jansky, a Bell Telephone Laboratory engineer, discovered that a surprising amount of unidentified radio "noise" seemed to be showering down from the sky in a steady stream. These were clearly no orderly radio signals transmitted from any human source on earth. Rather, they seemed to be a completely disorderly barrage of radio waves like the "disorganized static" that frequently interferes with normal radio transmission during electrical storms, for example—an annoying sequence of buzzes, hisses, crackles, pops, and squeals.

Radio communications experts already knew that such disorderly radio signals frequently emanated from our own sun during cyclical periods when the atmospheric disturbances on the sun that we know as "sun spots" could be observed. But the radio signals disturbing Jansky's work were not coming from the sun. They seemed, rather, to be coming in straight from space beyond our solar esystem.

For some years after Jansky's discovery, there was no concentrated effort made to try to identify the source of this cosmic radio noise. For one thing, radio receiving equipment in those days was relatively crude, and there was no known way of "focusing" such vagrant signals in such a way even to be certain which quadrant of the sky they seemed to be coming from. But during World War II radio and electronic communications devices and techniques were developed and improved on an emergency basis, particularly in the area of radar sending and receiving equipment. Radar was essentially a technique for generating a stream of electromagnetic waves and firing them off in a controlled beam in a specific direction, so that any of those waves striking some distant object such as a flying airplane would rebound and could be picked up again by means of a dishlike parabolic antenna and focused at the receiving point, We all know the importance of radar during World War II, and even today, in identifying and locating flying objects which have no business in the sky. But after the war was ended a great deal of surplus radar and radio equipment became available to astronomers in various parts of the world, so that in the late 1940s a number of workers begun attempting to track down the sources of the cosmic radio noise Jansky had first noticed.

## THE DISCOVERY OF "RADIO GALAXIES"

At first glance we might think the identification of such radio sources might be a comparatively simple matter, but the men who dug into the problem found it becoming more and more complex the more they worked at it. During the 1940s and 1950s first hundreds and then thousands of separate sources of radio signals were identified and catalogued in the skies. Many of these sources seemed to be located along the plane of the Milky Way galaxy, but many more appeared to arise from areas in the halo of stars surrounding the hub of our galaxy, or even in areas completely beyond it.

Naturally, astronomers were eager to identify these radio-wave sources with specific visible stars or other light sources in the sky, so that these sources could be studied more carefully with the great telescopes available. But this proved a bit more easily said than done. Radio receivers turned toward the night sky—instruments which came to be known as "radio telescopes"—might identify a strong source of radio signals in some relatively small area of the sky, perhaps even an area as small as one second of arc in diameter; but a study of astronomical photographs of even this small area would reveal perhaps tens of thousands of stars, nebulae, or galaxies! How could only *one* source be isolated from so many possibilities? It might have been easy if the radio signals emanated only from the largest and brightest stars, but unfortunately it did not seem to work that way. Many large and brilliant stars maintained complete "radio silence," while exceptionally faint and insignificant-looking stars poured out a veritable barrage. In some cases radio signals seemed to be coming from areas with no identifiable optical source whatever!

The problem of isolating radio sources thus required the development of techniques for "focusing in" on a radio source, and gradually these techniques—some of them very crafty indeed—began to improve. For example, two radio telescopes located at a distance from each other on the earth's surface might focus on the same radio source in the sky, and by measurement of the slight difference in time that the signals reached the two telescopes (a technique called "interferometry") the source could be pinpointed more closely than by a single radio telescope working alone. Another technique used the disk of the moon moving across the sky as a means of pinning down these elusive radio sources. Most optical astronomers generally considered the moon to be more of a nuisance than anything else, since it obscured their viewing of the area of space around it, and burned out their photographic plates. But radio astronomers found that the edge of the moon acted like the edge of a knife blade in the sky. The moon's disk would occlude or block off radio signals from a given source when it was in the way, and then suddenly "release" that source, so to speak, as soon as the moon's disk passed by. Since the precise location of the moon could be determined quite accurately at any given time, radio astronomers had available still another clue to the exact location in the sky where the source of invisible radio signals might be found.

At last, in 1949, radio observers succeeded in directing optical astronomers to one of the three strongest of all sources of radio signals that had thus far been discovered. This powerful radio source, a veritable broadcasting station in the sky, proved to be an old familiar friend to optical astronomers. A steady stream of strong radio signals seemed to be pouring forth from the exact spot in the sky occupied by the Crab Nebula, the great expanding cloud of gas and dust that remained from an enormous star explosion or supernova that had occurred in our own Milky Way galaxy thousands of years ago and was first observed by early astronomers in A.D. 1054.

As we saw earlier, that nebula had once been an ordinary star much the same as millions of others in our galaxy. Now it was nothing more than an enormous cloud of cosmic rubble. But radio astronomers were able to prove that that nebula was pouring out not only great quantities of visible light but a huge quantity of other kinds of electromagnetic radiation as well, including radio waves, the only kind of electromagnetic wave other than light which could penetrate the earth's atmosphere.

Bit by bit other radio sources were pinpointed, and a pattern began to appear. Characteristically, it seemed that radio signals emanated from regions of space where huge roiling gas clouds filled with high-energy atomic fragments in a highly excited state were to be found. Within a short time optical astronomers working hand in glove with radio astronomers located a visible source for powerful radio waves that lay far outside our Milky Way galaxy. This radio source was, in fact, two galaxies which were colliding with each other some 50 million light-years away.* The radio waves seemed to arise not from any violent collision of the compact matter of the stars in these two galaxies, but rather from the collision of atoms in great clouds of gas that lay between the stars of the onrushing galaxies. Astronomers calculated that when two such galaxies were involved in a collision—an exceptionally unusual event at best—the chances of any given star in one galaxy actually colliding with a star in the other was exceedingly small; most of the stars in both galaxies would simply pass along their way undisturbed. But atoms in the clouds of gas swirling between the stars of these colliding galaxies would be bound to undergo innumerable collisions with other atoms. As these gas clouds whirled together at velocities of hundreds of thousands of miles per second, heat would be generated. Atoms would be ionized into charged particles, and powerful radio signals would then be generated as these electrified particles interacted with each other again and again.

## THE DISCOVERY OF QUASARS

Thus over a period of years a few radio sources were clearly identified with visual objects—in most cases unusual cosmic events in the sky. But as the number of new individual radio sources that had been discovered rose to the hundreds and then to the thousands in the late 1950s, astronomers began to become more uneasy about them. There seemed to be far more of these sources of cosmic radio signals than there were supernovas or distant galactic collisions to account for them. And search as they would, astronomers could find no apparent identifiable sources for the greatest majority of powerful radio sources. Often these enigmatic signals seemed to emanate from areas of blank sky, areas, that is, which seemed to hold no celestial object any more exciting than the ordinary sprinkling of ordinary unremarkable stars and galaxies one would expect to find. If some radio sources could be clearly identified with peculiar galaxies and unusual cosmic events, why couldn't some of these other radio sources which were equally strong or stronger than those already identified also be associated with something unusual? And if these radio signals were *not* coming from sources that could be visualized, then what was going on out there to cause them?

Finally, in 1960, an answer to these questions was found by Dr. Allen Sandage and Dr. Thomas Matthews, of the California Institute of Technology, using the 200-inch telescope at Mt. Palomar. But welcome as it was, their answer was even more puzzling than the question. Searching

* As we discussed earlier, it is perfectly possible for the movements of individual galaxies in a localized area to bring two galaxies into collision, even though the over-all general pattern of motion of all galaxies is ultimately to draw away from each other in a vast universal expansion.

one of the "blank" areas of the sky that was clearly the source of a powerful radio signal, these men discovered that the enigmatic signals appeared to be emanating, not from a colliding galaxy or from the remains of a supernova, but rather from what appeared to be a small and totally undistinguished star—a star which had been photographed before, all right, but had never been studied carefully because it had always appeared to be just one of the multitudes of minor stars at the outer edge of our own galaxy. In short order three more such humble-appearing visual objects were also identified as powerful radio sources. Uninteresting as they had seemed to be, feeble as their visible light seemed to be, these "stars" were in fact pouring forth more invisible radio energy than would be expected even from whole colliding galaxies!

Puzzled and curious Sandage and Matthews took the next step—analyzing the feeble light from these objects with the spectroscope—and the mystery deepened. To their chagrin, they found that the faint lines that appeared on the spectrographs of these curious stars were completely different from the spectrographic patterns of any other known stars in the universe! Whatever else these light sources were, they were obviously *not* stars of any sort that had ever been discovered before in the sky. But all the same, they seemed to mimic the appearance of ordinary stars. For lack of anything better to call them, these puzzling objects were dubbed "quasi-stellar radio sources" or "quasars" for short.

The discovery of celestial objects that were neither stars nor galaxies nor nebulae nor gas clouds nor supernovas nor anything else that had ever been seen or identified before in the history of astronomy was really a bit too much. What were these strange "radio stars"? For that matter, *where* were they—within our galaxy or outside it? One of Sandage's associates, Dr. Jesse Greenstein, immediately set to work trying to figure out what kind of objects could possibly have this sort of visual and spectographic appearance and at the same time be throwing out such energetic streams of radio signals. Other astronomers began to find more such strange objects, now that they knew what to look for. No hypothesis made any sense at all, and for over two years, as more and more quasars were discovered, Greenstein found no plausible answers.

Then another astronomer at Cal Tech, Dr. Maarten Schmidt, dug into the problem, studying a particularly bright radio source identified only by a catalogue number 3C–273 (simply meaning the 273rd radio source listed in the Third Cambridge Catalogue, one of the most complete listings of radio sources that had been compiled by astronomers up to that time). When Schmidt analyzed the light coming from 3C–273, he also found the unexplainable lines on the "spectrograms"—the tiny photographic plates of the spectral lines created by the dim light of these quasars, so small that the shadings of black and white could only be distinguished under a microscope. Like Sandage and Greenstein before him, he could find no familiarity in the pattern of the spectral lines. It was wildly frustrating; Schmidt was convinced that there was an answer, that there *had* to be an answer, but it eluded him. And then, in one exciting night in February, 1963, after struggling with the problem steadily for weeks, Maarten Schmidt had an inspired idea and found the amazing solution to the two-year-old puzzle of the curious spectrographic patterns of these quasars opening up before him in a matter of a few hours.

It had seemed to Schmidt from the first that at least four of the faint spectral lines of the six he had identified from 3C–273 had a pattern that was somehow vaguely familiar. And familiar they were; they were simply located *in the wrong part of the spectrum altogether!* These four lines were usually associated with hydrogen, most commonplace of all elements present in stellar objects, and were usually found in their characteristic sequence near the blue end of the spectogram of the light of any given star. But in the spectogram of light emanating from 3C–273 these four characteristic hydrogen lines were shifted far to the *red* side of the spectrum, a fantastic 16 per cent shift from their normal positions! This was precisely the same "red shift" of spectral lines that Hubble had described in demonstrating that distant galaxies were moving rapidly away from each other in an expanding universe. But the red shift in the spectogram from 3C–273 was far, far greater than any that had ever before been measured. A 16 per cent red shift of the light from this radio source—if red shift was what it was—could only mean that this object was one of the most fantastically distant objects ever before visualized in the universe, lying out near the fringe of visibility even with the 200-inch telescope. And if that were the case, it could only mean that this faint and insignificant "star" at such a remote distance from our own galaxy had to be pouring forth light and radio energy with a fantastic brilliance, greater than that of a hundred galaxies of a hundred billion stars each, all taken together!

Hurling its light and radio waves toward us across a staggering distance of 2 billion light-years, and racing away from our galaxy at an unbelievable 27,000 miles per second, the quasar 3C–273 had to be one of the most brilliant objects ever observed in the universe. Excitedly rechecking the spectograms of other quasars photographed two years earlier, Schmidt and Greenstein found that one, identified only as 3C–48, had an even more marked red shift; this one was over 4 billion light-years away from the earth, and was receding faster than twice the speed of 3C–273.

The breakthrough had begun.

## THE ENIGMATIC BEACONS

Schmidt's electrifying discovery was made early in 1963. In the years since, quasars have been the object of intense and excited study in the world of physics and astronomy. A great deal has been learned about their characteristics, but the mystery of what they really are and how their presence in the universe must be interpreted has steadily deepened. Astronomers and physicists alike feel confident that the ultimate answer to the great cosmological puzzle, the question of where the universe came from and what is ultimately going to happen to it, must be intimately bound up with these enigmatic radio beacons in the sky, and most astronomers feel that at long last some kind of answer may be within reach. But so far no answers have been found. Indeed, virtually everything about these strange objects today remains a question mark. Practically nothing is certain.

Quasars look like dim nearby stars, but they are not stars, and the preponderance of evidence suggests that they are certainly not nearby. Their outpouring of radio energy is fantastic. Believed to be more remote from our galaxy than any other celestial objects ever known, their outpouring of visible light must also be fantastic in order for that light to appear even as dimly as it does on our telescopic photographs. No single star (as we normally think of stars) located 2 billion light-years away from our galaxy could conceivably produce enough visible light to register at all, even in our giant 200-inch telescope. A whole huge galaxy located at that remote distance would at best appear as a barely perceptible fuzzy light spot, a mere shadowy wisp that might only be found upon microscopic examination of a photographic plate—nothing more than a few flakes of silver precipitated in the emulsion. Yet quasars located at that distance away pour out enough radio energy to stand out like veritable radio beacons, and enough visible light to appear on photographic plates to be as bright as moderately sized stars located in our own galaxy.

Nor are the quasars numbered 3C–273 or 3C–48 by any means the most distant quasars that have been found. Those two, located respectively 2 billion and 4 billion light-years away, and apparently receding from our galaxy at velocities of 15 per cent and 30 per cent of the speed of light respectively, were far enough away to be incredible stellar objects. But as the search has continued, well over a hundred quasars have been found, all definitely associated with strong radio sources. Spectographic red shifts have been laboriously measured for over sixty of these. Many of them reveal red shifts indicating recession velocities of 30 or 40 per cent of the speed of light. Many are receding even faster. One, identified and studied by E. Margaret Burbidge, held the record for a while: its red shift indicated that it lies somewhere between 5 and 10 billion light-years away (no more accurate estimate can be made) and is slipping away from us at a whopping 148,000 miles per second—a velocity 80 per cent of the velocity of light. This made it the most remote celestial object ever observed by astronomers, until Maarten Schmidt himself recently reported another quasar receding at 81 per cent of the velocity of light!

But aside from the unresolved mystery of what in the name of creation these strange remote objects are, their discovery and study has a special significance to astronomers and physicists. In studying them, astronomers are not only probing farther out into the universe than our telescopes have ever before permitted, but they are, in a sense, looking farther back into the remote past of the universe than was ever before possible. The light now reaching our telescopes from the remotest of all quasars must have left on its journey toward the earth up to 10 billion years ago, perhaps before our own galaxy had even condensed out of its primeval gas cloud. What might have happened to that quasar in the course of that immense time period can only be guessed. For all we know, it might not even exist any more. And we might well have to wait 8 billion or 10 billion years in order to find out. More likely, we might never learn the answer, even if we waited, for if such a remote quasar continued receding from us faster and faster during that time interval, it would at some point have reached or exceeded a velocity of recession beyond the speed of light and thus moved forever beyond the "curtain of darkness" and out of the observable universe altogether, never again to be detectable in any way.

But considerable doubt exists that these remote quasars whose light is now reaching our telescopes could ever have actually survived anything like 5 billion to 10 billion years. Much larger than individual stars, yet smaller than galaxies, it seems inconceivable that any such aggregation of matter or energy in any form could continue to "burn" and radiate light and electromagnetic waves in such huge quantities in a stable fashion for anything like that period of time. Here again the question is unresolved. No one really knows the answer, and no one can any more than guess what, exactly, these celestial objects are, or where they get the unimaginable energy that they are pouring out.

There are speculations, of course. One possibility is that quasars represent some sort of super-supernova, involving a multitude of stars within a galaxy in a titanic chain reaction of supernova explosions. Yet quasars do not seem to be large enough in volume to be comfortably explained on that basis. Attempts to measure the distance across a quasar have been fraught with confusion and possible error, but at least one observation seems to indicate unmistakably that none of these quasars could be anywhere near as far from one side to the other as the diameter of any known galaxy. This conclusion was drawn from the observation that some of the brightest quasars can be seen to vary sharply in their brilliance and radio emanations over surprisingly brief periods of time. In the short three or four years that they have been studied, some quasars have been observed to vary as much as 50 per cent in their brilliance, flaring up and dying down, so to speak, in much the same manner as the Cepheid variables that we have discussed. In some quasars this cycle of variation occurs in as brief a time span as a single month. Astronomers argue that if these objects were anywhere near as large in diameter as a galaxy is, no such short cycles of variable brilliance could possibly be observed; it would take so long for the light from one side of the quasar to pass through to the other side that, as far as our telescopes could see, light coming from a "flaring up" period at one time would be mingled with the dimmer light from a "dying down" part of the cycle at another time, so that any observable variability in brilliance would be canceled out as far as our instruments could detect.

But whatever the structure of these quasars may be, they are almost certainly smaller and denser than galaxies; and whatever the source of the energy pouring forth from them may be, they are unstable members of the celestial family. It is quite possible that they might have a life span of as little as a few million years, a mere blink of the eye by astronomical standards.

## THE MYSTERY OF THE QUASARS' ENERGY

Among other things about quasars that have disturbed astronomers is the simple fact that practically everything that has been observed about these enigmatic celestial bodies hinges upon the interpretation of a single piece of evidence: the apparently extreme red shift observed in the quasars' spectra.

The assumption is that such an extreme red shift can only mean that these objects are extremely remote and moving away from us at extremely high velocities, but what if that assumption were incorrect? Suppose some other explanation for these extreme red shifts were to be found? Since the first discovery of these extreme red shifts by Maarten Schmidt, astronomers have made concerted efforts to find some other plausible explanation to account for this finding, and to find some other plausible explanation for the apparently staggering amounts of energy that quasars seem to be pouring forth.

Unfortunately, in this search scientists have been impaled on the horns of a dilemma. If quasars are truly as remote and distant as they seem, there is no really satisfactory explanation of where all their energy comes from; but if they are in fact much closer celestial objects than we think they are, then there is no explanation for the observed red shift. But a number of interesting alternative explanations have been proposed in order to escape this dilemma.

In a sense, it would be a massive relief to scientists to be able to prove that quasars are not really as remote as they seem, in other words, that their spectra show red shifts for some other reason than remoteness and extremely high recession velocity. According to one alternative explanation, the red shift of quasars may indeed reflect high-velocity movement of the light-emitting body away from the earth at the time the light left its source, but this does not *necessarily* have to imply that the recession of the light-emitting quasar has actually been going on for any prolonged period of time. According to this theory, quasars might actually represent some very local phenomenon, something which occurred in our own galaxy comparatively recently in the cosmological time scale, and comparatively close by, perhaps only a few hundred light-years away. Whatever it was that occurred, however, resulted in the quasars moving away from us at a very high velocity *from within our own local neighborhood in space.*

This idea would account for the presence of red shifts indicating high recessional velocities, and would at the same time imply that the *total* energy of the quasar (whatever its source) might actually be far less than it seems to be. But it would also mean that the energy emitted *per gram of mass* by such a "nearby" quasar would have to be fantastically high—far, far higher than if the quasar were remotely distant. So far, no one has the foggiest idea what kind of "incident" might have occurred so close by and so recently to cause such an outpouring of energy, or to cause a "nearby" quasar to move away from us at a velocity high enough to account for its red shift. Granted that some kind of titanic explosion or celestial collision very close by in space *might* have resulted in a great mass of luminescent matter racing away from us at velocities approaching light speed, but considering the *number* of quasars that are known to exist with extreme red shifts, this explanation gets a little sticky.

It fact, most astronomers and cosmologists feel that the "local-origin" hypothesis creates more questions about quasars than answers. If such an uncommonly remarkable event occurred within our own galaxy or in neighboring space, then why don't we see other evidences of such an event, or observe more such events occurring, or find evidence of them occurring in other galaxies that we can study closely? The fact is that we *don't* observe these things, so this theory is really hypothesizing not merely some kind of unusual event but indeed a truly and literally *unique* event in cosmic history. And astronomers and physicists are very nervous about theories or explanations which depend upon truly unique natural phenomena. We recall the basic assumption that underlies all the study of physics: that whatever happens in the universe happens in an orderly fashion. And the necessary corollary to this assumption, physicists feel, is that if some phenomenon—*any* phenomenon—can and does occur once at some time in the universe, then it can and must occur repeatedly at other places and at other times. A truly *unique* event or phenomenon is anathema.*

* This, of course, is merely a restatement of the scientific and philosophical reasoning behind the contention that the universe contains other forms of life similar to that which has developed on our own planet. According to this reasoning, the peculiar conditions and chain of events which led to the development and evolution of life on our planet could not be unique, but must have prevailed (or be prevailing) on multitudes of other planets of other suns in our own galaxy or elsewhere. And if such conditions did lead to the development of life in one place, then the same or similar conditions not only could but *would have to* lead to the development of life in other cases.

Another argument against the thesis that quasars might be the result of some local phenomenon in which some sort of starlike object is being expelled at high velocity from the Milky Way or some other nearby galaxy is just as fundamental: There is serious question that there is enough energy, *in toto*, anywhere in the Milky Way to produce an explosion great enough to hurl local objects away from us at such high velocities. Furthermore, there *is* some evidence—not a lot, but reasonably convincing—to indicate that quasars are indeed very remote from the earth. For example, the light from at least one quasar which has been carefully studied shows changes that suggest that it must have been filtered, before it reached earth, through a huge cloud of hydrogen gas which is known to lie near the Virgo cluster of galaxies some 40 million light-years away from the earth. Thus, however far away that quasar may really be, it seems certain that it *must* be farther away than 40 million light-years, a strong argument against the notion that quasars result from some phenomenon that has occurred close to our own galaxy.

On the other hand, some interesting theories have arisen in an attempt to explain what kind of reaction might conceivably create the amount of energy that quasars seem to have. British astronomer Fred Hoyle, of Cambridge University (whose unorthodox cosmological theories command serious scientific attention in spite of their obvious vulnerabilities), and his colleague William Fowler, of the California Institute of Technology, recently suggested one possible explanation. Quasars, they contended, might be huge superstars, extremely massive ones, which had depleted their hydrogen fuel sometime in the distant past and begun to contract under the force of their own gravity. The enormous amount of energy released by matter falling into a massive star's core, they maintained, might be sufficient to explain a quasar's almost unbelievable volume of radiation.

This was not entirely a new idea; for decades astronomers have speculated that such "gravitational collapse" of a burned-out star might occur if the star were big enough and if it reached a certain critical degree of massiveness due to the normal contraction that occurred during the expenditure of its nuclear fuel. But the Hoyle-Fowler proposal with regard to quasars raised two questions, neither of which have really been satisfactorily answered. First, the gravitational collapse of a star sufficient to generate an amount of energy of the magnitude that a quasar seems to possess would have to be occurring on an unimaginably massive star—a star with a total mass hundreds of times greater than the mass of any known star in existence. Many astronomers doubt that a star of such huge mass could ever actually have formed at all, for they speculate that it would have broken up into fragments during its early formative stages.

The other question raised by the Hoyle-Fowler gravitational-collapse theory is even more mind-boggling. Critics of the theory have pointed out that even if such a massive star could have been formed in the first place, the collapse of matter into its core during the gravitational-collapse catastrophe suggested by Hoyle and Fowler might well create such overwhelmingly powerful gravitational forces that even light and radiation would be unable to escape from the star and move out across space. Such a situation, it is argued, would not result in a massive "superstar" shining brilliantly and radiating radio waves the way a quasar does. Under these extreme circumstances, such a superstar would simply *vanish from view altogether*, leaving a black, lightless hole in the sky, as far as telescopic instruments could detect.

Still another explanation for the amount of energy radiating from a quasar has been proposed by a Swedish physicist named Hannes Alfven. Alfven speculates that quasars could be the results of a meeting, deep in space between the normal matter of a star as we know it and some great accumulation of contraterrene matter—a mirror-image form of matter in which the subatomic components that we normally expect to bear positive electrical charges bear negative charges, and vice versa. The existence of such contraterrene matter—so-called antimatter or antiparticles—has actually been demonstrated in modern physics laboratories. These antiparticles seem to be generated naturally in certain kinds of interactions between subatomic particles. What is more, it has been demonstrated that when a "normal" particle and its contraterrene counterpart or antiparticle happen to meet, they annihilate each other instantly and completely, with the transformation of the entire mass of both particles into huge amounts of energy.

The idea is intriguing, and not entirely unreasonable, either. If there were some great aggregation of contraterrene matter somewhere in the universe which encountered a massive object composed of "normal" matter, the resulting explosion would be titanic indeed and might be expected to pour forth radiant and electromagnetic energy in quantities of the order of magnitude that quasars seem to pour forth. The trouble with Alfven's theory, unfortunately, is that so far no one has ever been able to demonstrate any significant aggregation of contraterrene matter anywhere in the universe. When antiparticles are generated under conditions of nature they seem to be generated one particle at a time, and are subsequently annihilated one particle at a time instantaneously as they encounter their normal terrene counterparts.

Thus, although antiparticles *do* come into existence one by one, they manage to survive no longer than an incredibly tiny fraction of a second before they come into contact with terrene matter and are wiped out, at least so far as physicists have been able to discover. It is hard to see how any accumulation of such antiparticles could ever occur. Still, the idea cannot be ignored. As we will see later, physicists working in the microworld of nuclear particles believe the universe *ought* to contain as much antimatter as "normal" matter. They simply have never been able to determine just where in the universe such a quantity of antimatter might be found.

But suppose that *all* these alternative theories are incorrect, and that these enigmatic and brilliant beacons in the sky are indeed as unimaginably distant as they appear to be, and are racing away from our own galaxy still farther all the time. If this does ultimately prove true, there would be real reason to hope that quasars might very soon provide us with vital clues to a number of the mysteries which have been puzzling cosmologists, physicists, and philosophers for decades or centuries. If astronomers indeed *are* peering at the uttermost limits of the observable universe in their study of quasars, then what they ultimately see may help provide answers to the question of whether the universe began with a "big bang" some 10 billion to 13 billion years ago or not. If the expansion of the universe is indeed a continuation of the aftereffects of some such titanic explosion, then there is hope that within a very few years we will have definite evidence to indicate whether such a big bang occurred as a single, unique episode or whether the entire universe is expanding and contracting in a vast pulsating cycle.

Of course, it is also possible that further study of quasars may reveal new and totally unsuspected support for the steady-state model of the universe, the picture of a universe which has been expanding since the beginnings of time and will continue to expand forever as new matter is created piecemeal between the galaxies, so that new stars and galaxies are continually forming to push the old ones aside. So far, the bulk of the evidence seems to support the notion of a pulsating universe best of all. As we have seen, there is a gathering body of evidence to indicate that even now, 10 billion years after the hypothetical big bang occurred, the rate of expansion of the universe is slowing down slightly. The evidence is hardly overwhelming; it hangs mostly upon the interpretation of astronomical observations and upon complex mathematical analyses of available data—but if this interpretation is correct, then the universe has already begun very slowly to reduce its rate of expansion and will presently slow down that expansion to a stop, and then begin a slow but ever-increasing era of contraction which will go on through the span of another 40 billion years before the next big bang occurs.

Not all physicists and astronomers are convinced that such a future big bang could ever occur even if such a period of contraction *did* begin and proceed to an end point. It can be argued that if such a process of contraction took place, sooner or later all the mass in the universe rushing together would reach a certain critical stage of denseness and massiveness so great that the gathering mass would begin to undergo a swift process of gravitational collapse until all the matter in the universe would reach a volume so small and so incredibly dense, with gravitational forces so stupendously great that no possible explosion of any sort could conceivably counterbalance it and hurl the matter of the universe out into another phase of expansion. Some theorists have speculated, in fact, that this might be the place where antimatter comes into the cosmic picture—that only the total annihilation of a mass of normal matter colliding with a mass of antimatter could create sufficient energy to "escape" from such stupendous gravitational forces. Some even speculate that time might also slow down its passage in one direction to a stop and then reverse its direction, so that any cosmic picture of the universe would have to take into account not only antimatter but also "antitime" as a basis for another big bang followed by another period of expansion.

Granted that all of this is speculation, and in some ways very wild speculation. But as evidence accumulates, however tenuous, old ideas are constantly being reconsidered and re-evaluated. In recent years Fred Hoyle, the strongest proponent of the steady-state picture of the universe, has been forced to reconsider certain aspects of his theory in light of a gathering body of evidence that seems to contradict it. He now concedes that there is convincing evidence that at least the part of the universe that we can observe may exist as a huge expanding and contracting bubble of matter and energy which lies embedded in an even more huge over-all universe, carrying on its cyclic expansion and contraction according to "local"(!) physical laws which might differ from the over-all physical laws governing the larger whole of the universe. In this way, Hoyle contends, our "local" part of the universe could be conforming to a cyclic or pulsating big-bang type of behavior at the same time that a much larger "outer universe" might be conforming to the steady-state model that he has proposed, with a steady and gentle creation of matter proceeding throughout all time, past, present, and future.

## THE CONTINUING SEARCH

One thing is certain: no cosmologist, physicist, or astronomer today is too eager to make any very positive declarations as to what answers may ultimately be found, or as to which (if any) of these models of the universe may ultimately prove correct. The very fact that a totally unpredictable and unexpected phenomenon, such as the discovery of the existence of quasars, can suddenly be made, and can completely revolutionize our understanding of astronomy in a half a dozen years or so is enough to give any scientist pause and make him very cautious about which theory he accepts and which he does not. Far too many rigid and dogmatic stands have been wiped out overnight by unexpected discoveries.

Nor is an occurrence like the discovery of quasars by any means unique in the history of science; such revolutionary discoveries have turned up far too often in the past for anyone to imagine that more such discoveries will not be made in the future. This, of course, is one of the great excitements of the study of science, an excitement shared by every one of us, and not merely by a few learned men working on the outermost frontiers of science.

Another reason that no one wants to be too dogmatic in his views about the expanding universe is simply that scientists all over the world recognize far too painfully the limitations of the tools and instruments with which they have to work. Astronomers recognize that the ultimate answers must be found in the study of objects and events that can be observed only poorly, at best, at the uttermost limits of observation that are allowed by our present-day telescopes and radio signal receivers. Those limits must be probed and expanded still further before we can have any real hope of finding positive answers. At the present time, measurements at such enormous distances are necessarily so very crude that even the tiniest human or instrument error might make the difference between coming up with one answer or another.

And even given more accurate means of measuring and observing, and granted the accumulation of more and more data, there are still astronomers who suspect that the ultimate and fundamental answers will continue to elude us. Men like Dr. Jesse Greenstein who are actively pioneering on the frontier of modern astronomy and cosmology have deep-seated doubts about the ultimate success of their work. They feel that it is possible that modern-day scientists are simply side-stepping or even totally ignoring questions which they must ultimately face squarely and deal with—such questions as why matter exists in the first place, why there is as much of it as there is and not more or less, or questions of how matter came to be in the first place. As Dr. Greenstein recently expressed it: "Such questions leave astronomers philosophically unhappy." He feels that cosmologists may sometimes have to face the possibility that the ultimate answers to such questions might well lie beyond the comprehension of science at all.*

No doubt all scientists involved in this modern cosmological search, as well as many thoughtful nonscientists who follow the progress of the scientists' work with interest, must have such doubts and reservations from time to time. And if the ultimate answers do indeed lie outside the scope and comprehension of science, then scientists engaged in the search will indeed ultimately have to face this fact. But this does not mean that these men are tossing in the towel and saying, "Oh, what's the use?" Their attitude is quite the contrary. They feel that the really great work of science is to continue diligently to root out new knowledge about the nature and behavior of the universe to the uttermost limit of which human intelligence is capable.

Indeed, astronomers and cosmologists are far more likely to complain

* See C. P. Gilmore, "The Birth and Life of the Universe," *New York Times Magazine,* June 12, 1966.

about the human limitations placed upon their quest for knowledge—the limits of existing instruments, the limits of money, the limits of time and personnel—than about the possibility that they may be grappling with the unknowable. They point out that no limit to new discovery has yet been reached; there is still an enormous amount that must be knowable that has not yet been reached or touched upon. They point out that the greatest telescope in the world, the 200-inch reflecting telescope on Mt. Palomar, however splendid and powerful an instrument it may be, is still not big enough to penetrate as far as men *could* penetrate into the cosmos. With a 400-inch telescope he could probe farther, and a 400-inch telescope is within man's technical capabilities. Such a telescope could be built, given enough money and incentive to build it. A great gaping hole in our knowledge of the universe exists simply because no telescope of major size exists in the southern hemisphere,* so that totally unexplored reaches of the universe could be opened up to us if only a few million of the billions of dollars we are devoting to our space program could be diverted to building a 200-inch or 400-inch telescope south of the equator. As Fred Hoyle expressed it in his book, *Galaxies, Nuclei and Quasars,* tens of billions of dollars are being spent "to set a man afoot on the ruined slag heap we call the moon," a project he feels can have comparatively little if any real scientific value, at the same time that we wonder whether we can afford the sum of $100 million for the construction of instruments which would make possible a breakthrough to final understanding of the workings of our universe.

Not only are existing telescopes not large enough, there are nowhere near enough of them, nor are there enough trained astronomers to use them. A Maarten Schmidt or an Allen Sandage may be limited to thirty days of work a year at the 200-inch telescope on Mt. Palomar, simply because of the multitudes of other astronomers and other projects that are clamoring for its use. Yet it can take an astronomer weeks or months of meticulous labor using that 200-inch instrument in order to obtain a reliable red-shift measurement for a single quasar!

But even with existing instruments and with all of their limitations, the frontier of knowledge in this exciting field is nevertheless being pushed back, slowly but steadily. Techniques of observation and measurement are being polished and refined so that results that are obtained are becoming more and more reliable. Furthermore, here and there new and better instruments *are* being constructed. In the spring of 1963, for example, one of the most remarkable astronomical instruments ever devised was completed and set into operation by the staff of the National Ionospheric Observatory near the town of Arecibo on the northern coast of the island of

* The largest telescope located anywhere south of the equator has a reflecting mirror with a diameter of only 74 inches.

Puerto Rico. Built by the United States government and operated under the joint auspices of the National Aeronautics and Space Administration and Cornell University, this huge radio-radar telescope is a modern miracle of engineering technology and scientific imagination. Its receiving and reflecting antenna is a huge hemispherical cup of wire mesh 1,000 feet in diameter which was built (out of ordinary chicken wire, of all things!) in a natural valley in a range of mountains in the lush tropical jungle a few miles inland from Arecibo. Built with such care that no single point in the entire antenna cup varies as much as one inch from a perfect spherical contour, this antenna is by far the largest "radio ear" in existence pointed toward the sky and is capable of receiving radar signals bounced off the surface of the planet Jupiter and of focusing radio signals from far out in space with previously unheard-of sensitivity.

Suspended at the focal point five hundred feet above the center of this great hemispherical "ear" and held up by cables from three towering pillars is the technological heart of the telescope: a receiving device for picking up radio signals from space and two spearlike sending antennas for directing radar pulses outward. This "business machinery" of the telescope is supported on a huge triangle of I-beams and suspended on a circular track so that the whole affair can be turned around a full 360 degrees and the sending antennas can be pointed anywhere within a 40-degree arc across the sky. This whole central mechanism, which weighs over five thousand tons (you can imagine the size of the cables that hold it suspended!) can be reached only by narrow catwalk strung out across the empty space from the rim of the reflector to the suspended cage—a heart-stopping walk for the astronomers who flock to use the telescope, but an everyday matter for the technicians who swarm about the cage keeping the working mechanism in order.

The Arecibo radio-radar telescope was a much needed addition to the world's meager collection of such instruments. Astronomers come from all over the world to use it, and in the first few years of its operation it has already contributed an enormous amount of new knowledge. As a great radar sending station, capable of beaming tight radar signals as far out into our solar system as the planet Jupiter and then picking up recognizable reflections of those signals, the telescope has already allowed us to make some sharp revisions in our knowledge of our planetary neighbors. The thickness of the fine layer of dust on the moon's surface was accurately predicted, for example, long before soft landings and transmitted close-up photographs were possible. The long-standing mystery of the apparent lack of rotation of Venus was solved, and a great deal of new information about the planet Mercury has been obtained.

On the other hand, as a radio telescope for receiving signals from great distances and helping to pinpoint their source, the Arecibo instrument is deeply involved in the search for hitherto undetected quasars, entering into the close-knit partnership effort between radio telescopy and optical telescopy that has made the great advances of the last three or four years at all possible. Nor is this great telescope the only new instrument to be available. Another radio telescope with a somewhat smaller "receiving ear" but with greater maneuverability is on the drawing boards and is planned for construction at Greenbank, West Virginia; and elsewhere other links in the world-wide net of radio telescopic equipment are planned or under construction already.

Thus, new knowledge is constantly accumulating in the great world of macrophysics. The universe of the inconceivably huge of thirty years ago has steadily expanded into a mega-universe even more inconceivably huge today. The quest for new knowledge goes on unabated, but many questions remain unanswered, and many new questions continue to appear. There is no telling what new observations may shake the foundations of what we know, or think we know, tomorrow, next week, next month, or next year, but the workers on this frontier are not waiting for new discoveries to drop in their laps. The exploration of the world of the inconceivably huge continues with dogged determination.

Yet, paradoxically, other scientists using other instruments and other techniques are searching for virtually the same answers to virtually the same questions in an exploration of the universe at the opposite end of the spectrum. The great majority of contemporary physicists are probing deeply into the puzzles and conundrums of the world of the inconceivably small. Today there seems no doubt that the ultimate structure of the universe of stars and galaxies is inextricably related to the question of the tiniest building blocks in the universe: protons, electrons, and the multitudes of other tiny subatomic particles and the energy with which they interact. In the study of this inconceivably tiny micro-universe lies another exciting chapter in the history of modern physics, and here also scientists today stand on the brink of breathtaking discoveries. In the next section of this book we will take a close look at this continuing study of the micro-universe and at some of the remarkable and incredible directions that it is taking today.

# Part V

## *The Universe of the Inconceivably Small*

# CHAPTER 23

## *Micro-Universe: The Earliest Explorations*

For most readers of this book, the ideas which we have encountered so far about the nature of the physical universe in which we live have at least one consistent quality in common. Many of these ideas may have seemed strange, difficult to comprehend, perhaps even quite contradictory to the way we ordinarily think of our everyday world, but for the most part they have at least been comprehensible.

It was not too difficult, for example, to understand the way things work in the everyday world of classical Newtonian physics, once we understood the over-all pattern of natural laws that were used to describe the structure and behavior of objects in that world. The inconceivably huge mega-universe of the astronomers and cosmologists is also at least comprehensible if we are willing to stretch our imaginations a little. Granted that we have been discussing unimaginably huge distances and unimaginably long periods of time—but at least the sun, the stars, the galaxies, even the mysterious quasars are all distinct and visualizable physical entities capable of being observed, described, measured, and interpreted. We may not understand everything about this huge mega-universe and the parts that make it up, and there may be no satisfactory answers for many of the questions it calls to our minds, but at least we can feel comfortable that there is nothing *imaginary* about it. And for most readers, even the bizarre ideas and apparent contradictions of the theories of special and general relativity are comprehensible and visualizable to some degree, as long as we are willing to "play the game" and follow out the rigorous trains of logic that led to these theories and the quite different picture of the universe that they force us to consider.

But in the world of the inconceivably small—the "micro-universe" which so many modern physicists have been exploring so rigorously in recent decades in hopes of learning the basic underlying submicroscopic nature of the universe and of the matter and energy it contains—we face quite a different problem as far as comprehension is concerned. For it is from the study of this micro-universe that ideas and discoveries have arisen in recent years that boggle the mind and seem to defy comprehension.

Fifty or sixty years ago the situation was quite different. Physicists at the turn of the century had succeeded in discovering and identifying two or three of the basic structural building blocks of matter—or so they thought—and had developed a simple but rational picture or concept of how material objects were constructed of those basic building blocks, and how energy was exchanged from one such "elementary particle" to another. There was even a period of time when physicists were beginning to feel quite pleased with themselves for having learned all there was to learn about the submicroscopic nature of matter, so that nothing was left to be done but a sort of mopping up exercise, a matter of tying up a few loose ends that were still unexplained.

That attitude of complacency was short-lived. Modern physicists today studying the micro-universe find themselves grappling with a bewildering array of allegedly "elementary particles," a veritable swarm of subatomic entities too tiny to imagine. It would be pointless even to try to catalogue all of the dozens upon dozens of distinct and separate particles, nonparticles, and almost-particles that are believed to have critical roles in the subatomic structure of the universe as we know it; to date physicists have either described, predicted, or actually identified more than a hundred such "particles," and there is no reason to think that this accelerating parade of newly discovered tidbits of matter—or of energy, or whatever they are—is going to slow down in the foreseeable future. Most detailed accounts of these multitudes of particles and their even more multitudinous interactions are both confusing and boring to anyone but the professional physicist, and there is no reason or need to add to the confusion in these pages.

But on the other hand, we cannot ignore the world of the inconceivably small, nor can we ignore the work of the atomic and nuclear physicists and the strange views of this micro-universe that they have come to hold, so completely alien to everything that we have ever observed with our own physical senses. Exotic and obscure as this work may seem at first glance, we cannot afford the luxury of ignoring it for the simple reason that it impinges heavily upon the everyday world in which we live and the kind of world in which we are going to live in the future. The micro-universe of physics touches our everyday lives in far too many important ways already, and it will touch our lives more and more insistently every day and every year to come. We may be less than entranced by the details of the beta decay of certain atomic nuclei, or feel somewhat indifferent to the behavior of a slow neutron when it collides head-on with the nucleus of a heavy atom, but we cannot ignore such everyday facts of life as the existence of hydrogen bombs capable of wiping whole nations off the map, or the enormous potential for good that would arise if some way could be found to harness and control the devastating violence of such a bomb and convert it into a sustained source of unimaginable quantities of useful power.

If such things are of concern to us—and they obviously had better be—

then the work of physicists exploring the submicroscopic nature and structure of matter in the world of the inconceivably small must also be of concern to us.

This is not to say that we need a physicist's detailed comprehension of that micro-universe. We don't. What we *do* need is an over-all picture of what has been learned about it, what remains unknown, and what direction our increasing understanding of it is taking. And in fact, the work of the men who are studying this micro-universe of physics is perhaps the most fascinating and exciting chapter in all the history of science. In the course of a mere fifty or sixty years more has been learned about the true underlying nature of our universe than scientists were able to learn in all the previous five thousand years of recorded history. This short period of discovery, a period that still continues, has not only witnessed the identification of totally unsuspected objects and processes, but has brought forth a wealth of strange and alien ideas—ideas which have forced scientists everywhere to revise quite radically their thinking about the way the universe works and why.

In fact, the real implications of some of these ideas are only today beginning to reveal themselves to the men working in this field of physics. We are living now in a period of scientific revolution every bit as fundamental to our everyday lives and our future as that earlier revolution that was begun centuries ago by Galileo, Copernicus, and Isaac Newton.

## THE GOLDEN ERA OF DISCOVERY

In order to understand the modern exploration of the world of the inconceivably small it will be necessary for us to adjust our sights slightly, and to approach what has happened and what is happening with a somewhat different viewpoint from that of the earlier pages of this book. In this discussion we will turn our attention more and more to the meaning of certain great revolutionary ideas that have arisen than toward specific discoveries, theories, and processes. Some things will seem to vary from, or even flatly contradict, things that we have considered and accepted earlier. We will find ourselves calling older concepts by different names, and discovering that certain old and familiar definitions must either be modified or thrown out altogether here, because in this strange micro-universe they simply do not apply. We will be discussing distances, sizes, and intervals of time so exceedingly tiny that we will have to introduce new terms and new units in order to describe them comprehensively. Whereas in earlier chapters we have often discussed matter and energy quite separately, even as we learned that these two "entities" were really only two manifestations of the same thing, we will now discover, more and more, that neither term taken by itself means anything at all in discussing the micro-universe. Each must *always* be considered in close connection with the other.

We will also encounter curious and paradoxical ideas, things which don't seem to make any sense at all on the surface. For example, we will discuss certain "particles" which spin around their axes like tops, yet have no discernible mass whatever. Although we have seen that matter-energy can never be either created or destroyed, we will see instances in which certain particles actually seem to appear suddenly out of nowhere, and other instances in which two particles annihilate each other instantaneously if they chance to collide. We have seen that any object with mass has inertia, and that no material object or signal can be accelerated to a velocity exceeding the speed of light; yet in the micro-universe we will see instances in which particles which make their appearance instantaneously are *already* traveling at the speed of light the very instant they appear! We will discuss certain kinds of "permanent" atomic or nuclear building blocks which may already have been in existence for an infinite period of time and which seem likely to continue in existence infinitely without alteration; but in the same breath we will talk about other subatomic structures or particles which exist for such a tiny interval of time that we would not have room enough on this page to write the number of zeros necessary to express that time as a fraction of a second—yet which are still considered by physicists to be relatively long-lived particles!

In many ways the study of the micro-universe during the last fifty or sixty years has been a matter of trying to fit together a huge and complex jigsaw puzzle from which a great many of the pieces have been missing. The challenge of this jigsaw puzzle has probably occupied more brilliant human minds than any other challenge in the history of science: The ranks of the nuclear physicists have been filled with an unbelievable succession of brilliant, perhaps even genius-level, thinkers. The names of some of these men will be familiar to us all; we may never even have heard of many others. They have come from no one country, or class, or school of scientific discipline, yet all have worked in splendid cooperation and mutual respect toward a common underlying goal in one of the most remarkable examples of enlightened international human teamwork in all history.

As we will see, the roster of these men and women includes an elderly German mathematician who recognized relativity before Einstein did, a proud and wealthy deposed French aristocrat, a British nobleman, a cheerful and gregarious Dane, a young German mathematician whose logic was more inexorable than Einstein's, a brilliant young American Nobel Prize winner, an ethereal British scientist-philosopher who dreamed up a weird picture of the universe which demanded the existence of an unheard-of and impossible particle, and a doggedly patient American who finally demonstrated that that particle actually did exist!

These and a multitude of others, famous and obscure, worked at the puzzle—and working with them all, challenging them all, as deeply involved in the puzzle and its progress as any of them, was the towering figure of Albert Einstein, with his clear guidelines of reason and logic and his awkward but unavoidable ideas of relativity.

The puzzle of the world of the inconceivably small is not yet fully solved, far from it. But a very great deal has been learned in the past few decades, and the things that have been learned have caused both scientists and laymen to look at the universe in which we live in quite a different way than we did before. Today, as in the case of cosmology and astronomy, it seems that most of the easy problems have been solved; physicists at last are now grappling with the basic and fundamental problems. And the story of how we have come to this stage of understanding is as exciting as any detective story.

## THE EARLIEST THEORIES

We have seen repeatedly that physicists and other scientists throughout history have always been primarily concerned with only two things. They have sought to discover *how things are constructed* in the universe, and at the same time they have tried to find out *how things work*.

It is obvious that knowledge about one of these things must necessarily be incomplete and unsatisfactory without knowledge about the other as well. Yet the structure of things in the universe, on the one hand, and their behavior and interactions, on the other, can be studied, observed, measured, and described quite separately. And more often than not knowledge has been gained a step at a time, first on one side and then on the other. Physicists, for example, first might come to understand some aspect of the structure of the physical universe without the vaguest idea of how that structure contributes to the way the universe works, or vice versa. And somehow, time and again, it has been necessary to find a bridge of natural law between structure and function. Characteristically, each time such a bridge has been discovered, science has then moved a step further ahead in its over-all understanding of the universe.

We can see how this works with a simple example. Imagine that a time traveler from ancient Rome was suddenly and miraculously transported to our modern times and confronted with a modern sports car. The ancient Roman might examine the vehicle in the minutest detail, taking it apart and examining each of its components. He might well be able to deduce from what he sees that this is some kind of a vehicle for transportation, vaguely reminiscent of a chariot or an ox cart in some respects. He might even learn how to duplicate its construction, and be shown how to make each of its individual parts and fit them together. Equipped with that knowledge, he then might return back through time to ancient Rome and make an exact duplicate of the sports car, faithful in every detail, yet still have no comprehension whatever of how it was supposed to function, behave, or operate.

Without some comprehension of the sports car's function, his detailed knowledge of the structure of the vehicle would not be much help. The best the ancient Roman could say to the other Romans would be, "Well, it seems to be some kind of a vehicle, and it is made of this part and this part and this part, but how it works is beyond me!" As a display piece it might be intriguing, but as far as using it was concerned, the best he could do would be to harness it up behind a horse.

Then imagine the reverse of this picture. Suppose the same Roman time traveler, arriving in our modern times, were to be given a detailed demonstration of how a modern sports car could be made to run and what it could be made to do, without so much as a hint as to how it was put together. As far as over-all usefulness was concerned, he would be in no better shape when he got back home under these circumstances than in the first case. He would have observed a wheeled vehicle putting on what would have appeared to him to be a truly miraculous performance, traveling at enormous speeds, performing a great deal of work with no apparent propelling force dragging it or pushing it. He might even have been taught to drive the car, and found that *he* could make it function just as well as anyone else. In such a case, he might take it home with him to ancient Rome and drive it all over the city, but it could never be any more to him than a "mysterious vehicle that performs in a remarkable fashion for reasons unknown," and his joy ride would come to an abrupt end the moment any single component part broke down or became disconnected, or the moment the last drop of gasoline ran through the carburetor.

Indeed, the only possible way that our ancient Roman time traveler could possibly develop any *truly useful* understanding of this mysterious vehicle would be by learning both the detailed structure of the car *and* the basic laws governing its function.

Ideally, then, an understanding of the structure of objects in the universe and an understanding of the way they behave ought to advance hand in hand. Of course we know that it has never been that simple. In most cases the scientist's understanding of how things work has generally always been a step or two ahead of his understanding of how things were constructed. Isaac Newton worked out the laws of universal gravitation—and more important, recognized them as *universal* laws—long before astronomers had any clear idea of where stars came from or how they were formed into galaxies and clusters of galaxies. Michael Faraday and James Clerk Maxwell evolved their theories of electromagnetic fields and the propagation of electromagnetic waves long before they had any idea, for certain, of just what an electromagnetic wave was. Indeed, it came as quite a surprise to them to learn from their own study of how things worked that a very common but apparently unrelated phenomenon in the universe, the phenomenon of light, was nothing more nor less than one form or manifestation of an electromagnetic wave. And again, as we will see later, scientists were able to work out useful and reliable natural laws regarding the behavior of gases under varying conditions of temperature and pressure long before the component atoms or molecules making up the gas were understood as anything but tiny hypothetical particles of solid matter.

Indeed, as scientists looked more and more closely at the world of the inconceivably small, they found their ideas of the structure of matter changing repeatedly in order to catch up with new and deeper understanding of the way things *behaved.* It is one of the great paradoxes of modern physics—and one of the most difficult things for nonscientists to comprehend—that our concept of what matter is actually composed of and how it actually behaves in the submicroscopic world of the very small has grown steadily farther and farther away from the concept or picture of the universe that is revealed to us by our own sensory experiences. Of course, we recognize that everything we see in the world around us must be explained by natural laws. But in the world of our sensory experience, matter is solid and enduring; time moves inexorably in one direction from past to present to future; and everything that we see happen occurs (or *seems* to occur) as a result of cause and effect; one thing happens because something else happened a moment earlier, and that happened because something else had happened still another moment earlier, and so on.

But the physicist today sees quite a different picture of the micro-universe than we see in the world around us. He sees a universe in which no single particle of matter is ever entirely "solid" or even necessarily enduring. He sees a world in which the very existence of "matter" as we think of it is seriously questioned. He sees a world in which time may, and probably does, flow quite as satisfactorily from the future to the past as from the past to the future; a world in which virtually nothing happens as a result of cause and effect, but rather according to the rules of a strange mathematical game of chance which we know today as the *laws of probability.*

This strange view of the micro-universe has not evolved by chance, but by a careful piecing together of observed phenomena and a thoughtful interpretation of multitudes of experiments. And it has grown up as a result of conclusions which appeared incredible but were nevertheless forced upon scientists inescapably by actual things that were observed.

The key idea underlying our understanding of the structure of the micro-universe is that the world as we know it is composed or built up of certain basic and indivisible building blocks. Every structure that we see in our everyday universe, every phenomenon that occurs, must arise ultimately from the way such elementary particles or building blocks are put together to form larger structures, and the way they behave or interact with each other.

We already know that this is not a new idea. It was regarded by the ancient Greeks almost as a philosophic necessity, and it had probably been considered as a self-evident truth by thinking men far earlier than the days of ancient Greece. After all, a piece of wood or a chunk of stone could be broken up into fragments, and those fragments could be broken into smaller and ever smaller fragments. It stood to reason that if one continued to break up a given piece of wood or stone into smaller and smaller bits, one would at last reach some "smallest possible" piece or particle that was the fundamental underlying structural unit of the wood or stone and which could no longer be split apart or further divided.

The ancient Greeks, notably Leucippus in the fifth century B.C. and Democritus, later in the same century, recognized perfectly well that these ultimate indivisible units of matter must be exceedingly tiny indeed—far too tiny to be seen with the naked eye, perhaps too tiny even to be imagined. But if one were able somehow to shrink down in size and thus examine ever tinier and tinier structures, the argument went, one would ultimately encounter a "smallest possible unit" of matter which could not be made any smaller by any means.

This idea may have been very satisfying philosophically, but scientifically it was not particularly useful. Democritus and other Greeks made no particular attempt to explain the structure and function of everyday objects on the basis of these tiny indivisible components. They used the word "atomos" to refer to them, but were not even particularly interested in their possible structure, their physical characteristics, or the way they might be joined together to form tangible and visible objects. These early philosophers considered the idea of such atoms mainly as an interesting abstract idea, not as a starting place for observation and experiment. In short, they simply did not think of atoms as real physical things about which specific truths could be learned.

They did, however, outline two characteristics of atoms which later proved to be useful guides for scientists who did begin to study them as real physical entities. First, the Greeks considered it self-evident that these atoms were far too small for men ever to observe them directly. But on the other hand, they assumed that they were of some specific, finite size. However finely matter might be subdivided, one would ultimately reach a particle of *some* size, however small.

Thus for centuries the idea of an elemental atom remained an intriguing philosophical idea which was generally accepted as such, but was essentially useless in any practical scientific study of substances making up the universe. With the great scientific awakening of the Renaissance, however, more and more observers and experimenters began reevaluating the idea of matter composed of basic individual building blocks and considering it from a far more practical point of view. As the metallurgists, alchemists, and later scientific chemists began to make close observations of a wide variety of kinds of matter, scientists became preoccupied with the idea that various substances with different physical properties might have differing underlying atomic structures as well.

One reason for this was that multitudes of strange and inexplicable observations were made which seemed somehow to be connected with the underlying structure of matter. Some substances seemed to have fixed physical properties, yet could be changed from one physical state to another and back under varying conditions. Other substances seemed, under certain circumstances, to undergo irrevocable changes—now known as *chemical changes*—in which apparently new and different substances were formed. If all substances were made up of atoms that were indivisible and indestructible, were those atoms themselves then changeable? Were different substances different merely on the basis of the numbers of atoms present and the configurations in which they joined together, or were their atoms themselves of different sizes, shapes, and colors? If there *were* different kinds of atoms, then *how many* were there? It seemed incredible that every single different substance known could be made of a different kind of atomic unit—but where was the line drawn?

Bit by bit observations began to suggest that there *were* numerous different kinds of atoms, but that there was still some limit upon the number of different kinds that existed. At the same time, these same observations raised new and puzzling questions. Water, for example, was known to exist in three different physical states, solid, liquid, or gaseous, depending upon its temperature. Each of these different physical states of water had certain characteristic physical properties; yet with variations in temperature water could be changed from one form to the other and back without altering its basic nature. This was naturally assumed to mean that water must be composed of multitudes of tiny "water atoms" which always remained the same, impossible to break up and impossible to destroy. But if so, what happened to those "water atoms" under changing temperature conditions to account for the three different forms that water could take? In the 1500s and 1600s no one had an answer, nor even a halfway decent theory to suggest.

Again, it had been observed that some substances seemed quite impervious to any environmental change, maintaining certain fixed physical properties under virtually all kinds of experimental conditions they were subjected to. Gold, for example, did not evaporate upon standing, nor did it melt readily, nor did it corrode, nor did it undergo any other change in its physical appearance. But other substances behaved quite differently. Wood thrown into a fire would begin to shrink, lose weight, and ultimately turn into powdery ash. Iron, when left exposed to the atmosphere, soon crumbled to a powdery red dust which was obviously quite a different substance from the original iron. It was found that some substances could be mixed together very thoroughly and then separated from one another again: sand and iron filings, for example. Other substances, such as salt and water, could be mixed together all right, but could not be separated into the original components except by laborious procedures such as boiling off the water to retrieve the crystalline salt while simultaneously attempting to collect all the steam and condense it back into water again. Still other substances when mixed together seemed to undergo basic changes of character, to combine and change completely into some new substance which could not then be changed back again, so that the mixture could no longer be "unmixed." Sometimes this happened with explosive violence, as when sulphur, charcoal, and saltpeter were ground vigorously together to form a very touchy substance we know today as gunpowder.

Of course the ancient idea of earth, air, fire, and water as the sole "elementals" was discarded as soon as anything approaching serious scientific investigation began. But slowly and painfully the idea evolved that certain substances *were* elemental in nature, composed of just one kind of atomic building block, while other substances were *compounded* of two or three or more elemental substances which could sometimes, with ingenuity, be separated from the compound they formed. Unfortunately it was not always possible to determine whether a given substance was elemental or in fact a compound of several elements just by examining it. Indeed, it came as quite a shock when an "obviously" elemental substance such as water was found in reality to be a chemical compound formed of two elemental substances, each of which by itself occurred in nature in the form of a gas.

But in spite of these problems of observation, the idea that a basic difference existed between elements and compounds continued to grow, and with it came the idea that each separate element must somehow owe its individuality—even its very existence as an element—to some special quality of the basic atoms of which it was composed. From that point it was not too much of a leap to imagine that each elemental physical substance was made up of its own unique kind of elementary atom and that these atoms, far from being all alike, each possessed its own peculiar physical characteristics. By the middle 1700s scientists were regarding these atoms more and more as actual physical objects—very tiny objects, to be sure, but discrete and separate objects with individual shapes and characteristics of their own—and to speculate that all kinds of physical and chemical changes that occurred in various substances came about as a direct result of some kind of interaction between the atoms of one element and the different atoms of another element.

## THE ATOM OF DALTON

Clearly the time was ripe for someone to come up with a concrete picture, or "model," of what these elemental atoms must look like and a plausible explanation for how they behaved. Numerous attempts were made in the last decades of the 1700s to formulate a coherent "atomic theory" of matter, but an English school teacher named John Dalton (1766–1844) is generally credited with propounding the first really satisfying theory and then working out its finer details.

Dalton's model of the elemental atom was a far cry from our present-day concept, but it was a good starting place and it provided a plausible explanation for things that had been observed to happen. Dalton thought of the atoms of various elements as a variety of tiny solid marbles of different sizes and colors, the atoms of each element differing from the atoms of all others, all of them exceedingly tiny and all of them indivisible. Some of these tiny atoms, Dalton proposed, were covered with some mysterious kind of glue that enabled them to stick to others, while the atoms of other elements were completely free of glue and thus unable to stick to any

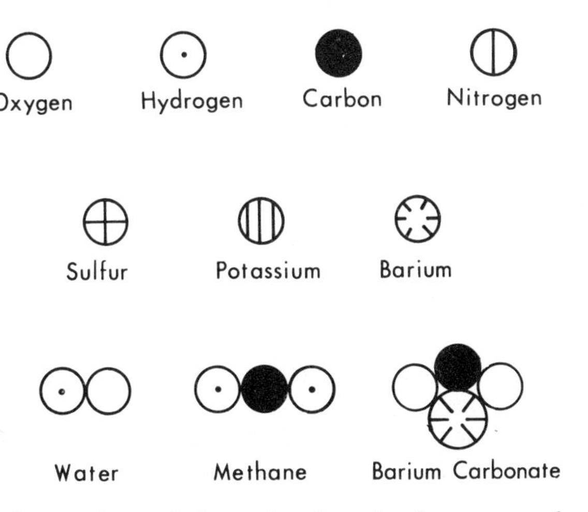


*Fig. 38* Dalton's atomic symbols, and a few simple compounds as he conceived them.

others at all. To emphasize the uniqueness of the atoms of each different element, Dalton devised a series of different symbols which he assigned to atoms of different elements. In Figure 38 we see Dalton's atoms represented diagrammatically together with some of the compounds that the "sticky" ones could form when they stuck together and a few of the structural configurations that he used to represent them.

Today Dalton's picture of the atoms seems simple and naive, but whatever its shortcomings it had one enormous advantage that the ancient Greek theory of atoms had lacked. Dalton's atomic theory represented a practical, down-to-earth attempt to explain observed phenomena on a physical basis. Right or wrong, it was at the very least *scientifically useful.* It provided a springboard for experiment and observation. Like any good hypothesis, it offered a possible explanation for a number of occurrences that had hitherto been unexplained. Furthermore, it was *capable of being tested.* By picturing atoms of various known elements as tiny solid marbles which sometimes become glued together, Dalton was able to explain, for example, how two gases such as hydrogen and oxygen might be able to join together to form a substance, water, which had quite different characteristics from either of the constituent elements. Using his own symbols to represent atoms of the various elements, Dalton was able to sketch out possible *structural diagrams* of many chemical compounds, and to show how they *could* be composed of atoms of a variety of elements.

Dalton's achievement was particularly remarkable when we consider the handicap under which he was working. For one thing, he had no way to determine the exact number of atoms in a given compound. Thus, he assumed that water was made up of one hydrogen atom and one oxygen atom glued together, whereas later experiment proved that a water molecule consists of *two* hydrogen atoms and one oxygen atom. He suffered from the problem, commonplace in that day, of getting compounds confused with elements and vice versa, but at least he showed what could be done and what could be learned with a well-conceived working hypothesis, even if that hypothesis were not entirely true.

Carrying his work further, Dalton proposed that the reason that a lump of one element was heavier than a lump of another element might be that the atoms of each element had a characteristic weight or mass. An atom of gold, he reasoned, must itself weigh more or have greater mass than an atom of nitrogen gas. Following up this idea, he organized the known elements into a table of elemental atomic weights, assigning a specific mass to the atoms of each known element. He reasoned that all atoms of any given element ought to weigh the same as all other atoms of the same element, but ought to differ in weight from those of all other elements. Granted this, then a given atom's weight would have to be a critical characteristic of the element that it composed. By assigning an arbitrary value of one unit of mass to atoms of the lightest known element, hydrogen, Dalton worked out the first complete table of atomic weights: primitive, frequently incorrect, but again extremely useful as a working scientific hypothesis.

One of the first beneficial effects of Dalton's new atomic theory was an enormous speed-up in the chemical description of a wide variety of substances, and the beginning of a whirlwind search to discover which substances were truly elements and which were not. In 1808, when Dalton's theory was first proposed, only twenty or thirty elements were known. (Lavoisier in 1789 had listed twenty-three that correspond to elements that we know today.) By 1858, fifty years later, the total number of known elements that had been identified and tagged with comparative atomic weights had reached sixty-five, and a Russian chemist named Dmitri Mendeleev had set up a chart of elements according to similarities and differences in their physical and chemical properties—the familiar "periodic table" of the elements that can be found in any high-school science classroom today.

Mendeleev's periodic table of the elements was a startling achievement and provided both chemists and physicists of that day with an invaluable tool in their study of the nature of elements and the structure of chemical compounds. Based as it was firmly on Dalton's atomic theory, Mendeleev's work suggested not only that the elements could be arranged in orderly progression according to the comparative weights of their individual atoms, but that there were also distinct groups or "families" of elements which shared many similar chemical properties and even certain physical properties among them, except that the comparative weights of the atoms of elements in such families were progressively greater.

Fluorine and chlorine, for example, were both acrid-smelling gases in their natural state, while bromine was a dense brownish liquid with an equally acrid vapor. All three of these elements, however, showed great chemical activity—that is, they entered into chemical reactions very readily with a variety of substances—and when placed in contact with certain light metals such as sodium and potassium, combined with these metals to form crystalline salts which had very similar chemical and physical properties. This family of so-called halogens was soon recognized to contain a fourth element, iodine, the heaviest of the four and the most sluggish in forming chemical reactions, but still closely similar to the other three in many properties.

In the same way, the metals lithium, sodium, and potassium were found to have closely similar physical properties and to enter into similar chemical reactions, differing from each other primarily in that sodium was eight atomic weight units heavier than lithium and that potassium was eight comparative atomic weight units heavier than sodium. Other metals such as magnesium, calcium, strontium, and barium formed yet another family with similar properties, less chemically active than the lithium-sodium-potassium family, but still entering into a wide variety of chemical reactions. On the other hand, the so-called inert gases helium, neon, and argon also lay eight comparative atomic weight units apart from each other in Mendeleev's table, and all shared the curious property of entering into *no chemical reactions whatever* with any other substances.

Perhaps the greatest early triumph of Mendeleev's periodic table, and one of the earliest confirmations of its validity, centered around a number of "holes" or blank spaces in the table. Mendeleev had been forced to leave these places empty right in the middle of his periodic organization of the known elements simply because the elements which were needed to fill those places were not even known to exist! By studying the physical and chemical properties of other members of the families of elements in which these missing elements apparently ought to belong, Mendeleev actually described what these truants from his periodic table ought to look like and how they ought to behave, if only somebody could find them. And within a very few years at least three of these missing elements were indeed discovered to exist in nature. When these new elements were studied, they were found to have precisely the physical and chemical characteristics that Mendeleev had predicted they should have.

This turn of events not only confirmed the validity of Mendeleev's work with his periodic table, but at the same time suggested strongly that Dalton's atomic theory, crude as it might be, was at least within striking distance of the truth. Indeed, for all its shortcomings, Dalton's atomic theory proved a real boon to the chemists and physicists of the 1800s and early 1900s. It provided a reasonable explanation for the behavior of elements as elements and for a wide variety of chemical reactions between elements which had been observed. Even more important, it permitted scientists to make surprisingly accurate predictions about what might happen if various substances were allowed to react with each other, or if the various chemical compounds could be broken down. Bit by bit it was found that these predictions could be confirmed by laboratory experiment, so that the atomic theory provided chemists with a whole new approach to studying the nature of chemical reactions. Finally, it helped them explain a great variety of occurrences which hitherto had seemed quite unexplainable.

But for all this, it soon become evident that Dalton's picture of the submicroscopic world was not *altogether* accurate, or at least was not by any means the entire picture. During this intense period of scientific investigation a number of phenomena were observed which seemed to be related to the basic atomic structure of matter, but which did not jibe with Dalton's picture of elemental atoms as tiny lumps of solid matter of varying sizes and shapes equipped with sticky surfaces so that atoms of some elements could stick to atoms of others. Gradually it became clear that whatever the true structure of elemental atoms might be, it had to be far more complex than Dalton's simple picture suggested.

## THE BROWNIAN MOVEMENT AND THERMAL ENERGY

For example, Dalton's concept of atoms as tiny solid particles like marbles implied that these particles, tiny as they might be, ought to behave like any other solid particles, according to Isaac Newton's laws of motion. If so, then it followed that a solid atomic particle at rest ought to remain at rest unless it was acted upon by some outside force. Such an atom ought to be no more able to jump up and run away in the absence of an outside force than a billiard ball is able to leap off the table and go flying through the air of its own accord. Yet considerable evidence seemed to suggest that atoms and molecules were seldom if ever completely at rest. Far from it; chemists and physicists were beginning to suspect that atoms and molecules making up all sorts of substances were, in fact, in a constant state of agitated motion. Things in the world which men could see and touch might appear to be in a placid state of rest, but their inner structure at the atomic level was continuously in a state of frantic activity.

What evidence was there that this might be true? The first actual observation of this rather astonishing state of affairs was made by an English botanist named Robert Brown in 1827. While using his microscope to examine some pollen grains floating on the apparently placid and motionless surface of a droplet of water, Brown observed that the pollen grains were not by any means at rest; they were jiggling and twitching around constantly like puppets on a string. Since there was no reason to believe that these pollen grains themselves had any intrinsic energy with which to hurl themselves around like Mexican jumping beans, it seemed apparent that they were actually being jarred into this irregular motion by some kind of submicroscopic activity going on in the water itself.

We can observe this same phenomenon without the aid of a microscope if we study the constant activity of dust particles in a ray of sunlight cutting through the still air of a darkened room. Similarly, we know that if we place a lump of sugar in a cup of tea, the sugar lump will disappear after a certain period of time, and presently the cup of tea will become uniformly sweet throughout, even though no one has stirred it. But what makes dust particles dance in the air? How do the molecules of sugar in a solid sugar lump make their way from the bottom of the cup of tea to an area several centimeters distant in sufficient quantity to sweeten the liquid?

The answer seemed to be that the atoms or molecules making up any physical substance, whether a lump of solid matter, a drop of water, or a volume of enclosed gas, *do* possess some kind of intrinsic energy sufficient to set themselves in motion. Far from sitting placidly at rest at any time, atoms or molecules are constantly hurling themselves about and bumping into each other, jostling any unwelcome neighbors such as pollen grains, dust particles, or even other molecules that happen to get in their way. Careful investigation of this phenomenon indicated that the amount of energy atoms or molecules possessed with which to throw themselves about at any given time and under any given circumstances was directly related to the temperature of the substance which comprised these atoms or molecules. Accordingly, this intrinsic or "built-in" energy of atoms or molecules came to be known as their *thermal energy*.

At the time Brown made his discovery neither he nor any of his colleagues had any idea just what the nature of this thermal energy was, or where atoms and molecules acquired it. It certainly seemed to suggest that something was wrong with Dalton's picture of atoms as tiny solid marbles, yet it did help immensely in explaining a number of phenomena which had been observed for years without any good explanation. For example, here was an explanation for changes observed to take place in the physical state of certain substances at varying temperatures. Water would slowly evaporate and disappear from a pan left sitting in a cool room because some molecules of water racing around in the pan possessed enough thermal energy to break loose from the surface and "escape" into the surrounding air. If the pan were placed in a much warmer room, the rate at which water molecules would escape would increase in direct proportion to the temperature. But even a block of ice left at freezing temperatures would lose molecules of water to the surrounding air, although far more slowly; we have all observed how ice cubes left in the tray under the freezing compartment in a refrigerator for a period of days gradually become smaller and smaller even though their temperature never increases as far as melting point. Similarly, poorly wrapped meat stored in a home freezer or freezing locker will gradually lose its moisture by evaporation, molecule by molecule, and will "burn" in the freezer even though it remains frozen solid at temperatures of 5 to 10 degrees below zero.

Thus water will evaporate under all commonly encountered temperature conditions, due to the thermal energy of water molecules. But that thermal energy becomes much greater and the evaporation much faster when heat is added to the water. In fact, if water is heated to a high enough temperature, 212 degrees Fahrenheit, a point is reached at which molecules have so much thermal energy that they fly apart with great vigor to form bubbles of steam that rise to the surface and escape—the phenomenon of boiling. On the other hand, if heat is removed from a pan of water (that is, if the pan is cooled sufficiently) the water will congeal to form solid ice, and the amount and rate of evaporation will be sharply reduced. In other words, atoms and molecules of water (and other substances) also seemed to possess a second intrinsic quality, related to thermal energy but quite different from it: Some kind of *force of attraction* seemed to hold them together in certain substances, an *intermolecular attractive force* which seemed to behave quite the opposite of thermal energy.

Thus when water is cooled sufficiently, so that the thermal energy of its molecules is at a low level, the intermolecular force of attraction between the individual water molecules holds them close together in a tightly packed mass, so that we find the water in the solid state of ice. When the ice is warmed, the thermal energy of its component molecules becomes greater and greater until it overbalances the intermolecular attractive force, so that the molecules can move around on their own more freely as water in the liquid state. Under these circumstances the intermolecular forces are still powerful enough to prevent the water molecules from flying completely free of each other in all directions, but further heating can increase the thermal energy of the molecules to such a degree that finally the intermolecular forces become quite unable to hold them together, and water in the form of gas (or water vapor) will expand to fill every nook and cranny of whatever container it happens to be in, or to dissipate into the air if it is not held in by a container.

If the water vapor is held in a container, however, with the individual molecules racing about in such a highly agitated state, there are bound to be innumerable collisions of the water molecules with the walls of the container. These collisions create a uniform pressure against the containing walls. If the containing walls are solid and unyielding, this pressure steadily increases as more heat is added, the situation we find in the case of a steam boiler or a hospital autoclave. If the containing walls are elastic enough, the container may expand under this increased pressure. On the other hand, if the container and the water vapor within it are cooled, the pressure of the molecular collisions against the containing walls will be reduced as the thermal energy of the molecules is reduced, and if that thermal energy is reduced sufficiently, then intermolecular forces once again can draw molecules together into droplets of liquid. We see this happening in the vicinity of a glass of iced tea on a very hot, muggy summer day: The water vapor in the air surrounding the glass is sufficiently cooled that we see it "condense" into droplets of liquid water on the outside of the glass and drip all over our pants when we try to take a drink.

We have been dwelling on the varying physical states of water simply because it is the most commonplace and familiar example we have of the relationship between the thermal energy of molecules making up a substance and the physical state of that substance. But precisely the same thing could be observed to take place in the case of a multitude of other substances under more extreme temperature variations. Iron, for example, can be converted from a solid into a liquid if heated enough, and iron liquid can even be turned into vapor at a still higher temperature. Similarly, oxygen or nitrogen gas can be condensed into a liquid state if cooled enough, and if cooled still further can be literally frozen. It has become customary to list as "physical properties" of any substance the temperature at which it can be changed from the solid state to the liquid state (melting point) or vice versa (freezing point) as well as the temperature at which it can be changed from the liquid to the gaseous state (boiling point). And as we saw in an earlier chapter, elements or compounds in the solid state are (with rare exceptions such as water) more dense or compact than in their liquid state and far less dense in their gaseous state.

By the early 1800s the twin concepts of thermal energy tending to tear atoms or molecules of a substance apart from each other and send them moving about in a state of high agitation, and of an intermolecular attractive force tending to hold molecules or atoms of a substance together, proved of enormous help to chemists and physicists in understanding physical and chemical reactions of all sorts. Certainly it helped explain *how* the differing states of matter came about. Molecules of a gas were able to move about freely and at random without being impeded by intermolecular attractive forces; those same molecules, robbed of some of their thermal energy by cooling, might be bound tightly to each other in the solid state by intermolecular forces so powerful that they could not move freely at all, yet still possessed enough thermal energy to remain constantly vibrating in place as though "trying" to break free.

Chemists by and large found this a highly satisfactory picture of the structure of matter; it was extremely useful as a working hypothesis, and it seemed to answer a great many previously unanswered questions. But there were still some disturbing questions that remained unanswered. What was the nature of the intermolecular attractive forces holding molecules together? How did molecules and atoms acquire the thermal energy with which they thrashed about in endless agitation? Were the atoms making up solid matter actually tiny lumps of material that were capable of being glued or stuck to each other, yet still mysteriously possessed a sort of internal life of their own which did not obey the laws of motion at all? It was a clumsy and incomplete picture, useful in its way but obviously limited. It became the great work of nuclear physicists in the late 1800s and early 1900s to seek out answers to such questions as these, and to come up with such a completely different picture of the structure of the micro-universe that Dalton would hardly have recognized it at all.

## THE QUESTION OF ATOMIC VALENCE

The history of science is full of examples of theories which proved far more valuable on account of the efforts that were made to discredit them than from their own intrinsic usefulness. Oddly enough, one of the most fruitful things about John Dalton's atomic theory was not the number of questions it answered but the number of unanswered questions it raised. Some of these questions fairly begged for answers, and although chemists and physicists found the theory to be a useful guide to experiment as it stood, they were not satisfied to let these questions remain unanswered.

For example, what was the nature of the "glue" that held Dalton's marblelike atoms together? What was the nature of the "intermolecular attractive force" that seemed to bind atoms and molecules together into rigid solid forms or even into somewhat less rigid liquid states? Clearly these questions were related to one another, and equally clearly "glue" was not a satisfactory answer. Dalton had learned, for example, that certain kinds of atoms bound themselves up only with a single atom of a different element in order to form a chemical compound, while other kinds of atoms seemed determined to "glue" themselves to two, three, or even more different atoms at once. In other words, atoms of certain elements seemed not so much to be coated with "glue" as to be equipped with little hooks that could grapple and hold onto similar hooks on the atoms of other elements, thus holding the atoms firmly bound. Atoms of some elements seemed to have only one such "valence hook" available, while others had two or three or even more.

This idea at least provided a way to visualize how atoms might join together to form molecules, but it did not explain why or how molecules of the same element or the same compound tended to cling firmly together unless torn apart by their thermal energy. It was hard indeed for tough-minded scientists to accept this concept of atoms as little marbles with hooks on them literally, but if this was not a true picture of atoms then what other explanation could there be for the observed fact that certain atoms or molecules had a distinct affinity for certain other atoms or molecules while in other cases no such affinity was observed at all?

Obviously some kind of unseen and unrecognized force was at work here. But it had to be a force unlike any other force that had ever been observed. For one thing, it was a choosy force: Sometimes it seemed to pull atoms together; at other times it behaved just the opposite and pushed them apart. Futhermore, this force seemed to be present only in unit quantities or multiples. A given atom might cling to one other atom, or to two, or to three, but never to one and a half atoms. What could be the nature of this force?

Of course there was always the possibility that this could be a force acting upon the atoms from the outside, just as we might apply an outside physical force to a billiard ball to make it roll across the table. But these attracting or repelling forces seemed to be intrinsic, a built-in characteristic of the atoms or molecules themselves. Physicists already knew of one such intrinsic force existing in nature: the gravitational force that was known to draw one massive object toward another. When at least one of two massive objects was *very* massive and close to the other, this force was quite strong, as in the case of the gravitational force holding a man's feet firmly to the ground. But when two objects with very small masses were placed close together—two ordinary-sized marbles, for example—the gravitational force between them was so weak that it could not even be measured. Thus it seemed highly unlikely that the force holding atoms together could be a gravitational force, and as for the force that seemed to push atoms apart, no one had ever observed or measured a force of gravitational *repulsion!* Clearly the force of gravity was not the answer here.

Magnetic force was another force acting at a distance which seemed to create very strong attraction even between very small magnetized objects, and which also exhibited an opposite and equally strong repellent force when two similar magnetic poles were brought close to each other. Closely allied and similar to magnetic forces of attraction and repulsion was yet another unseen and unexplained force which seemed even more promising as a possible explanation of the way atoms might attract or repel each other: the attracting or repelling force of electrical charges.

And here, it seemed, was a truly pregnant possibility; electrical charges seemed to fill the bill splendidly. And indeed, the first real experimental breakthrough in the attempt to explain the nature of interatomic forces was made by Michael Faraday in his laboratory study of a curious, newly discovered electrical phenomenon: the phenomenon of *electrolysis*.

## ATOMS, IONS, AND ELECTRICAL CHARGES

In an earlier chapter we saw that English-born Michael Faraday (1791–1867) was an early leader in the study of the behavior of electrical currents and magnetic fields in the middle 1800s. This brilliant experimental physicist was the first man to investigate in detail precisely what went on when an electric current was passed through solutions of various common compounds dissolved in water. Faraday had already found, of course, that electric currents could pass very readily through various substances such as copper or silver wire, but could not be made to pass through other substances such as glass or india rubber. Faraday classified substances either as "conductors" or as "nonconductors" or "insulators," according to whether they would conduct electric currents or not, and he found that various substances exhibited varying degrees of conductivity.

Thus copper was an excellent conductor of electricity, while lead or mercury, while capable of conducting current, did a very poor job of it. A block of glass might serve as a splendid insulator, permitting virtually no electric current to pass through it, whereas a short length of green sapling was a less-than-perfect nonconductor; it would not conduct *much* electricity very fast, but a trickle of current could be made to seep through.

When it came to classifying water as a conductor or a nonconductor, however, Faraday found himself in trouble. Absolutely pure distilled water seemed to be one of the poorest conductors of electricity known: An electric current could pass through it only weakly, if at all. But if very small amounts of certain chemical compounds such as acids or salts were dissolved in the distilled water, Faraday found that the resulting solution became an excellent conductor of electricity. Furthermore, when he passed an electric current through a solution of common table salt dissolved in water, he found that atoms of the two elements sodium and chlorine which make up the compound seemed to be wrenched apart by the presence of the electric current and drawn to opposite sides of the container.

Thus, when the positive pole of a dry cell battery was connected to a carbon rod electrode and immersed on one side of a container filled with a solution of salt in water, and the negative pole of the battery, connected to another carbon electrode, was immersed on the opposite side of the container, the current of electricity passing through the solution from one electrode to the other caused atoms of chlorine to gravitate to the positive pole and appear as free chlorine gas while sodium atoms gravitated to the negative pole. In short, the compound sodium chloride, which seemed a stable and firmly bound combination of sodium atoms and chlorine atoms in its dry crystalline state, appeared to be broken down in a water solution into two parts carrying opposite electric charges—the sodium atoms carrying positive charges and being drawn to the negative electrode, the chlorine atoms carrying negative charges and being drawn toward the positive pole (see Fig. 39).

Today such positively or negatively charged atoms or fragments of compounds are known as positive or negative *ions,* and we know that a great many chemical compounds break up into such charged ions when dissolved in water. We even know that water itself breaks up to a very limited degree into positively charged hydrogen ions and negatively charged "hydroxyl" ions each containing a fragment of a water molecule composed

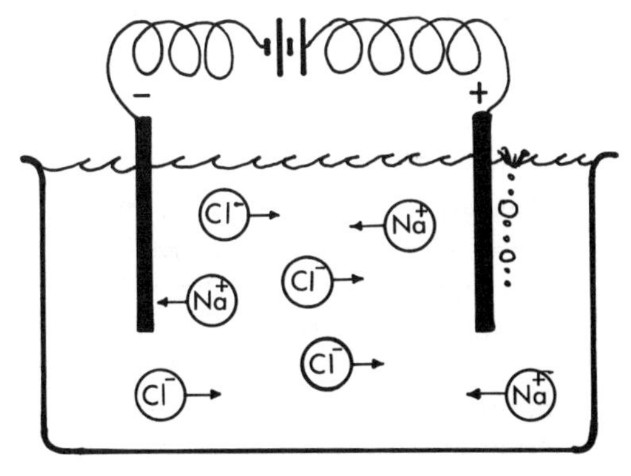


*Fig. 39* Electrolysis of sodium chloride solution. The chlorine ions are drawn to the positive pole, where free chlorine gas is released; positively charged sodium ions move to the negative pole, where free sodium metal is released to interact with water, forming sodium hydroxide and free hydrogen gas (not indicated on diagram).

of an oxygen atom and a hydrogen atom closely bound together and carrying a single negative electrical charge between them. These positively or negatively charged ions in solution behave exactly like any other object bearing positive or negative electrical charges: When an electrical current is passed through the solution, the ions with negative charges are attracted to the positive electrode and the ions bearing positive charges are drawn to the negative electrode. Because of the free movement of these charged ions in water solution, a current of electricity is conducted through the solution far more readily than it could be conducted through pure water; hence the radical change in the conducting properties of water into which small amounts of such ionizing compounds have been dissolved.

Michael Faraday did not know precisely what was happening when he began his study of the passage of electric currents through water solutions of ionizing compounds, but he realized that the "valence hooks" which bound atoms together into compounds had to be electrical in nature. Some atoms, he reasoned, must carry an intrinsic positive electrical charge, while others carry an intrinsic negative electrical charge. Under certain conditions, as when compounds are dissolved in water, the atoms of these compounds break apart into ions, either positively charged or negatively charged according to their intrinsic makeup, and then, since unlike charges attract and like charges repel, the positively charged ions are drawn toward the negative pole while the negatively charged ions are drawn toward the positive pole in the presence of an electrical current passing through the solution.

Furthermore, by carefully measuring the *amount* of the components of various compounds which collected at the positive and negative poles during electrolysis experiments, Faraday discovered that certain atoms carried exactly twice as much electrical charge as others, and still others carried exactly three times as much electric charge. By choosing as a measuring unit of electrical charge the atoms that seemed to carry the smallest possible amount of electrical charge—hydrogen atoms, for example, or the atoms of sodium or chlorine—Faraday found that other atoms always carried exact multiples of such a measuring unit of charge but never anything in between. A given atom might carry one unit charge, either positive or negative, or two or three charges, but never one and a half times the charge of a hydrogen atom.

From this observation Faraday concluded that there must be a *basic unit* of electrical charge of a very precise and specific magnitude. One positive unit of electricity or "positive charge" would attract one negative unit of electricity or "negative charge," no more and no less. An atom carrying two units of charge could join with one other atom carrying two units of the opposite charge, or with two other atoms each of which carried one unit of the opposite charge. Again, some atoms (such as those of the neon gas or argon gas in our atmosphere) seemed to carry no charge at all, but instead remained free and neutral, and these atoms, interestingly enough, were precisely the atoms which were never found forming any kind of compounds with other atoms.*

Faraday's discovery that compound-forming atoms always carried unit multiples of a certain "smallest possible" amount of electricity, and that these atoms tended to cleave together with other atoms carrying opposite charges, led scientists to quite a different idea of the nature of forces holding atoms together from the old unsatisfactory ideas of atomic "glue" or "valence hooks." It appeared that the forces holding atoms together or forcing them apart were nothing more nor less than the attracting or repelling forces of electrical charge. This proved to be a useful and extremely reliable explanation for the way that atoms bound themselves together into molecules, but typically it raised as many new questions as it answered. Why did some atoms always carry positive charges while others always carried negative charges? How could it be that still other atoms always re-

* True at least in Faraday's time. Recently the inert gas xenon has been found to form a compound with fluorine, one of the most furiously chemically active of all elements, and helium and argon have also been found to form a compound.

mained perfectly neutral at all times? Physicists and mathematicians, analyzing Faraday's work, found that they could calculate the exact amount of energy corresponding to Faraday's "smallest unit" of electrical charge, but what was the nature of the charge itself?

These questions could not be answered on the basis of Dalton's view of atoms as tiny solid particles of matter like submicroscopic marbles or billiard balls. Obviously the inner structure of the atom was far more complicated than that. But once they possessed the vital clue that the binding force between atoms might be electrical in nature, physicists soon began to discover just how much more complicated the picture really was than they had ever dreamed.

Faraday himself did not become deeply involved in studying the curious behavior of atoms and molecules in water solution, nor did he delve into the questions his discoveries raised. To him, his experiments with electrolysis were mostly a means to help him tie up the loose ends in his pioneering study of various electrical phenomena. But there were two of these loose ends which drew the fascinated attention of other investigators as well. First, the studies of electrolysis did *not* clarify just precisely what an electrical charge was. Was it an actual physical entity with characteristic physical properties such as size, shape, and mass, like a lump of butter or a hedgehog? Or was it merely another formless "entity" like a magnetic field, without any real physical substance and describable only by mathematical equations? The fact that charged atoms or ions in solution could carry electrical charges only in exact full multiples of some basic unit of charge suggested that an electrical charge had *some* kind of physical substance which could not be subdivided into fragments. As one scientist expressed it, electrical charges were like eggs in a bag: You might have one egg, two eggs, three eggs, or a dozen, but you could not have half an egg and still have an egg.

On the other hand, if the atoms of an element were indeed tiny particles of matter which could not be subdivided, what kind of a "particle" could an electrical charge be? This was a particularly difficult question considering that a given atom seemed to be able to take on such an electrical charge, or unload an electrical charge, under various circumstances (as in the case of a charged ion being drawn through a water solution to an oppositely charged electrode during electrolysis) without any apparent change in the atom's mass or in its integrity as an atom.

Difficult as these questions were, some remarkable and quite unexpected answers soon came to light as the result of four quite separate and startling discoveries that were made in a single brief period between 1895 and 1900, discoveries that were to force physicists to toss Dalton's comfortable picture of atoms as tiny solid marbles into the scrap heap and to revise completely their views of the ultimate structure of matter in the micro-universe. First came the detection of X-rays in 1895 by Wilhelm Roentgen. Next came the discovery of the phenomenon of natural radioactivity by A. Henri Becquerel in 1896. These achievements were followed in 1899 by the discovery of the ultimate "atom of electricity"—now known as the "electron"—by a British physicist named J. J. Thomson and his young assistant, Ernest Rutherford, just recently arrived from New Zealand; and finally, by the historic discovery of a previously unknown radioactive element, called radium, by Marie Curie in 1898. All unwittingly, these pioneer investigators started the world of physics off on one of the most fascinating detective stories in all the history of science. In order to appreciate the full impact of their discoveries on modern scientific thinking we must consider these people and what they discovered in more detail.

# CHAPTER 24

## *Micro-Universe: The Puzzle of Radioactivity*

When we read about scientific progress in the daily newspapers and magazines, we often come away with the impression that most of the really significant scientific discoveries arise from careful reasoning, shrewd guesswork, and the careful planning of ingenious experiments. It is easy to forget the role that sheer blind chance so often plays. Wilhelm Konrad Roentgen was indeed a shrewd and careful scientist with a flair for ingenious experiments—but Roentgen's discovery of X-rays in 1895, for which he was later awarded the first Nobel Prize in physics, resulted not from careful planning but from one of the most spectacular accidents ever recorded in the history of science.

As early as 1858 physicists had known that a current of electricity could be made to flow through an evacuated glass vacuum tube if wires were sealed into either end of the tube to form a "cathode" or negative terminal and an "anode" or positive terminal, and if the wires were then connected to some source of high-voltage electricity. Nor was a flow of electric current through such a tube the only thing that happened under such circumstances; at the same time the current was flowing, a fluorescent glow would appear on the glass wall of the tube.

Sir William Crookes, an English physicist working in the late 1800s, had experimented widely with this curious phenomenon without learning very much about it. He had been able to demonstrate that the fluorescent glow was caused by some kind of ray emitted by the cathode or negative terminal of such a "cathode ray tube," and that these rays traveled in straight lines; but that was about all that was known about them. By the early 1890s, cathode ray tubes devised by Crookes had become popular laboratory curiosities. They produced a fascinating natural phenomenon which everyone agreed must have some significance; but nobody had the vaguest idea what was actually going on inside a Crookes tube while it was operating.

Among those trying to puzzle out the curious electrical behavior of a Crookes tube was a physicist named Wilhelm Roentgen working at the University of Würzburg in 1895. Late one afternoon Roentgen had wrapped a Crookes cathode ray tube carefully with black paper, then darkened the room and thrown the high-voltage switch to send an electric current moving through the shielded device. To his amazement, a discarded piece of cardboard coated with fluorescent crystals which lay on a bench across the room began to give off a pale green glow whenever the Crookes tube was turned on. The glow vanished immediately whenever the current was cut. Although Roentgen could see no light escaping from the shielded tube at all, it was apparent that *some* kind of invisible ray was emanating from the Crookes tube, passing completely through the black paper wrapping, and striking the piece of cardboard across the room with some kind of wave that caused the fluorescent crystals to begin glowing.

Hardly believing his eyes, Roentgen held a book between the tube and the cardboard, and then a block of wood. In each case the mysterious rays appeared to pass through the intervening object without any obstruction. Then Roentgen held a sealed and wrapped photographic plate between the tube and the piece of cardboard. Again the cardboard gave off its eerie glow, but when the photographic plate was unwrapped and developed it was as thoroughly exposed as if he had held it up uncovered in the sunlight.

Indeed, only a very few materials seemed to obstruct these curious emanations in the slightest. A sheet of lead seemed to block them completely, as did platinum. When Roentgen placed his wife's hand on a wrapped photographic plate and exposed it to the tube, he discovered that he had produced a photograph unlike any ever seen before: The calcium in his wife's bones had obstructed the rays just sufficiently to leave a ghostly skeletal shadow on the plate, with the silhouette of a gold wedding band draped eerily around the bony ring finger!

Roentgen could not identify the nature of these invisible "X-rays" that emanated from his tube, nor could he guess why they appeared to behave in this fashion. But obviously here was a ray unlike anything ever before encountered, a ray capable of passing through surprising thicknesses of solid matter as if it were empty air, and capable of piercing flesh to reveal bony outlines on photographic plates. Within a very few weeks of the time of his discovery, Roentgen's new X-rays were being used by physicians to help them diagnose and identify fractured bones in human bodies, and physicists all over the world were racking their brains and climbing laboratory walls trying to figure out what these X-rays were and how they were generated in the cathode ray tubes from which they seemed to come. Although he did not know it at the time, Roentgen had thrown open a floodgate and released a whole new wave of discovery in physics. A discarded scrap of cardboard in his laboratory had triggered a search which is still going on full force sixty years later.

## THE DISCOVERY OF NATURAL RADIOACTIVITY

No one had ever observed Roentgen's mysterious X-rays anywhere in nature or under natural circumstances; these X-rays seemed to be purely a man-made phenomenon, even if no man was able to say precisely how man had made them. But if Roentgen's discovery of X-rays started an intellectual chain reaction in the world of physics, another chance discovery that followed just a few months later added fuel to the fire.

Like Roentgen's discovery, this new discovery made by A. Henri Becquerel, a French physicist, was the result of chance and accident, doubly so, in a sense, because it occurred as a result of Becquerel's own investigation of Roentgen's mysterious X-rays. As we mentioned earlier, a Crookes tube or cathode ray tube connected to a high-voltage electrical circuit not only produced invisible X-rays but an odd visible light as well: The glass wall of the Crookes tube could be seen to fluoresce with a ghastly glow. This quality of fluorescence (or phosphorescence, as it was sometimes called) was not associated exclusively with vacuum tubes or with high electrical potentials. A number of natural substances including the element phosphorus itself were known to develop a strong fluorescent afterglow as a result of exposure to sunlight. Once removed from the light, these substances would emanate a pale light for varying periods of time, sometimes even for hours. Fishermen for centuries had seen a similar fluorescent or phosphorescent glow in ocean water disturbed by their oars when they were returning home after dark. As word of Roentgen's discovery of X-rays spread, a number of investigators including Becquerel began to wonder if some of these naturally fluorescent substances might not also give off penetrating X-rays.

Planning to investigate this question when other experiments he was doing had been completed, Becquerel had collected a number of samples of various mineral ores and compounds and stored them in a desk drawer which also happened to contain several unopened packages of photographic plates. Perhaps he had an overzealous wife who had stowed away all the miscellaneous trash lying around on his desk just to tidy things up. But when Becquerel later went to use the photographic plates, he discovered to his chagrin that all of them were badly fogged, as if they had been exposed to light. This seemed particularly odd considering that the boxes of plates were sealed and wrapped in heavy black paper.

Becquerel might well have simply tossed out the plates and gotten some new ones, but scientists by habit tend to dislike curious and unexplained coincidences. Of course the plates *might* have been inadvertently exposed while they were being wrapped, but this seemed highly unlikely. Puzzled, Becquerel began checking out the ore samples in the drawer, placing each of them on a sealed paper-wrapped photographic plate for a period of time and then developing the plates. In short order he discovered the silent offender: a chunk of Bohemian pitchblende known to contain quantities of a compound of the element uranium known as potassium uranyl sulphate. Investigating further, he found this same incredible photographic effect produced by a number of other compounds of uranium. Indeed, the more uranium present in a given chunk of ore, the stronger the effect seemed to be.

It was clear to Becquerel that whatever was going on had nothing to do with phosphorescence. Neither did it have anything to do with high-voltage electrical currents. Ordinary, cold, untreated uranium ore in its natural state appeared to be continuously producing some kind of invisible ray that could pass through opaque material such as black paper as though it weren't there at all. Nor were uranium compounds the only ones to exhibit this curious behavior; natural ores of another heavy metal called thorium exhibited the same disturbing quality.

Becquerel called this newly discovered phenomenon "radioactivity." Further studies revealed that all uranium ores poured out their puzzling radiation at the same slow and steady rate, the amount of the radiation always proportional to the amount of uranium in the ore. There was no change in the rate of radiation whether the substance was heated to incandescence or frozen, ground up into a powder or treated with strong acids. The radiation appeared to be totally unrelated to any other physical or chemical property of the material and was completely unchanged by anything that was done to it. There seemed to be no way to increase the amount of radiation from a given sample in a given period of time nor any way to slow it down or turn it off; it just went plodding steadily along, radiating in the same fashion no matter what anyone did to it.

Obviously there was no ordinary chemical property involved here. It appeared that radioactivity had to be a sort of built-in quality of the atoms of these heavy metals—a phenomenon that occurred independently of all the other properties of the metals. But what was the nature of this radiation? Becquerel investigated it in a very simple way. He had found that a thick layer of lead seemed to block all the radiation emanating from uranium ore; so the French physicist placed a small quantity of the ore at the bottom of a deep hole in a lead block This way the only radiation that could escape from the ore was a thin beam emerging from the hole like steam from a geyser. Satisfied of this, Becquerel then arranged a magnet outside the hole in the lead block so that the beam of radiation, whatever it was, had to cut perpendicularly through a strong magnetic field (see Fig. 40).

To his consternation, Becquerel discovered that his uranium ore was pouring out not merely one but *three* totally different varieties of ray, each with different qualities and each separable from the others by the action of the magnetic field. One of the three rays thrown off by the uranium seemed to be a stream of particles which carried a strong positive electrical charge and was thus drawn to one side in the magnetic field. These particular particles seemed to travel only a few centimeters through the air; they were capable of passing through a sheet of tissue paper or a very thin sheet of gold leaf, but any objects much thicker than that seemed to block their passage.

A second emanation from the uranium formed a beam that was also deflected in the magnetic field, but in the opposite direction from the first, suggesting that it was made of particles carrying negative electrical charges. This beam traveled much farther than the first before it was dissipated, and could penetrate through a sheet of aluminum half a centimeter thick, or

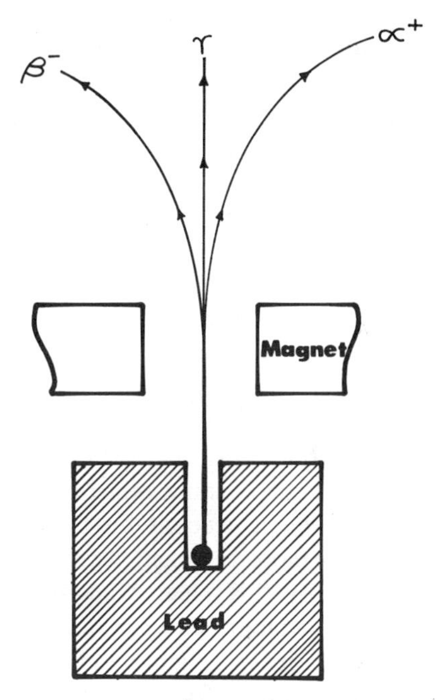


*Fig. 40* Becquerel's analysis of decay products from uranium decay. Powerful magnetic field through which decay particles passed pulled alpha particles (helium nuclei) in one direction, beta particles (electrons) in the opposite direction, but had no effect on path of gamma rays.

even through a few millimeters of lead. The third beam of rays went straight up, showing no magnetic deflection whatever. Becquerel thus concluded that these rays had no electrical charge. But they were far more penetrating than either of the other two kinds. In fact, these highly penetrating rays seemed from their behavior to be virtually identical to Roentgen's mysterious X-rays. For convenience in keeping them straight, Becquerel named these three streams of radiation respectively "alpha rays," "beta rays," and "gamma rays," and found that each consistently behaved in its own peculiar manner. One kind, for example, did not suddenly change into another kind.*

* Many readers are annoyed and irritated by the physicists' apparent delight in tacking Greek-letter labels on practically everything in sight. Couldn't Becquerel's alpha, beta, and gamma rays just as readily have been called A-, B-, and C-rays? Of course they could have, particularly since at the time Becquerel had no more idea what they actually were than the man in the moon. On the other hand, he

Becquerel's discovery of the phenomenon of natural radioactivity, following so closely on the heels of Roentgen's discovery of X-rays, served to intensify the explosion of excitement in the world of physics at the turn of the century. Here was a totally new and quite unsuspected phenomenon. What was more, it seemed quite unexplainable. No one had any idea why atoms of uranium or thorium should behave in this curious fashion, for radioactivity did not seem to fit into the accepted view of the atomic structure of matter at all. In fact, the phenomenon seemed in some ways flatly to contradict the laws of conservation of matter and energy. Yet there was an unmistakable similarity between the gamma rays pouring forth spontaneously from a lump of uranium ore and Roentgen's X-rays which were produced only under highly artificial circumstances when a high-voltage electrical discharge was passed through a vacuum tube. It was also arresting to discover that other emanations from radioactive metals very distinctively bore positive or negative electrical charges, much as atoms or molecules took on positive or negative electrical charges during the process of electrolysis.

Further fuel was added to the fire when a Polish-born chemist, Marie Curie, became convinced that the amount of uranium metal alone in Becquerel's pitchblende ore could not possibly be producing the *amount* of radiation that was observed emanating from it. It occurred to Madame Curie that perhaps some other unknown element that was mixed in the pitchblende might account for the fact that the ore seemed to have four to five times more radioactivity than it ought to have.

Obsessed by this idea, Madame Curie began what must surely have been the most laborious experiment ever before undertaken in the history of science, patiently refining and re-refining more than a ton of Bohemian pitchblende donated for the purpose by a bemused Austrian government as she sought to isolate just that part of the ore which contained radioactive material. The final result of her two years of incredible labor has, of course, become modern legend: In addition to quantities of uranium, Madame Curie managed to isolate a few milligrams of two previously undiscovered elements, one of which (called polonium, after her native country) was thousands of times more radioactive than uranium itself, and the second of which, radium, was so furiously radioactive that it poured off a million times more radiation, milligram for milligram, than uranium did. But like uranium, each of these new elements exhibited three distinct kinds of

---

could equally well have called them aleph, beth, and gimel rays (using the Hebrew alphabet) or alif, ba, and ta rays (Arabic alphabet). Then *everyone* would have been confused. For practical purposes throughout the international world of science, letters of the Greek alphabet are universally used as symbols, since physicists everywhere in the world at least recognize these letters even if they can't read a single word of Greek.

radioactive rays: positively charged alpha rays, negatively charged beta rays, and the highly penetrating gamma rays which possessed no electrical charge.

## THE "ATOM OF ELECTRICITY"

At the same time that Madame Curie was doing her monumental work in France, British physicist J. J. Thomson, working in the Cavendish Laboratory at Cambridge University, discovered the third piece in the three-cornered puzzle, the vital clue which, together with the discovery of X-rays and of the phenomenon of natural radioactivity, was to lead physicists to quite a new and totally unexpected picture of the basic structure of matter.

Thomson, assisted by a young New Zealander named Ernest Rutherford, became intrigued by the question of just how an electric current was able to pass through an evacuated cathode ray tube (the sort of tube which produced Roentgen's X-rays) in the first place. As we discussed before, when most of the air had been pumped out of a sealed glass tube to create a near-vacuum, with electrodes sealed into either end of the tube and connected to a source of high-voltage electricity, a current would pass from the cathode or negative electrode of the tube to the anode or positive electrode. But how did it jump the gap? Of course physicists realized that the vacuum in such a tube was never perfect. Some small quantity of gas always remained—a highly rarefied gas, perhaps, but still a gas. Thomson speculated that in the presence of a high-tension electric field between the cathode and the anode of such a tube the atoms or molecules of the gas remaining in the tube might be broken up into positive and negative ions very much the same as the molecules of salt in solution in water would dissociate into positive and negative ions in the presence of a high-tension electric field through the water solution. Thus, in the cathode ray tube, the positive ions would be expected to flock toward the cathode while the negative ions would stream away from the cathode toward the positive terminal or anode, forming a beam of "cathode rays." Investigating this matter, Thomson discovered that the positive ions that occurred in the rarefied gas in such a cathode ray tube were identical in every way to the positive ions that were found during the electrolysis of a salt dissolved in water. They were no more nor less than positively charged atoms of the rarefied gas inside the tube. But the negative "ions" which made up the beam of cathode rays seemed to be something quite different. These also seemed to be a stream of particles (Thomson first referred to them as "corpuscles") carrying negative electric charges much the same as, say, the chlorine ion carries a negative electric charge in a water solution of common table salt. But in the cathode ray tube these "corpuscles" seemed far, far smaller than any atom and moved far more swiftly than an ordinary atom could be expected to move.

Earlier experiments with cathode rays bombarding the anode of a Crookes tube had demonstrated that a stream of these particles could be bent or deflected from their straight-line course if a magnet were placed around the tube so that the rays had to pass through a magnetic field. Thomson applied electric fields and magnetic fields simultaneously along the path of cathode rays moving through a vacuum tube in order to measure the exact amount of bending or deflection that these external forces could bring about. And by studying what happened to these high-speed "corpuscles" when magnetic and electric fields were acting upon them simultaneously, Thomson made some startling discoveries.

First of all, he confirmed that these "corpuscles" were indeed negatively charged particles. In fact, they formed a veritable stream of moving electrical charges, nothing more nor less, in fact, than an electric current. Secondly, these particles were in fact *particles,* not just electrical charges, and as particles they seemed to have real physical substance and mass, although they were moving at a very high velocity. But Thomson's calculations indicated that the actual mass of each of these negatively charged particles had to be incredibly small, far smaller than the mass of an atom of even the lightest element known, hydrogen. It seemed as though the high-tension electric field between the cathode and the anode inside a Crookes tube somehow had the power of breaking off tiny, tiny negatively charged "chips" from the atoms of rarefied gas within the tube and could then send these negatively charged "chips" flying at high velocity toward the anode. At the same time the remainder of the gas atoms with the "chips" torn away but still containing most of the original mass of the atoms were left bearing positive charges and thus began moving toward the cathode, although much more slowly because of their much greater mass.

But if this were really what was happening when an electric current passed through a cathode ray tube, the implications were staggering. This could only mean that the supposedly indivisible atoms of Dalton—the tiny, solid marbles—were not really indivisible after all; "chips" could be torn from them, and if this were the case then those atoms must actually be made up of two separable parts, the largest part bearing a positive electrical charge and most of the atoms' mass, while a tiny fragment with only a fraction of the mass bore a full unit of negative electrical charge!

Because these tiny "chips" or negatively charged particles seemed so much more like "atoms of electricity" than any other kind of matter known, Thomson spoke of them as *electrons.* He could find no way to measure the exact mass of these tiny particles, nor was he able to measure the *amount* of negative electrical charge carried by each particle. The closest he could approach this was to determine a mathematical ratio between the charge of a single electron and its mass, without being able to specify the proper quantity for either.

This failure was not so much a reflection on Thomson's ability and imagination as it was upon the crudity of the instruments and techniques for study of these tiny particles that were available in his day. It was not until twenty-five years later that an American physicist, Robert A. Millikan, finally devised a method for measuring the electrical charge of a single electron, and thus to calculate that this tiny negatively charged particle had a mass of only 1/1,840 the mass of a single hydrogen atom. Millikan did a brilliant piece of work in achieving this measurement, and ultimately won a Nobel Prize in physics for it; but long before Millikan had even dreamed of his clever technique for measurement, J. J. Thomson, Rutherford, and a multitude of other physicists had come to recognize clearly what the discovery of the electron had to mean.

Just as the other high-energy and highly penetrating gamma rays emanating from uranium and other radioactive atoms had been found to be very similar in their properties to Roentgen's high-energy and highly penetrating X-rays, Thomson's electrons streaming toward the end of a cathode ray tube were found to be virtually identical in every way to the negatively charged beta rays that were given off by uranium. What was more, the massive and slow-moving alpha rays thrown off by uranium seemed to behave precisely like atoms of helium from which negatively charged electron "chips" had been stripped, thus turning what remained of the previously neutral helium atoms into positively charged helium ions. Today we know that these observations were precisely correct. Beta rays are indeed nothing more than ordinary electrons bearing negative changes, and alpha rays are much more massive and slower-moving ions of helium. It remained for Millikan to determine in precise detail what the mass of an individual electron was, and to prove that each electron carried a single unit of negative charge equivalent to a specific very tiny quantity of electrical energy, but as for the real major importance of the discovery of electrons, Thomson and Rutherford had let the cat out of the bag.

Atoms could no longer be regarded as tiny indivisible particles similar to wee billiard balls with hooks on them. All atoms, from the very simplest to the most complex, had to be made up of at least two different kinds of components which could be separated from each other under certain circumstances: a heavy or massive component bearing one or more positive electrical charges, and one or more far less massive negatively charged electron "chips." Considered in this way, one could think of the positive ions in an electrolysis solution as atoms which had been temporarily stripped of one or more electrons, thus leaving them out of electrical balance with one or more excess positive charges, while negative ions in the same solution were atoms which had temporarily gained or taken on one or more electrons in excess of their normal quota and thus had an excess of negative electrical charge.

Thus it became clear, bit by bit, that the atoms of radioactive elements were doing something far more specific than merely releasing some kind of mysterious emanation. In any lump of radioactive material there were, in fact, atoms spontaneously exploding from time to time—literally hurling part of their atomic structure away from them. Sometimes such a radioactive "explosion" resulted in the emission of a negatively charged electron. Sometimes a more massive fragment of an atom was emitted, indistinguishable from a positively charged helium ion. At still other times such a radioactive "explosion" merely seemed to unload excess energy, as it were, in the form of electromagnetic waves or gamma rays. And in the course of this process, radioactive atoms were one by one spontaneously performing, on a submicroscopic scale, precisely what alchemists had been seeking unsuccessfully to do for centuries. These radioactive atoms were actually transforming themselves from atoms of one element with certain specific physical and chemical properties into atoms of completely different elements with quite different chemical and physical properties, while converting a portion of their mass into pure energy in the process.

## CAUSE AND EFFECT VERSUS PROBABILITY

From this time in the late 1890s on, physicists began changing their views, almost on a week-to-week basis, of how atoms really were constructed, as more and more was learned about the behavior of electrons, of positively charged ions such as those of helium (alpha rays) or hydrogen (soon to be called "protons"), of gamma rays, X-rays, and other electromagnetic waves.

All of these subatomic particles were far too tiny ever to be observed with even the finest, most high-powered microscopes. A bit later we will try to visualize more clearly just how very tiny these particles are, and to see *why* no microscope can ever be expected to reveal them. But since these particles could not be viewed directly, ingenious devices were invented to enable physicists to observe them indirectly—quite literally by observing the wreckage they left in their path as they moved about. And with the invention of such devices, more and more was learned about what the internal structure of the atom had to be like. Brilliant and complex new theories were worked out to try to explain and correlate the observed behavior of subatomic structure with their physical structures as they were imagined. Indeed, in the early 1900s a veritable chain reaction of discovery began which is still continuing to this day.

But even before these new discoveries were made—in fact, on the very heels of the work of Becquerel and Madame Curie and the earliest investigators of radioactive phenomena—a new and alarming idea about the behavior of things in this micro-universe began to appear in the minds of physicists. It was actually an idea that was implied by one of the earliest observations that had been made of the behavior of radioactive substances, an idea so subtle that it was overlooked at first by many of those pioneer nuclear physicists even when it was staring them in the face, yet so far-reaching in its fundamental implications about the behavior of matter and energy in the micro-universe of subatomic particles that it was soon to challenge virtually everything that classical physics had ever discovered or taught.

The idea was simple and obvious, yet it was also inescapable. Stated baldly, it was simply this: Radioactive atoms did not "explode" (or in physicists' terminology, "decay") and transform themselves into atoms of new elements on a cause-and-effect basis predictable by any specific natural law. It was totally impossible for anyone to predict when, if ever, any single given radioactive atom in a mass of radioactive material would spontaneously undergo this transformation; when and if it happened, it happened as a result of random chance, not in a cause-and-effect sequence, and *was therefore predictable only on the basis of mathematical laws of probability.*

We should pause for a moment to consider this idea in more detail, because it challenges one of the basic premises on which all previous understanding of natural law had been based: the premise that everything that happens in the universe happens in an *orderly* fashion, with each individual event occuring as a direct result of some previous event.

What are "mathematical laws of probability" and how do they differ from the idea of cause and effect? In brief, the laws of probability are a means of making reasonable and approximate predictions of what event is likely to occur in the future when the multiple causes which might determine the event that occurs are too multitudinous and too variable to permit the calculation of a single specific predictable effect—or when those "causes" themselves arise as a result of random chance. Stated more simply, the laws of probability are a means of making a "best guess" of the likelihood of various possible events occurring when the causes leading to those events either cannot be calculated readily, or cannot be determined *at all* by any scientific means.

There is an apocryphal story of four women who were playing contract bridge one day when in a single deal each of the four was dealt a straight thirteen-card suit; one found all thirteen clubs in her hand, another had all thirteen hearts, and so forth. The story goes that the women were so flustered by this incredible turn of events that they spent all afternoon talking excitedly about their hands and completely forgot even to bid, so that the holder of the suit in spades lost the opportunity of a lifetime to bid a grand slam on the opening round of bidding with 100 per cent certainty of being able to make it!

What does such an event involve? An ordinary pack of playing cards contains fifty-two cards divided into four suits which are shuffled together in a random fashion before each deal. Obviously, it is *possible,* purely on the basis of chance, for such a shuffling to leave the cards so arranged that when they are dealt in sequence to four players each player receives an unbroken suit of thirteen cards. But there are so many possible alternative ways the cards might fall after a shuffling that the *probability*—that is, the *likelihood*—that such an incredible hand would be dealt after any given shuffle is ridiculously small. By knowing the number of cards in the deck and the number of cards in each suit we could calculate that there would be one chance in 158,500,000,000 that *one* person of the four would be dealt all thirteen cards of one suit, and that the probability that each of the other three players would also be dealt perfect thirteen-card suits in the same deal would be on the order of one chance in 24,900,000, 000,000,000,000,000,000,000,000,000,000,000,000,000,000,000,000. We could even say that, since this "perfect hand" is indeed one possible alternative, it would stand as good a chance of occurring in any given deal as any other possible alternative distribution of cards, and that if enough hands were dealt, sooner or later such a fall of cards would, in fact, occur. *But there could be no possible way to predict when.* A quartet of bridge addicts might sit down prepared to deal 5 billion successive hands, if necessary, in order to achieve this "perfect hand," but it could just as well be dealt the first hand around as the five billionth. On the other hand, it also might not fall until the ten billionth hand was dealt.

Thus if we were to try to say *when* such a hand would be dealt under these circumstances, or to estimate how long the players would have to sit there in order to achieve it, we would have no way to make such a prediction. There are no laws of nature we could call upon to help us do so. There is no way of examining or controlling the cause (the way the cards happen to be lying in the pack during the shuffle) in order to be able to predict when the effect (the fall of cards in a perfect hand) would occur. The very best that we could do would be to calculate the degree of likelihood or probability—we might say, the percentage chance—that the perfect hand would happen to be dealt within the first ten hands, or within the first twenty, or within the first thousand, or the first million.

But could there be any usefulness or value to such a prediction of probability? It couldn't possibly tell us when the anticipated event would occur. We couldn't lay plans for a grand party to celebrate. What value would an estimate of probability be, in this case? In point of fact, it would be of inestimable value in that it would show us coldly and unemotionally how exceedingly small the chances would be of seeing such a perfect hand dealt even if all four players lived to be two hundred years old and continued dealing cards without respite to the day they died. This information about the likelihood or probability of this desired event happening could then serve us as a basis for judging whether even to embark upon the experiment or not. In this case the probabilities of success within human lifetimes would be so very small that a recognition of those probabilities would lead any sane-minded human being to conclude that the effort would be ridiculous and that one's time might better be put to doing something potentially more productive.

It is not surprising that the mathematical laws of probability first thrust themselves to man's attention as a result of attempting to predict probabilities in various games of chance, nor is it strange that most of us today are still accustomed to thinking of probability in terms of games or gambling situations of one sort or another. The idea that probability might play any part at all in scientific prediction has only been seriously considered in the world of science for perhaps a hundred years, and never (until the first decade of the 1900s) was it seriously regarded as a useful or necessary scientific tool. What was it, then, about the discovery of the phenomenon of radioactivity that suddenly forced scientists' attention to the laws of probability—made them wonder, indeed, if the old cause-and-effect basis for natural law might not in the final analysis prove to be nothing but a false illusion of security?

To answer this we must consider how scientists had always regarded cause and effect in relation to the occurrence of events. It had always been assumed that if one could only accumulate enough information about all aspects of some natural process, no matter how complicated those aspects might be, one could then predict with absolute precision what the outcome of the process would be. Having made that prediction, one could then use it accurately to predict further the outcome of events that would occur still farther in the future. In other words, it was assumed that if all the forces at work in situation A were known at a specific point in time, then one could predict what end result those forces would bring about in situation A, and then use those predicted results as future forces in action in a future situation B in order to predict in advance what end result those forces would bring about in situation B at some later point in time. The classical laws of nature were thus nothing more than simple statements of relationships which allowed precise predictions to be made regarding future situations that would have to arise from present situations. Convinced of the validity of those classical laws, scientists therefore firmly believed that *nothing* could ever happen except as a fixed and inescapable result of some other thing, or series of things, which had happened before.

This might seem to us to be a rather sweeping assumption, but over the centuries there has been an enormous quantity of evidence, observation, and experiment to support and validate this cause-and-effect principle. Drop a steel ball down a long vertical tube from which all the air has been exhausted, and the ball will fall to the bottom of the tube with a steadily increasing velocity until it strikes the bottom. Not maybe, not sometimes, not even usually, but always and invariably—as long as the laboratory is located within a gravitational field somewhere, such as on earth or on the moon or within gravitational reach of the sun or some other massive body. If we know the precise mass of the ball, the precise length of the tube, the precise length of time it takes for the ball to reach the end of its fall, and the strength of the gravitational field in which the ball is moving, we can predict with perfect accuracy what the rate of acceleration of the falling ball will be the next time that it is dropped, or the next time, or the next, and we can tell in advance precisely what its velocity will be at any point in the course of its fall. If we had still further data, other things could also accurately be predicted. We might predict in advance, for example, the amount of kinetic energy that would be converted into heat energy at the moment the ball strikes the bottom of the tube, or the velocity and magnitude of the ball's rebound or bounce.

These are not questions of likelihood, probability, or chance. These are clearly matters of cause and effect. We are saying that as a result of *this, this,* and *this* factor or force, *this, this* and *this* predictably will happen—not maybe, not sometimes, but always.

This principle of cause and effect is not just observed in the physics laboratory alone. We live with it continuously in our everyday life, and unconsciously accept it as the underlying basis upon which things around us normally happen. Consider this example: A man driving an automobile at high speed on a busy freeway through a city has his attention distracted momentarily by a glaring neon sign reading DRIVE SAFELY which is mounted on the roof of a nearby auto-insurance company building. During the instant of inattention while the man reads the sign, the muscles in one of his arms contract slightly, pulling the steering wheel slightly out of line, so that the car veers into a neighboring traffic lane and a bloody collision occurs. "A dirty break," we say when we hear about the wreck, "bad luck. A tragic chance occurrence"—but is it? Not really, since everything that happened in sequence occurred as a result of something else that had happened. Did the neon sign cause the accident? Not directly; it obviously did not jump off the building and fall on the two unfortunate automobiles —although as an attention-getting nuisance, deliberately calculated to pull drivers' eyes away from the road, it was certainly a prime contributing cause.

Let us consider a case that is a bit more extreme. Some years ago western newspapers headlined an account of perhaps the most freakish automobile tragedy in all history. A man was driving his automobile over a western mountain pass highway when a huge fir tree, undermined by erosion at the side of the road, suddenly toppled over onto the highway at the instant the car was underneath it, pulverizing the car and killing the man instantly. Surely, we are tempted to say, *this* cannot be considered anything but blind freakish chance; no one could have predicted or forestalled such an improbable accident. Not so, however, from the scientist's point of view. He would argue that had *all* the facts been known in advance —the place from which the driver started driving that morning, the velocity and route of his car, the degree of soil erosion beneath the roots of the tree, the age of the tree, the amount and direction of wind, etc.—it would have been possible to predict accurately *before that man even started on his trip that morning* that four hours later his car would be at that exact point in space and time at which that tree struck the ground. Chance did not enter into it at all; the accident was a direct result of cause and effect.

Obviously, in practical terms, this is a ridiculous exaggeration. The most knowledgeable and competent scientist in the world cannot predict future events in this fashion, for the simple reason that so many totally unknown or incalculable variables enter in. The driver of that car might have decided to stop for a cup of coffee along the way because of a sudden feeling of sleepiness. One of his bald tires might have blown out two minutes before he reached strike point. He might have forgotten to kiss his wife that morning before he left, and turned back to correct this oversight. Any one of a million other unpredictable variables might have conspired to save the man from his fate.

But when we speak of "unpredictable variables" here we are speaking of alternative events that cannot be predicted in advance not because prediction is impossible but simply because we don't have enough data upon which to base such a prediction. Scientists had assumed for centuries that, at least theoretically, if there were enough data at man's disposal about conditions in the universe, available in great enough detail, there might be no such thing as an "unpredictable variable"!

What about the case of the true random chance? Whenever probability is discussed, the example of the flipping of a coin almost invariably comes up, usually to demonstrate that there are some kinds of events which happen only by chance and can be predicted only on the basis of likelihood or probability. Surely whenever a coin is flipped there is always an even 50–50 chance that it will fall heads up, and an equal 50–50 chance that it will fall heads down. But is this really a matter of chance? Perhaps not. Suppose that we knew the precise weight of the coin, the precise force with which it was flipped, and the precise angle of rotation with which it was sent spinning, together with data about the gravitational field in which the coin and the coin-flipper are performing, and such other things as air resistance, air movement across the path of flight of the coin, elasticity of the surface upon which it is to fall, etc. If we had *enough data in great enough detail* to begin with, and knew which side of the coin was facing up before the flip, we should, at least theoretically, be able to predict which side of the coin would be up when it landed, and make that prediction *before the coin was ever flipped.*

We might even conceivably devise a mechanical coin-flipping machine designed under carefully controlled conditions to deliver precisely the same force and spin to a coin with every flip. Suppose the machine were carefully designed to rule out all external forces of any kind, and then were calibrated in such a way that if the coin is facing heads up before the flip it will always and invariably land heads up on the table after the flip. Would the result of coin flipping by such a machine be a matter of chance? Of course not. The element of chance enters into ordinary coin-flipping only because of our inability accurately to regulate all the forces at work in the process of the coin-flipping. We are not talking about a matter of chance; we are talking about a lack of data, of physical control, and of engineering know-how.

It was precisely such thinking to the ridiculous extreme which had long since convinced scientists of the validity of the assumption that all things in the universe had to occur as a result of cause and effect. In the last analysis even the dealing of the perfect hand at bridge would have to be considered the result of cause and effect, even though the precise forces at work while the cards were being shuffled could never accurately be calculated. We had to settle for the laws of probability in that case only because of our inability to secure enough detailed data upon which to make cause-and-effect predictions. With such a traditional and deeply entrenched conviction of the operation of cause and effect in all natural phenomena, we can imagine the shock and consternation that arose when, for the first time, a natural phenomenon was encountered which simply could not be predicted on the basis of cause and effect because it appeared to occur totally as a result of random chance.

The decay of atoms of a radioactive element was precisely such a phenomenon, and try as they would physicists could not find any other way of explaining what they observed to happen.

What is it about radioactive decay that is so totally different from any other natural process ever observed? Part of the answer lies in what actually happens when a radioactive atom emits a fragment of itself or a burst of energy and transforms itself into another element. Another part lies in the question of *when* in the natural history of a radioactive element this convulsive change takes place. Atoms of radioactive elements such as uranium, thorium, or radium differ radically from other atoms (such as lead, for example, or oxygen atoms or atoms of calcium) in that they possess as an inherent property a continuing, built-in instability. After remaining in such a state of instability for an interval of time, an atom of such a radioactive element may undergo a spontaneous internal "explosion" or change, hurling fragments of itself away at high velocity, dissipating some of its mass in the form of energy, and ending up transformed into an atom of a different element.

When such an event occurs, it seems to occur for no discernible reason and its occurrence seems to be truly spontaneous—meaning, of course, that it does *not* occur as a result of some outside force acting upon the atom. On the contrary, it is obvious that some internal force within such an atom itself has to account for this continuing instability and the chain of events which may result. What, then, triggers this chain of events and leads to the result?

If there were some event or change discernible within the atom which acted as a triggering mechanism to the radioactive "explosion" we could then relax, confident that the "explosion" has after all occurred as a result of cause and effect. But the very uncomfortable fact is that there is *no* apparent or discernible change or event occurring within such atoms that has ever been identified as such a triggering mechanism. It is as though one has a ball dangling from a string, and then without any change in the ball, in the string, in the gravitational field, or in any other condition, the string suddenly parts and lets the ball fall to the floor.

But then, *when* would a given radioactive atom transform itself in a radioactive "explosion"? The answer is that there is absolutely no way to predict when. And therein lies the rub.

This unpredictability of radioactive decay became clear very early in the study of radioactive atoms. But physicists found one thing about radioactive phenomena which could be observed and accurately predicted. This was the *rate of decay* occurring in a large aggregation or lump of radioactive atoms of the same element. Whatever triggered any given individual atom to undergo a radioactive transformation, and however unpredictable that event might be in the case of any given atom, the over-all *average rate* at which atoms seemed to decay in a given mass of a radioactive element could be very accurately measured.

For some radioactive elements this average rate of decay seemed to be very slow. In a milligram of such an element, containing millions upon millions of individual atoms, radioactive transformations of atoms might occur comparatively infrequently, so that the total number of unchanged, or "unexploded," radioactive atoms that remained would decrease at a very slow and stately pace. In the case of other elements, the rate of decay was observed to be far, far faster, so that in an equivalent milligram of such an element individual atoms might be exploding constantly in a veritable volley and the total amount of unaltered, or unexploded, element remaining from the original amount could be seen to decrease very rapidly.

In fact, this rate of decay of each radioactive element that has ever been discovered seems always to be a highly unique characteristic of the particular element. No two radioactive elements decay at the same rate, and the varying rates of decay of various radioactive elements can be measured with extreme accuracy. This property of radioactive elements is, in fact, so reliable that it is one of the best ways physicists have of determining precisely what element (or what radioactive fragment, or what unstable particle) is involved in any radioactive interaction.

Thus within a few years of the first discovery of radioactivity, physicists knew that the rate of decay of radium atoms was such that in any given quantity of pure elemental radium, half the original number of radium atoms present would have decayed and transformed themselves into something else within a period of 1,620 years. If one were dealing with the same quantity of another radioactive element, radon, half the original number of atoms of that element would have decayed and transformed themselves into something else by the end of 3.82 days. For an equivalent number of atoms of uranium, however, some *5,000 million* years would have to pass before half the original number of atoms had transformed into something else—a very slow rate of decay indeed!

It became the custom to measure and describe the rate of decay of any given radioactive element in terms of that element's corresponding "half life," the length of time that would be required for half the original atoms present in a given quantity to have changed or decayed into another element. But what about the "whole life" of the radioactive substances? If half a given quantity of pure radium would have been transformed into something else within a period of 1,620 years, there would at that time obviously still be half the original atoms remaining untransformed. During a second period of 1,620 years, then, half of that remaining half of the atoms would have been transformed. During a third period of 1,620 years, *another* half of one-half of one-half of the original number of atoms present would have been transformed—but at the end of that third period there would still be half of one-half of one-half of the original atoms that were present to begin with, *still present and unchanged,* 4,860 years after the beginning of our measurement.

As we can see in the graph in Figure 41, a steadily decreasing number of the original atoms would always still be present after each succeeding half-life—in each case half of the number of atoms that had been present at the beginning of that half life. This means that even after a long succession of half-life periods, a small fraction of the original radium atoms present in the original lump would still be in existence and untransformed even if we measured successive half lives for 5 million years. By then the fraction remaining of the original lump of radium would be very small, true, but it would still contain a huge number of original, still-unchanged atoms. An even smaller fraction—although *still* a very large number of individual atoms—would still be present and unchanged 100,000 years from the time we started measuring; and so it would go until the number of radium atoms remaining becomes so small that we can no longer use statistics.

But if this is so, then *how long will any individual radium atom in our original lump live before it decays and changes itself into something else?*

The answer is that nobody knows and there is no possible way that anyone can ever find out. There is *simply no way to predict* when a given individual radium atom will decay. That atom might "live" for no longer than a 1 millionth of a second, but it might just as readily "live" for a period of 150,000 years, or for any other interval of time in between these extremes. For that matter, it might "live" on, unchanged, for 5 billion years, improbable as that may be. Accurate predictions can be made *on the average* about the behavior of a *large number* of radium atoms *all taken together*: We can predict accurately that at least half of them will be transformed within 1,620 years, and that this will hold true no matter how large or how small the total number of atoms we start with may be; but there is no way we can make any prediction whatever about precisely when any single given radium atom will undergo this transformation.

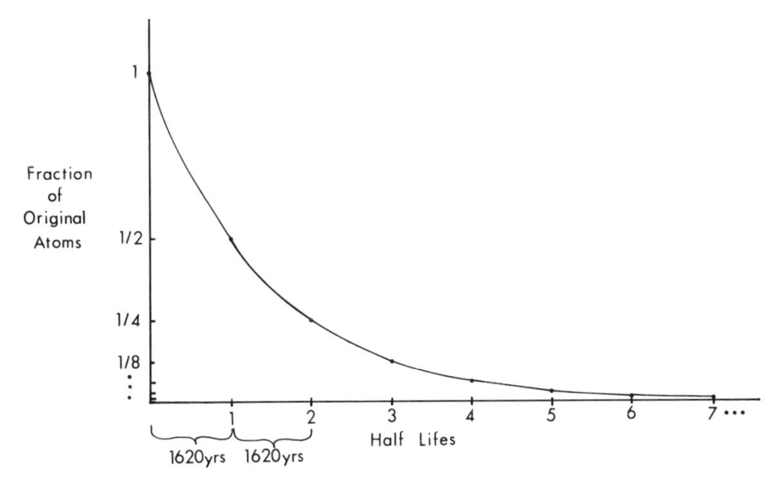


*Fig. 41* Graph depicts the decay of radium atoms in any given sample according to half-life periods.

The very best that we can say is that there is a certain *probability* or likelihood—in this case, a 50–50 chance—that this particular atom we are interested in will have undergone decay at some point in time within a period of 1,620 years. If our atom survives that half-life period untransformed, then we could again say only that it has a certain probability—again a 50–50 chance—of surviving another 1,620 years, and so on, and so on, to the point where only a few atoms are left—perhaps 140,000 years. And even then we could not predict when any one of those remaining atoms would decay; according to the laws of probability, they, too, will have decayed within a few thousand years more, but any one of them might *conceivably* survive for 5 billion years or 10 billion years or whatever other interval of time we might choose to guess.

## PROBABILITY AND UNCERTAINTY

It was this simple implication of the observed behavior of radioactive elements—the discovery that the decay or transformation of any given radioactive atom could never be predicted on the basis of cause and effect, only on the basis of mathematical laws of probability—that shook the minds and convictions of physicists at the turn of the century.

It was an ominous and frightening observation, for here was a pattern of behavior obviously associated with the most fundamental inner structure of matter itself which now suddenly appeared flatly to contradict a basic assumption upon which the whole body of natural law had been built up for centuries before. Many physicists found the idea so upsetting that they wholly rejected it. Of course they didn't just say, "I don't believe it because I don't want to believe it," and go off to sulk; rather, they began searching for some cause-and-effect explanation—some hitherto unsuspected bit of data about the interior of radioactive atoms—that would demonstrate that the time of decay of any given radioactive atom *was* preset and predetermined by previously existing conditions inside the atom and thus occurred as a result of cause and effect after all. Perhaps changes were going on in the core of a radium atom all the time, a predetermined and discoverable chain of events which led stepwise to the inevitable result of radioactive decay of that atom after a certain fixed period of time. If this were so, it might mean that all radium atoms actually did live precisely as long as all others, or if not, it might then mean that there was some predictable reason why one atom decayed early in the game while another survived longer.

Unfortunately, as time passed, it became increasingly clear that such an idea was little more than wishful thinking. The awkward but incontrovertible fact was that there were no such cause-and-effect changes going on in the heart of a radium atom or any other radioactive element. Subsequent exploration of the form and nature of the interior of atoms has, in fact, produced an increasing number of situations in which mathematical laws of probability are the *only* laws that seem to govern natural events. Brilliant mathematicians have been able to demonstrate mathematically that there are situations in the micro-universe of subatomic particles in which laws of probability are the *only laws that can possibly apply.* Of course this was not known with any certainty in the years when radioactivity was first discovered, but the awkward and uncomfortable idea had presented itself nonetheless. And it is quite possible that this idea—the idea that some of the most basic of all phenomena in the universe cannot ever be predicted in advance with accuracy—contributed more than anything else to the electric sense of urgency felt by everyone working in atomic and nuclear physics at the beginning of the century.

One final simple analogy will help us see more clearly the enormous distinction between a natural phenomenon governed by cause and effect and a natural phenomenon governed by the laws of probability. All of us at one time or another have watched popcorn popping on the stove. Up to a certain point, what we observe happening to a given kernel of popcorn in the popper seems remarkably similar to what happens to a radium atom. In either case, after a period of time a sharply defined fragment of matter undergoes a sudden change and ends up transformed into something quite different from what it was before. Just as the radium atom "explodes" in radioactive decay, we might say that the hard dense inedible kernel of popcorn suddenly "explodes" and is transformed into a large, white, yummy, crunchy morsel.

Furthermore, just as all the radium atoms in a lump of radium do not decay and transform themselves at once, so all the popcorn in a popper does not pop at once, either. As heat is applied and as the popper is shaken, first one kernel pops, then another, then another, and another, with increasing rapidity. If we knew the exact number of kernels we put in the popper to begin with and if at some point in the course of the popping procedure we could count the number of kernels that were popped and the number that were still unpopped at that instant, we might conceivably calculate a half life for popcorn in the popper under these conditions of heating. And no matter how careful we are to get all the popcorn popped before we take it off the stove, we will invariably discover a few kernels left in the bottom of the bowl that never did get popped at all.

But is the popping of a pan of popcorn an accurate analogy to what occurs during the process of radioactive decay? Actually it is not, not even remotely so, for the simple reason that a few crucial points have been overlooked. For one thing, popcorn does not pop spontaneously. A marked change in its environment must occur before any popcorn will pop at all: External heat must be applied. Without that "outside intervention" the kernels of popcorn could sit forever in the pan until the mice got them, and not one kernel would "explode" spontaneously. In sharp contrast to this, we have seen that radioactive decay occurs without any prompting.

In addition, once external heat *is* applied, there is a specific and predictable point at which any given kernel of popcorn will pop. Some force intrinsic to the kernel of corn—the moisture trapped inside a very hard and unyielding shell—responds to an external change—the application of heat—until the pressure of expanding water vapor inside the kernel exceeds the containing strength of the hard outer shell and the kernel of corn goes blooey. If all the kernels of popcorn in our pan were physically identical in their composition, then we might accurately predict a "trigger point" at which any given kernel would explode. Heat a kernel to such and such a temperature at such and such a rate of speed for such and such a period of time, and at *this specific instant* in the future it will pop. The reason we see the kernels of corn popping at a variable rate of speed is not because they pop on a basis of chance, but only because we are not controlling external conditions well enough; we allow some kernels to get hotter faster than others do.

We might even imagine a perfectly ideal situation in which a whole panful of popcorn kernels, each physically identical to every other one in every way, was warmed up slowly and uniformly in an oven so that each kernel heated up at precisely the same rate as every other one. Under such ideal circumstances, we might then observe the grand spectacle of every kernel of popcorn in the pan—*every* kernel, without exception—suddenly popping at precisely the same split second in time: instant popcorn, if you will. But the real significance of this "improbable" idea is not that it might conceivably happen. The significance is that if those specific circumstances and conditions could be achieved, and if each kernel of popcorn was indeed identical to every other kernel, instant popcorn *would* happen. It would *always* happen under those circumstances; in fact it simply *could not fail to happen.*

What we are saying, in effect, is that the popping of popcorn is a phenomenon which occurs as a result of cause and effect, not of chance. At the same time, we know that radium atoms possess an intrinsic instability which can cause them to "pop," and we know that in a mass of pure elemental radium each atom must possess the same identical physical structure and properties. If the same laws of cause and effect applied to the decay of radium atoms as apply to the popping of popcorn, then there ought, at least in theory, to be conditions achievable in which instant decay of all the radium atoms in a lump would have to occur. But since we have no way to arrange such conditions, nor even any way to imagine what conditions might conceivably be arranged to make this possible, no such circumstances have even been observed with radioactive atoms, and early nuclear physicists found more and more evidence to suggest that no such circumstances ever could exist, even theoretically.

The best that could be said was that a certain *excessively* small probability existed that each and every atom of radium in a given lump *might* decay at the same instant, transforming the whole mess into something else all at once, but there was no conceivable way that circumstances could be arranged to increase the minuscule probability one whit. Nor could circumstances be arranged to decrease the probability one whit. Fortunately or unfortunately, it is possible to calculate that such an event, while conceivably possible, is so massively unlikely that we would not be well advised to hold our breath until it happens.

Thus it was that physicists in the early years of this century began to realize that the underlying assumption of cause-and-effect predictability of events in the universe might not necessarily obtain at all in the microuniverse of subatomic particles. One could not predict with certainty the time when a given radioactive atom would decay, so that any time the decay of radioactive atoms was at issue there was always a degree of *uncertainty* about what precisely could be expected to happen when. Even in the earliest years after the discovery of radioactivity, this principle of uncertainty was an exceedingly awkward and uncomfortable idea for nuclear physicists. Later we will see how other even more awkward examples of this principle became manifest as the wave of discovery moved on, and review some of the evidence, both experimental and mathematical, that forced physicists more and more to make peace with a micro-universe in which uncertain and unpredictable events were the rule.

It was not an idea greeted with joy in all quarters. Many brilliant men found the concept philosophically uncomfortable, even repugnant, and sought valiantly to overcome it. But to modern physicists the principle of uncertainty is a key idea in our present-day understanding of the structure and behavior of the micro-universe. In the next chapter we will see how this idea assumed greater and greater importance as physicists all over the world began doggedly working to probe the inner structure of atoms and to try to develop a valid "model" of atoms and other elementary particles in the micro-universe—to visualize them as they really exist.

# CHAPTER 25

## *In Quest of the Atom: Measurements and Tools*

With the discovery of the phenomenon of radioactivity in the late 1890s, and with the subsequent discovery of the thoroughly unorthodox and unsettling manner in which radioactive atoms decayed and transformed into other atoms, physicists all over the world considered it imperative to develop a better picture of what an atom really was and how it was really constructed than had previously been achieved. Up until then, vague and hazy pictures and crude models of the structure of atoms had at least been serviceable, but now there was a pressing need for detail. With uranium and radium atoms known to be spontaneously decaying and hurling fragments of themselves out into space, it was not enough to speculate that such an atom must be constructed of positively and negatively charged segments arranged together more or less loosely in a ball. Physicists needed to know *exactly* what the internal structure of the atom was. And so a valiant effort was mounted to try to visualize a model or imaginary descriptive picture of the atom's detailed inner structure which might still be consistent with the actual laboratory observations of atoms and their behavior that had begun pouring in.

J. J. Thomson and Ernest Rutherford were, of course, among those on the front lines of the search. Their discovery of the existence of tiny, negatively charged "chips" of atoms—which they called electrons—was a great step forward. This fact alone made it clear that atoms must be composed of separate or different positively and negatively charged components which could be wrenched apart under certain circumstances. This, of course, meant that Dalton's picture of atoms as simply tiny, solid, indivisible marbles had to be discarded and some other picture of what atoms were like substituted in its place.

Thus in 1898 J. J. Thomson speculated that atoms might really consist of a central dense "droplet" of positively charged material which had tiny negatively charged "chips" or electrons embedded in it like cherries in a fruit cake. He suggested that there ought to be just enough negatively charged electrons in any given atom perfectly to balance the number of positive charges present, under ordinary circumstances, so that the atom would normally be electrically neutral. But if something came along to pick one or more of the cherries out of the fruit cake. Thomson speculated, then the atom that remained would not be neutral, but would have become a positively charged ion. He considered that the smallest possible portion of the atomic component that carried a positive electrical charge, the so-called proton, would have to be the part that carried exactly one unit of positive charge. Thus the smallest possible atom, the hydrogen atom, would have to be composed of a single proton "droplet" with a single electron "chip" embedded in it like the plum in Little Jack Horner's pie. We see this early concept of the shape of an atom diagramed in Figure 42a.

In 1903 an alternative model atom was proposed by Philippe Lenard, who suggested that atoms might perhaps be made up of pairs of negative

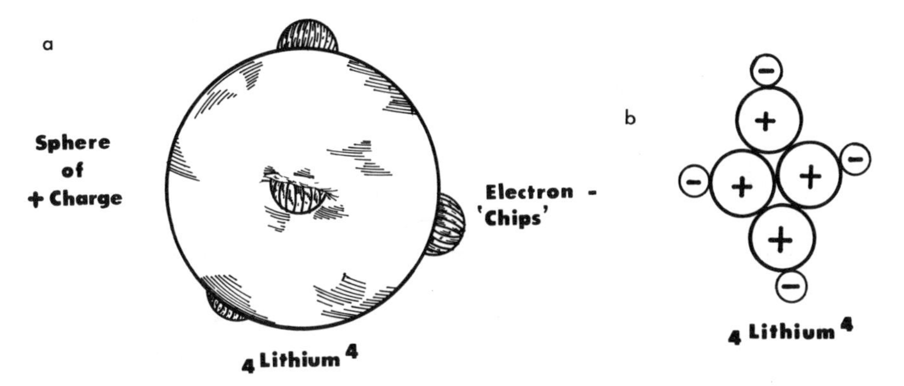


*Fig. 42a and 42b* Diagrammatic representations of two early "models" of the atom. In J. J. Thomson's model (42a) the electrons were considered to be small negatively charged "chips" embedded in a large spherical "droplet" of positively charged material. Philippe Lenard's concept (42b) reflected an earlier hypothesis that all atoms might be composed of aggregations of hydrogen atoms (i.e., proton-electron pairs).

and positive charges floating, as it were, hand in hand in space. Interestingly enough, this model of the atom fitted in surprisingly well with an idea first proposed over a century earlier by a Britisher named William Prout, who had suggested with uncanny prophetic insight that all atoms of elements heavier than hydrogen might simply be composed of various numbers of hydrogen atoms (now known to be proton-electron pairs) packed together into a ball (see Fig. 42b).

Prout's contemporaries had scoffed at his idea because they had found that if they took the measured atomic weight of a single hydrogen atom as 1.000, virtually none of the other heavier elements whose atomic weights were compared with hydrogen came out to exact multiples of 1.000, as one would expect if those atoms were merely aggregations of hydrogen atoms. The atomic weight of neon, for example, had been measured at 20.13. Chlorine had an atomic weight of 35.46; and copper, an atomic weight of 60.54 (using oxygen at 16.000 as the standard of measurement). Since it seemed unlikely that 18/100 or 46/100 or 54/100 of a hydrogen atom could exist, Prout's idea did not seem to jibe with observed facts and had almost been forgotten by the early 1900s.

Then J. J. Thomson and another British physicist, F. W. Aston, made a discovery which explained how the atomic weights of various elements might have fractional values when measured in comparison to hydrogen atoms. In their early investigation of the inner nature of atoms, these men found that many elements were made up of mixtures of atoms having the same atomic numbers (that is, the same location on the periodic table of the elements, in which elements were listed according to their order of increasing massiveness) and also had precisely the same chemical and physical properties, but had slightly differing masses. In such mixtures of atoms of the same element, the atomic weight of each individual atom was an exact integral number, but when the *average* atomic weight of such a mixture was determined, it would come out as a fraction.

Thus neon gas found in nature was discovered to be a mixture of three slightly different forms or "isotopes" of neon atoms, all identical in their chemical nature, but some with an atomic weight (i.e., comparative mass) of 20, some with an atomic weight of 21, and some with an atomic weight of 22. Because of the ratio in which these three isotopes were found to be mixed in nature, the *average* atomic weight of neon atoms came to 20.13, which simply meant that most of the atoms in a natural sample were neon 20 isotopes mixed with a few atoms of neon 21, and with a rare individual neon 22 atom thrown in for good measure.

In any event, this suggested that each individual atom of neon *did* have an atomic weight which was a whole-number multiple of the atomic weight of hydrogen, so that Prout's idea that all atoms were made up of aggregations of hydrogen atoms began once again to look more respectable. This, then, was the model of the atom championed by Lenard in 1903 (see Fig. 42b).

Unfortunately Lenard's model very soon ran into trouble as a result of some new experimental techniques that Thomson's assistant, Ernest Rutherford, had devised for studying the inner structure of atoms. It was obvious from the behavior of natural radioactive elements that intrinsic or internal forces within certain atoms could split away fragments of those atoms, hurling out either electrons or even larger and more massive chunks such as alpha rays (helium ions). Powerful electric fields also seemed able to split apart negative and positive charges in atoms, creating positive ions and negative ions, as we saw in the Crookes tube. It seemed to Rutherford that it should be possible to produce artificial fragmentation of atoms if some way could be devised to fire some sort of atomic "bullet" at them; any atoms that could be "smashed" or fragmented in this way could then be carefully studied and would perhaps supply new knowledge of just exactly what the internal structure of an atom really was.

But what could the physicist use for his atomic "bullet"? It occurred to Rutherford that precisely such an atomic bullet already existed: the alpha particles that were expelled spontaneously from decaying radioactive atoms. After much work Rutherford finally designed an ingenious device, actually remarkably similar to Becquerel's lead-block device for "aiming" radioactive particles thrown off by uranium, which would allow only a beam of alpha particles to escape from a lump of radioactive matter in one specific direction. Using this device to aim alpha particles, Rutherford

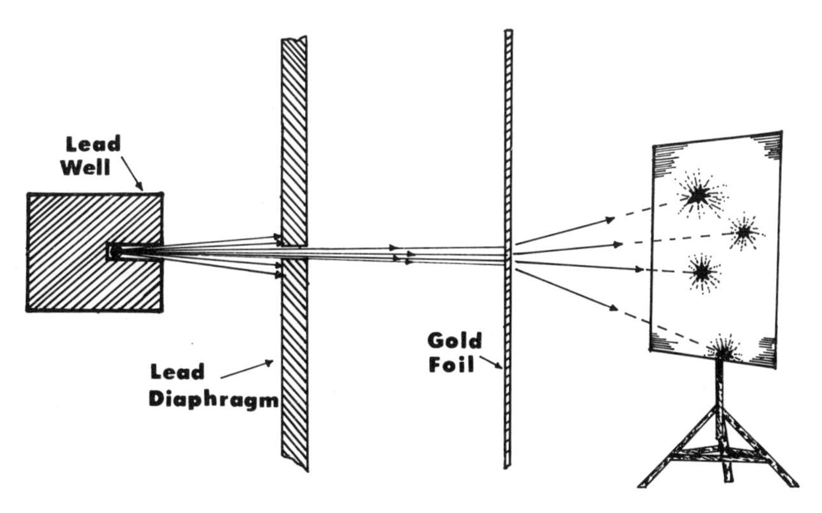


*Fig. 43* Diagram of Rutherford's early bombardment experiments. Alpha particles screened and "focused" through a lead diaphragm struck gold foil. Many passed through in straight line to register on fluorescent screen, but others were sharply deflected from their straight-line paths.

began using these massive and comparatively slow-moving "decay particles" to bombard such things as the atoms of gold in a thin piece of gold foil (see Fig. 43). Later such alpha particle bullets were "fired" through tubes of gas known to be easily ionized, as well as through other materials.

The results of Rutherford's experiment were little short of astounding. Most of the alpha-particle bullets moved serenely through the target substance to strike the fluorescent screen at the far end of Rutherford's device. These alpha particles were themselves undisturbed in their path, but left streams of ionized atoms behind them, apparently knocking electrons off and sending them flying in all directions. It was much as if an alpha particle bullet were a twenty-ton truck plowing blindly through a collection of tricycles: The tricycles would be scattered all the way from here to Guinea, but the truck would continue to run serenely on its way without even recognizing that it had hit anything. Rutherford was not overly surprised at this; alpha particles were known to be relatively huge in comparison to the tiny electrons, so that in any encounter with such minuscule tidbits of matter as electrons the alpha particles would not be expected to be jostled very much. The thing that was *not* expected, however, was that some few of the alpha particles *were* thrown off their courses, some veering off slightly, some veering at much sharper angles, and some indeed actually being hurled back in the direction from which they came!

The only plausible explanation for this strange behavior on the part of a few alpha particles was to assume that they had actually collided (or appeared to collide) with the main body of atoms in the target material. Indeed, from the violence with which some of these atomic bullets were thrown off course, it seemed as if they must have been actively repelled or hurled away by close encounters with the much more massive atoms in the target material.

But these were not classical collisions in the mechanical sense of billiard balls colliding with each other, with the results of those collisions governed by Newton's laws of motion and inertia. Rather, these errant alpha particles behaved as if they had been actively shoved aside or even thrown back in the direction from which they came by some kind of force far more powerful than the force of any mechanical collision. These positively charged alpha particles behaved as if they had encountered large and compact bundles of positively charged material, and instead of colliding head-on had been forcibly thrust aside by some powerful force such as that of electrostatic repulsion.

These experiments led Rutherford to a quite new and different view of how atoms must be constructed. If atoms were merely a collection of heavy, positively charged particles and very light, negatively charged particles all scrambled up together, they ought normally to be electrically neutral, and thus would not be expected to repulse a positively charged particle approaching them on collision course. Yet this was precisely what seemed to be happening in the alpha particle bombardment experiments. Rutherford realized that this strange behavior of alpha particles on collision course with much more massive atoms could only be explained if the parts of the atoms bearing negative charges and positive charges were considerably separated from each other in space.

Thus Rutherford evolved a model of the atom in which all of the positively charged particles making it up were packed together into a dense and compact "nucleus," with the whole nucleus carrying a very strong positive electrical charge. This positively charged nucleus was then surrounded at a significant distance by a swarm of far tinier negatively charged electrons, which spun around the nucleus in orbital paths at just such a distance and with just such a speed as to perfectly counterbalance the force of electrostatic attraction seeking to tug them into the nucleus of the atom (see Fig. 44).

This Rutherford model of the atom, with its compact nucleus of positively charged protons and its so-called planetary, negatively charged electrons moving in orbit around the nucleus at a distance, was the first

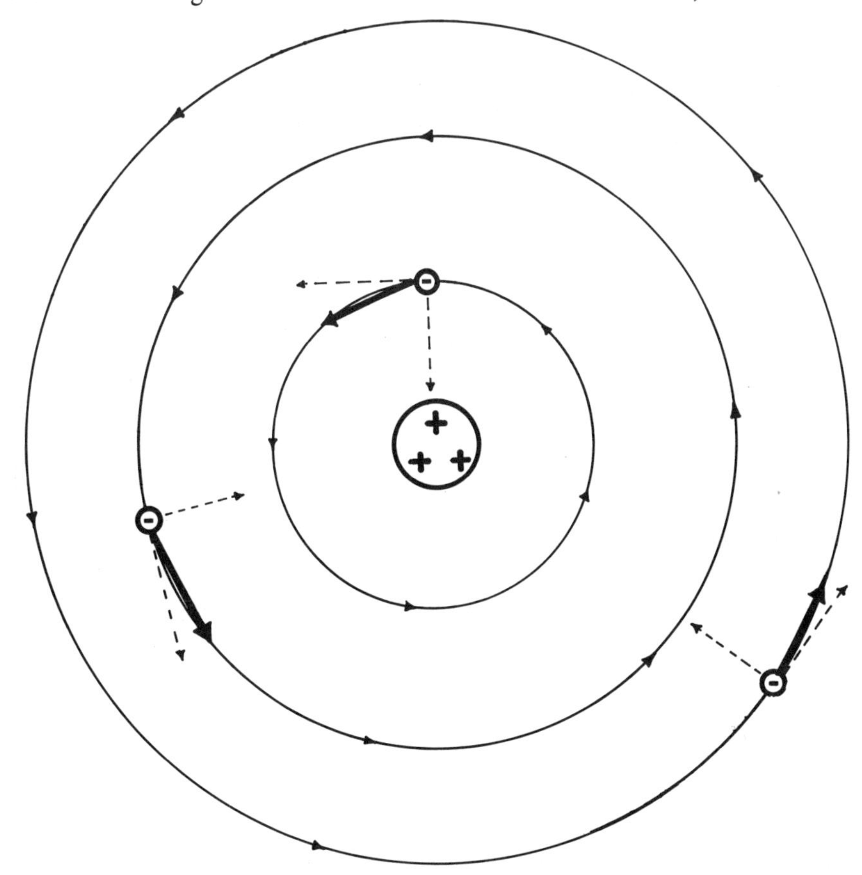

*Fig. 44* Rutherford's concept of the atom as a positively charged nucleus with negatively charged "planetary" electrons. Electrons were thought to remain in orbits much as the earth remains in orbit around the sun, as a result of balance between in-pulling attraction and inertial (tangential) forces, as shown by arrows.

visualization of the structure of the atom that even came close to the modern view held by physicists today. Even now, in many ways, it remains the most useful *visual* representation of what atoms would "look like" if we had any way of visualizing them directly. In the ensuing fifty years, the *details* of the structure of Rutherford's atom have been modified repeatedly, and most modern physicists are quite certain that the components of Rutherford's atomic model are, in fact, not even remotely like what Rutherford imagined. But even so, the over-all picture of Rutherford's atom still is a useful visualization to which we will return again and again as we see the ways in which it has been modified by new discoveries, and why. Historically, it was a great jumping-off place which led physicists through a series of startling and sometimes flatly incredible discoveries to their present-day understanding of what an atom is "really" like.

But even in these earliest studies of the internal structure of atoms, Rutherford and other physicists were plagued by two recurring headaches. First, they were handicapped by the difficulty of trying to visualize and comprehend the actual *size* of subatomic particles that were really so tiny that they were beyond man's power to comprehend. To work with these particles and study them at all, some comprehensible basis for *measurement* of tiny physical dimensions and tiny intervals of time were badly needed. At the same time, physicists desperately needed some kind of tools or techniques that would permit them at least *indirect* observation of what was going on in the micro-universe they were studying. If things were happening to particles so tiny they could never be directly visualized, then there had to be some sort of device to observe these particles indirectly.

Thus, the early years of study in nuclear physics evolved a system of measurement of tiny dimensions which is often unfamiliar and uncommonly confusing to the nonscientist, and those early years saw the development of some ingenious devices for observing indirectly what was going on in the micro-universe. To understand the exciting wave of discoveries that began in the early 1900s, we must consider for a moment this means of measuring dimensions and the tools that were used for observation.

## HOW TINY IS TINY?

Earlier in this book we found ourselves in trouble trying to visualize and comprehend the enormous sizes and distances in the macro- and mega-universe—the sizes of stars, the distances of galaxies, or the great reaches of space which seem to separate us from the mysterious quasars. In discussing the micro-universe we face precisely the same problem but at the other end of the scale of magnitude. So far we have used terms such as "extremely small," "tiny," or "minuscule" in discussing such structures in the micro-universe as molecules, atoms, protons, or electrons. But these words give us no real comprehension of just how unimaginably tiny these structures really are.

In point of fact it is doubtful whether even the most brilliant and experienced nuclear physicist can really comprehend the scale of smallness in which he is working; the human mind is simply not equipped to imagine structures in the micro-universe on any kind of realistic scale. We could use any number of imaginative comparisons to help us visualize the dimensions of this world of the inconceivably small, but such attempts invariably fail. It may well be that it is this very failure which makes the micro-universe of nuclear physics seem to be such an incomprehensible Never-land to most people.

We can easily see why this is so. In our everyday life we depend constantly upon measurements and comparisons of all kinds to maintain our orientation. We are not confused when a mountain that we know is 20,000 feet high appears to be tiny at a distance, for we automatically compensate for distance and perspective when we look at it. Nor are we ordinarily confused about whether it was a mouse or an elephant that we just saw dashing across the living room floor; one is big and the other small, and all we have to do is look at the animal—or the wreckage it leaves behind—to tell which was which.

But in our everyday life there are limits to our ability to comprehend size and magnitude at both ends of the scale. These limits are set by our senses and our experience. We can tell at a glance whether there are one, two, three, or even four or five people in a room without counting them, and some can even correctly identify six people in a room at a glance, but to see seven people in a room is to see "a crowd" which we must count to be sure how many are present. In the case of larger numbers our impressions become more and more inaccurate without actually counting. Few of us could estimate whether an auditorium holds 200 people or 400 people, and when it comes to guessing the number of beans in a five-gallon jar, offhand estimate will vary all the way from several thousand to several *million!* By using some kind of computational scale (such as the seating capacity of a large football stadium) we can comprehend how many people 70,000 or 80,000 are, but when we get up to 100,000 or so we reach an order of magnitude that has no real meaning to us.

Things are just as bad at the other end of the scale. We know that a fly is smaller than a bumblebee, and a gnat smaller than a fly, but a gnat's whisker is a little too small for us to comprehend realistically. As we get into the realm of the very small, devices such as magnifying glasses, microscopes, or electron microscopes can create enlarged images for us to see, so that a single streptococcus may appear to be a fairly husky fellow, but to all intents and purposes our normal sensory experience tells us that a streptococcus is totally invisible (that is, unimaginably small), and we believe that it exists at all only because doctors can prove that these "invisible" fellows can cause very uncomfortable throat infections. Most of us cannot really comprehend the difference in size between an amoeba (which can *almost* be seen with the naked eye), a streptococcus (which can't), and an Asian flu virus particle (which can barely be seen at all, even with the finest electron microscope in the world). All three are "invisibly small" entities as far as our senses are concerned, yet on a comparative scale an amoeba, a streptococcus, and a virus particle might compare roughly in size to a woolly mammoth, a caveman, and a flea.

When it comes to discussing atoms and elementary particles, we are even further at sea. These objects or entities are so much tinier than even a virus particle that there is no comparison whatever. We know that there will never be any kind of visual instrument invented that will allow us actually to observe such tiny entities. The only proof of their existence that we have at all is based upon the effects such particles can have upon things that we *can* observe, and enormous effort has gone into the design of ingenious devices to enable physicists to observe these particles *indirectly*.

What do we mean by "indirect observation"? When a doctor examines a 250-pound piano-mover who is laid out flat in bed with a high fever, a pounding headache, and aching muscles, this might constitute fairly good circumstantial evidence that *somewhere* inside that piano-mover some kind of virus is lurking, even if nobody can see it. Similarly, when the physicist observes and photographs a tiny row of water droplets appearing in a cloud chamber quite suddenly, he can be confident that these droplets have condensed around atoms of gas which have suddenly been stripped of electrons and thus ionized, and he considers this good evidence that *some* invisible particle has just rushed by, knocking electrons off atoms. But just as examining the 250-pound piano-mover is the closest that the doctor is likely to get to seeing the flu virus itself, the line of water droplets in a cloud chamber is the closest the physicist can come to seeing the actual particle that caused the water droplets to form. He has to deduce from what he observes precisely what particle it must have been that rushed by earlier and caused the droplets to form.

But even if there is no way we can comprehend or visualize the actual smallness of things in the microworld of physics, we still need some scale of measurement for comparing the size of one microworld particle to that of another. Ordinarily scientists use grams and kilograms as units of weight, centimeters and millimeters for units of linear distance, and seconds for units of time. These work splendidly in our everyday world, but in the micro-universe we are invariably speaking of inconceivably minute fractions of grams, tiny fractions of millimeters, or incredibly small fractions of seconds, and we need some way to gauge (comparatively) how large or small those fractions are.

Of course, in the everyday world there is little practical difference between 1/10,000 second and 1/1,000,000,000 second; either interval of time is altogether too short to imagine. But in the micro-universe, time differences such as these can mean the distinction between an elementary particle which exists for a fantastically "long" period of time and another particle which has hardly any period of existence at all before it changes spontaneously into something else.

It would be perfectly possible to express these tiny measurements as ordinary fractions or decimal fractions of larger units. The trouble with this is that every time we expressed such measurements in fractions we might end up with page upon page of zeros—a great deal of bother counting them, and dull reading into the bargain. To simplify things—really just to get away from writing all those zeros—scientists commonly use *negative exponents,* standing for fractions of multiples of ten, to represent tiny fractions of larger units of distance, mass, or time. Thus we could contrast 1/10,000 second and 1/1,000,000 second not by writing down a confusing number of zeros in an ordinary fraction, but simply by comparing the number of times that ten is multiplied in the fraction. This method is simple, but it has one built-in source of confusion: we always have to bear in mind that an increase of one in a negative exponent represents a *tenfold* increase in the smallness of the fraction.

How does this actually work in practice? First consider how exponents can be used to compare degrees of largeness of things. Ten people ($10^1$) might be a comfortable number of people to have in for a summer barbecue. Ten times those ten people or 100 people ($10^2$) is the number we might find watching an obscure art film in some very small theater on a rainy evening. Ten times $10^2$ people or $10^3$ people (1,000) might be the number on that same evening watching a Broadway show that has recently opened to good reviews. We might find 10,000 people ($10^4$) filling a hall at an appearance of the Beatles, whereas 100,000 people ($10^5$) would be the number of people watching the Army-Navy game at Municipal Stadium in Philadelphia. The number of people packing St. Peter's Square in Rome during the election of a new pope might be 1,000,000 ($10^6$), while 10,000,000 ($10^7$) represents the total number of people in all of greater New York City and part of metropolitan New Jersey put together.

The total combined population of Germany and France is 100,000,000 people ($10^8$); 1,000,000,000 people ($10^9$) is approximately the combined population of all North and South America taken together, with all of Western Europe thrown in for good measure. And 10,000,000,000 people ($10^{10}$) is the total *expected* population of the entire world in 1975 at current rates of population growth.

Obviously there is a great deal of difference between $10^9$ people and $10^{10}$ people, even though the two numbers seem deceptively close to each other. Even at the level of $10^{10}$ we are not really running into trouble with zeros, but we can see that the difference between $10^{20}$ and $10^{21}$ is enormously greater than the difference between $10^9$ and $10^{10}$, and also that we would begin bogging down in zeros if we were to try writing such huge numbers out at full length without the assistance of exponents.

Precisely the same thing is true of *negative* exponents, which are used to represent successively smaller fractions of a given unit. One-tenth of a centimeter can be written as $10^{-1}$ centimeter. A hundredth of a centimeter then becomes $10^{-2}$ centimeter, one thousandth of a centimeter becomes

$10^{-3}$ centimeter, and so on. And as the negative exponent increases unit by unit the comparative smallness of the fraction increases by tenfold leaps.

## MEASUREMENTS IN THE MICRO-UNIVERSE

We can see how useful negative exponents of ten are in expressing measurements in the micro-universe with a simple example. We know that the basic unit of length or distance commonly used by scientists is the meter—approximately three inches longer than an ordinary household yardstick. For measuring small objects in our everyday world a meter is divided into 100 equal segments called centimeters, a distance of approximately ⅜ inch, or one inch is equal to approximately two and one-half centimeters. Measurements of length or distance in the micro-universe are thus ordinarily expressed in fractions of centimeters.

Now if you will press approximately 150 pages of this book tightly together and measure the thickness with a centimeter scale, you will find that 150 pages are very close to one centimeter thick. Fifteen pages held together would have a thickness of 1/10 centimeter ($10^{-1}$)—commonly called a millimeter or 1/1,000 meter. A single page in this book would then have an approximate thickness of 1/15 of 1/10 centimeter. We could express this thickness as 1/150 centimeter if we chose, or we could call it $1.5 \times 10^{-2}$ centimeter.

Of course this book is printed upon comparatively thick paper. Imagine a dictionary or a Bible with very, very thin paper, in which ten pages would be no thicker than a single page of this book. Each of those thin pages would then represent 1/10 of 1/15 of 1/10 centimeter—1/1,500 centimeter or $1.5 \times 10^{-3}$ centimeters. At this point we are beginning to get down to microscopic measurements: If we could somehow manage to shave away 1/10 the thickness of one of those thin pages, the resulting shaving would be 1/15,000 centimeter or $1.5 \times 10^{-4}$ centimeters.

This measure of $10^{-4}$ centimeters (1/10,000 centimeter) is a particularly useful unit for the measurement of microscopic objects and has a special name: the *micron*. With careful technique it is possible to etch glass slides for high-power microscopes with lines that are exactly 1 micron apart; such grids can be used in doing accurate blood counts, or in measuring the length of bacteria. This is obviously a very small unit of measurement, but we can see graphically the enormous difference in smallness between $10^{-1}$ centimeters (the thickness of a sheaf of fifteen ordinary book pages) and $10^{-4}$ centimeters (a shaving so thin that a high-powered microscope is needed to see it at all). An object $10^{-5}$ centimeters long is 1/10 again smaller than a micron—the diameter of certain very large virus particles.

But in the micro-universe of subatomic particles, atoms themselves vary from $10^{-8}$ to $10^{-9}$ centimeters in diameter. The nuclei of atoms are from $10^{-11}$ to $10^{-12}$ centimeters in diameter, while electrons are some $10^{-14}$ centimeters in diameter. Again, the difference between something $10^{-8}$ centimeters in diameter and something else $10^{-9}$ centimeters wide seems deceptively small, but we are speaking of a *tenfold* decrease in magnitude with each unit increase of negative exponents of ten.

The measurement of tiny fractions of seconds can be expressed in the same fashion. Virtually all other scientific measurements (weight, mass, distance, etc.) in the world of science are based on a decimal system but our measurement of time, interestingly enough, is not. Our time measurements are based upon the earth's orbital course around the sun, and the commonest unit of time, the second, is not a decimal but a rather awkward fraction of a day, which is an even more awkward fraction of a year. Nevertheless when we get down to measuring intervals of time in fractions of a second, using negative exponents of ten, time measure then *becomes* a decimal system based upon the second as the unit so that shorter and shorter intervals of time are expressed as increasing negative exponents of ten exactly as in the case of other scientific measurements.

## DIRECT AND INDIRECT MEASUREMENTS

Obviously in studying objects and events in the micro-universe, scientists face a major problem in detecting the presence of such tiny particles or measuring such tiny intervals of time, for there is no means of direct observation or measurement of either. Invariably such measurements can only be made indirectly, for human senses are simply not keen enough to do the job, and no measuring or magnifying instruments exist which are capable of sharpening man's senses enough to make direct observation and measurement possible.

No ordinary high-power light microscope, for example, can ever enlarge or "resolve" any object smaller in diameter than the wavelength of visible light; light waves would simply leap over or pass around such an object as if it were not there. At the very best we might be able to distinguish rather fuzzily that *something* very small was probably there because we could observe that some light waves were being disturbed and others were not, but we would never be able to distinguish precisely what the object was nor precisely what it looked like.

If we use an electron microscope which employs a beam of electron waves of much shorter wavelength than light, then far smaller objects can be clearly visualized. Large molecules of certain kinds of protein, for example, can be individually seen by means of electron microscope technique. Virus particles can be brought into clear "focus," as can various structures within the nuclei of living cells. But when we get down to the inconceivably tiny level of individual atoms and subatomic particles, there is no longer any possibility of direct visualization at all, for these particles are too small for even the very shortest wavelength electromagnetic waves to resolve, and we face the increasing problem that any attempt to use shortwave electromagnetic waves to observe these minuscule particules would cause increasing degrees of alteration or change in the particles themselves, so that the closer that we came actually to visualizing them the more distorted the picture that we might see.*

Thus physicists had no choice but to settle for indirect measurement, and much of the most exciting work that has been done to probe the interior of atoms has been made possible only with the use of ingenious devices that could provide indirect evidence of the presence of very tiny particles, or very tiny quantities of energy. In order to see more clearly how some of the amazing discoveries of the last fifty years were made, we should pause to see what some of these devices are and how they actually provide the information physicists need about the micro-universe.

Obviously one of the earliest devices used was the ordinary silver emulsion photographic plate. High-energy particles or microwaves striking molecules of a silver compound in the emulsion of a photographic plate stimulate tiny local chemical reactions in which free silver atoms are formed. Thus, the fact that such particles or microwaves were present can later be demonstrated by the presence of tiny flakes of precipitated silver on the film; the film appears exactly as if it had been exposed to visible light. Since photographic plates were already available at the beginning of this century, they were used from the very first as a means of detecting atomic and subatomic particles, a somewhat crude method, but still useful. Even today workmen and scientists likely to be exposed to deadly penetrating radiation wear "badges" containing small bits of photographic film which are removed and developed from time to time to determine if the person wearing the badge has unwittingly encountered more radiation than he thought he had.

Another early instrument for detecting the presence of elementary particles, designed and used by Ernest Rutherford, was a device called a scintillator. Essentially a scintillator is nothing more than a screen covered with some material such as zinc sulphite, which tends to "fluoresce" or give off a glow of visible light whenever its atoms are "excited" by being hit by free-moving particles such as alpha particles or electrons. When such a collision occurs on the scintillator screen, a tiny pinpoint of light appears momentarily, and this pinpoint glow or "scintillation" can be detected with the use of a microscope. Rutherford found the scintillator to be extremely useful in tracking down various results of particle bombardment experiments that he conducted, as well as providing a means of measuring the ability of magnetic or electrical fields to deflect moving particles from

* We will consider this extremely important problem of observation in greater detail in a later chapter.

a straight-line path, and the essential principle of the scintillator (with modifications, of course) is still useful today.

One of the men who helped Rutherford design the scintillator for use in his laboratory was a young German physicist whose name has become a household word today on account of another device that he invented for measuring and counting high-speed elementary particles. Hans Geiger's instrument is essentially a tube with an insulated wire running down the center and a high electrical tension between the wire and the wall of the tube. When an alpha particle or other elementary particle shoots through the tube, knocking electrons off atoms of gas in the tube and thus creating ions, the multitudes of ions so created join together to create a pulse of electrical charge in the wire. Thus when the wire is connected either to a mechanical counter or to a loudspeaker, such electrical pulses can easily be detected. Each time an ionizing particle enters the tube and acts as a trigger, another electrical pulse passes through the wire. These pulses from the "Geiger counter" can then be made to turn the dials of a counting machine just as the usher at a church door clicks a hand counter, or they can be made to light up a small neon light or to cause an audible "click" in a loudspeaker.

The number of pulses occurring per second is an accurate measure of the number of radioactive particles passing through the tube of the counter, a useful enough form of measurement in itself. But with certain modifications, Geiger counters can even be made to measure the original energy of each particle that causes a pulse to occur. Thus, this ingenious device has proved to have a wide variety of uses. It can be used, for example, to detect and locate sources of radioactivity and to gauge the amount of radioactive material in a given vicinity. Small and portable Geiger counters are invaluable field instruments for uranium hunters. In one form or another Geiger counters can be used to monitor the results of bombardment experiments, and to provide an immense amount of useful information about the atomic and subatomic interactions going on.

Two other devices have improved even further the physicists' ability to "see how elementary particles behave," that is, to observe the effects these particles bring about and then, by studying these effects, help the physicists to piece together what the particles must have been, where they must have come from, and so forth. One of these instruments was developed in 1911 by British physicist C. T. R. Wilson and is known as a "Wilson cloud chamber." The operation of the cloud chamber depends upon the fact that ionized gas molecules can act as "seeds" around which water vapor in a cold saturated atmosphere can condense into droplets of water. By filling a closed container with a transparent lid full of water vapor and then chilling the chamber and reducing the pressure inside it by means of a tightly fitting piston, Wilson created in effect a "goldfish bowl" filled with cool moist air—exactly the same sort of circumstances that leads to the condensation of water vapor into clouds and rain in our own atmosphere. Then when alpha particles or other radioactive particles capable of causing ionization of gas pass through the vapor-filled chamber, they leave behind a trail of condensed water droplets in their wake, seeded by the gas ions they have created along the way. These long, thin streaks of fog within the cloud chamber can easily be illuminated by side lighting and then photographed through the transparent lid of the chamber, providing permanent photographic records of the trails that ionizing particles leave behind them.

Wilson's cloud chamber became a commonplace tool in the nuclear physicist's laboratory from the time the first one was built, and these devices are still in use today. But the cloud chamber also has a number of drawbacks. Because it is filled with gas, the behavior of radioactive particles in collisions with atoms or other particles in the gas chamber are often hard to study, since such collisions are relatively few and far between. The fog trails are visible only during the fraction of a second in which the air pressure in the chamber is being reduced, and the whole device has always been clumsy to set up and operate.

In 1951 Donald Glaser, of the University of Michigan, devised an improved kind of observation chamber that worked on the same general principle as Wilson's cloud chamber, except in reverse. He filled a glass chamber full of a volatile liquid, ethyl ether, under pressure, and then intermittently reduced the pressure by means of a tightly fitting piston. Here again an ionizing radioactive particle passing through the chamber can be seen to leave a track behind it as it travels through the liquid—but in this case it leaves a track of gas bubbles in the liquid instead of water droplets in a gas. Modern physicists use such "bubble chambers" to great advantage in studying the behavior of charged subatomic particles, setting the bubble chambers up in powerful magnetic or electric fields that can make the particles bend away from straight-line flight in various ways according to their charges, their velocities, and their masses.

Finally, it became increasingly clear that in order to study what was going on in the nucleus of atoms, atomic projectiles were going to be needed that had far greater energies and traveled at far higher velocities than the alpha particles produced by the decay of natural radioactive materials. Thus much time, thought, and money have been invested in the design and construction of a variety of "particle accelerators," huge machines capable of trapping charged particles such as alpha particles, protons, or electrons in powerful electromagnetic fields and accelerating them to extremely high velocities before releasing them to bombard target atoms at high speed and with high energies.

One of the earliest of these "atom smashers" (so called because they were designed to bombard the nuclei of atoms with such high-energy atomic bullets that the target nuclei would be literally split apart) was the "cyclotron" built by Ernest O. Lawrence at the California Institute of Technology and set in operation in 1939. In the cyclotron the atomic bullets were accelerated in an expanding spiral to very high velocities by means of shifting magnetic and electric fields until they were finally released at high energy at the outer rim of the spiral and aimed at the target materials. Subsequent to the construction of the first cyclotron, a whole armamentarium of increasingly powerful particle accelerators have been devised and built, some working on a principle of cyclic acceleration, some accelerating particles in a straight line, and so forth. But nuclear physicists, never satisfied, even today become dreamy-eyed at the thought of all the things that they *could* accomplish in their studies if only someone would build them bigger and more powerful particle-accelerating machines.

In this chapter, we have looked at some of the devices that were invented as the working tools of the earlier nuclear physicists. As we go along we will discuss other devices invented for very specific purposes; the problem of trapping subatomic particles into revealing their presence, as we will see, often required the highest ingenuity on the part of physicists. But as these tools and instruments were developed, the search was progressing and the fascinating tale of nuclear discovery was unfolding in the world of physics.

# CHAPTER 26

## *The Puzzle of Energy Quanta*

In many ways the concept of the structure of an atom that arose from Ernest Rutherford's work in the early 1900s was tidy and comfortable. By using the alpha particles emitted from natural radioactive substances as "atomic bullets" with which to bombard a variety of target atoms, and then studying the results of these bombardments indirectly by the use of such devices as the scintillator, Rutherford had found conclusively that the older concept of the atom as a loose intermixture of positively and negatively charged particles could not possibly be valid. Atoms were indeed made up of two different particle components—positively charged protons and negatively charged electrons—in such groupings that the numbers of positive and of negative charges were in balance under ordinary circumstances, so that atoms were normally electrically neutral. But it seemed clear from Rutherford's work that all of the protons in a given atom must be packed tightly together into a dense and tiny core or "nucleus," while still tinier negatively charged electrons were busily racing around the nucleus in orbits at quite some distance away. Indeed, if Rutherford's model of the atom were correct, it appeared that the "solid" material of the universe was in fact anything *but* solid, if one could visualize it on a submicroscopic scale; most of any given atom, even an atom of a dense and heavy metal such as gold, was empty space.

Of course there were also some rather awkward aspects to this picture of the atom. Although protons and electrons seemed to carry individual minimum electrical charges of opposite nature but identical magnitude, the masses of these two atomic components were greatly out of proportion. Protons had almost two thousand times the mass of electrons, so that virtually all the *mass* of a given atom lay concentrated in the densely packed nucleus. There appeared no ready explanation for this curious disproportion.

Even more perplexing, no one could explain how an atom with several positively charged protons in its nucleus could avoid being split apart by the electrostatic force of repulsion between those protons lying so close to each other. Clearly there had to be some kind of "nuclear binding force" holding these similarly charged particles together that was powerful enough to overcome electrostatic repulsion, but no one had ever heard of any such nuclear binding force before. And again, no one could explain why the "planetary electrons" in their orbits around the nucleus were not inexorably drawn into the nucleus by the electrostatic attractive force between the positive charges on the protons and the negative charges on the electrons. Those electrons must be moving in their orbits very feverishly indeed in order to create sufficient centrifugal force to balance the inward-tugging electrostatic attraction of the nucleus, and even then it seemed unlikely that they could hold their position in orbit when the mutually attractive electrical charges were in such close proximity.

But if Rutherford's atom presented shortcomings on the one hand, it also explained some things very nicely on the other. For one thing, it provided a better explanation of how atoms might join together to form compounds. Atoms of very active elements—that is, of elements that tended to form many differing kinds of compounds easily—could be visualized as coming closely enough together with atoms of other elements so that two atoms might actually share one or two or more of their roving electrons between them. Normally such atoms joined in partnership might be held in close proximity to each other, for example, in the form of a crystal, if the compound they formed was a crystalline solid such as a salt. But if crystals of such a compound were dissolved in water, the partner atoms might tend to pull apart somewhat, so that the shared electron or electrons might be held closer to one atom of the pair than the other. Then the atom with possession of the extra electron or so would have an excess of electrons and become a negatively charged ion, while the atom bereft of the shared electron or so would have an excess of positive charges in its nucleus and become a positively charged ion.

Rutherford's concept of the atom also explained why certain marauding high-speed particles such as alpha particles could knock electrons off atoms so readily as they traveled through a gas but would so rarely actually approach the solid nucleus of an atom. In Rutherford's atomic model, the electrons were covering large areas of space around the atomic nucleus, with their orbits going in all different directions, so that an alpha particle might easily run into one, so to speak, as it went along its way, whereas the nuclei of target atoms were compressed into such small volumes of space and were actually so far distant one from another that an alpha particle might have to travel a comparatively enormous distance before ever coming anywhere close to one of them.

Naturally, Rutherford's work set off a vigorous chain reaction of investigation as alpha particles and streams of high-energy electrons were used to bombard the nuclei of many different kinds of atoms and physicists sought to learn more about the kind of forces holding the atomic nuclei together. It was soon proved that the nuclei of atoms could actually be broken apart by such artificial collisions, with identifiable chunks being split off according to certain definite patterns. It also became clear that atoms could be changed or "transmuted" from atoms of one element to atoms of another simply by breaking off part of a nucleus by collision with a bombarding particle, or by adding a chunk to the nucleus. Soon it had become abundantly clear that the physical and chemical properties of any atom depended greatly upon the number of proton particles in its nucleus and the number of electron particles swarming around it in space. These two subatomic particles somehow seemed to hold the key that distinguished atoms of one element from atoms of another, and atoms of one element could be altered into atoms of another merely by rearranging the number of protons and electrons they possessed.

But while Rutherford and many other physicists proceeded with this work, there was quite a different type of investigation under way with regard to the question of how atoms were made. In fact, a totally different and quite unexpected "elementary particle" was in the process of being identified concurrently with Rutherford's work—a particle so tiny that to all intents and purposes it seemed to have no mass at all; a particle that had no electrical charge, but which carried with it fantastic quantities of energy and traveled through space at incredible velocities. Indeed, this strange particle traveled through space with the velocity of light, for the simple reason that this particle was nothing more nor less than a particle of light, soon to be known as the "photon." And just as Rutherford's work brought about sharp changes in men's concept of the shape and the form of the atom, the study of these light particles or photons brought about a sharp change in physicists' thinking about the nature of the forces holding atoms together.

## THE ULTRAVIOLET CATASTROPHE OF JAMES JEANS

We will recall that the idea that light might actually be composed of a stream of tiny particles or bullets had long since been studied very carefully and finally, reluctantly, discarded. Centuries earlier, Sir Isaac Newton had contended that a beam of light was *discontinuous,* made up of a stream of tiny, indivisible units or chunks of light, just as philosophers had speculated that matter was made up of tiny, indivisible units or chunks of matter, the classical "atoms" of the Greeks.

But Newton's "particle" concept of light had run into serious difficulties. Multitudes of studies in the 1700s and 1800s had demonstrated that light simply did not behave like a stream of particles. Indeed, it behaved so much like the newly discovered electromagnetic waves that by the late 1800s physicists had virtually abandoned the idea of units or particles of light. Light was indisputably a wave phenomenon, proved by all sorts of diffraction and diffusion experiments. Different wavelengths were associated with light of different colors, and the wavelength of visible light had been found to blend into a broad spectrum of electromagnetic waves ranging from the very long radio waves at the one end to the moderately long wavelengths of red light, on through the color spectrum to the quite short wavelengths of violet and ultraviolet light, and finally on into the realm of still shorter wavelength microwaves and X-rays.

This was a comfortable and satisfying way of considering the phenomenon of light, but unfortunately physicists kept coming up with awkward theoretical and mathematical predictions of things which ought to be happening if light was composed of electromagnetic waves, but which steadfastly refused to happen under experimental conditions. One of these trouble-makers was a British physicist named Sir James Jeans (1877–1946) who caused all of his colleagues a headache when he predicted a dreadful sort of catastrophe that ought to occur any time light got trapped in a closed container, but which had in fact never been observed to occur at all in real life. It is just as well that this "catastrophe" didn't happen, since we would all have been fried in our own juices long since by deadly radiation if it did, but physicists were intensely annoyed that they couldn't explain *why* it had never happened.

All that Sir James Jeans was trying to do, really, was to apply the mechanical principles that governed the exchange of heat energy between molecules of a gas in a closed box to what seemed to him the similar case of the exchange of energy between light waves of differing wavelengths in a closed box. As we saw earlier, molecules of a gas are in constant motion. When the gas is heated, the molecules individually have increased kinetic energy and bounce around the interior of a closed container more and more furiously the more the gas is heated. If the gas is cooled the molecules have less kinetic energy, and move at much slower velocities.

But if a sample of hot gas and an equivalent sample of cold gas are introduced at the same time into a closed container, an exchange of kinetic energy will take place between the molecules. The high-energy molecules of the hot gas will thrash around wildly in the container, banging into its sides and into other molecules, sending the slow-moving and low-energy molecules of the cold gas spinning. Such encounters, while transferring kinetic energy from the "hot" molecules to the "cold" ones, would also cause the hot ones to lose kinetic energy and stop moving around so fiercely. Very soon enough hot molecules would have collided with enough cold molecules, with resulting exchanges of kinetic energy according to the classical Newtonian laws of motion and inertia, that the distinction between hot and cold molecules would begin to disappear. Very soon, in fact, the container would have very few hot or cold molecules left at all; instead it would contain a large number of "lukewarm" molecules. The total kinetic energy of both the hot gas and the cold gas originally introduced into the container would have been *equally distributed,* so that all the molecules of gas in the container (on the average) would have the same kinetic energy and the same velocity as all the others, lower than that of the original hot molecules and higher than that of the original cold molecules.*

This distribution of kinetic energy among hot and cold gas molecules had been worked out in mathematical detail by James Clerk Maxwell, the man who had also worked out the field equations describing the behavior of electrical and magnetic fields. In the late 1890s Sir James Jeans tried to treat the exchange of radiant energy between different wavelengths of light in a box in the same way that Maxwell dealt with the kinetic energy of gas molecules in a box. Jeans imagined a cubical container lined with curving mirrors that would reflect 100 per cent of any radiation striking them. He reasoned that if a ray of light of a single wavelength were introduced into this box with, perhaps, a few dust particles floating around in it to absorb energy from some of the light waves and then re-emit it at different wavelengths, the radiation reflecting back and forth inside the box would soon contain vibrations of many different wavelengths.

Jeans considered that just as the kinetic energy of the hot and cold gas molecules would soon be distributed more or less evenly among all the gas molecules in the container, so also the total amount of radiant energy inside his light-reflecting box soon ought to be distributed equally in the form of electromagnetic waves of all possible different wavelengths.

But here a problem appeared. In the case of the gas molecules, there might be a very large number of molecules to take into account, but the number was still finite. No matter how large the number of molecules, there would always be just so many and not more. But the number of possible wavelengths of radiant energy in Jeans's box was infinite. If each different wavelength were to receive an equal share of the total amount of radiant energy admitted to the box, more and more of the total available radiant energy would become concentrated in waves with shorter and shorter wavelengths. Thus, Jeans reasoned that if long-wavelength red light were admitted to his reflecting box to begin with, the light inside the box would rapidly become blue and then violet and then ultraviolet and would then be transformed into X-rays or even shorter wavelength gamma waves, and so on.

Indeed, *any* wavelength of light that was originally introduced into the box would presently be transformed exclusively into waves in the ultraviolet range of wavelengths or even shorter, if the laws of classical physics

* We say "on the average" because we realize that some of the hot molecules might well have collided with other hot molecules so as to have *increased* their velocity and kinetic energy, while some of the cold molecules might have been struck in such a way as to be stopped dead and become much colder. We might have to wait an awfully long time for *all* of the molecules to acquire *exactly* the same energy level as each and every one of the rest, but this would have happened to the vast majority of molecules within a very short period of time.

were applicable to radiant energy. The same thing would happen any time any object were heated enough to radiate light—a piece of coal in a furnace, for example. Instead of heat (infrared) waves greeting the fireman as he opened the furnace door, he would be faced with a blast of ultraviolet waves, X-rays, and deadly gamma rays!

Now it was quite obvious to Jeans's colleagues in physics that this sort of so-called ultraviolet catastrophe did not ever actually happen at all, and never had. Jeans was not proposing that it was likely to happen, either. He was merely proposing it, quite seriously, as a sort of theoretical nightmare which *ought* to happen every time anyone turned on the stove, *if the classical laws of physics actually applied to the behavior of radiant energy.*

When we recall that the classical laws of physics were the only laws of physics that were known in the 1890s, we can see what consternation this paradox caused. It seemed that either Jeans's mathematical analysis had to be wrong (unlikely, since no one could find any error in it) or else something had to be wrong with the classical laws of physics as they were understood. But no one could imagine what could be wrong with them, since they seemed to be perfectly good laws which applied splendidly to multitudes of other phenomena in the universe. Certainly nobody was willing to scrap them out of hand, for fear of throwing out the baby with the bath water.

What was needed was a new way of looking at things. And in December, 1899, just a short time before the beginning of the new century, the necessary revolutionary idea was supplied by one of the least likely people imaginable. Max Planck, a 31-year-old German physicist of the old school, was a solid and conservative citizen in the world of physics, steeped in the classical traditions—a sort of Old Stonebottom of the Newtonian school. Planck by his very nature abhorred the very idea of revolution in physics in any form, but he had courage enough to recognize that as far as radiant energy was concerned the classical laws had to be modified somehow. So he set about to find the smallest possible modification that would explain why Jeans's ultraviolet catastrophe did not and could not actually occur.

## THE CONCEPT OF ENERGY QUANTA

The solution that Planck came up with was deceptively simple. Once physicists had become convinced that light, radio waves, and other electromagnetic waves were indeed waves and not streams of "bullets" as Newton had thought, they then naturally assumed that the radiant energy contained in those waves traveled in a smooth and continuous flow. After all, if you are convinced that water is a completely homogeneous liquid, you do not expect it to come out of a bucket in separate cupful chunks when you empty the bucket. But Planck suggested that possibly the radiant energy in electromagnetic waves did *not* occur in a continuous flow but, rather, was broken up into chunks or packets. For each form of electromagnetic waves, including light waves, Planck suggested that there must be some tiny, indivisible parcel or chunk of energy. Thus the radiant energy of sunlight pouring in at a window was not a continuous flow of energy but a vast stream of tiny packages of energy, with a different-sized package for each different wavelength of the light.

Indeed, Planck contended that the amount of energy in a given energy package would be inversely proportional to the wavelength of the radiation. Thus long-wavelength radiation such as radio waves, infrared light, or visible red light would be carried in very, very tiny "energy packages," while the energy packages of shorter-wavelength green light would be much larger, and those of still shorter-wavelength ultraviolet light even larger yet.

Planck named these packets or bundles of radiant energy "quanta," from a Latin word meaning "a part or portion." He proposed that each individual wavelength of radiation had its energy bound up in quanta of a fixed and unchanging size which could never be subdivided in any way. Thus if radiant energy were to be exchanged from radiation of one wavelength to radiation of another wavelength, that energy could only be given up by the one wavelength in exact multiples of its quantum bundles, and could only be accepted by the other wavelength in exact multiples of *its* quantum bundles. There could be no way of giving up one-third of a quantum of energy, or taking on one-half a quantum. In each energy exchange, the exchange had to be an all-or-nothing affair.

It was this concept of radiant energy composed of discontinuous quanta, with the provision that the smaller the wavelength of the radiation the larger the quantum of energy that would be associated with it, that explained why Jeans's ultraviolet catastrophe did not happen when a beam of red light, for example, was introduced into his box of mirrors. True, an infinite number of different wavelengths of light might be possible in that box, all the way from infinitely long to infinitely short, and the radiant energy of the red light in the box might be divided up into any combination of those infinite number of wavelengths. But if the *total amount* of radiant energy present in the box was divided up into a *huge but finite* number of quanta of red light, that meant that only a finite number of quanta could be distributed among all the other possible wavelengths. Red-light quanta contained only a very small amount of energy each, since the wavelength of red light was very long. But other wavelengths—ultraviolet wavelengths, for example—were very short and could not accept any of the available radiant energy at all if it could not get a huge number of quanta at a time, and X-rays or gamma rays would require even huger numbers of red-light quanta in order to form a single one of their superhuge quanta.

We can see this more clearly if we can imagine the predicament of a mother at the dinner table trying to distribute a panful of noodles as equitably as possible among some two dozen children of various sizes and appetites. If each child would accept and eat whatever number of noodles (red-light quanta) that he was served from the available supply, the mother could distribute the available noodles perfectly equally just by dividing the total number of noodles available by the total number of mouths and then serving equal quantities to each child. If the pan happened to be small, the number of noodles given to each child would also be small, and if there happened to be forty children instead of twenty the number of noodles per child would be still smaller, but at least equal distribution could be managed.

That would be fine—but suppose that each of these children (that is, each wavelength of radiation) has a different appetite and needs a different minimum amount to satisfy its hunger (that is, different quantum sizes). Suppose further that no child will eat *anything at all* unless he gets the full amount of his minimum demand, however large his minimum demand may be. In such a situation, some children (representing long wavelengths of radiation) might be satisfied with very small amounts of food; two or three noodles might do. But another child might want a spoonful of noodles or none at all. Still another child might want a whole cupful or none at all. Still another child (representing very short-wavelength ultraviolet light) might demand *the whole pot* or none at all.

How, then, would a fair-minded mother distribute the noodles?

Obviously, she would not want to use up all of the noodles in the pot by distributing them all to the large number of children requiring only two, three, or four noodles apiece, while ignoring the plaintive cries of the ones who really need a spoonful to be satisfied. Neither would she give the whole potful to one greedy pig and deprive all the rest of any at all. If she decided that the fairest approach would be to accommodate the largest number of children possible with what she had available and let the unreasonable ones go hang, she would send the greedy ones away from the table and distribute the noodles to the rest as equitably as possible. Probably most of the noodles would go to the undemanding ones (long wavelengths, small quanta). Most of the spoonful-types would also get their fill (medium wavelengths, larger quanta). Perhaps one or two of those demanding a cupful (short wavelength, very large quanta) would be accommodated, but the pigs (extremely short wavelengths, huge quanta) would go hungry.

Clearly, the more modest the child's demands (that is, the smaller the quanta) the more *likely* the child is to have his demand fulfilled. The larger the demand, the less chance of fulfillment, and no significant chance at all for an ultraviolet catastrophe—all the available noodles going to a few big demanders—because the requirement of any single one of these big demanders (short wavelengths) would in itself use up all the available noodles and there would always be another bigger demander (shorter wavelength, larger quantum) waiting next in line.

Planck's concept that radiant energy was divided into "smallest possible" parcels or quanta, with the size of each quantum proportionate to the wavelength of the radiation, certainly provided a solution to the paradox of the ultraviolet catastrophe. Unfortunately, this idea also seemed uncomfortably reminiscent of Newton's idea that light was not a wave phenomenon but was broken up into little "bullets," and classical physicists had already proved beyond question that light waves *had to be* waves. They were diffracted by diffraction gratings, they would diffuse when passing around a corner or through a slit and not make a sharp shadow, and they would behave in a multitude of other ways as good waves ought to behave. Thus Planck knew that his suggestion of energy quanta was indeed radical, coming from one schooled in classical physics, and once he had published the idea he was, as George Gamow described it, "scared to death" of the idea.* He immediately began trying to figure out some way that energy quanta might not arise from the properties of light waves themselves, but rather from some mysterious internal properties of atoms. In other words, he wanted to argue that those atoms that would absorb and then re-emit radiation must have something wrong with them so that they could only transfer radiant energy in certain discrete quantities, rather than to agree that it was the nature of light or other radiation itself that gave rise to quanta of energy.

But Planck's timidity did not win; just five years after the quantum idea was out of the bag another man published unmistakable evidence that energy quanta did indeed exist as physical entities, and not as a result of some perverted mechanism by which atoms dealt with radiant energy. The man was Albert Einstein, the year was 1905, and the evidence was based upon studies of another mysterious and previously unexplained phenomenon that seemed flatly to contradict the classical laws of wave mechanics: the phenomenon known today as the photoelectric effect.

## ELECTRONS AND PHOTONS

For some years physicists had known that certain kinds of metals, when exposed to light of various wavelengths, seemed to become agitated or "excited" by such exposure and as a result would start hurling away electrons.

* *Thirty Years That Shook Physics,* New York, Doubleday, 1966.

These electrons, apparently torn or thrown away from atoms of the metal that lay near the surface, could be detected, their velocities measured, and their kinetic energies calculated. It appeared that the radiant energy of certain wavelengths (particularly light of very short wavelength such as violet or ultraviolet light) was somehow absorbed by the surface atoms of the metal, and that this absorbed energy was then sufficient to "kick out" electrons from those atoms with various velocities and kinetic energy. It was as if an overwound clock spring within those atoms had suddenly broken, with the result that electrons from those atoms were hurled in all directions.

Assuming, as the classical physicists did, that this ejection of electrons was due to some sort of electromagnetic force generated when the light waves struck the atoms, early observers of this so-called *photoelectric effect* naturally assumed that the electromagnetic forces causing the ejection of electrons ought to increase smoothly and steadily as the intensity or brightness of the light increased. It certainly seemed reasonable that the stronger the beam of light hitting the vulnerable metal, the more electrons ought to be ejected, and the higher the velocity with which they were hurled away. It also seemed reasonable to expect that if a given metal exhibited the photoelectric effect at all, any wavelength of light ought to throw out just as many electrons as any other wavelength of light of the same intensity, and that the number and velocity of electrons ejected by the surface atoms of the metal ought to increase only as the intensity of the light increased —that is, as the total *amount* of the light that was striking the metal increased.

According to the view of light waves that prevailed in the mid-1800s, all these things seemed reasonable to expect, and experimenters who set out to measure the number of electrons ejected, as well as their respective velocities and kinetic energies, had every reason to expect their experimental results to confirm the predictions of the classical laws of physics. We can imagine their chagrin when they discovered that their actual measured results did not, in fact, tally at all with what they had expected.

For one thing, they found that when light of a single wavelength—so-called *monochromatic* light—was beamed at a metal surface, first in a faint beam and then with increasing intensity, increasing numbers of electrons were indeed ejected, but there seemed to be no increase whatever in the amount of energy with which they were ejected. Each came flying off the metal with exactly the same velocity as all the others. If monochromatic light of a different wavelength were used, then and only then would the electrons be ejected at a different velocity; if the wavelength of the light were longer (toward the red end of the spectrum) the velocity of the ejected electrons would be lower, while if the wavelength were shorter (toward the violet or ultraviolet end of the spectrum) the electron velocity would be higher. But for any given wavelength of light, only the total *number* of electrons ejected would increase as the intensity of the light was increased.

Once again physicists found themselves faced with experimental results that did not make sense. No matter how carefully their observations and measurements were made, the results simply did not agree in the slightest with the predictions of established natural laws which had predicted other things so accurately. These electrons hurled off by atoms of metal which had absorbed light waves simply did not behave as they ought to.

Characteristically, the experimenters first questioned the accuracy of their own measurements, but repeated experiments with more and more accurate measuring devices turned up precisely the same experimental facts, time after time. And in a case like this, the facts had to be accepted. So many electrons coming from a metal surface were so many electrons, no more and no less, wherever they may have come from. If a given electron was found to have such and such a velocity, then it had such and such a velocity, no matter how awkward this fact might be. A few centuries before, physicists might have elected to ignore experimental observations that conflicted with "natural laws" which they were certain were correct, but in the 1890s and 1900s science had become far too sophisticated for that. If experimental fact contradicted accepted "natural laws," then there was no choice but to face the fact that the "natural laws" were incorrect as they had been understood.

Thus, when Einstein began studying the paradox of the photoelectric effect, he did not start from the premise that because the experimental data failed to agree with the facts, the experimental data must be wrong, as scientists of an earlier age might have done. His premise was that if the observed facts contradicted the rules, then something must be wrong with the rules. But how could the rules be changed to account for the odd observations that had been made of what actually happened with the photoelectric effect?

Einstein was aware of Max Planck's suggestion that all kinds of electromagnetic radiation might exist only in discrete quantum-sized bundles, with the size of the quantum getting larger the shorter the wavelength of the radiation. Basing his thinking upon Planck's hypothesis, Einstein published a famous paper in 1905 in which he proposed that the "paradox" of the photoelectric effect could be explained if Planck's quantum idea were in fact literally true. Suppose, Einstein said, that light traveling through space does not travel in continuous waves but rather in the form of individual energy packets or quanta, and suppose that upon encountering an electron in a surface atom on a metal surface such a quantum of light—whatever its energy content, depending upon its wavelength—simply *disappears,* giving up its entire energy to the electron and thus causing it to fly out of the atom at whatever velocity the energy content of the light quantum makes possible.

If this were truly the case, then a tiny quantum of long-wavelength light (monochromatic red light, for example) with comparatively little energy content might just not have quite enough energy available to push an electron out of an atom at all. In such a case, it might be possible to flood atoms of that metal with any amount of red light imaginable, no matter how intense, and never find a single electron ejected no matter how long one waited. But take light of a shorter wavelength—say, monochromatic violet light—made up of much larger energy quanta, each with a much greater energy content, and then even the dimmest beam of that light would make an encounter between a single one of these more robust energy quanta and a single electron likely. As soon as that encounter came about, the energy content of this larger quantum would be sufficient to send the electron flying out of the atom, but since such a quantum was still only of moderate size with a moderate energy content, the electron might fly away at a comparatively low velocity and with a low kinetic energy.

Thus Einstein suggested that there might be a quantum threshhold involved in the photoelectric effect. Very small light quanta from long-wavelength light just wouldn't have steam enough to throw an electron out of an atom, while the quanta of shorter-wavelength light which were just above some baseline threshold of quantum size and energy content would always kick off an electron any time the quantum and the electron met. If only a few of these quanta were available for encounters with electrons (very dim or low-intensity light), then only a few electrons would be thrown off. If a great many of these quanta were available for encounters with electrons (very bright or high-intensity light) then a flood of electrons would be thrown off—but no matter how many were thrown off as a result of the intensity of the light, each electron would be thrown off as a result of an encounter with one and only one quantum of that light, and would be thrown off with only the precise amount of energy available from that one quantum, no more and no less, so that each and every electron thrown off by quanta of that size and energy content would have to fly out of their atoms with the same velocity and kinetic energy as all the rest.

This concept of one-to-one encounters—one particle interacting with one and only one energy quantum—was very important, and carried to its logical conclusion, explained very neatly why the photoelectric effect behaved as it did. If instead of using a moderately short-wavelength light with quanta just barely exceeding the necessary threshold of energy content to kick an electron out of an atom at all, we switched to a *very* short-wavelength light made up of *very large* quanta with *very high* energy content we would again find that the total number of electrons emitted by the metal surface would increase as the intensity of the light striking the metal plate increased; but in this case each time one of these *very large* quanta happened to encounter an electron, it would throw the electron out of its atom at a *very high* velocity and with a *very high* kinetic energy.

But again, in the case of each electron, the effect would still be the result of a one-to-one encounter between one electron and one single energy quantum, which in this case happened to be not only above the necessary energy content threshold but happened to pack with it a whale of a lot more energy than the bare threshold amount, so that the electron departed at a frantic velocity and carried an enormous amount of kinetic energy with it. And again, so long as all these very large quanta were of the same size, each electron so ejected would buzz off with the same high velocity and the same high kinetic energy as every other one.

Planck's quantum theory fitted the observed facts of the photoelectric effect so perfectly that there was little serious question but that it was indeed true. But Einstein took the quantum theory a step further than Planck had dared to. Einstein put an end to any hedging about the true nature of light quanta or other energy quanta—the quanta of invisible radio waves, for example, which were extremely small quanta because of the very long wavelength of radio waves; or the energy quanta of extremely short-wave X-rays and gamma rays, which were large quanta, the size being always in inverse proportion to the wavelength.

It was nonsense, he said, to claim that such quanta were merely some function of the atom that is absorbing the light or other radiation. Rather, Einstein contended, light itself is *composed* of a stream of energy packets, rather like a stream of tiny sand grains pouring from a microcosmic sand-blasting machine toward whatever surface the light touches. Infrared or red light (long wavelength, small quanta) consists of a stream of extremely fine, powdery quanta like a shower of finely ground flour. Green or blue light (shorter wavelengths, larger quanta) is more "grainy," like a shower of sand and gravel on the microcosmic scale, while violet and ultraviolet light (very short wavelength, very large quanta) is like a shower of boulders. In those terms it is easy to see that a red light quantum—a grain of flour—striking an electron in an atom might not contribute enough energy to knock the electron free at all. A quantum of green light—a good-sized chunk of gravel—striking an electron might be just barely large enough to jar it loose and send it moving sluggishly away with very low kinetic energy, while a "boulder-sized" quantum of ultraviolet light encountering the same electron might knock it from here to Guinea. If the ground rules were such that only one quantum, be it flour-grain size or boulder size, could hit any given electron at any given time, then no red-light quanta could ever hope to dislodge an electron; but on the other hand, no ultraviolet light quantum could knock any given electron any farther or faster, so to speak, than any other ultraviolet light quantum.

But the really important point to Einstein's hypothesis regarding the photoelectric effect was the basic concept: that whether light quanta or other electromagnetic wave quanta were comparable to a shower of flour grains or to a shower of boulders, depending upon the wavelength of the radiation, *this was the nature of radiomagnetic waves of any wavelength,* and every single quantum, whether small or large, was a physical entity which was a built-in property of the radiation and could not be split into halves, thirds, or quarters, or piled together into combinations of two and one-half, three and three-quarters, four and one-third, etc. A quantum was a quantum, and it either possessed sufficient energy content to be capable of electron-knocking or else it did not. Two or three could not gang up on one electron, nor could one ever knock out two or three electrons at the same time.

This was a jolting notion, shaking apart the whole carefully developed classical concept of light and other electromagnetic radiation as waves in one sweeping blow. But before his colleagues could gather their wits together and start protesting, Einstein carried his concept one logical step further. All electromagnetic waves, he said, were divided into energy quanta as a part of their very nature. But since mass itself was merely one manifestation of energy, it followed that all massive bodies were also basically composed of energy quanta, according to the same rules: the larger the wavelength, the smaller the individual quantum. Thus an electron had to be composed of quanta. A proton had to be composed of quanta. A man walking on the street was also composed of quanta. In the case of the man on the street, his "wavelength" was so extremely long that the quanta were tiny to the point of insignificance—the reason that a man walking on the street behaves more like a "particle" than like a "wave."

But moving in the other direction, Einstein insisted that a quantum of light of a given wavelength, although possessing a certain specific energy content, was still as much a "particle" as an electron or a proton was, excepting that in the case of such a "particle" or "energy packet" it was constantly in motion at the velocity of light. Light quanta or other radiant energy quanta are never at rest; they are either moving through space at light velocity or else they are encountering and interacting with particles such as the electrons of surface atoms of metals. Thus if a light quantum "particle" happens to be "brought to rest," so to speak, by such an encounter, it exists at that instant entirely in the *energy form* of matter, and therefore can have no mass at rest. We say that it "disappears" and "gives up its energy" to the electron which flies away frantically as a result of the gift—but we might more accurately say that the light quantum that is "brought to rest" by encounter with an electron simply instantaneously becomes that electron's energy and ceases to exist as a "particle" of light. At rest it has no mass, but while in motion at the velocity of light some of its energy content *is* actually in the form of mass.

Thus Einstein predicted that light ought to exert a certain pressure against any surface which it struck. This prediction was later confirmed when delicate measurements demonstrated that light does have a very minute but measurable pressure on surfaces it encounters. The closest that a quantum of light racing along at light velocity could come to a state of rest was at the instant of collision with an electron, at which instant it became all energy and all of that energy in one fell swoop became a property of the electron. The reason that the electron flying out of the atom did not depart at the same velocity of light that the light quantum was traveling at the instant of encounter was simply that the electron *does* have a certain very small but significant mass at rest, and therefore has inertia which some of the light quantum's energy has to overcome in order to get the electron moving at all.

## THE PARTICLE-WAVE CONFLICT

From the very beginning Einstein and those enlarging upon his work deliberately referred to a quantum of radiant energy, whatever its size and energy content, as a photon in order to identify it firmly as a particle with just as much claim to particlehood as an electron, a proton, or any other particle. But the idea of an energy quantum as a physical entity—as a particle—was hard to swallow. Max Planck, with his entrenched conservatism, had known in 1900 that the "corpuscular theory" of light had been as dead as the proverbial coffin nail for years, and had come up with his revolutionary idea of quanta of light rather by accident and quite in spite of himself. He must indeed have been in quandary, for the idea made him so nervous he tried thereafter to apologize for it and soften it so that physicists (including himself) would find it more palatable.

Albert Einstein was more of a revolutionary from the start and had no apologies to make, but even he was hesitant enough about the revival of this corpuscular theory of light that he did not at first come out for it dogmatically and flatly, but hedged it a bit, saying in effect that light unquestionably had wavelike properties, but that in the photoelectric effect, at least, it seemed to behave more like a stream of particles than like waves. It was only later that he took the bull firmly by the horns and insisted that light existed *only* in the form of particlelike quanta or photons and always had.

In this respect it was worth emphasizing that Einstein was a *mathematical theorist* in physics, not an experimentalist. He had never actually seen a light quantum or photon demonstrated in a laboratory. He himself had never even attempted to find experimental evidence for his theories; it is likely that if he ever happened to wander into a physics laboratory in his life it was by sheer accident. All he really did was to say, in effect, "Look, fellows: photons exist. They have to exist. Since they exist, all you have to do is go looking for them and you're bound to find them." Thereupon, looking infinitely sad, Einstein went back to his theoretical work on another shy and modest project like the special theory of relativity with which he was about to jar loose still another foundation stone of classical physics.

Einstein's challenge to other physicists to go find and demonstrate the existence of photons as particles in the laboratory was not easy to take up. For almost twenty years after Einstein's first publication of his work on the photoelectric effect, the idea that photons of light or other radiations were actual physical particles had to be assumed without experimental proof. Then in 1923 an American physicist named Arthur Compton finally found clear-cut laboratory evidence that such particles did exist and did indeed behave like *particles*. Being of a practical turn of mind, Compton realized that if photons really were particles they ought to behave like particles when they collided with other particles such as free electrons (i.e., electrons that were not tied down in the structure of an atom). He realized that if he could induce a photon to strike such a free electron, the result of the collision ought to be precisely the same as an ideal, perfectly elastic collision between ivory billiard balls, obeying the laws of conservation of mass, energy, and momentum.

By using the very high-energy quanta of high-frequency X-rays as his "bullets" and electrons as his target, and then measuring the energy of the X-rays deflected from the collision between electron particles and X-ray photons, Compton proved that such collisions behaved just as he predicted. In a direct head-on collision, a high-energy X-ray photon would deliver all of its energy to the electron and cease to exist, while the electron would be catapulted away, carrying with it all of the energy of the collision. In more glancing type collisions, the X-ray photons and the electrons would bounce away from each other at angles, with the electrons taking on part of the energy of the photons, and the X-ray photons deflected or scattered by the collision and carrying smaller amounts of energy and therefore longer wavelengths (see Fig. 45).

The results of Compton's experiments confirmed in every detail Einstein's theoretical concept of light and other radiation as being composed of streams of energy packets or quanta which had a real physical existence as particles or photons. Thus Compton established experimentally that photons were indeed just as much bona fide elementary particles as electrons or protons were, differing from them solely in the fact that the photons had mass only while they were in motion, and when brought to rest instantly became tiny massless packets of pure energy.

Thus it seemed that the work of Planck, Einstein and Compton had brought physicists back full circle to the old discredited view of light and other radiation as a stream of "bullets" such as Isaac Newton had championed. What then about all the body of evidence indicating that light traveled as waves? Once again physicists found themselves struggling with the question of which light really was, wave or particle. Einstein replied that the question was not only unanswerable but that asking it was just as

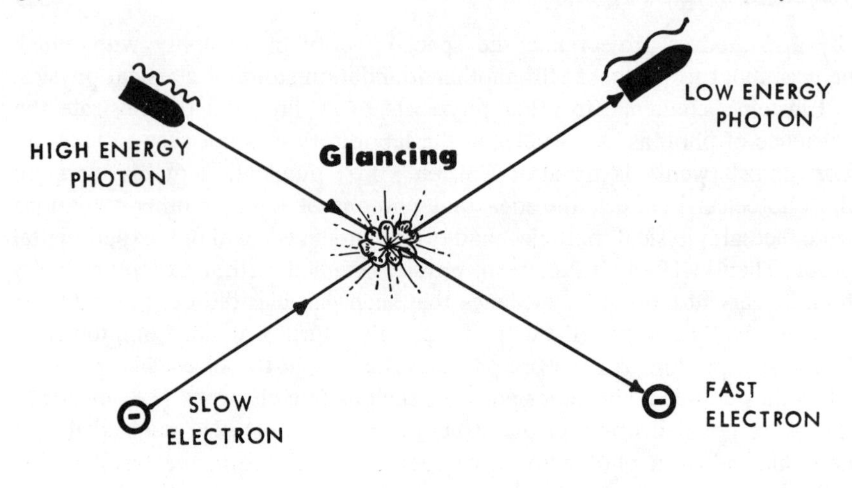


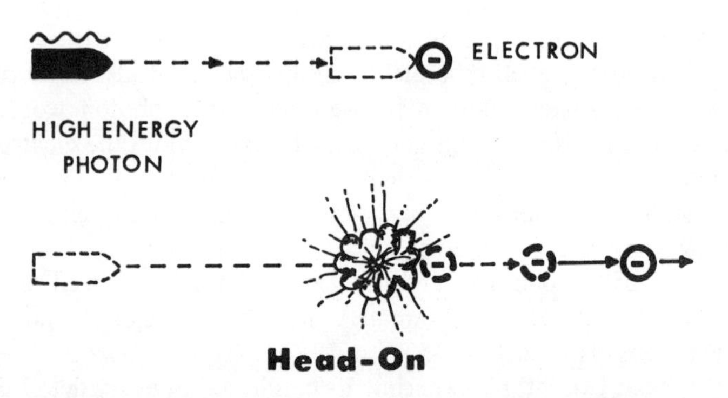


*Fig. 45* The "Compton effect" when photons and electrons collide. In a *glancing* collision between a high energy (short wavelength) photon and a slow electron, the electron takes on part of the photon's energy while the wavelength of the photon is reduced in an encounter much like a classical "billiard ball" collision. When a photon collides with an electron head-on, however, the photon ceases to exist, giving up all its energy to the electron, which therefore moves from the collision scene at a faster velocity than it had prior to collision.

inappropriate as the legendary blind men asking themselves whether an elephant was a tree trunk or a wall. To the man gifted with sight the elephant was obviously neither, and yet it was like both in certain respects: Its leg was like a tree trunk, its side was like a wall. According to Einstein, light simply could not be considered exclusively either wave or particle, nor was it necessary to insist upon one or the other. Light could never be accurately visualized or described as exclusively either one or the other. It merely had certain of the qualities of one under certain circumstances, and of the other under other circumstances.

Thus, from one point of view and in one kind of experiment the wave-like properties of light might be demonstrated; in a different kind of experiment, the results might emphasize more strongly the particle side of its nature. Even in circumstances of nature, light sometimes behaved in wavelike fashion, as when it passed through a diffraction grid and developed an interference pattern, while under other natural circumstances (as when it struck the electrons in the surface atoms of certain metals) it behaved in a more particlelike fashion.

In short, Einstein was saying that physicists who tried to categorize light and other radiation as *either* wave *or* particle were, in effect, like old-fashioned movie-makers, trying to categorize all the people in the world as either "good guys" or "bad guys." Granted that it made the plot easier to follow, but the nature of people is such that they cannot generally be crammed exclusively into either a good-guy or a bad-guy mold. All people, by their nature, have both good and bad qualities; they merely reveal more of their good-guy nature under certain kinds of circumstances, and more of their bad-guy nature under other circumstances.

The quantum theory introduced by Max Planck in the 1900s and expanded by Einstein in 1905 set off a wide-ranging chain reaction of investigation and study in the following decades. Physicists who were first confused by it, or suspicious of it, or both, soon came to realize that the concept of energy quanta was not merely a radical and revolutionary hypothesis. In a very practical scientific sense it was a rich and fruitful hypothesis. Throughout the history of science hypotheses and theories of all sorts and varieties have arisen, some clearly impossible, some merely poor, some good, and some truly great. The measure of the poorness or greatness of any given theory is not necessarily a matter of whether the theory ultimately proves correct or not. Granted that a great theory tends to stay around longer and proves more fruitful in the long run than a poor theory, and that any theory is strengthened more and more as it proves to stand up to experimental verification. But the real importance and the real "greatness" of a theory depend even more upon its *usefulness* in opening up doors of understanding. The phlogiston theory of oxidation, for example, widely supported two hundred years earlier, had ultimately proved about as completely incorrect as any major theory in all the history of science; yet it was a great and important theory for the very reason that it lent itself so splendidly to experimental invalidation, and in the process of being invalidated led to a clearer understanding of oxidation than would have been possible without it.

The concept of energy quanta qualified in both respects. Not only was its theoretical validity confirmed repeatedly by experimental observation as time went on; it also proved to be a veritable gold mine of usefulness, a key that unlocked the doors to a multitude of other baffling mysteries in physics. The quantum theory became a sort of "missing link" without which physicists had been helpless to probe the nature of the structure of atoms beyond the crude picture that Rutherford developed in the first decade of the 1900s. Without quantum theory it is doubtful indeed that any deeper understanding of the nature of atomic nuclei could ever have been achieved. But thanks to this powerful theoretical tool, a groundswell of discovery began which is still continuing to this day.

## THE ELUSIVE ELECTRON

One of the first places in which the concept of energy quanta proved invaluable to a deeper understanding of the structure of atoms was in suggesting answers to certain nagging questions surrounding the behavior of electrons as a part of basic atomic structure.

At the time Einstein's paper on the photoelectric effect was published and physicists in general began exploring quantum theory, Ernest Rutherford's experiments in bombarding atomic nuclei with alpha particles from natural radioactive materials were well underway, and his picture of the atom as a dense, compact nucleus of positively charged protons surrounded by a cloud of tiny, negatively charged planetary electrons was well established. But as we have seen, there were some problems involved in Rutherford's model of the atom. Why didn't all those similar positively charged particles compacted into the nucleus push each other apart by the force of electrostatic repulsion? Why weren't some of these electrons gradually pulled into the nucleus of their respective atoms by the force of electrostatic attraction?

One of the physicists struggling unsuccessfully with these questions was a young and gregarious Dane named Niels Bohr. It seemed to him, as it did to everyone else, that if Rutherford's model of the atom really were valid, then the electrons of a given atom must clearly be moving in some kind of circular or elliptical orbit at some distance away from the nucleus, much as the planets circled the sun. If an electron with its negative electrical charge were *not* moving, the attractive force of all the positive charges concentrated in the nucleus would certainly tug on it until it "fell down" into the nucleus. Only if such an electron were moving very swiftly indeed would it be possible for such a tiny negatively charged particle to have sufficient centrifugal force out and away from the nucleus to counterbalance the inward tug the nucleus exerted upon it.

There certainly were at least superficial similarities in this concept to our sun's planetary system. The earth, in order to remain in orbit and not fall into the sun or drift out farther away from it, must move with just sufficient angular velocity that its centrifugal force outward exactly counterbalances the sun's gravitational force pulling the earth inward. But even if one thought of an electron as a tiny "planet" in orbit around a nuclear "sun" in an atom, the electron had one known property which a planet such as the earth did not have: *It carried an electrical charge.*

Now the work of Faraday, Maxwell, and others had long since established that objects with an electrical charge had certain peculiar properties related to their motion. If some object bearing a negative electrical charge was perfectly motionless, it would exert a certain electrical potential in a field surrounding it. But the moment it was set into motion at all, it would begin to generate electromagnetic waves. Niels Bohr reasoned that this effect ought to come about even if the charged object *were* as tiny as an electron, and bore only a single unit of charge. It seemed to him that an electron moving around the nucleus of an atom ought therefore to be generating electromagnetic waves constantly, and thus dribbling away its energy into the space around it. As its kinetic energy was gradually dissipated, the angular velocity with which such an electron was moving in orbit ought gradually to decrease, so that the electron would slowly spiral downward and ultimately join the nucleus, pulled in by the action of electrostatic attraction. In fact, it seemed that this very thing *ought* to have happened long since to all the electrons of all atoms in existence if Rutherford's planetary model of the atom were the whole story of its structure and behavior.

This, of course, was assuming that the kinetic energy or energy of motion of electrons could in fact be converted to the radiant energy of electromagnetic waves and drained off in a continuous stream. It seemed to Bohr and other physicists studying this problem that there was no possible way that Rutherford's atom could ever remain a stable entity with a central nucleus and planetary electrons swarming in orbit around it at a distance if such an assumption (based on classical laws of physics) were correct. Yet it was abundantly clear that at least the atoms of most elements were indeed stable entities. How could the puzzle be resolved?

The concept of energy quanta provided a splendid explanation. According to quantum theory, a charged body such as an electron of an atom moving in orbit around its nucleus could indeed convert some of its kinetic energy to electromagnetic waves, but *not* in a smoothly continuous manner for the simple reason that radiant energy didn't exist as a smooth stream. It could exist only in quantum-sized bundles. Thus an electron could not dribble away its kinetic energy by generating a steady stream of electromagnetic waves and as a result spiral in toward the nucleus; if it were going to convert its kinetic energy into radiant energy at all, it would have to do so only by ejecting a bare minimum of one quantum of radiant energy all at once, since there was no such thing as a half quantum, or a third of a quantum, or a sixteenth of a quantum.

If this quantum theory of radiant energy were correct, Bohr pointed out, then an electron moving in orbit at a great distance from its nucleus would in effect have two "choices." It could either continue moving in its stable orbit without giving up *any* radiant energy at all, or it could emit a whole quantum of radiation at once, in a bundle so to speak, and drop instantly down closer to the nucleus, where it would again take up a stable orbit but with less kinetic energy. Its new orbit, closer to the nucleus, might be thought of as a "lower-energy orbit" than the "higher-energy" orbit it had been traveling in before it emitted a quantum of radiation. Once established in that lower-energy orbit the electron would again have a "choice" of remaining where it was in a nice stable orbit around its nucleus, or of converting part of its remaining kinetic energy into another quantum of radiant energy and hurling it away, thereby dropping down into a still-lower-energy orbit that was yet closer to the nucleus. Thus, if the electron could only emit radiant energy in quantum-sized bundles it could only fall toward the nucleus in a series of sudden jumps, rather than in a smooth spiral.

This was fine as far as it went: But if this were true, what prevented the electron from ultimately falling into the nucleus anyway, even if by leaps and bounds, so to speak? Again the answer was provided by the quantum theory. Electromagnetic waves of different wavelengths consist of different-sized quanta. Suppose, Bohr said, that at each successive energy level the electron might reach, jump by jump, the possible wavelengths of radiant energy the electron could emit at that level grew shorter, so that with each successive jump to a lower-energy-level orbit the electron would have to hurl out a larger and larger quantum of radiant energy. Yet with each quantum of energy the electron emitted, its remaining kinetic energy would be less. Thus, one might say, it would "become more difficult" at each successive energy level for the electron to muster up enough enthusiasm to emit the necessary quantum-sized bundle of radiant energy to drop to a still lower level. At each level the electron would be "more inclined to choose" just to remain in the stable orbit where it was without emitting any more radiation. And finally, there might be an innermost limit—some very-low-energy-level orbit quite close to the nucleus at which any possible radiation emitted would have to be of such very short wavelength with such enormously huge quanta that the electron just wouldn't have enough kinetic energy left to make the grade even if it "wanted" to. Such an electron might remain in that innermost orbit throughout all infinity without ever being able to emit the quantum of energy large enough to let it get "home" to the nucleus (see Fig. 46).

Of course in his proposal Bohr did not endow atomic electrons with such human capabilities as freedom of choice, ability to make an effort, inclination to do something, or other human attributes that they quite certainly do not possess, as we have in the above. We can get away with such literary devices in the interests of clarity, whereas a physicist cannot. But Bohr did use the quantum concept to develop a startlingly different picture of the atom than Rutherford's model presented: an atom whose electrons circled its nucleus not in just any old way but at specific levels in fixed layers or orbital shelves, like the layers of an onion.

Bohr reasoned that these orbital levels would have to be the *only* areas in which electrons *could* orbit, because in order to get from one level to a lower-energy level closer to the nucleus the electron had to give

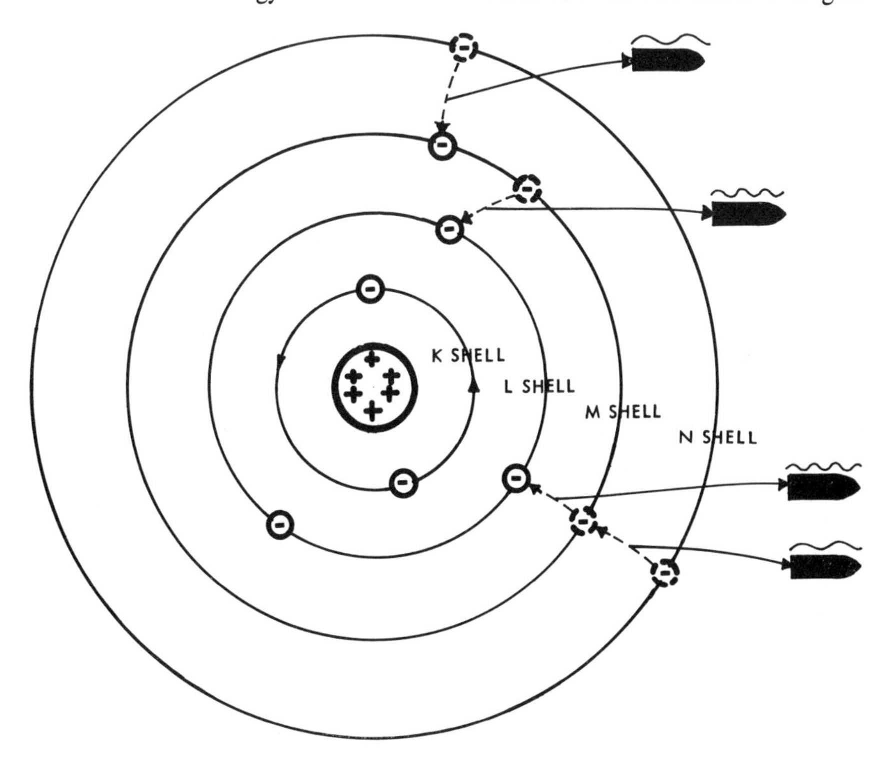


*Fig. 46* A diagram of Bohr's atom, showing the electron shells. Electrons can make "quantum jumps" from outer shells to shells closer to the nucleus by emitting energy quanta. Here the quanta are represented as bullets, with the energy of each indicated by the wavelength associated: high energy-short wavelength, low energy-long wavelength. For an electron to make a quantum jump from the L shell to the K shell (assuming there was room) would require emission of a still higher energy (shorter wavelength) quantum than shown here.

up a fixed quantum of energy in the form of radiation of a fixed wavelength that was associated with that particular "quantum jump" and no other. And since the size of the quantum to be emitted as "toll," so to speak, before each successive inward quantum jump could be made was larger in geometric progression as the wavelength of the radiation became shorter, it became just that much more difficult for an electron to emit the necessary-sized quantum for the next inward leap at each successive level.

Bohr calculated just how far from the nucleus each of these successive "electron layers" would have to be, what angular velocity the electrons would have to have at each successive energy level, what quantum-sized bundle of energy would have to be emitted in the form of radiation of what wavelength for each successive inward leap, and how stable the electron would be in each successive orbit. Clearly, the electron occupying the innermost possible energy level would be the most stable. Electrons in higher-energy-level orbits farther out from the nucleus would be far more unstable, far more readily able to emit radiation and drop into a lower-energy-level orbit.

But Bohr also pointed out that electrons should be able to behave in the opposite fashion as well. If a photon of electromagnetic energy—a photon of light, for example—should encounter an atom and be absorbed by one of its electrons, thus increasing the electron's kinetic energy, that electron could make a quantum jump *away* from the nucleus into a higher-energy-level orbit farther out. Electromagnetic radiation composed of long wavelengths (small quanta, such as photons of red light or green light) could only affect the electrons in the outer orbits where the size of the "energy toll" was comparatively small for a quantum jump from one orbit to the next. Radiation of very short wavelengths (extremely large quanta) would be needed before the electrons on the more stable inner energy-level orbits could be affected.

If this picture were true, however, there were still some unresolved problems. Since all the electrons of an atom had the *capability* of emitting energy quanta and hopping down toward the base-line low-energy orbits, and since the electrostatic attractive force of the protons in the nucleus ought continually to be encouraging electrons to do just that, so to speak, why didn't *all* an atom's electrons ultimately congregate in the lowest possible energy-level orbit? Bohr speculated that there must be some other hitherto unrecognized force or influence at work *limiting* the number of electrons that could orbit at the same energy level. He even calculated how many electrons could be permitted to "coexist" at each successive level: two at the innermost energy level, eight at the next higher energy level, eight at the third energy level, sixteen at the fourth energy level, and so forth.

Superficially, of course, Bohr's new model of the atom seemed very similar to Rutherford's, but on close examination we see that it is very different indeed. Just as Planck and Einstein before him had re-examined the classical concept of radiant energy as a smooth, continuous flow of waves and then "quantized" radiant energy into discontinuous bundles or energy packets called quanta, Bohr had re-examined one form of *mechanical* energy—the kinetic energy of electrons in their atomic orbits—and "quantized" this mechanical energy.

Planck and Einstein had not discarded the classical electromagnetic wave theory of Faraday and Maxwell altogether. They had simply found that in one critical area—the area in which electromagnetic energy is *exchanged* from one material body to another—the classical theory did not cover all of the ground. The way that radiant energy could behave was restricted by the fact that it could only be exchanged in unit packets or quanta, with the size of each quantum being proportionate to the frequency of the radiation concerned. Einstein went one step further and contended that such radiant energy actually *existed* only in the form of quanta which he called photons.

In almost identical fashion, Niels Bohr had found that the classical Newtonian laws of motion, however well they described exchanges of mechanical energy in the everyday world, simply did not apply in the tiny micro-universe of the atom. Electrons in orbit around an atomic nucleus did *not* behave like tiny solid orbiting marbles ought to have behaved if they followed Newton's laws of mechanics.

But Bohr's hypothesis did not throw out the rules of classical mechanics wholesale. It simply said that in the case of atomic electrons moving in their orbits around the atomic nucleus the classical rules did not *fully* explain what was observed to happen; certain restrictions had to be added to those classical rules so that out of an infinite variety of possible paths that electrons might otherwise follow in their orbits only certain special ones were permitted.

The mere fact that Bohr's theory explained what had been observed to be happening to atomic electrons did not in itself necessarily mean that his theory was true. It didn't even mean it was a very good theory, to say nothing of a complete one. It had, in fact, been worked out for the specific purpose of explaining precisely what it explained; so it was not remarkable that it did its job. After all, if a tailor cuts cloth and then carefully sews it together to a customer's measurements, the customer should hardly be surprised if the finished suit fits well!

Indeed, the real strength of Bohr's theory was not that it explained what he had set out to explain with it, but that *in addition* it also perfectly dovetailed with Einstein's presentation of radiant energy existing only in quanta, and then went even further than that to predict certain things that ought to be demonstrable if his theory was correct. For example, Bohr's theory predicted that since electrons could make quantum jumps from one energy-level orbit to a lower-energy-level orbit only by emitting a photon, these photons should in some cases be detectable as visible light with a spectroscope, and should appear in a spectrogram as spectral lines of predictable color (that is, predictable wavelength) and at predictable intervals from each other. Bohr was able to calculate mathematically that each different kind of atom should demonstrate its own characteristic spectral lines which should appear in very precise places.

Of course, as we have seen, it was already known that different chemical substances, when heated and examined with the spectroscope, showed patterns of spectral lines that were very distinctive and characteristic for each substance—previously a totally unexplained phenomenon. Now when the spectral lines of various substances were compared in their color and location with the location of the spectral lines predicted by Bohr's quantum theory hypothesis of electron energy levels, it was found that Bohr's predictions coincided very closely with what had already been observed in the case of atoms of a variety of pure elements. And this kind of confirmed prediction, together with others, began to make Bohr's hypothesis look like a very good theory indeed.

## WOLFGANG PAULI AND THE EXCLUSION PRINCIPLE

Thus it was soon evident that Bohr's theory of the electronic structure of atoms had a great deal going for it, and a number of physicists began examining his work and the whole question of atomic structure with great enthusiasm. In fact, a sort of feverish excitement began to pervade the world of atomic physics, beginning with the publication of Bohr's theory in 1913. Part of this was a direct result of Bohr's own unique personality. Of all the many investigators in the history of physics, here was a man who possessed not only a brilliant, inquiring, and imaginative mind but also a high sense of humor, an outgoing nature, an aura of overflowing enthusiasm, and an enormous quality of personal magnetism. When a proud Danish government granted him extended facilities for his work in Copenhagen following publication of his theory, Bohr became a sort of magnetic focus for a wide circle of students, associates, friends, and colleagues, all of whom were excited by this revolutionary new era of investigation in physics.

Unlike so many scientists who tend to guard their work jealously against encroachment, Bohr was outstandingly generous in offering time, attention and solid help to others who began coming up with additions, extension and modification of his ideas and pointing out implications of his theory that even Bohr had not thought of. A German physicist named Arnold Sommerfeld, for example, extended Bohr's ideas (which had been developed on the assumption that electrons make circular orbits) to the case of elliptical orbits, helping to explain certain minor discrepancies between Bohr's predictions of where a given atom's spectral lines ought to be found and where they were actually found. Sommerfeld also pointed out that electrons orbiting around an atom's nucleus need not necessarily be oriented in the same plane in space the way planets orbiting around our sun are oriented in the plane of the ecliptic, thus making the solar system like a great flat wheel. Electrons could occupy orbits in virtually any plane in space, so that the atoms of heavy elements with dozens or even hundreds of associated electrons (if they could be seen) would not be found to appear as flat, disklike planetary systems but, rather, more like three-dimensional bodies.

Another German physicist named Wolfgang Pauli, living in Switzerland, tackled a different problem arising from Bohr's theory. The Bohr model of the atom worked splendidly in the case of hydrogen, the lightest of all elements, with a single proton in its nucleus and a single electron which could occupy only certain permitted orbits which represented specific integral or "quantized" energy levels. But what happened in the case of heavier elements with their larger atoms? If the outermost "permitted" orbits represented high-energy levels and the innermost "permitted" orbit represented the lowest possible energy level, there seemed to be nothing in Bohr's theory to prevent all the electrons of a given atom from emitting their excess kinetic energy in the form of radiation and one by one hopping down the ladder of permitted orbits until all were huddled together at the innermost orbit possible.

But if this actually happened, Pauli reasoned, then the negative charges of all the electrons crowded into the innermost permitted orbital level would greatly increase the intensity of the field of electrostatic attraction between negatively charged electrons and positively charged protons in the nucleus. The electrons and protons would tend to pull toward each other, and the larger the atom, with the more electrons in the innermost possible orbits, the more this force of electrostatic attraction would tend to squeeze down or contract the volume of the atom. Thus one would expect atoms of the heavier elements to occupy a much smaller volume of space for their size and mass than would be the case if electrons remained in a number of the outer orbital levels as well.

But this was known not to be the case at all. In fact, experimental findings suggested strongly that the atoms of the heaviest elements actually had approximately the same total outside diameter as atoms of much lighter elements. Furthermore, if all the electrons of an atom actually did accumulate on the innermost permissible energy level, it would become more difficult to dislodge an electron from a large atom than from a small one, and again experiments had demonstrated that this was not true. In bombardment experiments with alpha particles, for example, it had been found that electrons were jarred loose from large and massive atoms such as those of gold just as readily as from the much lighter atoms of neon.

From this line of reasoning Pauli concluded that some additional restriction of natural law must be acting upon electrons in atoms, which Bohr had not taken into account, a restriction that prevented all the atomic electrons from cramming themselves into the lowest possible quantum state. Pauli's solution to the problem was neat and simple. He suggested that each quantum level or orbital level of an atom must have a characteristic "quota" of electrons which it could accommodate at any one time, and that as soon as the quota of the innermost orbital level was filled up, any other electrons would be "excluded" and would have no choice but to occupy the next higher orbital level. When the quota for that level had been filled, further electrons would be "excluded" from it and forced to occupy a still higher orbital level until a "vacancy" occurred in one of the lower-energy levels. This might occur, for example, in a case in which an electron in a low-energy level absorbed a photon of radiation coming in from outside and consequently "jumped" out to a higher-energy level, leaving a lower-level "hole." Then if there were some higher-level electron waiting out there just twitching to throw off a photon and drop back to a more restful spot at a lower-energy level, it could occupy the "hole" that had been vacated.

This notion that only a predetermined number of electrons could occupy any one quantum level or energy state in an atom at the same time, so that other electrons would be "excluded" from that quantum level unless a vacancy occurred, became known as the *Pauli exclusion principle*. Working out its theoretical details, Pauli concluded that the exclusion principle fitted experimental data best if *only two* electrons were permitted to occupy any single quantum state. He contended that the eight electrons Bohr mapped out for the second electron shell of an atom, for example, were actually four pairs of electrons, each pair in its own quantum state, although the four quantum states were so closely similar that the eight electrons appeared to be all in one shell. The exclusion principle soon proved to be an extremely fruitful expansion of Bohr's original theory. Using the Pauli principle, Bohr, Pauli, and others were able to construct detailed imaginative models of the individual atoms of all the elements from hydrogen on up to uranium.

These models dovetailed perfectly with what was already known of the chemical properties of various elements, and of their ability—or inability—to form compounds with other elements. For the first time there was a really sound physical explanation of why the atoms of various elements fell into a periodic system such as had been described earlier by the Russian chemist Mendeleev. This expanded concept of the structure of atoms explained why elements in certain well-known families (such as the halogen family, which included fluorine, chlorine, bromine, and iodine, or the alkaline metal family, including lithium, sodium, and potassium) had such similar chemical properties. It explained why other families of elements, such as the inert gases neon, argon, krypton, and xenon, appeared to have no chemical affinity whatever for atoms of other elements, nor to form any kind of compounds.

Later on two other physicists, G. E. Uhlenbeck and S. A. Goudsmit, found it necessary for mathematical reasons further to modify the Pauli exclusion principle when more delicate experimental observations had indicated that electrons in orbits, in addition to having mass and electric charge, also had a certain angular momentum—that electrons were spinning like tiny tops about their axes. Pauli's original exclusion principle, which stated that only two electrons could occupy any given quantum orbit, was modified to say "only two electrons *which spin in opposite directions* could occupy any given orbital level." This modification was then still further modified when new studies indicated that the orbits occupied by two such electrons spinning in opposite directions would have to be *slightly* different from each other, so that Pauli's original idea of two electrons in a single orbit was changed in favor of the idea of two electrons with opposite spins each occupying a unique orbital level all its own but so closely similar to that of its partner that it seemed to be the same orbital level. None of these modifications, of course, undermined Pauli's basic idea in the least, nor reflected on its usefulness in understanding the behavior of atoms and their electrons. These were merely the sort of super-refinements which became important to physicists who were finally getting down to an understanding of the basic fine detail of how an atom was constructed. It was Pauli's essentially valid concept that made this sort of refinement of understanding possible.

## THE PILOT WAVES OF LOUIS DE BROGLIE

The Pauli exclusion principle was essentially a direct extension of Niels Bohr's original theory of quantum orbits. It was a very simple but necessary modification which proved extraordinarily fruitful in fitting physical theory to observed experimental fact, and thus deepened physicists' understanding of the construction of the atom and of the new rules of *quantum mechanics* that limited the behavior of atomic electrons. But a few years later, in 1926, a far more radical modification of Bohr's theory was introduced by a French physicist named Louis de Broglie.

Bohr's theory, as modified by Pauli, had already introduced some very sharp restrictions and changes in the classical Newtonian laws of mechanics, at least insofar as the behavior of atomic electrons was concerned. These restrictions and modifications explained how certain things that were known to occur within the atom could occur in spite of the classical rules which would—if fully valid in the micro-universe—have made things happen quite differently. When these restrictions and modifications were later confirmed by experiment and observation it became clear that they were not merely clever stabs in the dark but were in fact absolutely necessary if the basic structure of atoms was to be understood. But no one had been able to suggest any reason *why* these restrictions should exist in the special case of electrons orbiting around their atomic nuclei.

A startling reason why was introduced in 1926 by a young French aristocrat, a wealthy and refined man with a taste for chamber music and an almost accidental acquaintance with the behavior of electromagnetic waves, derived from his experience in the French army during World War I. Louis de Broglie grappled with the question of why Bohr's electronic orbits were as they were as the basis for a doctoral thesis, and came up with such an extraordinarily peculiar explanation that many physicists roared with laughter when they heard it. But the laughter dropped off sharply when experimental data began appearing to support de Broglie's weird idea, and soon those same physicists were trying to find relief from the headaches that resulted from it. For this young man's ideas opened up a Pandora's box of troubles that no one (with the possible exception of Einstein) had even halfway anticipated.

It had been Einstein who had first pointed out the basic interchangeability of solid matter with various forms of energy. He had insisted that neither mass nor energy could be considered to exist alone and separate one from the other, but that both are different manifestations of the same thing. Accordingly, in any interaction of matter in motion the mass may be more predominant than the energy in some cases and the energy more predominant than the mass in other cases, but the conversion from one to the other and back again was not only possible but actually did occur under a variety of circumstances.

Einstein had also contended that radiant energy, which had previously been thought to exist in a smooth and continuous unbroken stream of electromagnetic waves, actually existed as particlelike quanta or photons which not only behaved like solid matter under certain circumstances but had a certain amount of mass associated with them when they were in motion. He had emphasized the necessity of accepting a two-headed dragon, so to speak, regarding radiant energy under certain circumstances not so much as energy but as "elementary particles" that were as real and particlelike as electrons, for example. We have even seen that Einstein suggested that other and different units of mass-energy, such as electrons, might also under certain circumstances have wavelike qualities and behave more like electromagnetic waves than like solid massive bodies.

Later Bohr had "quantized" the mechanical energy of electrons, but Bohr and his followers had continued to regard electrons in their atomic models primarily as tiny chunks of mass—tiny solid billiard balls, if you will. Louis de Broglie reviewed the work of these men and began to wonder if this way of regarding electrons was really necessary or even desirable.

In trying to understand why atomic electrons had to be limited to certain permitted orbits at certain permitted energy levels or quantum states, it occurred to de Broglie to take the Einstein concept of the duality of mass-energy at face value. He speculated that possibly atomic electrons, or any moving electrons, might be inseparably associated with waves of some specific wavelength, with the associated wavelength varying for each orbital energy level, so that the electrons in an atom were able to move only along specific paths laid out for them by their accompanying waves and along no others. In a sense these waves might act as pilots, going on ahead of the electrons and creating a "path of least resistance" for the associated electrons to follow. But they might at the same time act as restraints, forcing their associated electrons to follow *only* the specific paths made possible by the waves.

Indeed, de Broglie speculated that the electron might well *not really exist at all* as the sharply defined chunk of matter it had always been thought to be. Perhaps electrons had no such sharp physical definition at all, but were really nothing more nor less than diffuse wave patterns filling their entire orbits, each with a rapidly moving concentration or nidus of energy at some point along the track which presented slightly more of the characteristic of mass at that point than anywhere else in the orbit. For that matter, de Broglie speculated, even electrons that were *not* part of the integral structure of an atom but were traveling freely through space—the electrons making up a cathode ray beam, for example—might have this diffuse wavelike quality and follow specific paths of motion that were directed by their accompanying pilot waves.

At first many physicists of the day found this a very shifty idea, a little more than they could readily stomach. They could agree, perhaps, that radiant energy might exist in particlelike chunks and even, perhaps, demonstrate mass under some circumstances. But to stand by and watch age-old concepts of matter as discrete, solid, massive stuff begin to dissolve away into a sort of limbo of wavelike fuzz was a bit too much. Some men, however, found the idea intriguing. Just a year after de Broglie published his theory of pilot waves, a Viennese physicist named named Erwin Schrödinger obligingly worked out the mathematical equations that perfectly described and matched de Broglie's strange pilot waves and showed how they could be perfectly applied to all kinds of electron motion. But even at that the idea was considered a somewhat wild-eyed speculation until 1927 when two American physicists came up with some thoroughly disturbing experimental data.

These men, C. J. Davisson and L. H. Germer, working at Bell Telephone Laboratories, reasoned that if electrons actually *were* accompanied by "pilot waves," or if they were themselves wavelike in nature, then it should be possible to make them demonstrate certain kinds of wavelike properties in the laboratory rather than always behaving stubbornly like solid massive particles. Davisson and Germer tried aiming a beam of high-energy electrons at a diffraction grating in hopes of being able to demonstrate a wavelike interference pattern, just as earlier physicists had demonstrated interference patterns when a beam of light was aimed at a diffraction grating. Of course the "wavelength" of high-energy electrons would have to be so very, very short that any ordinary etched diffraction grating would be far too coarse to serve the purpose. Instead of trying to find a way to etch a finer grating artificially, these men decided to use the surface of a crystal for their grating, reasoning that neighboring molecular layers of the crystal would act as "lines" in the grating.

Incredibly enough, the idea worked. Davisson and Germer were able to demonstrate that electrons were diffracted by their crystalline "diffraction grating" and formed a perfectly characteristic diffraction pattern on a screen or a photographic plate, with the diffraction bands widening or narrowing as the velocity of the electrons was increased or decreased. This astounding discovery could only have one meaning: that electrons did indeed demonstrate wavelike properties. Solid particles did not bend around corners, and streams of solid particles could not form interference patterns. De Broglie's pilot waves were proven a physical reality although no one knew what they were, and the always-elusive electron suddenly became far less solid and far more elusive than anyone had ever dreamed before.

## HEISENBERG AND THE UNCERTAINTY PRINCIPLE

By this time it was becoming clear that something very odd was happening in the pursuit of the elusive electron. It seemed that the more that physicists studied the nature and behavior of atomic electrons in detail, the farther away they were moving from a clear-cut, visualizable picture of what the electron was, of how it moved, of where it might be located at any given time, and in what shape it might be found if it could be found at all.

This was an alarming departure in the history of the study of physics. Always before, more careful, detailed, and refined study of things had led to a *clearer* and *more understandable* picture of how things worked. In this case, however, it seemed that the more refined the study became, the more *confusing* the picture became. And this was not even the end of the study. The worst was yet to come. Physicists had always assumed that by studying the elusive electron more and more closely and in greater and greater detail they should ultimately be able to pin down specifically and precisely what it was and how it behaved. As they waded deeper into the study of quantum wave mechanics, particularly after de Broglie's contribution, that assumption began to look slightly doubtful.

Then, in the same year that Schrödinger published his mathematical analysis of the mechanics of de Broglie's pilot wave, another German mathematician named Werner Heisenberg struck a fatal blow at the hopes of those who were determined to pin down precisely what an electron was and where it might be found at a given instant in time. Heisenberg came up with a mathematical analysis which indicated that physicists never *would* come close to pinning down an electron in minute detail. Heisenberg's analysis suggested that the very best one could ever do would be to discover the *probability* (within certain limits) that an electron would be one place doing one thing at a given instant of time rather than some other place doing something else, and that the very act of attempting to examine that electron any more closely in order to be *more certain* of where it was and what it was doing at a given instant *would itself alter where the electron was and what it was doing at the instant in question.* Heisenberg, in effect, was saying that in dealing with the behavior of electrons and other elementary particles the laws of cause and effect do not and cannot apply, that all we can do is make predictions about them on the basis of probability, and not a very high degree of probability at that.

Heisenberg's contention that the behavior of particles in the micro-universe can never under any circumstances be defined with any degree of certainty has come to be known as the *uncertainty principle.* It is of such enormous importance to the understanding of modern physics that we need to see clearly just what it means and how it is applied. A couple of simple examples will be illuminating.

Suppose that an elephant wants to see what is going on inside an anthill at a given moment; so he pushes it over and turns it upside down with his trunk. What happens? Does he find out what was going on inside that anthill at some moment before he started meddling with it? Of course not. All he really learns is what goes on inside an anthill that has been turned over by an elephant, with ants fleeing in all directions. He is no closer to satisfying his original curiosity than he was to start with, and he probably ends up with a trunkful of stinging ants into the bargain.

Again, suppose that a clumsy scientist wants to find out how fast a mouse's heart beats; so he starts picking up mice to see. Unfortunately every time he picks up a mouse the foolish thing dies of fright and its heart stops beating. Does the scientist find out what he wants to know? Obviously not; all he gets for *his* trouble is a growing collection of dead mice.

Some years ago a clever science fiction story was written about a scientist who set up an elaborate closed-room experiment with which to investigate how a monkey behaves when it doesn't know it is being observed. The room was arranged so that the only way anyone could see into it was by peeking through the keyhole. The scientist put the monkey in the room and settled back to wait for the monkey to get used to his new surroundings before beginning to observe him through the keyhole. But to the scientist's disgust, every time that he peeked through the keyhole to observe what the monkey was doing, all he was ever able to see was the monkey peeking back through the keyhole at him!

Each of these fables for frustrated physicists is a crude example of the uncertainty principle in action. Each, in one way or another, illustrates the same thing: that in any experiment there is always some risk that the observer's attempt to observe something may in itself alter the phenomenon he is trying to observe so drastically that he ends up failing to observe what he wants to at all. All he sees is some phenomenon that has been altered by his own attempt to observe it. His own presence, or the presence of his instruments, in themselves, alter what he wants to study so that he can't study it. Yet obviously he also can't study what he wants to study without introducing himself or some kind of instruments in order to make observations. In such a case, it seems, the observer is damned if he does and damned if he doesn't.

But isn't there some way that he can succeed in studying what he wants to study without running into this awkward dilemma? If what he wants to study is the behavior of subatomic particles in the micro-universe, the answer is: No, there is no way that he can avoid the dilemma. In 1926 Werner Heisenberg published a paper proving mathematically that there was indeed no way to escape damnation, at least when it came to gathering detailed information about the nature, behavior, and location of any individual electron in an atom at any given instant of time. If one *insisted* upon trying, he might, figuratively speaking, end up with a trunkful of red ants, or a pocket full of dead mice, or a curious monkey looking out the keyhole at him—but he would *not* find out what he wanted to find out.

But why not?

Heisenberg's contention was based entirely upon theoretical mathematical analysis. Thus his conclusions might conceivably have been shrugged off (at least for a while) on the grounds that they had no basis in reality; just because it works out that way on the blackboard, we might argue, doesn't mean it has to work out that way in the laboratory. But the idea that lay behind Heisenberg's mathematics was unfortunately not so easy to shrug off.

We can understand why more clearly by following a very simple imaginary experiment through to its logical conclusion. Suppose we are serving hot buttered rum on a cold winter's day in a chilly room, and someone complains that the drink is too hot while someone else complains that it's too cold. To settle the dispute we attempt to measure the temperature of the toddy using an ordinary laboratory thermometer. Very quickly a temperature is registered on the thermometer. "Aha!" we say. "The temperature of this hot buttered rum is exactly 120 degrees Fahrenheit—not too hot, not too cold, but just right."

Now for all practical purposes our experiment is perfectly satisfactory, the result is useful, and the dispute is settled. But was our temperature measurement *accurate?* That is, did it express the *exact* temperature of the toddy at the instant we stuck the thermometer in? The answer of course is no. The temperature registered on the thermometer may have been *useful,* but it was not *accurate* by any manner of means.

Why not? Because since the room was cold the thermometer itself was considerably cooler than the toddy at the time we began to measure the temperature. The thermometer had a certain volume, as did the toddy, and as soon as it was inserted into the fluid, heat from the hot buttered rum began to flow into the thermometer until the thermometer had warmed up to a point that the toddy temperature and the thermometer temperature were in equilibrium. But this had to mean that the final equilibrium temperature, which was the temperature we read from the thermometer, was actually *slightly lower* than the exact temperature of the toddy the instant we stuck the thermometer in it. What we ended up with was not the exact temperature of the toddy at the moment it came into question but a *measured approximation* of the temperature which we could have foreseen would have to be slightly lower than the exact temperature we were seeking to determine.

But how *much* lower was the approximation than the exact actual temperature before the measurement was made? The answer is that it depends. If we had known the exact temperature of the thermometer to start with and its exact volume, both of the part that was to be immersed in the drink and of the part that was not to be immersed (but which would also absorb *some* heat by conduction), and if we knew numerous other physical characteristics of the thermometer such as the coefficient of expansion of the glass and of the mercury in it, and if in addition we knew in minute detail the exact volume of the hot buttered rum that we were going to examine and the exact temperature of the room air to which that drink was steadily radiating heat—if we knew all of these things and a few dozen others we will not bother to itemize, we could then conceivably calculate *approximately* how much error our act of measurement introduced and come up with an *approximate* correction of the measured temperature to account for this error. But why just *approximately?* Couldn't we calculate the error *precisely* and thus correct our end result to come up with an absolutely accurate temperature?

The answer is still: No, we could not. Each of these items of data entering into our calculation could be obtained only approximately rather than exactly. We might measure the volume of the thermometer accurately to five decimal places, but we would still not have the *exact* volume. We might calculate the coefficient of expansion of mercury to ten decimal places if we chose, but even with that figure used in our calculations we would still come up with only an approximate correction, not an exact correction. Granted that our correct results would be *a much closer approximation* than before, but it would still be only an approximation. In other words, with a great deal of labor we might *approach* the exact temperature of the hot buttered rum at the instant we wanted to know it, perhaps even approach it very closely, but we could not completely reach it.

Again, from a practical standpoint, the answer we did come up with would ordinarily be quite close enough to satisfy the most fussy man—except that by now we are getting annoyed. We don't *want* an approximation. We want an *exact* temperature. Clearly a major hang-up here is the comparative size of our measuring instrument and the interaction that is occurring between it and the toddy; so we attempt to get around this problem by using a much smaller thermometer.

This thermometer, too, will interact with the toddy, but to a much lesser degree. The resulting measurement would be a closer approach to the exact answer than we could possibly accomplish with the larger thermometer. If we picked a still smaller thermometer we would end up with a still closer approach, but even then *only an approach.*

In fact we could go through a whole succession of smaller and smaller thermometers, calculating the degree of alteration of temperature brought about by each one, and find ourselves getting closer and closer to the exact temperature we wanted. We could plot our results on a graph, and find that the result with each thermometer would fall somewhere on a curve of a geometric progression. The strike point we are seeking to reach would lie at the very end of that curve. If we had an *infinitely small* thermometer, we could then get the *exact* temperature end-point we sought. Or so it would seem. Surely, then, the only thing preventing us from obtaining an absolutely accurate answer is the large and clumsy nature of our equipment. Surely we could at least *imagine* circumstances which would provide us our absolutely exact answer, couldn't we?

Once again the answer is: No, we could not.

If all the laws of classical physics were completely valid for all aspects of the universe, large and small, then we could indeed imagine circumstances that would provide the answer we wished. But as we have seen, those classical laws are not entirely valid for all aspects of the universe. In the micro-universe of physics, they become flatly invalid. Descending into that world, we would ultimately reach a point at which *we could no longer reduce the size of our imaginary thermometer any more.*

Bear in mind that in measuring the temperature of a fluid we are essentially using an indirect means of measuring *interchanges of energy*—heat energy, kinetic energy, and radiant energy. These interchanges of energy occur between our measuring instrument and the substance that we are measuring. And these are not just effects, they are *interchanges,* exchanges of energy back and forth, so that the measuring instrument is changing the thing that is being measured just as the thing that is being measured is changing the measuring instrument. Granted that the more we reduce the size of the measuring instrument the more we reduce the *amount* of interchange of energy between that instrument and the substance being measured. If we regard the energy being interchanged as composed of quantum-sized bundles, then we might say that fewer and fewer quanta of energy are interacting. But when our thermometer has finally been reduced to such a size that only *a single quantum of energy* remains to interact with the substance being measured, we have blown the works.

We still have only an approach—a very close approach, but still just an approach—to an absolutely exact temperature reading, and we have reached a point where we simply cannot come any closer to an "infinitely small thermometer" than one that has but a single quantum of energy with which to interact with something else.

In other words the nice curve on our graph, which ought to show us zero deviation from the exact answer at the point that the thermometer has become infinitely small, actually has to stop somewhere short of that imaginary point, as if it had run into a brick wall. As Heisenberg expressed it, just as the velocity of light is the upward limiting velocity with which any material object or signal can travel through space, a single quantum of energy is the lowermost limiting unit at which any kind of measuring interchange can take place. We can go no further, and we still do not have the answer we want.

Of course we have emphasized right along that this fanatical insistence upon absolutely accurate measurement is completely pointless and unreasonable in the case of measuring the temperature of a hot buttered rum. As with a vast multitude of other situations in our everyday life, a reasonably close approximation to an exact measurement fulfills our needs just as well as the exact measurement itself would. We don't really *need* to know the *exact* temperature of the hot buttered rum. If we are sawing boards to build a door frame, we may need the accuracy of our measurement of the various pieces to be within a tolerance of an eighth or a sixteenth of an inch; we don't *need* accuracy to 1/10,000 inch even if we could achieve it, because the door will hang just as well without it. In building components for an Apollo rocket measurement tolerances of 1/10,000 or 1/20,000 inch may well be absolutely necessary, whereas tolerances of 1/1,000,000 or 1/1,000,000,000 inch might be extremely difficult if not impossible to achieve and would still yield no more useful results.

Thus in most areas of our everyday world the classical laws of physics provide us with perfectly satisfactory answers. If those answers are slightly incorrect, the tiny degree of inaccuracy doesn't make any difference. We can disregard such inaccuracy for the simple reason that it isn't significant.

But when we are exploring in the micro-universe of physics, trying to measure and predict the exact position of a given electron revolving around the nucleus of a given atom at a given instant of time, and seeking to measure its momentum at that very same instant, we find ourselves in serious trouble. On this microcosmic scale a measurement tolerance of a billionth of an inch is far too coarse and crude to be of any use whatever. And in this microcosmic universe *anything we do* to try to pin that particular electron down, *any* kind of observation or measurement whatever, is in itself going to alter either the location of that electron or its momentum, one or the other. The closer we try to come to an accurate measurement of one property, the more we will be altering the other property. We are not only unable to predict with any certainty *where that electron will be,* or *what it will be doing, at some specific instant in the future;* we cannot even describe where it actually is and what it is actually doing *right now* with any accuracy or certainty whatever!

The best that we can come up with is the *probability* that that electron is somewhere in such and such a general area, and the *probability* that it has a momentum somewhere between this limit and that limit, right now. As for the future, all we can predict again is a degree of *probability* that the electron will be found somewhere between this area and that area, and the probability that it will have a momentum between this limit and that limit at the instant in the future that we specify. We can even calculate the *degree* of probability with regard either to the electron's position or to its momentum; it is precisely this sort of calculation that was made possible by Werner Heisenberg's mathematical analysis and equations. And this, it begins to appear, is something of considerable value.

But in the last cold analysis probability is *not* certainty. The more we try to reduce the limits of probability—that is, the more certain we try to become about a given electron's *position* at a given instant, the wider the limits of probability we must accept with regard to what its *momentum* is at the same time, and vice versa. The more closely either one property of the electron or the other is examined, the more closely we approach certainty with regard to one property or the other, the more wildly uncertain the other property becomes. And since an electron can really only be fully described in terms of *both* its position and its momentum at any given instant, *it becomes utterly impossible to describe an electron at all* in terms of absolute certainties. We can describe it only in terms of uncertainties or probabilities.

This concept was a blockbuster, appearing as it did in the world of physics in the late 1920s. It meant that the whole concept of how the micro-universe could be explored and what might actually be discovered about it had to be changed radically. A great many physicists of the day were perfectly appalled at the idea that virtually everything happening in the micro-universe, everything happening in the world of electrons, photons and atomic nuclei, could only be known in terms of probability. Gone was the hope of ever coming up with a clear-cut visualizable model of an atom that described all of its properties accurately. Gone was the hope of ever describing *precisely and for certain* where a given electron actually was and what it was actually doing at a given time.

Even more appalling, gone was the classical and hitherto unchallenged concept of a totally orderly universe in which every event, large or small, occurred as an effect of some previous event, and in which every event went on to become a cause of a subsequent event. If events in the micro-universe followed laws of probability rather than laws of cause and effect, then it surely seemed, as Einstein emotionally expressed it, that "God was playing with dice."

Even to this day there are physicists who are still trying to contend with Heisenberg's uncertainty principle and demonstrate that it is invalid. Einstein himself could never accept Heisenberg's conclusions, and he spent the latter years of his life in an unsuccessful search for some way to avoid the implication of Heisenberg's equations. Many physicists saw all hope of learning more about the inner structure of the atom vanishing before their eyes. But Heisenberg himself took no such gloomy view.

True, he said, there could be no closer approach to an accurate physical model of the atom than had already been obtained. True, more careful physical observation would not produce new knowledge. But Heisenberg maintained that all this was unimportant and beside the point anyway. The uncertainty principle was merely telling physicists that the path they had been following in the study of atomic particles was in fact a blind alley and that they might just as well toss aside once and for all their foredoomed efforts to make exact determinations and specific measurements at the atomic level and lower. The uncertainty principle implied that, in order to learn more, a whole new approach to investigation had to be developed. The classical physical methods of measurement and observation had to be discarded, if more were to be learned, in favor of a theoretical mathematical approach which would simply ignore exact description of individual particles and concern itself only with the probable activities of those particles.

And already, Heisenberg pointed out, other physicists had explored this new mathematical approach and found it fruitful. De Broglie's concept of electrons as wavelike particles and Schrödinger's mathematical equations for those waves had already laid the groundwork for this new mathematical approach to investigation of the micro-universe.

For the nonscientist, the reader of this book, Heisenberg's uncertainty principle has a somewhat different significance. In a sense it marked the end of the clearly understandable and the easily comprehensible in the progress of modern physics. As we have seen in the foregoing pages, the pathway explored by Planck, Einstein, Bohr, Pauli, de Broglie, and the others was becoming increasingly more tortuous and confusing; a simple understanding of the nature of the electron and the structure of atoms seemed to become *more* complex and elusive rather than less so. But at least nonscientists could follow lines of reasoning, and concepts and conclusions based upon a visualizable, measurable physical approach to the atom.

And now, here was a man who claimed that this approach had to be discarded, that further progress could only be made with the aid of complex mathematical constructions and analyses far beyond the grasp or comprehension of ordinary mortals and considerably beyond the grasp of many physicists.

This does not mean that we must stop at this point and toss in the sponge, for some of the most exciting developments in the world of modern physics are yet to be discussed. What this *does* mean is that to comprehend these advances at all we must look more to general concepts and generalized descriptions of the implications of those concepts and less to concrete demonstrable example. It means, in effect, that to comprehend what has been happening in physics since 1930 or thereabouts we must become more inaccurate in what we say, from the scientific point of view, and settle for analogies that fall farther from their mark than those we have employed so far.

But why did physicists have to accept Heisenberg's decrees in the fashion they did? Strange and confusing as Heisenberg's ideas may seem to us, and much as great physicists resisted the implications of those ideas and fought against junking time-tested techniques of study and methods of thinking in favor of a new and unexplored approach, the uncertainty principle had two enormously potent forces going for it.

First, like Mt. Everest, it was there. There was nothing wrong with Heisenberg's mathematics, and his conclusions drawn from his mathematics had to be considered just as valid as any conclusions drawn from observation in the laboratory. They could not be ignored, nor shrugged off, nor explained away, although they have indeed been challenged again and again, and quite effectively by certain modern-day physicists, as we will see.

But perhaps more important, as the new mathematical techniques demanded by the uncertainty principle were slowly worked out, they began opening doors that had never opened before. The indisputable fact was that *Heisenberg's mathematical approach worked.* In one particular area in which physicists had literally been stumbling in the dark—the study and investigation of the structure and behavior of the nucleus of the atom—these new mathematical techniques were literally lifesavers; they were the *only* techniques of investigation that led to any progress whatever. By using the complex mathematical groundwork of wave or quantum mechanics, physicists during the 1930s and in subsequent decades were at last able to probe deep into the nucleus of the atom, seeking to identify the true basic elementary particles of which atoms of all matter were composed. Those investigators did not find just the few simple particles they expected to find there; instead they found themselves discovering a veritable volley of "elementary particles" of all sorts and varieties, particles that no one had even suspected of existing. Some of those particles were so fantastic as to defy credibility, but all are related in one way or another to the way in which the nuclei of atoms are constructed, the way those nuclei break apart, and the way they interact with each other.

If Heisenberg's work marked the bitter end of an era of scientific investigation, it also marked the beginning of a new and even more fascinating one.

# CHAPTER 27

## *Into the Heart of Physical Matter*

In the previous chapter we turned our attention to just one of the exciting areas of research and discovery in modern physics: the origin of the concept of energy quanta, the evolution of the quantum theory, and later the development of the complex and paradoxical concepts of quantum or wave mechanics arising from increasingly close and critical study of atomic electrons.

This work took place largely between 1900 and 1930, and focused largely on the study of the tiny negatively charged particles making up the outer shell of atoms. It laid the basic groundwork for practically everything that was subsequently discovered about the structural nature of matter. But this was not the only work that was progressing during this revolutionary thirty-year period. So far we have said very little about the dogged investigation of the compact and massive core or nucleus of atoms that was going on at the same time, other than to point out Ernest Rutherford's conclusion that atomic nuclei were compact gatherings of elementary particles that were almost two thousand times more massive than the planetary electrons, but each carrying a positive electrical charge equal in magnitude to the negative charge of an electron. And indeed, Niels Bohr, Wolfgang Pauli, Louis de Broglie, and many other pioneering physicists were concentrating their attention and their thinking so fiercely upon problems that seemed related to the electron shells of atoms that it is easy to get the impression that they regarded the nuclei of atoms rather indifferently as vague concentrated lumps of positively charged matter and nothing else.

Of course this impression is not accurate. From Rutherford's time on, physicists were quite as deeply concerned about the structure of the nuclei of atoms as with the behavior of atomic electrons. It was obvious that this part of an atom's structure was every bit as complex as the nature and behavior of the electrons that surrounded it, and all the more mysterious and inaccessible to investigation because it was so difficult to approach the nucleus of the atom in any way. Electrons, at least, could be knocked flying from an atom relatively easily if suitable techniques were used. Beams of free-moving electrons could be created at will, boiled off the heated metal filament of a cathode ray tube. Thus, in effect, electrons could be separated, counted, weighed, and measured. They could be stripped away from one atom and added to another with comparative ease. Their interactions with quanta of radiant energy such as light photons, X-rays, or gamma rays could be brought out into the open and studied in a variety of ways.

In other words, electrons were more or less freewheeling characters, seldom tied up so tightly in the structure of atoms that they could not be separated, examined, and counted, at least within limits. But the matter in the nucleus of an atom was something else altogether. Whatever its true nature and composition, it was *not* easily accessible for investigation. Whatever the constituent parts of atomic nuclei were, they were clamped tightly together by forces that no one understood—forces far stronger than the forces of electrostatic attraction and repulsion. In order for physicists to invade the nuclear core of an atom at all, it was necessary to bombard it with atomic bullets of enormous energy, and even then it was difficult to get such bullets even close to the nucleus, much less to penetrate it.

Furthermore, since all the positively charged matter in atomic nuclei was clustered in a tight little bundle, concentrated in one tiny area in space instead of being spread around diffusely the way electrons were, it was exceedingly difficult even to approach the problem of separating one constituent part from another. To tear apart this tiny atomic fortress and examine what made it tick seemed an almost unattainable goal to physicists; yet it was far too exciting a challenge for them to leave alone.

## THE MODERN ALCHEMISTS

At the time of Rutherford's pioneering experiment in which he bombarded various kinds of atoms with alpha particles from decaying radioactive material, at least one constituent of the atomic nucleus was quite well acknowledged. Whatever else atomic nuclei might contain, it seemed clear that they always contained one or more massive positively charged particles, the *protons,* which seemed to be identical in every way with the nucleus of a hydrogen atom from which the electron had been stripped away.

Physicists of Rutherford's day were well convinced that any atom, if it were to be electrically neutral, ought to contain in its nucleus at least enough positively charged protons to equal the number of electrons in its periphery. Thus it seemed reasonable that atoms of the lightest of all elements, hydrogen, should have a single proton in the nucleus with a single electron in the periphery, the one positive charge perfectly balancing the one negative charge in spite of the difference in masses of the two particles. Atoms of the second lightest element, helium, then ought to have two protons in the nucleus to balance electrically two planetary electrons, and this contention seemed to be confirmed by the fact that alpha particles, which were nothing more or less than the nuclei of helium atoms from which the electrons had been stripped off, did have two positive charges—or at least, a total positive charge twice the magnitude of the positive charge of a hydrogen nucleus. Carrying the same reasoning on, atoms of lithium, the third lightest element and the lightest of all the metals, was assumed to have three protons in each nucleus and three planetary electrons.

It was obvious then, that if the number of electrons in a given atom was always matched perfectly by the number of protons in its nucleus, then atoms of each successively heavier element should contain one proton (and one electron) more than atoms of the element preceding it, and this succession of heavier and heavier atoms should follow the periodic succession of elements that had been worked out by Mendeleev in his periodic table. The number of electrons in any given atom (and thus, the number of protons in its nucleus) was called the *atomic number* of the element, and elements with increasing atomic numbers could be lined up in the periodic table in a neat succession.

But already, by Rutherford's time, two problems had appeared that made this neat succession somewhat less neat than it seemed at first glance. Bohr and his associates had explained satisfactorily why the electrons in an atom's periphery were not pulled down toward the positively charged nucleus in a steady spiral: Electrons could only drop down from outer quantum levels to inner quantum levels by emitting photons of energy, and the photons that had to be emitted became progressively larger the lower the quantum level to which the electron dropped. Once an electron had reached the innermost quantum level it could no longer have the energy necessary to emit the high-energy photon it would have to emit to make the final jump down into the nucleus.

Thus a hydrogen atom with its single positively charged proton in the nucleus and a single negatively charged electron revolving around it might represent a fine stable system, with perfectly balanced electrical charges. But what about the helium atom with two positively charged protons apparently sitting side by side in the nucleus, packed together into a very small space, as Rutherford's experiments had indicated? According to any number of experimental observations two objects with similar electrical charge should *repel* each other, with that force of repulsion becoming increasingly violent the closer the particles were brought together. Why, then, didn't the two protons in the helium nucleus fly apart from each other explosively? *What made them stick together?*

Atoms of lithium were just as puzzling, only more so. Here *three* protons were packed together in the nucleus. All three ought to burst apart from each other in violent repulsion. And what about atoms of uranium, in which no fewer than ninety-two positively charged particles had to be crammed together into a very tiny area of space? What kept all these positive charges from flying apart violently? For that matter, why hadn't the nuclei of all atoms of solid matter long since exploded as a result of electrostatic repulsion of the protons in their nuclei, disintegrating all solid matter into hydrogen atoms?

In Rutherford's time no one had answers to these questions, but it was obvious that these disastrous explosions of atomic nuclei were not occurring. Clearly *some* kind of hitherto unrecognized, powerful intranuclear force had to exist that counteracted the electrostatic repellent forces of similarly charged particles in the nuclei of atoms, so that protons in those nuclei remained firmly bound together in spite of themselves. But what sort of force was this "intranuclear binding force"? Could it be related to gravity? Unlikely: The masses involved in atomic nuclei were far, far too small for any significant gravitational attraction whatever to exist between their constituent particles. Could some unrecognized electromagnetic force be at work here? Possible, but again unlikely: Even electromagnetic forces could have only strength enough barely to counteract the mutually repellent force that must exist between these protons, so that at the very best any atom would be excessively unstable and ready to fly apart at the slightest jostling if electromagnetic forces were binding its nucleus together. And since this obviously did not appear to be the case, the electromagnetic forces were pretty well ruled out. Indeed, it became increasingly clear that this mysterious nuclear binding force, whatever it was, had to be far more powerful than either gravitational or electromagnetic force.

But this was not the only problem with the idea that atomic nuclei were composed only of protons. As we saw earlier, physicists and chemists of the late 1800s had observed that atoms of all elements heavier than hydrogen actually weighed considerably more than they should be expected to weigh if their nuclei were really made up only of enough protons to balance the number of electrons and thus to occupy a given atomic number on the periodic table. In fact, atoms often had measured atomic weights that were two or three times more than they should have been if they had only protons in their nuclei.

A helium nucleus, for example, obviously had only two protons in its nucleus since it had the positive charge of only two protons, but when actually weighed it was found to have a mass of at least *four* protons. Where did this extra mass come from without any additional electrical charge? Atoms of oxygen, the element occupying the eighth lightest place in the periodic table, have an atomic number of eight—that is, they contain eight protons and eight electrons—yet the actual atomic weight of oxygen atoms is more nearly sixteen. We have also seen that all the atoms of a given element do not necessarily have exactly the same atomic weights as all other atoms of that element; different *isotopes* of a given element have differing atomic weights, although all isotopes of a given element have *exactly* the same number of protons in their nuclei. Again, where does the extra mass come from?

In the search for answers to these questions Rutherford remained a vigorous experimental pioneer. He knew that the atomic nucleus was a well-defended fortress, but his own early bombardment experiments had proved that it was not necessarily impregnable. Even in nature, radioactive decay of heavy atoms such as uranium or radium resulted in the spitting-out or emission of chunks of the nuclei of those atoms (alpha particles) as well as electrons (beta rays) and short-wavelength high-energy radiation (gamma rays). By using the same techniques of atomic bombardment with high-speed alpha particles which earlier had proved that atoms had nuclei in the first place, Rutherford set about attempting "heavy bombardment" of various atomic nuclei, hoping to break them apart just by hitting them hard enough, so to speak.

It was painfully slow work. Most of the alpha particles were hurled away as they approached atomic nuclei. Even in rare cases of a head-on collision with the nucleus, an occurrence about as unlikely as Robin Hood's legendary feat of splitting an arrow lodged in the target with his own arrow from five hundred yards away, alpha particles just didn't seem energetic enough to fight their way through into the nucleus in most cases. Rutherford tried a wide variety of different target atoms; in case after case even those alpha particles scoring "direct hits" were merely thrown back in the direction from which they had come.

Then Rutherford, in the year 1911, tried bombarding a tube filled with nitrogen gas, and at last was rewarded with success. The long-awaited direct hits were rare, but when they did occur the scintillator screen revealed fragments flying off that appeared to have significantly longer range than repelled alpha particles would. These fragments were, in fact, hydrogen ions—positively charged protons—which were being expelled when the nuclei of nitrogen atoms were smashed open like rotten melons from the impact of an alpha particle. Furthermore, a quite different atom remained after the impact than had been present before. For the first time in history the nucleus of an atom had been smashed artificially, with a small chunk (a hydrogen ion) torn away, leaving a residue which, having "captured" the bombarding alpha particle, had become the nucleus of a different atom altogether. The original nitrogen nucleus had become the nucleus of a "created" oxygen atom! Ernest Rutherford had achieved what alchemists throughout the centuries had failed to achieve: he had artificially *transmuted* atoms of one element into atoms of another. What happened in the course of this transmutation can be represented diagrammatically as in Figure 47, or written in a simple shorthand form as follows:

$$ {}_2He^4 + {}_7N^{14} \rightarrow {}_8O^{17} + {}_1H^1 $$

Rutherford's experiment was a spectacular success. It demonstrated beyond question that atoms *could* be transmuted from those of one element into those of another by means of artificial tampering with the nucleus. But this modern alchemy was slightly irrelevant to Rutherford; he was not, after all, primarily interested in transmuting base metals into gold. Far more important, his experiment demonstrated beyond question that *the nucleus of an atom could be breached,* given the right kind of battering ram with the right amount of energy behind it.

Nor was this some kind of special magic that Rutherford alone could achieve. Following the publication of his work in 1919 physicists all over the world scrambled to repeat his experiments in a multitude of different forms. The awkward scintillator screen was discarded in favor of more sophisticated cloud chambers with which to observe the results of collisions

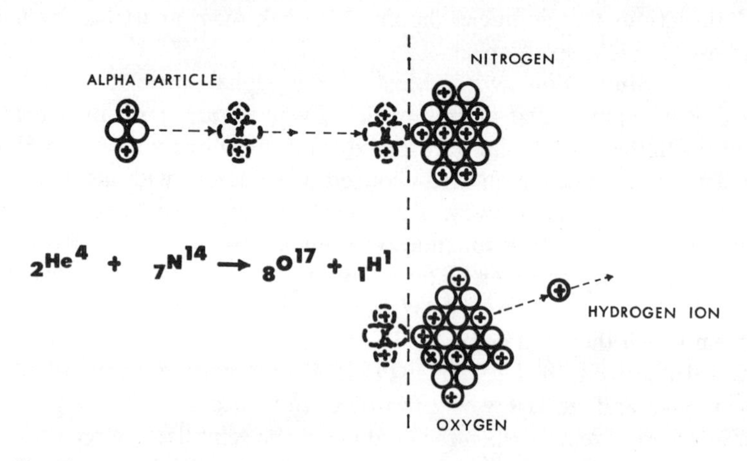


*Fig. 47* Transmutation of a nitrogen atom into an oxygen atom by bombardment with an alpha particle (helium nucleus). A proton (hydrogen ion) is also produced in the reaction. (In physicist's shorthand the reaction is written out as a formula, using the letter symbol for the given elements; the number before the symbol indicates the atomic number of the atom—i.e., the number of protons in the nucleus—and the superscript after the symbol indicates the atomic weight —i.e., the number of protons and neutrons combined).

between atomic bullets and their nuclear targets. By 1925, six years after Rutherford's first atom-smashing party, a technique had been devised whereby the paths of charged particles revealed in a cloud chamber could actually be photographed, so that when an atomic nucleus was shattered the event could literally be caught on film and preserved for leisurely study and comparison later.

## THE DISCOVERY OF THE "NEUTRAL PROTONS"

Once it was established that atomic nuclei could be broken up under the right circumstances, and that high-energy atomic "bullets" were the key to the process, physicists all over the world began to search for ways to provide more energetic projectiles with which to bombard nuclei. Alpha particles were natural radioactive breakdown products, endowed only with the relatively low energy that God gave them. Now work began to develop machines which could accelerate various bombarding particles to higher and higher velocities before hurling them at the targets. One of the earliest such devices was a high-tension particle accelerator built by John Cockcroft and E. T. S. Walton which was capable of accelerating bombarding bullets to energy levels of up to one million electron volts.*

A few years later a man named Robert Jemison Van de Graaff devised a generator which could accelerate particles up to energies of many millions of electron volts. After a succession of these "small fry" particle accelerators had been devised, the world's first truly big atom-smasher was built: the famous cyclotron at the University of California in Berkeley, capable of driving atomic projectiles up to energy levels of 40 million electron volts, almost five times higher than that of the energy of the fastest natural particles.†

With the construction of bigger and better machines to deliver bombarding particles at their targets, it was not long before atomic nuclei of all kinds were being broken, altered, and transmuted in experimental physics laboratories. Then in 1932 another major breakthrough was made in the study of the composition of these nuclear fortresses, a discovery that shook the world of nuclear physics. This discovery was the work of James Chadwick, one of Rutherford's students. Rutherford himself had never been able to explain why nuclei of atoms weighed so much more than they ought to, judging from the number of protons they contained. These nuclei certainly contained protons, but it also seemed clear that they must contain some other unknown constituent that no one had yet detected or noticed, some kind of particle with approximately the same mass as the proton but without any electrical charge. Even though these mysterious particles had not been detected, Rutherford had become so certain that they had to exist that he began speaking of them as *neutrons* and started a search for proof of their existence.

* The "electron volt" is a unit of energy used by physicists in describing the velocity and energy with which various particles strike their nuclear targets and is defined as the energy an electron gains when moving in an electric field between two points with a potential difference of one volt between them. Thus a single electron volt is a very tiny amount of electrical energy. A million or a billion electron volts is something else again. In the physicists' shorthand one electron volt is indicated by the symbol eV. A million electron volts is indicated by MeV, and a billion electron volts by BeV. It is because of this terminology that atom-smashing machines capable of delivering bombarding particles with a billion or more electron volts are commonly referred to as "bevatrons."

† We get some notion of the extremely rapid development in this field of physical research when we consider that the famous Berkeley cyclotron, first set in operation in 1939, was torn down in 1962, less than twenty-five years after it was built, because it had become obsolete.

In 1932 the search was rewarded when his student Chadwick discovered some kind of mysterious ray that was thrown out by atoms of the element beryllium when it was bombarded by alpha particles. By using these rays in turn as the "bullets" with which to bombard nitrogen atoms, Chadwick obtained cloud chamber photos which showed nitrogen nuclei recoiling as if struck by some kind of unseen heavy particle. These unseen particles were not repelled by the nitrogen nuclei the way alpha particles were; on the contrary, the nitrogen nuclei seemed particularly vulnerable to direct collisions with these new particles and were bounced around by them like sticks in a stream. What was more, these unseen particles did not leave a stream of ionized gas in their wakes the way positively charged alpha particles did, and seemed not to be influenced by electrical or magnetic fields. Their presence could only be assumed from the way the nitrogen nuclei behaved when they had been hit.

By measuring the amount of recoil of these nitrogen nuclei Chadwick found that the unseen particles emitted from the beryllium perfectly matched the theoretical description that had been worked out for neutrons. These were particles of a mass almost the same as that of a proton, but with no electrical charge at all. Beyond doubt these heavyweight particles had to be the "silent partners" that existed as integral parts of atomic nuclei, along with protons, in most atoms and accounted for the difference between the atom's atomic number in the periodic table and the same atom's measured atomic weight. Neutrons were bound together with protons by some mysterious nuclear attractive force to form the massive nuclei of atoms.

Having no charge themselves, neutrons could not affect the electrical balance of atoms in any way, but they did affect their weight. Thus helium nuclei with their double positive charge and their mass equal to four times the mass of a single proton were really composed of two protons and two neutrons, bound together in some mysterious fashion and traveling as a unit. The nuclei of all oxygen atoms contained eight and only eight positively charged protons, but various isotopes of oxygen might contain seven, eight, nine, or ten neutrons as well. And the nuclei of uranium atoms, which were known to contain 92 protons also contained, in addition, 143 or more neutrons!

## PARTICLES IN PROFUSION

Chadwick's experimental proof that neutrons really did exist as definable, detectable elementary particles within the nucleus of atoms did not exactly come as an unmixed blessing in the world of nuclear physics. Not that it was not an exciting discovery, as well as an extremely useful one. It appeared to clear up one persistently nagging question: the mystery of where an atom's excessive weight was coming from. What was more, the neutron was an extremely useful particle to use for further nuclear bombardment. Having no electrical charge itself, it could penetrate the electrical barrier of a nucleus easily, and could be used to disrupt nuclei even when it was traveling at relatively low velocities.

But there was also something slightly ominous about the appearance of this newcomer on the scene. What was the neutron itself composed of, and how did it get to be a part of the atomic nucleus? Its presence there did not help in the least to explain the nature of the nuclear binding forces holding protons closely packed together, and the appearance of yet another elementary particle at a party that was already getting a little crowded made physicists slightly nervous. They couldn't help but wonder just what *else* might come spilling out of atomic nuclei if they kept bombarding them with projectiles of higher and higher velocity and energy.

These worries of the experimental physicists were compounded by some bizarre and upsetting predictions that the theoretical physicists were beginning to come up with as a result of some of the highly rarefied mathematical analyses arising from the study of quantum mechanics. The experimentalists liked things that they could get their hands on or point to in cloud chamber photos; they were often suspicious of this new brand of "blackboard physicist" who depended upon complicated mathematics for his discoveries and predictions. The trouble was, some of the most wild-eyed mathematical predictions imaginable had proved to be right on target as subsequent experimental evidence of their validity piled up in the laboratory.

One such theoretical gadfly was a British theoretical physicist named P. A. M. Dirac, a thin and ascetic young man who appeared more like a vaguely absent-minded philosophy professor than a phvsicist. As a result of some extremely abstract mathematical considerations Dirac, in 1929, came up with the notion that electrons bearing single *positive* charges ought theoretically to exist in the universe just as well as the "ordinary" electrons known to have negative electrical charges. Dirac further predicted that such positive electrons should have a number of curious properties unlike those of any known particles in the universe.

For example, he predicted that they should have what he called "negative inertial mass"—that is to say, when they were pushed in one direction by a physical force they would respond to the force by moving in exactly the opposite direction from which the force was applied.* Dirac's mathematics indicated that such positive electrons could exist only in matched pairs with ordinary negative electrons, but that since such a positive electron and an ordinary negative electron would annihilate each other instantaneously and become converted to pure energy the moment they came in contact, there had been no way that such positive electrons heretofore could have been experimentally detected.

* This prediction turned out to be wrong, in case anyone was thinking of using this strange "antiparticle" as a basis for an antigravity machine!

This idea seemed so utterly fantastic and so far removed from anything that had ever been observed to happen in the micro-universe or anywhere else that few physicists took it very seriously at first. Dirac's work was subjected to heavy criticism as living proof that the newly developed mathematical techniques were actually coming up with results so totally useless and meaningless that they had no relationship to reality at all. But the criticism stopped abruptly a couple of years later when the "impossible" was actually found to happen in the laboratory. In 1932 an American physicist named Carl Anderson photographed a particle track in a cloud chamber which was identical to the track one would expect an ordinary electron to leave behind it, but which *curved in the opposite direction* on its path through a magnetic field. Whatever strange particle left the track behind only survived for an instant, but there was no doubt that this was indeed the "positive electron" (later called a *positron*) that Dirac's mathematics had predicted. There was simply nothing else that it could be.

The positron, when it was discovered, was not found as a part of any atomic nucleus. It appeared to be formed when the air in a cloud chamber was bombarded by extremely high-energy gamma rays striking the earth from outer space—the so-called *cosmic rays* that had been discovered as early as 1913. True to Dirac's predictions, a positron always made its appearance paired up with its opposite number, an ordinary negative electron, and could be detected only if mutual annihilation of the pair were prevented by pulling them apart at the instant of their formation by means of powerful magnetic fields. Under the influence of high-energy cosmic rays bombarding atoms in the air or the surface atoms of metal plates, these positron-electron pairs gave every appearance of being created out of nothing. In truth, however, such pairs of particles were really "created" from the energy carried by the gamma rays. Over 1 million electron volts of radiant energy is required to produce one such positron-electron pair, with part of that energy being converted into the actual mass of the two particles and part transformed into the kinetic energy with which they speed away from the point at which they are created.

The electron that is created in such an event is precisely the same as any other "ordinary" electron, and becomes a permanent citizen on the scene, with a capability of existing infinitely just as long as it does not have the misfortune of running into a positive electron anywhere. Ordinarily, once freed from its partner-in-formation, the encounter of such an electron with another positron anywhere in the universe at any time would be exceedingly unlikely. After all, negatively charged electrons are the standard kind of electron that exists, at least in our part of the universe; it is the positron which is the intruder in the normal order of things.

But the positron that is formed in our part of the universe as part of a positron-electron pair, whether by natural happenstance or by human intervention, does not fare so well. Surrounded by a virtual sea of negative electrons, it can hope to exist only for a split second before encountering its nemesis and joining it in a mutual suicide pact in which both positron and electron cease to exist and a new photon of radiant energy exactly equivalent to the combined rest masses of the positron and the electron together appears in their place and speeds away at the velocity of light.

Indeed, under ordinary circumstances and without the help of a powerful magnetic field to pull these particles apart at the very instant of their formation, positron-electron pairs would inevitably annihilate themselves just as fast as they were formed, instantaneously throwing out a new photon so identical to the photon that created them in the first place that there would be virtually no way to detect that anything had even happened. This, of course, was precisely why there had been no hint that any such particle as a positron might conceivably exist, previous to Dirac's mathematical predictions. Considering the rapid development of increasingly sophisticated apparatus for studying nuclear bombardment, it now seems likely that the positron would have been discovered experimentally sooner or later whether Dirac had predicted it or not, and we can imagine what a highly embarrassing customer it would have been if it *had* been so discovered experimentally without any warning! As it was, however, the mathematical techniques of quantum mechanics had scored a major victory, validated as a tool for frontier exploration in physics, and another perplexing member had been added to the increasingly crowded elementary particle zoo.

Dirac's positive electron or positron was not by any means the only bizarre and peculiar elementary particle that was first "discovered" as a mathematical necessity by theoretical physicists and then only later confirmed by experimental physicists to have real existence. This was, in fact, becoming more and more a *pattern* of discovery as the ranks of the theorists and the experimentalists gradually began to diverge during this period of almost breathtaking progress. One of the great changes brought about by the development of the quantum theory and mathematical analysis of quantum mechanics was the gradual disappearance from the scene of the "compleat physicist" who was fully competent to deal both with the theoretical (that is, mathematical) aspect of research and the experimental aspect with equal facility. It was getting to be too much for any one man to master both sides of the fence; more and more it was becoming a case of the left hand of physics not knowing (or not quite understanding) precisely what the right hand was doing.

Theoretical physicists like Einstein, Bohr, Pauli, de Broglie, Schrödinger, Heisenberg, and Dirac were on one side of the fence—men who never went near an experimental laboratory if they could help it, and (according to George Gamow's account) invariably smashed all the equipment in it if they did. On the other side of the fence were experimental physicists like Rutherford, Chadwick, Millikan, and Carl Anderson, who were brilliantly effective in working out experimental devices and then using them to make profound observations of things occurring in nature, but who were either uninterested, unwilling, or unable to deal with the complex mathematics of the theoretical side. Ernest Rutherford was probably one of the most brilliant experimental physicists of all time, but it is said that he was so poor in mathematics that he had to have a mathematically inclined associate work out the formulas to describe what was happening in his famous alpha particle bombardment experiment. Only a very few modern physicists, Enrico Fermi, for example, have had sufficient breadth to do really outstanding work in *both* theoretical and experimental physics.

Fortunately this division of physicists into the ranks of the theorists and the experimentalists seldom led to the sort of conflict that might seem inevitable. Rather, a singularly fruitful sort of partnership began to evolve in which workers on one side of the fence began to respect and pay attention to what those on the other side had to say and workers on both sides found areas in which their work dovetailed splendidly. Sometimes the experimentalists needed the theorists to figure out what the devil it was they had just discovered; sometimes the theorists needed the experimentalists to demonstrate that particles or interactions they had predicted on the blackboard really did have physical reality. And between the two divergent approaches to physics an exceptionally powerful two-edged weapon was forged by which to attack the basic problems of the structure of matter.

We can see how fruitful this kind of partnership approach was in the discovery of one of the most bizarre and unlikely of all elementary particles as yet known to physics—a particle that was first predicted in the 1920s but proved so elusive that it was not actually trapped and identified in the laboratory until 1953. As far back as 1914 experimenters such as James Chadwick had noticed a peculiarity in the manner in which naturally radioactive atoms decayed. Chadwick had found that while the alpha particles and the gamma rays that were hurled out of radioactive nuclei were always emitted with exactly the same energy each time, the beta particles or electrons which were emitted could have energy levels that varied widely from particle to particle. For some reason or other some electrons emitted from such radioactive atoms seemed to fly out with far more energy than others did, and as a result the mass-energy balance sheet before and after the emission of these electrons didn't quite add up properly.

Niels Bohr was one of the theorists who tackled this problem, in a typically bold and radical way. Bohr, who had seen so many of the "laws" of classical physics crumble with the appearance of relativity and quantum theory, became convinced that here was a case in which the classical law of conservation of mass-energy was violated. If so, it was the first time such a violation of that venerable and powerful law had ever been observed; it had survived the Einstein revolution as one of the most firmly established and unshakable of all natural laws. Assuming that here was a case in which it did *not* hold true, Bohr labored to work out a mathematical "proof" of his contention, but the best proof that he could come up with proved so clumsy that even he didn't like it very much.

Then his friend Wolfgang Pauli suggested an alternative solution to the problem. Pauli pointed out that the mass-energy balance before and after the emission of electrons from natural radioactive nuclei could be maintained if the electrons that were emitted were accompanied by the emission of one or more of a completely unknown particle—a particle that had no electrical charge and practically no mass at all, but which carried with it a very significant quantity of energy. Pauli maintained that these mysterious particles could be emitted along with the electrons in such a way that the sum total of the energy of the emission was always the same, thus keeping the energy books in balance, so to speak.

Of course, the trouble with this handy little "unknown particle" of Pauli's was simply the fact that if it actually had no electrical charge and little or no mass it might well be able to buzz along through thousands of miles of solid matter without ever once either hitting or otherwise affecting a single atom in any way, just as a small space vehicle might be able to travel in a straight line through millions upon millions of light-years of space without ever running into a star or a planet. But if this were true, it would mean that Pauli's unknown particle might well be practically undetectable in any way. Many physicists of the 1920s and 1930s thought this convenient little particle of Pauli's was really nothing more nor less than a "solution" that had been cut out of whole cloth in order to answer an awkward problem; they didn't really think such a thing existed at all. Enrico Fermi humorously gave Pauli's imaginary particle the name "neutrino" meaning "little neutral one," and even half-believed it might exist, but saw no way anyone could ever prove whether it did or didn't.

But the *idea* of Pauli's neutrino continued to exist, and indeed, as time passed more and more theoretical and circumstantial evidence began piling up to suggest that such a particle really ought to exist. As one physicist expressed it, "If it *hadn't* existed, somebody would have had to invent it." Fortunately physicists were spared that labor of creation; in 1953, some thirty years after Pauli first suggested the existence of his hypothetical neutrino, two physicists working at Los Alamos finally succeeded in trapping a few neutrinos escaping among the radioactive decay products from an atomic pile. They found that neutrinos, like electrons, spun on their axes like little tops and otherwise behaved very much like electrons excepting that they did not carry the burden of any electric charge at all. As more and more has been learned about them in the ensuing years, it appears increasingly likely that neutrinos play an absolutely critical key role in the maintenance of the mass-energy balance throughout the entire universe.

Another theoretical prediction of a previously unsuspected elementary particle provided the long-sought answer to a major problem in nuclear physics: the problem of what kept protons and neutrons so tightly glued together in the nucleus in spite of electrical forces tending to force them apart. At the same time, the prediction of this particular elementary particle opened a veritable Pandora's box of other peculiar particles upon the world of physics. As we saw earlier, the discovery of the neutron as an integral part of atomic nuclei was no help in explaining the nature of the forces holding atomic nuclei together. No one could understand why they did not simply explode as a result of the mutual repulsion of all the protons packed so tightly together.

Nor, on the other hand, did the discovery of neutrons help in the least to explain why nuclei of certain heavy radioactive atoms were so unstable that they spontaneously emitted part of their mass and energy from time to time in the process of radioactive decay. Enrico Fermi, the brilliant Italian-born physicist who later became one of the leaders of the Manhattan Project which developed the first nuclear fission bomb, tackled the

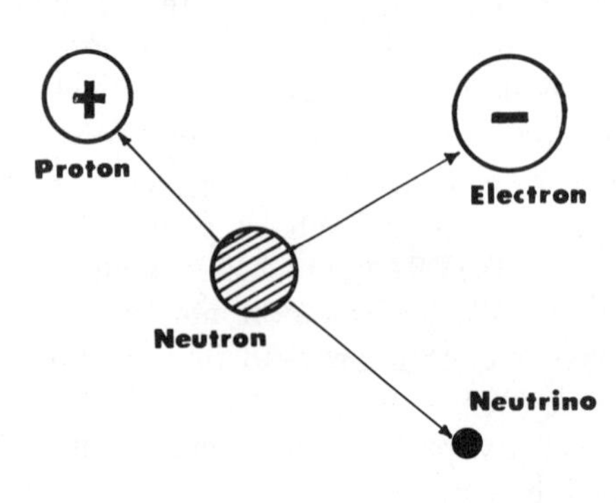


*Fig. 48* A neutron spontaneously decays into a proton and an electron-neutrino pair. In this drawing the electron is shown exaggerated in size to emphasize its massiveness in contrast to the massless neutrino. If the diagram were to represent size of the particles in true proportion to their masses, neutrino would not appear in the drawing at all and proton would be 1800 times larger than electron.

problem of what was happening within radioactive nuclei. Fermi's approach to the problem was to try to determine what energies had to be involved in radioactive emissions, and in the nuclear binding forces holding protons and neutrons together. From his work Fermi came up with the theory that the neutrons in atomic nuclei, unlike the protons, are basically unstable particles and that in radioactive nuclei in particular a certain number of neutrons tend spontaneously to transform themselves into more stable protons by expelling matched pairs of electrons and neutrinos. It was these electrons, Fermi maintained, that could be detected as beta rays being emitted by radioactive materials. This process is illustrated diagrammatically in Figure 48.

Fermi spoke of such particle transformations as "weak nuclear interactions" because they seemed to occur reluctantly and only with comparative difficulty, or more accurately, they seemed to occur with a low degree of probability. Using the mathematical techniques of quantum mechanics, Fermi computed the probability of such a neutron-to-proton transformation with the resultant emission of an electron-neutrino pair, and then calculated the energy that would have to be involved in such interactions. He arrived at the series of formulas which could be applied theoretically to all transformations of one nuclear particle into another, whether spontaneously or because of contact or collision.

These theoretical considerations, far out as they seemed, actually agreed splendidly with what was observed to occur during the decay of radioactive nuclei, and inevitably physicists began to wonder if these same formulas might not also explain the nuclear binding forces holding neutrons and protons so tightly together. Unfortunately, they did not. Fermi's weak nuclear interactions might theoretically provide *some* degree of binding energy between protons and neutrons, but further mathematical analysis indicated that the binding energy arising from such weak interactions would have to be far too little to accomplish the job. Protons and neutrons clearly had to be bound together by some type of interaction far more powerful than this weak one.

In 1935 a Japanese physicist named Hideki Yukawa came up with an alternative proposal. Yukawa suggested that if the internuclear binding forces could not be explained on the basis of weak nuclear interactions in which electron-neutrino pairs, for example, might be jumping to and fro between a neutron and a proton, thus binding them together, then perhaps an entirely new and unsuspected nuclear particle might be present to do the jumping to and fro. Having proposed this purely theoretical possibility, Yukawa then embarked upon the ambitious task of developing a theoretical "description" of what such a purely theoretical particle would have to be like if it were really to be able to provide sufficient internuclear binding energy to do the job.

Again, Yukawa's approach was mathematical and theoretical, and the hypothetical particle he came up with was a strange one indeed: it had to be a particle that was two hundred times heavier than an electron, but some ten times lighter than either a proton or a neutron. Since this particle theoretically would have to occupy a sort of limbo halfway between a proton and an electron in mass, Yukawa called it a mesotron or *meson.*

One of the most obvious curiosities about this hypothetical particle, of course, was simply that there had never been any kind of elementary particle even remotely like it either observed or even suspected before. If it *did* exist, it would have to be quite a different breed from anything anyone had ever yet heard of. Yukawa was undaunted by the fact that this "made up" particle of his had no similarity whatever to any known elementary particle—that it had, in fact, been "built up logically" on the blackboard—and proceeded to make predictions as to how mesons would have to behave if they existed. He theorized that these particles would have to undergo very strong interactions between protons and neutrons, interactions involving such powerful forces that they could serve as the "nuclear glue" that kept protons and neutrons packed tightly together inside atomic nuclei.

Having reached this goal, Yukawa more or less dusted off his hands and said to the experimental physicists, in effect, "Okay, boys, now it's up to you."

The experimentalists, needless to say, were somewhat nonplussed by Yukawa's mesons. None had ever been observed; indeed, mesons had been concocted solely by means of mathematical computations based on theoretical considerations. If they really existed at all, they were particles so different from any other elementary particles known, and so deeply buried in the nucleus of atoms, that it seemed ridiculous to hope that they could ever actually be observed or identified in any possible way.

But once again the high-energy cosmic rays raining down upon the earth's atmosphere came to the rescue. In 1937 Carl Anderson detected certain kinds of particles that seemed to shower constantly down on earth as a result of cosmic ray bombardment of the gas in the earth's outer atmosphere. In fact, he detected not one but two different kinds of particles which were unlike any kind of elementary particles hitherto known and which seemed to have the sort of "halfway" intermediate mass predicted by Yukawa for his mesons.

Further laboratory study of these odd particles indicated that they were constantly present in the earth's atmosphere and that one of the two kinds of particles not only demonstrated *precisely* the mass of Yukawa's predicted particles but also demonstrated the very type of *strong nuclear interaction* between protons and neutrons in the nuclei of atoms that Yukawa had predicted that his mesons ought to demonstrate. Today these two kinds of particles are known as "pi mesons" or *pions* and "mu mesons" or *muons;* and while the role of muons in the infinite scheme of things still remains a riddle today, there is no longer any serious question that pions are the identical particles that Yukawa "created" on his blackboard in 1935, and that they play an exceedingly important part in binding the particles in atomic nuclei together.

## RADIOACTIVE DECAY AND NUCLEAR CATASTROPHES

The discovery of pi mesons or pions and their identification as the source (or at least *one* source) of strong nuclear interactions that bind protons and neutrons together was a staggering achievement. It answered a multitude of questions about the nature of the structure and behavior of atomic nuclei. For one thing, the discovery of pions helped to clarify the phenomenon of natural radioactivity. The nuclear binding force generated as a pion hopped back and forth between a neutron and a proton (as diagramed in Fig. 49) was powerful indeed—far more powerful than gravitational forces; weak nuclear interactions such as the four-way interaction that can occur among a neutron, a proton, an electron, and a neutrino; or even electromagnetic forces. We might imagine this "hopping back and forth" as somewhat akin to a man staying afloat on the water by hopping back and forth between two logs, neither of which was big enough to support him for any length of time but either one of which could support him momentarily. As long as he keeps moving from one to the other, he stays afloat, but each hop from one to the other involves an energy exchange between the man's foot and the log. A pion hopping back and forth between a neutron and a proton in an atomic nucleus was in effect a rapid succession of strong nuclear interactions which kept the neutron and proton bound together, as long as the pion kept moving, with a remarkably powerful force of attraction.

That intranuclear binding force generated by strong nuclear interactions within an atom's nucleus is indeed powerful—but it is not *all-powerful.* It is far more powerful than the electrostatic force of repulsion tending to push

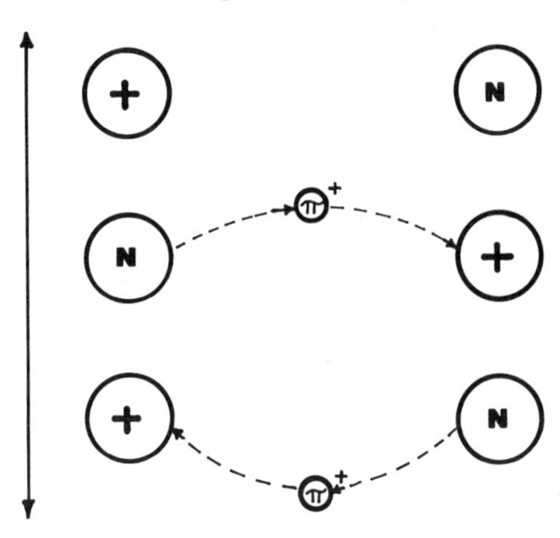


*Fig. 49* A strong nuclear interaction, in which a proton approaching a neutron spontaneously transforms itself into a neutron and a positive pion, and the neutron absorbs the pion to become itself a proton. Thus, the two original particles "change place." The pion shuttling back and forth between such a "changeable" pair of particles creates a powerful binding force between them—a force now called an *exchange force.*

protons apart from each other in the nucleus but it has, so to speak, an Achilles' heel: It cannot exert its power over any great distance. Within a certain very small radius it can completely overcome the electrostatic force and keep the protons and neutrons in the nucleus of a comparatively light atom tightly bound together, by virtue of pions leaping back and forth unremittingly. But in the case of a very heavy atom with a large nucleus containing a great many positively charged protons and still more electrically neutral neutrons, the nucleus is intrinsically unstable, and a curious thing can happen: If an interacting set of neutrons and protons near the outside of such a nucleus should chance to move just far enough away from the rest of the nucleus to escape the powerful but short reach of the nuclear binding force, electrostatic force would instantly act to expel this chunk of nuclear material from the nucleus en masse.

Granted that such an event is not very likely to happen in any given heavy nucleus. Certainly it never happens *predictably,* even in the nuclei of the largest, heaviest, most unstable or radioactive atoms known. But in any such radioactive atom *there is always a certain small probability* that just such a chunk of nuclear material will at some time be found far enough away from the rest of the tightly bound nucleus to be "unprotected" by the intranuclear binding force and thus to be hurled out of the atom altogether by the force of electrostatic repulsion. And as we have seen earlier, an alpha particle hurled out of the nucleus of such an unstable or radioactive atom is nothing more than a pair of protons tightly bound together with a pair of neutrons (together with their pions hopping back and forth)—a relatively stable structural unit which periodically gets thrown out of the interior of an unstable nucleus. Thus, we could think of the emission of an alpha particle by the nucleus of a uranium atom as a sort of "step toward stability"—getting rid of two protons from a nucleus that was already overloaded with them to the point of instability.

With this event, of course, the nucleus of the uranium atom changes spontaneously into the nucleus of an atom of thorium, an element that has two fewer protons in its nucleus than uranium does. But there is also a slight difference in the neutron-proton ratio in the new nucleus. Both uranium atoms and thorium atoms normally have about 1.6 times as many neutrons as protons in their nuclei. But the newly formed atoms of thorium will have two more neutrons in their nuclei than they would normally have; so the thorium atoms formed from radioactive decay of uranium will be heavy isotopes of thorium. Since alpha particles always have an equal number of neutrons and protons, two of each, as successive emissions of alpha particles take place from decaying radioactive nuclei, the neutron-proton ratio grows larger and larger; there are progressively more neutrons in proportion to the protons with each alpha particle emission. But for each element there is an outer limit to the number of neutrons that can be tolerated in proportion to the protons. When this ratio is exceeded, one of the neutrons has to go, one way or another, and this is usually accomplished by a simple and straightforward process: One of the "excess" neutrons spontaneously emits an electron and a neutrino, and thus becomes a proton. Energy imbalances within the newly formed nuclei may also be resolved by the emission of a quantum of radiant energy from the nucleus in the form of a gamma ray. Some of these adjustments in electrical charge, neutron-proton ratio, and energy imbalance in radioactive nuclei are illustrated in the diagram in Figure 50.

What does this add up to? Obviously the stepwise decay of an unstable radioactive nucleus is not as unpredictable and sometime-maybe as it first appeared to earlier investigators. Rather, it is a remarkably orderly and logical chain reaction of events, with the stage for each event set by another event that happened before. When a radioactive nucleus takes the "major step" of emitting an alpha particle in order to move toward greater stability, a different kind of instability is usually created by this event. That new instability can be resolved only by emission of electron-neutrino pairs (beta rays) and/or quanta of radiant energy in the form of gamma rays. Ultimately, stage by stage, protons within the unstable radioactive nuclei are unloaded, the energy books are tidied up temporarily, excess neutrons discard electrons and neutrinos to become new protons, and new imbalances must then be corrected. The result of this is that the radioactive nucleus follows a zigzag and changing path, transforming itself from the nucleus of one element to the nucleus of another element, and then another, but always

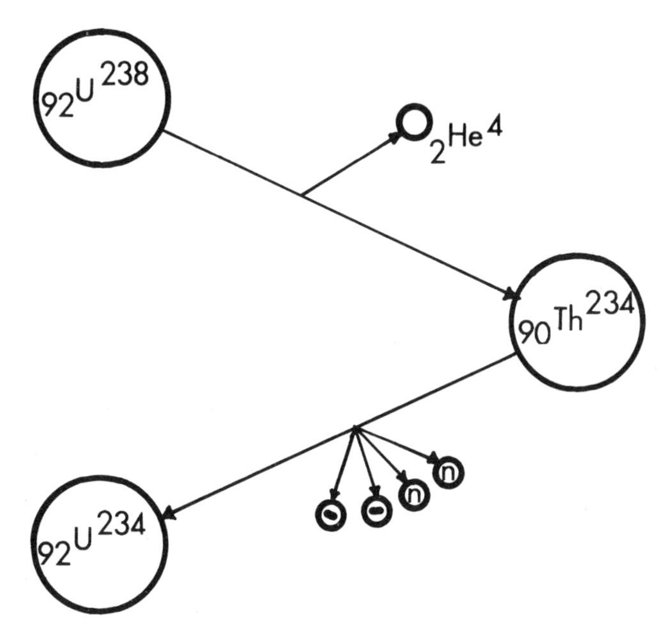


*Fig. 50* The first two steps in the radioactive decay of Uranium 238. First an alpha particle is emitted (2 protons and 2 neutrons lost from nucleus) leaving an extra-heavy, unstable isotope of thorium, $Th^{234}$. In seeking stability of its unstable ratio of protons to neutrons, the thorium nucleus in turn decays, but in different fashion: 2 of its excess neutrons expel an electron (—) and a neutrino (n) each and become protons, thus changing the thorium nucleus into a nucleus of uranium again, this time $U^{234}$. Subscript before the atomic symbol represents *atomic number* (i.e., number of protons in nucleus). Superscript following atomic symbol represents atomic weight (i.e., total number of protons and neutrons together).

steadily decreasing in mass, steadily decreasing in energy, and steadily working toward a newer, more permanent stability as the total number of protons in its nucleus is whittled away and discarded.

It is an uneven and hesitating path, sometimes a matter of taking two steps forward and one step back; but gradually the total number of protons in the unstable nucleus is reduced in a long and orderly sequence of transformations until at last the heavy, unstable radioactive nucleus has been "whittled down" or reduced to a permanently stable, nonradioactive nucleus in which the nuclear binding force holding the protons together can permanently counterbalance the electrostatic repulsion seeking to force the protons apart.

In Figure 51 we have diagramed the stepwise radioactive decay process

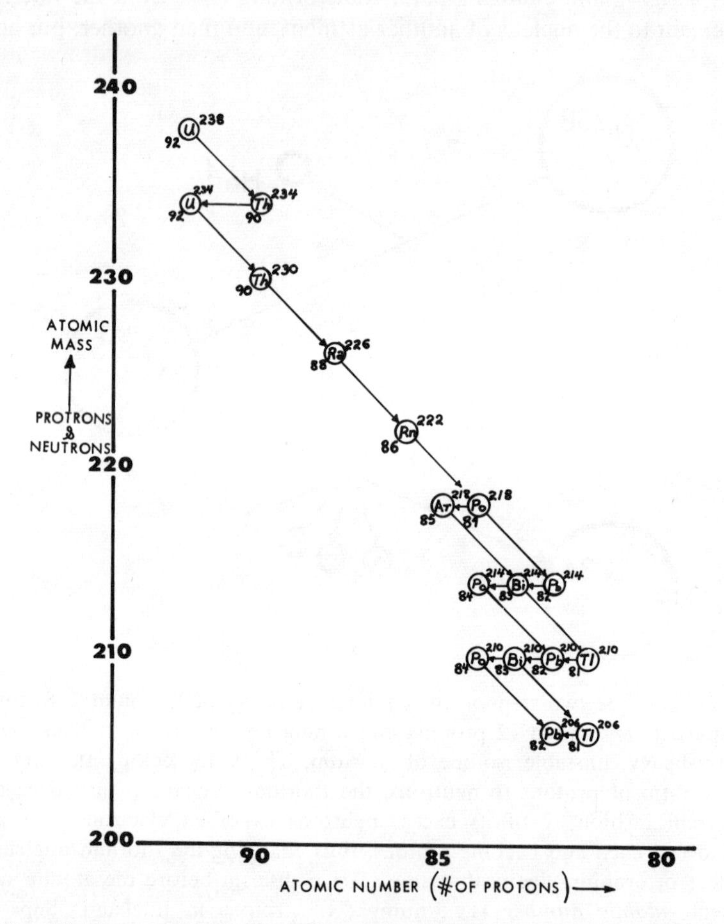


*Fig. 51* The long trail of radioactive decay as $U^{238}$, with its 92 protons, alters spontaneously in a stepwise, zig-zag fashion to become, finally, $Pb^{206}$ with 82 protons, a stable non-radioactive isotope of the element Lead.

that occurs in the case of one kind of radioactive atom, the natural atom of uranium 238. Other heavy radioactive elements follow different zigzag courses, making different elemental way-stops along the route, and reaching somewhat different ultimate permanent and stable goals, usually one or another stable isotope of the element lead. But if the sequence of steps in the radioactive decay process is logical, fixed, and unvarying, with each step predetermined by the nature of the instability left by the occurrence of the last step, the timing of various steps may vary widely, since the occurrence of any given step is a matter of probability rather than predetermined cause and effect.

Thus, no one can say how much time will be required for any given step to be taken in any given stage of the radioactive decay of any given nucleus: A given atom may be "hung up" at one stage for years, decades, centuries, or millenniums even though that particular next necessary step *on the average* only requires a split second to occur. The best one can say is that each step in a given decay process requires an average time—each step has a specific, calculable, and unvarying half-life—yet even the average times or half-lives for each of the steps may be totally different, varying from an average time of a billionth of a second in one step to an average time of thousands upon thousands of years for another step.

At points in the process where energy imbalances are acute, or where instability is excessive, the transformations are likely to occur very quickly, as if the nucleus were sitting on a hot stove at such-and-such a stage and wanted to get out of there into some other more comfortable state in a hurry. Thus the half-life of some highly unstable, newly formed nucleus may be very short, with the probability that any given nucleus will remain in that state for longer than, say, one billionth of a second, being very small. Other "way station" nuclei formed in a given progression may be far more stable, so that the probability of transformation into the next state at any given instant becomes much less; such "way stages" may have average half-lives of thousands of years. But in each and every radioactive decay process, the over-all end result is the transformation of an unstable radioactive nucleus by means of gradual disintegration into a nucleus of an isotope of lead—a permanently stable nucleus—and once this goal is reached the long, difficult road of radioactive decay for that atom comes to an end.

Thus we can see that the discovery of neutrons, neutrinos, and pions led physicists to a far clearer and more rational understanding of the natural radioactive phenomena that had been puzzling investigators since they were first discovered in the 1890s. But the discovery of these particles and the development of new techniques of investigation also brought about a long series of totally unexpected discoveries as well, some of which had grave implications indeed for the future of human society in the universe of everyday life.

As we pointed out, once neutrons had been discovered and identified, they were soon found to be extremely useful as "atomic bullets" for the bombardment of all kinds of atomic nuclei. Carrying no electric charge themselves, even very low-velocity and low-energy neutrons could easily pierce into the very heart of a nucleus, providing a new and comparatively easy means for transforming lighter elements from one into another, or even for creating isotopes of various elements which were found in nature very rarely if at all. By means of neutron bombardment and other newly developed bombardment techniques, dozens of such new isotopes were created.

Some were unstable radioactive forms of commonplace elements which never ordinarily had any radioactive isotopes at all in their natural state, and some of these man-made radioactive isotopes were found to have some remarkably useful properties. Naturally occurring iodine, for example, is composed almost entirely of stable, nonradioactive atoms with atomic weight 127, with a few equally stable atoms with atomic weight 126 thrown in, the heavier isotope differing from the lighter only in having one more neutron. But by bombarding slightly heavier atoms such as xenon or tellurium with neutrons, and then allowing the resulting radioactive atoms to decay, physicists were able to produce an unstable, artificial, radioactive isotope of iodine with an atomic weight of 131, bearing four more neutrons in its nuclei than natural iodine atoms do.

These man-made radioactive $I^{131}$ atoms behaved exactly like natural iodine atoms in every way except that they were radioactive, with a half-life of approximately eight days, and their remarkable usefulness was immediately apparent to medical researchers. It had long been known that natural iodine in small amounts was essential for human life, since the iodine enters into a complex chemical combination to form the thyroid hormone produced in the thyroid gland. Thyroid gland disorders and diseases have tormented mankind since time immemorial; yet there had been no good way to examine thyroid indirectly or find out what was going on in it as regards thyroid hormone production, so that many thyroid disorders were extremely difficult to diagnose before they had already done their damage.

Researchers knew that iodine taken into the body by mouth was very quickly concentrated in the thyroid gland, and when radioactive $I^{131}$ was produced, it soon became widely used as a "radioactive tracer" by which the inner chemical activity of the thyroid gland could be examined and studied quickly, painlessly, and harmlessly. And as a result, early, life-saving diagnosis of thyroid disorders could be accomplished.

Other man-made radioactive isotopes also proved of great medical value. Radioactive phosphorus, for example, provided the first effective means of treating such disorders as polycythemia vera, in which huge excesses of red blood cells are produced in the bone marrow. $P^{32}$ administered as a medicine to a victim of polycythemia was quickly concentrated in the patient's bones and survived long enough before decaying into a stable element to whittle down the bone marrow's excessive cell-making activity.

Research into artificial and naturally occurring radioactive isotopes led in other useful directions as well. It was discovered, for example, that a great many commonplace and normally stable elements actually contain small quantities of radioactive isotopes occurring in the natural state. Of all known elements, carbon, with an atomic weight of 12, is one of the most common and most stable to be found on earth. But among every hundred carbon atoms a few oddballs are found with an atomic weight of 14. These atoms, "overstuffed" with neutrons, so to speak, are radioactive and decay spontaneously with a half-life of approximately 5,700 years.

Archaeologists immediately recognized that $C^{14}$ could provide them with a crafty sort of radioactive time clock with which they could estimate accurately the age of ancient artifacts containing materials compounded of carbon. Animals living millenniums ago, for example, ate vegetable products throughout their lives and collected carbon compounds, including some radioactive carbon 14, in their bodies and bones.

Once those animals died, the rate of intake of carbon ceased for all time, but the carbon 14 already stored in, say, an animal's skull, continued gradually decaying for millenniums after the animal's death. By analyzing the skull of such an animal, the exact amount of carbon 14 that still remained in it undecayed could be measured. This amount could then be compared with the amount of carbon 14 present in the skull of a similar animal still living today. By this technique, archaeologists could determine how much of the carbon 14 that had originally been in the ancient skull had so far decayed, and knowing the half-life of carbon 14, could pinpoint to within a dozen years precisely how long ago that ancient animal died. Willard Frank Libby first pioneered such "radioactive dating" using carbon 14 in 1946, and was awarded the Nobel Prize for this work in 1960. But by now, a wide variety of other radioactive dating techniques have been developed as well, and their accuracy demonstrated beyond question. Understanding of the radioactive decay process has thus provided scientists with a powerful tool for accurate dating of events known to have occurred in earth geological ages ago.

Unfortunately, all the studies and discoveries in this area of physics were not so fruitful and beneficial to mankind. From the very first, the bombardment of atomic nuclei had revealed one thing very clearly: that fantastic quantities of energy are bound up in the nuclei of atoms, and that tampering with those nuclei could result in the release of that energy, often in very dangerous or even violent form. When neutrons were used to bombard the nuclei of lightweight atoms, transmutation of those atoms from elements of one atomic weight to another could occur in a variety of ways, generally either by chipping away part of the existing nucleus or by adding neutrons to the nucleus. When certain nuclei were struck by a neutron, the bombardment might result in the expulsion of a proton or an alpha particle from the nucleus. In other cases such a neutron bombardment would cause another neutron to be expelled from the target nucleus without the bombarding neutron itself being captured at all, while in other cases the neutron "bullet" might be caught and held by the nucleus. But in many such bombardments it was found that energy was released from the target atoms in the form of gamma rays, and sometimes unstable radioactive isotopes of elements would be formed which subsequently regained stability by means of the emission of electrons or of energy in the form of gamma rays.

These kinds of reactions became well known and predictable eventualities expected by physicists during the bombardment of light atoms. But the bombardment of nuclei of heavier atoms which were, by virtue of the large number of protons held in their nuclei, intrinsically more unstable, did not always lead to such predictable results. In the late 1930s a number of nuclear physicists who were engaged in such bombardment experiments had already begun to speculate that some kinds of neutron bombardment might conceivably disrupt the nuclei of certain heavy atoms far more radically than had previously been observed, perhaps even tearing away great chunks of nucleus with the release of large quantities of energy and the formation of any number of strange and unexpected fragmentary by-products.

Then in 1939 a German chemist named Otto Hahn performed an experiment which proved how grimly prophetic these speculations had been. Hahn found that when the nuclei of uranium atoms were bombarded with relatively slow-moving, low-energy neutrons, the result was not the expected splitting-off of a tiny fragment of the nucleus such as an alpha particle or a proton. Instead, the uranium nucleus was literally split in half, divided into two approximately equal fragments which proved to be abnormal isotopes of the much lighter element boron. This so-called fission of uranium nuclei was accompanied by the release of energy—but not of just a few quanta of energy. As much as 200 million electron volts of energy could be released in the fission of a single uranium atom.

Furthermore, in addition to the main fragments that were formed in Hahn's fission experiment he discovered that a number of free neutrons were also emitted in the course of the breakup, and that when the fission occurred in the nuclei of atoms of uranium 235, enough of these excess "fission neutrons" were produced that some of them could proceed to collide with other uranium nuclei, again causing the nuclei to split and pour out energy and produce further excess "fission neutrons." In short, the fission of a single uranium 235 nucleus could trigger an expanding chain reaction of nuclear fission, so that in the space of a very short time all the $U^{235}$ nuclei present would be broken up with the violent liberation of a perfectly staggering amount of energy.

The overwhelming significance of Hahn's discovery was recognized in a matter of hours by physicists all over the world. This was no mere chipping away of fragments of atomic nuclei; this was a release of energy such as had never been observed before. Obviously if some way could be found to control the speed of the chain reaction taking place in a given quantity of uranium 235, an incredibly rich source of usable power would be within human grasp. Equally obviously, if some way could be found to trigger an uncontrolled chain reaction in a critical mass of uranium at the precise instant and place desired, the explosive release of nuclear energy which would occur could be used as a bomb of such violent destructive force that the ordinary chemical explosion of TNT or nitroglycerin would be like the explosion of a firecracker by comparison.

Unfortunately, the discovery of uranium fission was made on the eve of the outbreak of the Second World War, so that primary attention was turned to exploring the destructive use of nuclear chain reactions. Study of Otto Hahn's uranium fission phenomenon was immediately buried in secrecy in the major warring countries and a race to the death was on for the development of the most horribly destructive weapon yet known to man.

Fortunately for the world, Adolph Hitler had a low opinion of scientists in general and physicists in particular; Hitler's government in Germany responded indifferently to the potential weapon power that lay within its reach. In the United States, thanks to the urging of respected scientists who clearly saw the handwriting on the wall, development of a controlled and self-sustaining chain reaction in uranium 235 became a crash research program in the ultrasecret "Manhattan Project" under the over-all direction of Enrico Fermi. At the time there was no method known for separating the tiny amount—less than 1 per cent—of fissionable $U^{235}$ out of the great mass of unfissionable $U^{238}$ in natural uranium; but with his powerful mathematical ability Fermi figured out a geometrical configuration for a "uranium pile" which would give a chain reaction even in such $U^{235}$ as there was in the natural metal. Thus the world's first nuclear reactor was constructed secretly in a basement under a squash court at Chicago University, and was first set into successful operation on December 2, 1942.

Although no publicity whatever was given to this milepost achievement until years later, it was an absolutely essential first step in the development either of a uranium bomb or of controlled production of atomic energy. The problem was to gather together a sufficient amount or "critical mass" of uranium 235 into a single "pile" so that a steady, slow chain reaction of fission would begin to occur among the uranium nuclei. To do this chunks of uranium were interspersed with chunks of graphite which acted to slow down the velocity of emitted neutrons so that fission could take place. Control of the speed of the chain reaction lay in control of the numbers of "fission neutrons" that were created as a result of $U^{235}$ nuclei splitting apart. It was found that the metal cadmium had the singular property of absorbing great numbers of free neutrons and thus taking them out of business, so to speak, so that they could not proceed to strike other $U^{235}$ nuclei and feed the chain reaction. By inserting long rods of cadmium at carefully calculated intervals through the whole structure of the nuclear pile, and then literally adjusting the number of neutrons available to feed the chain reaction of uranium fission by pushing those cadmium rods deeper into the pile or pulling them further out, it was possible to "heat up" or "cool down" the rate of uranium fission in the pile.

Once physicists had learned to control the nuclear chain reaction in this fashion, the next step was to devise some way to build and store a potentially explosive mass of uranium (or its even more fissionable man-made first cousin plutonium) until a desired detonation time and then be able to detonate it at will. Development of a uranium fission bomb was undertaken in a highly secret gathering of physicists, technicians, engineers, and other personnel at Los Alamos, New Mexico, and in 1945 the first prototype of such a fission bomb was tested on a nearby stretch of New Mexican desert. Later the same year the Japanese city of Hiroshima was demolished by such a bomb, and Nagasaki was similarly shaken. The rest of the story is too well known to require repetition. The long and terrible war was brought to an end, and the weaponry for another much shorter and far more hideous one was already in hand.

## A NO MAN'S LAND OF PARTICLES

Thus the increasing knowledge of the structure of atomic nuclei during the last thirty or forty years quite inadvertently thrust modern atomic physicists into a role that fitted them poorly, the role of scientific weapon makers. Once the fission bomb was an accomplished and proven item in the arsenal, subsequent research led to an even more horrendous source of concentrated quantities of raw energy: so-called thermonuclear reactions, in which nuclei of lightweight atoms were fused together, with the resultant conversion of tiny amounts of their mass into vast quantities of energy under conditions of extreme temperature and pressure. Intensive research went into ways and means of developing such thermonuclear reactions, closely related to the hydrogen-to-helium fusion reactions that accounted for the tremendous outpouring of heat and radiant energy from the stars, under sufficient control that they could be used as the basis of controllable weapons without blowing the user himself to kingdom come. But this was a matter of controlling the *triggering* of such reactions, not of controlling the reactions themselves once they were triggered.

To date these studies have resulted only in the development of massively destructive "hydrogen bombs"—thermonuclear weapons which require such very high temperatures to start the explosion going that uranium fission bombs must be used as triggering devices, much as percussion caps are used to trigger the explosion of gun powder in a shotgun shell. So far no way of controlling or even beginning to control the resulting thermonuclear reactions has been found; once triggered, they result in explosions of massive violence. But those engaged in this research have never lost sight of the staggering quantities of useful constructive power that would be available to mankind if some way of control could be found; and as we will see later, one of the most important areas of modern physics today is concerned with a vigorous search for means of producing "slow" or controlled thermonuclear reactions so that mankind can have available for its benefit the same source of power that fires the sun.

For most physicists the role of weapon maker has never set well; men in physics have come to dread their identification with the terrible weapons of mass destruction that their knowledge of the micro-universe of physics has made possible. Many who would like to have divorced themselves from the aspects of nuclear research that led to the development of war instruments found that they could not do so, no matter how much they desired to. The more they probed the nuclei of atoms the clearer it became that here was a source of unimaginable energy which could be released as a result of their work. They could not study the one and escape the other.

And sadly enough, many nonscientists have gained the impression that physicists all over the world have been largely preoccupied in the last twenty-five years with building bigger and better bombs. This picture of modern physics and the work of modern physicists is not, of course, even remotely fair or true. Nuclear weapons were an inevitable side effect of the great discoveries about the interior of atoms that evolved during the early years of this century; but side by side with the development of modern weaponry, a far more basic and ultimately far more important series of discoveries was being made about the nature of matter on a microcosmic scale.

Following the trails blazed by theoretical physicists in their work in quantum and wave mechanics, and using their newly developed instruments and technological procedures, experimentalists in the 1940s and 1950s began producing a wide variety of different "atomic bullets," capable of moving at higher and higher velocities and energy levels, to fragment the nuclei of their subatomic targets with ever increasing degrees of refinement, and then studying the results of those high-powered bombardments.

With particle accelerators such as the giant Synchrotron that was built at the Brookhaven National Laboratory on Long Island, physicists were able to use high-energy protons as their projectiles, accelerating them faster and faster around a circular track by means of multiple synchronized bursts of electrical energy, and finally hurling them on collision course toward target nuclei at energies up to 33 billion electron volts, while standing by with instruments to help isolate and analyze the multitudes of fragments scattered by such collisions. Machines such as the "linear accelerator" now operating at Stanford University made it possible to use electrons as the atomic bullets in bombardment experiments. This machine has an acceleration chamber two miles long that is capable of driving electrons up to energy levels of 20 billion electron volts, with provision made for additional length to be added to the chamber in the future in hopes of ultimately pushing the energy output up to an incredible 40 billion electron volts.

Other devices were designed to accelerate alpha particles and other subatomic particles as projectiles, while studies utilizing neutron bombardment have continued. Physicists have even developed techniques for using nature's own high-energy cosmic rays as bombardment projectiles in order to learn more about the inner nature of atomic nuclei through study of cosmic ray collision products.

As a result of these increasingly ambitious atom-smashing endeavors during the last three decades, physicists have discovered a bewildering profusion of new and peculiar elementary particles hidden within the nuclei of atoms. Some of these particles, like the pions and muons, had been predicted theoretically before they were demonstrated experimentally in the laboratory; the existence of others had not even been suspected by the theorists. Some were clearly particles that had been created artificially by the impact of high-energy particles striking nuclei and causing strange and abnormal intranuclear interactions to take place. Others are now believed always to have played a natural and normal part in a variety of interactions continuously taking place within atomic nuclei, but simply never recognized before because the interactions in which they play their critical roles take place with such incredible rapidity that it was practically impossible to pin these particles down before they had changed themselves into something different. Some of these particles were found to have comparatively long lives, at least on a micro-universe scale in which one millionth of a second, for example, may represent a whale of a long time. Others, more recently discovered, are so incredibly short-lived, existing perhaps for no more than one hundred thousandth of a billion billionth of a second, that physicists to this day cannot be entirely certain that they ever do exist as particles at all but only as strange kinds of "resonances" that change spontaneously into something else virtually as fast as they are formed.

Indeed, spontaneous degeneration, disintegration, or decay seems to be one of the most universal and striking characteristics of virtually all of these strange subatomic particles that have been discovered, of which now a total of almost a hundred different ones have been identified in some stage of existence or almost-existence. Most of them have been identified at all as independent citizens of the subatomic particle clan only because they interact with other particles in very individual and characteristic ways. Some are believed to exist under normal circumstances deep in the nuclei of atoms, always present but exceedingly hard to pin down, while it is suspected that others may never be present in atomic nuclei excepting under the most unusual circumstances. But no matter how strange the very strangest of these bizarre particles may seem, the very fact that they can be identified at all as a possible form of matter-energy even under the most extreme artificial circumstances—say, in the course of atomic bombardment with extremely high-energy particles—suggests that such particles must have had *some* natural function at *some* time and under *some* natural circumstances in the universe.

And it is not hard to find extreme and violent circumstances occurring naturally in one part of the universe or another. We know, for example, that the violence of the most high-energy atom-smashing techniques ever devised by man cannot even begin to match the violence of high-energy cosmic ray bombardment of atoms in our own outer atmosphere, nor can it begin to approach the kind of violence that the cosmologists of the big-bang school of thought believe must have occurred in the first few microseconds or even minutes of the inconceivably violent cosmic explosion they believe must have set all the matter in the universe on its age-long course of expansion.

The study of this apparently endless proliferation of strangely behaving particles has become all the more complex and confusing as physicists have gradually come to realize that a startling and almost fantastic theoretical idea which appeared in the 1920s was actually valid: that virtually all elementary particles (with the single exception of photons or light quanta) have their own corresponding antiparticles, just as real as the positive electrons or "positrons" that were found experimentally to be closely associated with ordinary negative electrons.

P. A. M. Dirac was the one who first suggested, on the basis of his mathematical studies, that such antiparticles ought at least theoretically to exist. But ever since experimentalists confirmed in the laboratory that Dirac's positron existed in reality as well as in theory, both theoretical and experimental physicists began searching very alertly for other such antiparticles corresponding with "normal" particles that were known to exist.

There seemed, for example, to be no mathematical reason to doubt that antiprotons should exist, corresponding to "normal" protons, or that antineutrons might not be found corresponding to "normal" neutrons. Starting off with that sort of "why not?" approach, theorists soon began to find innumerable mathematical indications that such antiparticles, some 1,800 times heavier than positrons, not only *could* exist but *should* exist; indeed *had to* exist. Theorists of the late 1920s and early 1930s speculated that the antiproton, if it could ever be identified under circumstances in which it could be studied, would be found to be identical to the normal, positively charged proton in mass and other properties, differing from it only in that it would bear a negative charge instead of a positive charge.

Actually finding such a particle, however, proved to be a far more difficult task than that of finding the tiny low-mass positron. Since a proton was 1,800 times heavier than an electron, and since the antiproton would have to have the same mass as a "normal" proton, calculations indicated that antiprotons could only be formed if it were possible to bombard atomic nuclei with atomic bullets having energies in excess of 4.5 billion electron volts—and no one in those days had the foggiest idea how to go about achieving such energy levels. The early atom smashers could not deliver an atomic projectile with anything approaching this much energy, and even more sophisticated later models were still not powerful enough.

Little impediments like these, however, did not dampen the enthusiasm of theorists and experimentalists alike for the idea that antiparticles did exist, and the search for them continued with gathering intensity. Finally, in 1955, physicists, using a giant electron accelerator at the radiation laboratory at the University of California in Berkeley, managed to hurl electron projectiles into their nuclear targets with energies exceeding 6 billion electron volts, and at last were able to identify beyond question a few negatively charged protons or antiprotons that were thrown out of the target nuclei as a result of this bombardment.

Just as predicted, the antiprotons so produced had masses identical to protons but carried negative electrical charges instead of positive charges. Furthermore, they met with abrupt and violent ends bare microseconds after their formation, as a result of collisions with normal protons. In such collisions both particles were annihilated with the total conversion of all their combined mass into radiant energy. Nor was it long before other antiparticles had also been produced: About a year after the antiproton was identified, antineutrons were also discovered. In this case, however, the antiparticles were exceedingly difficult to distinguish from their normal counterparts, since neither had any electrical charge whatever. In fact, the only way that neutrons and antineutrons could be told apart was that when two normal neutrons collided with each other they merely recoiled from the collision and went their various ways, whereas when a neutron and an antineutron collided they were both annihilated.

Subsequent to the discovery of the antineutron, experimental evidence bit by bit was found to demonstrate that virtually every kind of particle known had its corresponding antiparticle. The discovery of these antiparticles actually in physical existence in the "real world" of nuclear physics was of course a great triumph for the soothsaying capabilities of the theoretical physicists. From that point on practically nobody in physics could live with them, much less understand how they arrived at their conclusions, but everybody listened to them and paid attention to their predictions. In a larger sense, the confirmation of the existence of antiparticles seemed to many scientists to be a very comforting confirmation of the existence of a basic symmetry in nature. But the discovery of antiparticles also raised a perplexing question for which even today a satisfactory answer is lacking: the question of where in the universe antiparticles might normally be found.

Clearly, in that part of the space-time continuum we know the best there is no "comforting basic symmetry" in evidence, as far as particles and antiparticles are concerned. Where we live, antiparticles are literally sudden death to their unfortunate counterparts and themselves alike whenever they meet. Where we live, the protons that make up the nuclei of atoms have positive charges; the electrons have negative charges; and it is only through massive effort under highly artificial circumstances that antiparticles can be formed at all—and even when they *are* formed under such circumstances, they don't last long. They get clouted in collisions with their infinitely more numerous counterparts before they have a chance to turn around.

Yet if the presence of antimatter, whether it is artificially created or not, really represents a basic symmetry in nature, then it would follow that there ought to be as many antiparticles in existence somewhere in the universe as there are normal particles. But if this is so, then *where are they*? They don't seem to be around these parts; but then, what parts *are* they around? Some physicists have speculated that we happen to be living in a part of the universe in which virtually all of the physical matter around us is composed of "our kind" of particles, while in other parts of the universe virtually all of the matter in existence there must be made up entirely of antiparticles.

Some men have even argued that certain of the great galactic explosions that astronomers have observed might conceivably be the result of an entire galaxy composed of antimatter that happened to come into proximity with a galaxy composed of "our kind" of matter. But in many respects such an idea is a sort of desperate reach for an explanation, and not even a very satisfactory reach at that. Among other things, it does not explain why more such contacts between "normal" galaxies and "antigalaxies" are not observed; the kind of galactic explosion that might conceivably represent such a disastrous contact seems to be exceptionally rare. Nor does this idea offer much comfort to those who wonder why "normal" matter and antimatter should happen to be distributed in such irregular globs here and there throughout the universe, or those who wonder how it got segregated in this fashion in the first place.

An even more fantastic speculation has arisen from the mathematical considerations of theoretical physicists who have found certain existing relationships in quantum mechanics impossible to explain mathematically without at the same time assuming that a negative or retrograde flow of time is just as possible, or even as necessary, as our familiar flow of time. What these workers are suggesting, in essence, is that "normal" matter and antimatter may very possibly exist simultaneously and correlatively in a sort of side-by-side relationship in the three linear or spatial dimensions, but never actually come into physical proximity under normal conditions because they are moving in opposite directions in time. According to this view a positron or positively charged electron is nothing more than a "normal" negatively charged electron moving backward in time instead of forward in time.

Now such an idea, of course, sounds like the sort of stuff science fiction writers might have explored in some of their more wildly speculative stories. But today it is only the incautious physicist who laughs when he encounters such a story. Others are more likely merely to complain that they thought of the idea first. And when we consider the mounting volume of evidence lending credence to a cyclical big-bang theory of the formation of the universe, and at the same time consider the notion that antimatter in the universe may be nothing more than "normal" matter moving through the time dimension in a direction opposite to the direction our "normal"-matter universe is moving, there is certainly room for serious speculation. It is fascinating to consider, for example, that the big bang believed to have set our universe into its present state of expansion might have occurred at a point at which normal matter moving forward in the time dimension and antimatter moving backward in the time dimension met at some zero point, bringing all the normal matter and all the corresponding antimatter in the universe into an instant of motionless proximity both in space *and* time. We could imagine that only the briefest of inconceivably brief instants of such time-and-space proximity would have been needed to trigger an annihilative explosion in which all the matter in the universe was instantaneously converted into raw energy, with new normal matter and new antimatter then reforming and racing away from each other in the time dimension once again as "we," on "our side," entered another phase of matter-building and expansion and "they" on "their side" did the same.

Fantastic? Of course. Maybe not even remotely probable—but antiparticles *do* exist and this is the sort of mind-wrenching speculation that their very existence forces upon the most mundane working physicist, or thinking nonscientist.

Unquestionably the discovery of the existence of antiparticles to correspond to every known "normal" particle vastly increased the scope and complexity of the elementary particle search, an already complex enough game as it was. But having found them, they could not be ignored; a place for them had to be worked into the over-all scheme of understanding that physicists were trying to work out.

It is easy for nonscientists who are not themselves deeply involved in the technical aspects of nuclear physics to become totally confused trying to sort out the multitudes of new and bizarre subatomic particles that have so recently been discovered, and are still so poorly understood, even by the physicists who have discovered them. Nor is there any particular point in trying to sort them out individually in these pages. As time passes physicists may learn which ones are really important and significant enough that nonexperts need to know about them and which are not; in the meantime, there is nothing quite so dull as the long tables of meaningless statistics and characteristics of these particles that we find turning up even in the pages of daily newspapers from time to time.

What *is* important and meaningful here is to recognize the broad general classes or families into which these particles fall, and to consider the significance of some of the over-all properties they seem to display. These categories or families of particles are distinguished by a number of identifiable characteristics. Their masses help fit them into a pattern, for example. So does knowledge of the direction of their spin. So also does the length of time that they are able to endure as independent particles, to say nothing of the particular patterns of interactions they demonstrate in encountering other particles or in disintegrating or changing from one to another. Many are excessively short-lived particles that exist only for incredibly tiny fractions of seconds before either disintegrating, changing, or interacting in some way that alters them. Others are far more stable and long-lasting, capable of entering into other kinds of interactions. All these characteristics assist physicists in finding categories in which to place both the old and relatively well-understood particles and the newly discovered and obscure ones.

The two most familiar of the particle families, of course, are the *electron* family and the so-called *nucleon* family. The electron family includes the familiar negative electron, the positron, the neutrinos associated with the movement of negative electrons, and the antineutrinos associated with the movement of positrons.

These, as we have seen, are the "tiny-mass" or "almost-no-mass" family of particles; electrons and positrons are arbitrarily assigned a comparative mass unit of one, since they are the smallest or lightest-mass particles that are clearly understood to exhibit true particlelike properties—all other particles that truly behave like particles (at least some of the time) have greater masses than these featherweights. The neutrino and the antineutrino are believed to have *no mass at all*—they have been proved to have less than 4/10,000 the mass of their corresponding electrons or positrons —but are included in the family because of their constant close association with these particles. Electrons and positrons each have a single unit of electric charge, negative and positive, respectively; neutrinos and antineutrinos have no electric charge. All four of these particles demonstrate characteristic spin, with particle and antiparticle spinning in opposite directions, as we might expect.

The family of nucleons are so called because they seem to constitute the basic stable building blocks of atomic nuclei. The major members of this family of particles are the familiar proton, the antiproton, the neutron, and the antineutron. Like electrons and positrons, protons and antiprotons each bear a single unit of electrical charge, in this case the proton bearing the positive charge; the antiproton, the negative charge. Neutrons and antineutrons bear no electrical charge whatever. The striking contrast between nucleons and members of the electron family is in their much greater mass: if the electron's mass is considered the basic unit of one, protons and antiprotons have masses over 1,800 times greater. Neutrons and antineutrons have similarly great masses, with the mass of the neutron roughly equivalent to the combined masses of a proton and an electron.

For all their differences in mass, the members of the electron family and of the nucleon family share one enormously important characteristic in common: *They are extremely stable.* Among them they constitute the reliable workhorse team that makes perpetuating physical matter possible at all. Neutrons and antineutrons are the shortest-lived of both these families; when moving free of an atomic nucleus, a neutron has a half-life of approximately 19.5 minutes before it spontaneously emits an electron and a neutrino and transforms itself into a proton. This may not seem like a very long life expectancy, but when we consider that almost all other known particles except those in these two families have life expectancies of only thousandths, millionths, or billionths of a single second, the neutron is a veritable Methuselah by comparison.

Electrons and protons, on the other hand, can exist infinitely and have no known pattern of spontaneous degeneration whatever. Thus most of the electrons and protons in existence today may well be precisely the same ones that existed at whatever instant in the remote past matter was first formed as such. Neutrinos also have infinite lifetime, as long as they can avoid collisions with other particles. They do not degenerate spontaneously, and the universe is believed to be full of neutrinos which have been in existence ever since they were formed, perhaps billions of years ago.

Even the neutron, when associated with a proton in the nucleus of an atom, becomes a building block of infinite life expectancy, probably because the neutron and the proton associated with it are in constant interaction, with one or more pions providing the nuclear binding forces that hold them together. Thus the neutron may remain stable under these circumstances because it is constantly engaged in high-speed interactions with pions hopping back and forth between it and the proton, changing from a neutron to something else and back into a neutron again at such fantastically rapid speed that it never actually exists as a "free neutron" long enough at any one time for there to be any significant probability whatever of its 19.5-minute half-life ever catching up with it.

The third most familiar family of elementary particles comprises those intermediary in mass between the electron family and the nucleon family. These are the family of mesotrons or *mesons.* This clan includes the muon, with a mass of approximately 206 electron masses; the pion, with 273 electron masses; their corresponding antiparticles; and the neutrinos and antineutrinos known to be closely associated with their movement, recognized to be different in certain complex ways from the electron's neutrino and the positron's antineutrino.

Actually, this particular family categorization is awkward for a number of reasons. Although physicists by now are quite convinced that the pion is the particle responsible for the strong nuclear interactions that hold protons and neutrons together, there is considerable evidence to suggest that the muon is really a quite totally different class of particle altogether and has gotten thrown into this family only because it seems to have a superficial similarity to the pion. Unfortunately, no one yet has any very clear idea of precisely what the muon's place in the infinite scheme of things may be.

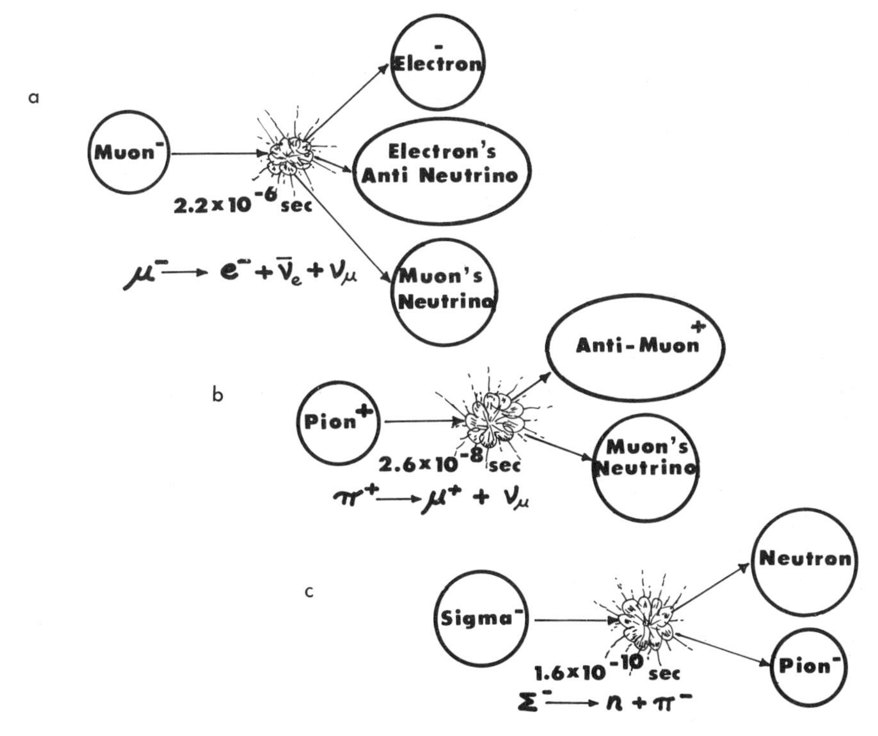


*Fig. 52* Typical disintegration patterns of particles in the meson family (muon− and pion+) and of one of the hyperon family (Sigma−). Half-life of the respective particle is listed in each case, and the nature of the reaction is written in physicist's shorthand beneath each diagram.

It is known that both pions and muons differ sharply from the more stable electron or nucleon family in that they have extremely short half-lives and tend to disintegrate spontaneously into other particles almost as soon as they come into existence in the free state (see Fig. 52). But here their similarity ends. The pion undergoes strong nuclear interactions with neutrons and protons, but the muon undergoes only electromagnetic interactions much as an electron does. In fact, apart from the facts that the muon has a mass some two hundred times greater than an electron, and is unstable and disintegrates spontaneously, it behaves *exactly* the same as an electron, even to the point of joining with atomic nuclei to form strange "mesic atoms." The energy levels of the muons in such atoms can be obtained from Bohr's formula if the mass of the electron is replaced by the mass of the muon!

An even more obscure and peculiar family of particles includes a number, just recently discovered and differentiated, which seem to have masses considerably *greater* than the mass of the proton or neutron—masses running in the neighborhood of 2,400 to 2,700 electron masses compared to the proton's 1,840 or so. Because of their apparently excessive masses, these particles as a group are referred to as "*hyperons*," and their observed behavior is so very odd that physicists have been referring to them quite candidly as "strange particles" ever since their discovery.

The family of hyperons is numerous and prolific: There seem to be dozens of different kinds, each differing from all the others in one obscure but identifiable fashion or another. Physicists separate them into three main groups or types, and for lack of anything more descriptive to call them, merely refer to the groups by Greek letters: lambda particles, sigma particles, and xi particles. In each of these groups some members bear negative charges, some bear positive charges, and some are neutral. Each individual particle in each group has its own corresponding antiparticle. All of these particles have been discovered and identified as components of the atomic nucleus, whether they have been artificially created by human tampering, or are normal and natural residents of those domains. It is known that they have only been discovered at all under conditions in which atomic nuclei are excited or disturbed in some way, but no one has any clear idea precisely *why* they are there in the nucleus (when they are there) or why they are the way they are.

Universally, hyperons are among the shortest-lived of any of the particles yet known that are ever actually in existence long enough to be clearly and unmistakably identified as particles and not merely resonances. They have half-lives as short as $10^{-20}$ seconds, and seem to have a very strong tendency to disintegrate spontaneously just as fast as they can. In Figure 52c is diagramed the disintegration pattern of a negatively charged sigma particle, believed to be typical of the disintegration of hyperons, if anything about them can seriously be considered "typical."

Finally, within the last decade or so, yet another class of particles has been discovered. These are known as K-particles or kaons, and are both sharply different and closely similar to hyperons. The masses of kaons and antikaons are in the vicinity of 975 electron masses, as compared to 273 for pions, 1,836 for protons and 2,100 to 2,500 for various hyperons. As in the case of the hyperons themselves, it is not entirely clear whether these strange K-particles which seem to exist as a whole cluster of separate and individual members with identifiable differences among them, are in fact real, live, individual customers, or merely a multitude of differing forms or manifestations of the same particle appearing under different circumstances. Kaons are even harder to distinguish one from another than hyperons, and like hyperons, they don't often hang around long enough for physicists to get much of a look at them; they have half-lives on the order of $10^{-8}$ seconds, which is getting down about as close to no time at all as you can get. Also, like hyperons, nobody knows for sure why kaons exist at all, what their purpose is, or whether they ever exist as "natural" components of the nucleus or merely appear under extreme conditions of human tampering.

## TOWARD THE ULTIMATE PARTICLE

Where do all these multitudes of particles, plain and fancy, fit within the over-all structural pattern of matter? What parts do they play, and how important is their existence? There is little doubt that the long-lived, stable work-horse particles such as the electrons, neutrinos, protons, and neutrons are the fixed and basic building blocks of physical matter, but are these building blocks themselves built up of other still more basic underlying particles?

There are no good answers to these questions yet; only guesses. But setting these work-horse particles aside, precisely what part do the vast multitudes of *really* strange particles play? This remains today one of the deepest and most profound mysteries of modern physics. Nobody knows. Some of the men who know the most about them admit that they don't even have any very good guesses. Physicists feel that the fact that these strange particles do exist indicates that they do—or at least that they once did—play an important part in the structuring of the nuclei of atoms. Perhaps they play important or even critical roles under certain special kinds of circumstances, or did so at some time long past in the course of the universe's evolution. Perhaps they will again play such a critical role at some future time under similar circumstances, or conceivably they may still be merely waiting in the wings for their moment on stage to arrive.

But whenever physicists think about these strange particles they cannot forget that most of them seem to be born of violence during atomic bombardment, or to arise from violent and powerful interchanges of mass-energy components in atomic nuclei. It is conceivable that they appear under these circumstances but not under normal stable low-energy circumstances at all. Perhaps their existence was needed or is needed only at the times and in the places where matter is being created from energy, or being changed from one form to another under the circumstances of violent mass-energy interchanges that occur in nature, under such circumstances, for example, as the nova explosions of stars, or the circumstances which might have prevailed in the midst of massive big-bang explosions which may have been a part of the evolutionary history of our present universe 10 billion years ago.

Clearly these strange particles cannot be ignored, for their very existence may provide to physicists the vital clues required to unlock the deepest secrets of mass-energy relationships. And just as modern physicists can no longer think of even the stable workhorse electron as a tiny solid particle but only as a particlelike entity with wavelike properties, physicists are equally convinced that all the profusion of particles and antiparticles that have been discovered, strange or otherwise, must also be regarded as wavelike energy manifestations as well as in the role of solid massive particles. And throughout all the investigation of these particles, the conviction has been growing that many of the basic answers, if they are ever found at all, are going to be found in the mysterious mathematical half-world of quantum mechanics.

Thus it is no accident that men working in the micro-universe of nuclear physics are all keeping their eyes sharply upon the work of other men studying the universe at the extreme opposite end of the magnitude scale, the work of the astronomers and cosmologists studying the mega-universe around us. Essentially, nuclear physicists are searching for simple, universal, natural laws describing the behavior of matter and energy within the nuclei of atoms under circumstances of rest and circumstances of violence. So far no such simple laws have been discovered. In very recent years men such as Dr. Murray Gell-Mann, one of the frontier explorers of the universe of strange particles, working at the California Institute of Technology, have been raising serious questions about the real nature of even those elementary particles which seem best known, best understood and most stable. Anticipating the time soon to come when new and more powerful atom-smashing machines will be available to produce atomic bullets with energies as high as 40 or 50 billion electron volts, Gell-Mann has been one who has begun considering the possibility that all of the so-called elementary particles we have been discussing, ranging all the way from the tiny electron and almost massless neutrino to the most massive hyperons, may in fact prove to be complex composites of still tinier and still more basic elementary particles, held bound together in their apparent form just as atoms are bound together in *their* apparent forms, even though they are all composed of a variety of different kinds of subatomic components.

Not all physicists agree with Gell-Mann that such subelementary particles actually exist, but enough suggestive evidence has accumulated by now that it would not surprise anyone if such extra-tiny or fragmentary particles were suddenly to be discovered and identified in a physics laboratory tomorrow. Physicists speak somewhat irreverently of these tiny hypothetical tidbits of mass-energy as "quarks," a made-up name suggesting a sort of entity that nobody is really convinced exists at all. But if quarks do exist, there must also exist enormously violent forces acting to glue them together into more complex particles such as those we know today as protons, neutrons, or hyperons.

Nor is it surprising that Murray Gell-Mann and others working in this hyper-refined area of physical research are looking not only to their own experiments with atomic bombardment and mathematical analysis, but are also looking to the work of the cosmologists and astronomers—watching the feverish study of quasars, for example, with sharply interested eyes. Whatever quasars are, it seems entirely possible that they might represent the sort of violent energy storms in which quarks might conceivably be found or formed, later to be driven into more complex forms. From one point of view, it would be a splendid irony (as well as a splendid step toward unity in our understanding of the structure of the universe) if the study of the mysterious quasars which seem to offer astronomers and cosmologists clues to the size, duration, and life history of the universe, might also provide the key to an understanding of the innermost structure of nature's tiniest basic building blocks. The final answers have not by any means been found yet, but there is an increasing sense of imminence in the world of physics today, a growing sense of conviction that at least *some* of the most basic answers may not any longer be too far away.

## OLD LAWS, NEW MEANINGS

But of all the results that have come rolling in in the study of atomic nuclei and the elementary particles making them up, one of the most fascinating things which has come about has not been so much in the nature of a new discovery as a new way of regarding the great laws of nature which seem to govern the way things work in our universe. The search for the basic elementary particles of which matter is constructed has in essence become a search for simple, powerful, and exclusive natural laws governing the basic relationships of mass and energy in the micro-universe of the atomic nucleus. And out of that search a new way of regarding the meaning of certain existing natural laws and a new idea of how those laws may apply has slowly been evolving—a strangely different view of natural laws than any ever considered before.

Throughout this book we have said a great deal about what "laws of nature" really were, where they came from, and why they have been so important in the study of science. We have discussed natural laws with considerable respect, suggesting that the discovery of natural laws is really the major underlying goal of all scientific investigation in the past or in the present.

Yet at the same time in this book we have encountered repeated examples of apparently adequate, well-established, and powerful natural laws which have had to be discarded, modified, expanded, or changed as a result of scientific study. These laws were found not to be "laws" at all, merely approximate descriptions of relationships which seemed to hold true in the everyday world in which we live, but which often went awry in the mega-universe of galaxies and quasars, in the micro-universe of the atomic nuclei, or in the universe of very high velocities.

Of course, it is only right and proper that natural "laws" which do not adequately explain things that are observed to happen ought to be discarded, modified, expanded, or changed. We have seen from the very first that natural laws are not a matter of cosmic decrees which, once enunciated, take on an authority of their own. They are merely statements of how things behave in relationship to each other. They can be confirmed only by repeated experimental examination. And as we have seen, the most powerful natural law imaginable can only be considered valid so long as it accurately describes everything that can be observed, not just usually, not just most of the time, but *always* under all circumstances, everywhere in the universe, no matter who or where the observer may be.

Thus many such powerful "laws" which seemed for centuries to be completely adequate to describe perfectly all examples of natural phenomena as yet known were later found to be clearly inadequate, as new knowledge accumulated, and as areas of natural phenemona were found that they did not describe well at all. Newton's laws of motion and inertia, for example, were found inadequate under certain circumstances: circumstances involving the motion of objects at extremely high velocities. Thus those laws had to be modified. Those modifications did not discard the old laws, nor even change them, so much as they *extended* them and *broadened* them so that they applied more widely and covered newly discovered phenomena which the unmodified old laws did not describe at all.

Einstein's special theory of relativity, as we have seen, was essentially a modification of Newton's laws of motion. Later, Newton's laws of mechanics, so splendidly adequate for describing objects in motion in the everyday world in which we live, had to be modified further because they failed to explain the newly discovered nature of motion of electrons around the nucleus of their atoms.

Curiously enough, of all the laws of nature that we have discussed there is one group beyond all others that has stood the test of time, of repeated investigation and study, and of multitudinous challenges without ever needing any modification. These laws have proved to be so simple, yet so universally applicable and so powerfully strong, that repeated efforts to overthrow them, extending right down to the present day, have invariably failed. Thus, these particular laws have become the closest approach we yet have to ideal, complete, universal expressions of natural relationships which we feel sure enough are true that we say with confidence, "Yes, because of these laws we are dead certain that this particular thing or that particular thing will always happen any place in the universe under any circumstances." These most powerful of all known natural laws are the *conservation laws*: the laws of conservation of mass-energy, the law of conservation of momentum, and other simple laws which state essentially that certain qualities of the universe seem to remain forever conserved or unchanged no matter what happens under any circumstances.

Granted that physicists have never regarded the conservation laws as being in any way sacred or untouchable. They have always been considered fair game for anyone who could find a demonstrable instance in which they did *not* hold true. Probably there has not been a physicist of stature in the last five hundred years who has not at one time or another succumbed to the challenge and pitted his mind against the conservation laws in one way or another; but even the most brilliant minds have failed to overthrow them. When Niels Bohr, in his study of quantum mechanics, found a situation in which he felt that the laws of conservation of mass-energy were violated, he did not hesitate to consider that those conservation laws might be wrong. As it turned out, it was Bohr who was wrong, and Wolfgang Pauli's concept of the neutrino, based upon the purely tentative scientific assumption that the mass-energy conservation laws were *not* really violated, turned out to be the right answer to the conundrum. Yet in that particular case, those conservation laws were in serious jeopardy right up to the day that the neutrino was actually proven, by experimental observation, to be a real entity in the universe.

Encouraged by the apparent strength of the conservation laws, nuclear physicists in the past forty years have consistently looked for ways to apply the laws of conservation to a number of new qualities or properties demonstrated by the flock of new elementary particles that were being discovered. We mentioned earlier, for example, that electrons, protons, neutrinos, and other particles were found to have a quality known as "spin" among their basic properties. These particles behave in their movements in orbit as though they were tiny tops spinning on their axes. Furthermore, some particles were found to spin in a right-handed or clockwise direction while others demonstrated left-handed spin.

As soon as this property was recognized, physicists began to speculate that the spin of interacting particles might be a quality that was always conserved in any interactions that might occur. In other words, if two particles with an equal amount of spin in opposite directions were to interact to form different particles, the new particles that were formed after the interaction would have to have the same net total of spin that the original particles had entering into the interaction.

We can see this concept of "conservation of spin" illustrated quite simply in Figure 53. We see that if the net spin of the two original particles was zero (that is, one particle with right-handed spin and the other with left-handed spin), then the net spin of the resultant particles after the interaction would also have to be zero: The interaction would either have to produce two particles with opposite spin, for example, or perhaps only one new particle with no spin at all, but if conservation of spin were valid, the interaction could *not* result in two particles with right-handed spin and none with left-handed spin. Experimental evidence soon began to appear, strongly suggesting that spin was indeed a quality that was conserved in all interactions of elementary particles, and to date the "new" laws of conservation of spin always seems to obtain in all experiments that have been conducted to test it. In fact, spin seems so consistently and doggedly to be conserved in inter-particle interactions that the law of conservation of spin has already begun to take its place among the other "strong" conservation laws.

Not all qualities of elementary particles, however, fall under the aegis of conservation laws, even though they may perhaps seem to for a time. In-

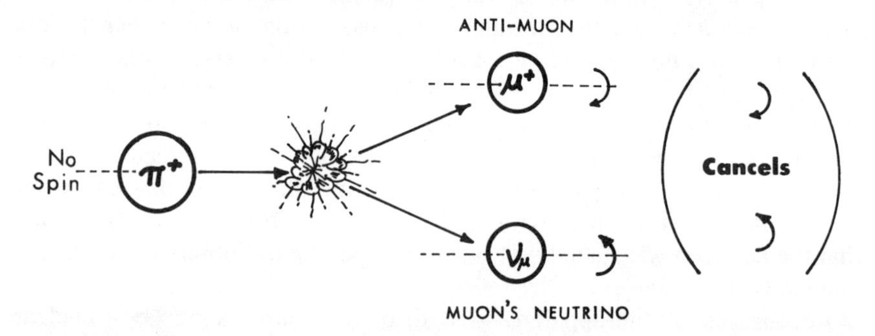


*Fig. 53* Conservation of spin illustrated in the spontaneous decay of a pion$^+$ particle. The pion, which has no spin, yields two particles, each with spin but in opposite directions, so that the *net spin* of the disintegration products is the same as that of the pion before disintegration, namely zero.

deed, in recent years some evidence has been found, too complex to detail in this discussion, that suggests that in the course of certain kinds of interactions between elementary particles even some of the very "strongest" conservation laws may become temporarily suspended, as it were, for very, very brief intervals of time during which the interactions are taking place. This disturbing possibility first came to the attention of physicists not because any cases had been observed in which the laws of conservation of mass-energy, for example, were found invalid, but because physicists could not see how, for extremely technical reasons, certain steps that must have ocurred in the course of certain interactions could have taken place at all the way they did, or in the sequence they did, if those conservation laws had been "in full force" all the time from the beginning of the interaction to the end of it.

One thing is clear: The great conservation laws are beyond doubt the most fundamental and widely significant of all descriptions of natural phenomena that scientists have ever succeeded in working out, and it is not hard to understand why physicists today almost unanimously agree that these laws in some way are at the very root of any possible profound over-all understanding of the behavior of matter and energy in the universe. Where other paths have met blind walls, forked, meandered, or wandered in all directions, the conservation laws have remained straight and above all else trafficable. But in recent years many physicists have become convinced that the conservation laws actually represent something quite sharply different from what they have always been thought to represent, and that the universe described by those laws may be quite different from the universe it has always been assumed to be.

How, after all, have the conservation laws always been regarded? From the earliest beginnings of classical physics, when the first conservation laws were recognized, physicists have always heretofore considered them as rigid templates or guidelines, so to speak, describing certain ways in which all mass-energy interactions always had to occur. Just as a moving train is confined and forced to go precisely where its track leads it and nowhere else, on pain of disaster, so events in nature have been regarded as confined to certain specific tracks or channels and no others. The conservation laws were regarded as the confining "railroad tracks" laid out to guide, direct, and force the direction of natural phenomena. These laws seemed to be saying, in essence, "*This* and *only this* may happen under these circumstances: nothing else can or will be permitted." And indeed, for centuries, this seemed to be precisely the relationship that existed between the conservation laws and the occurrence of all sorts of natural phenomena in the universe.

But recently a growing number of physicists have begun to wonder if science has not always been regarding these conservation laws in a completely backward fashion. Suppose these powerful laws are *not* fixed directives at all. Suppose they are *not* specific, direct orders which natural phenomena are forced to follow. Suppose, rather, that the conservation laws are really nothing more than a series of fairly broad extreme limitations within which all natural phenomena must occur. Suppose, rather than saying, "*This* and *only this* can happen under these circumstances," the conservation laws are really saying, "*Anything* may happen under these circumstances just as long as whatever happens does not exceed this or that outer limit." Suppose the conservation laws, rather than railroad tracks going only in certain places, are really merely confining walls on either side within which the train can and—to a certain degree of probability will—go in any direction it chooses.

Consider a simple analogy. Suppose a mother wants the evening dishes done after dinner, and assigns her daughter to the task. She wants the dishes sparkling clean, and she wants the job done before 11:00 P.M. because that is her daughter's bedtime. There are two quite different ways that the mother could approach the problem.

First, she could say to her daughter: "I want the dishes done and done now. Use *this* pan, *this* water, *this* soap, *this* towel, and *this* brush. Wash the dishes in *this* way to *this* degree of cleanliness, rinse them *this* way and dry them *this* way; and if you deviate from this method anywhere along the line, you'll be sorry . . ."

If her daughter follows this directive, Mother would get her dishes done sparkling clean, and before 11:00 P.M., too. Every step of the job has been rigidly prescribed in rigidly ordered fashion. Thus, no other result could be possible.

Alternatively, the mother could say to her daughter: "I want the dishes done, and done well, before 11:00 P.M. I don't care *when* you do them in the time you have available; I don't care *how* you do them; do them any way you want. Call in your friends to help if you want to; coerce your little brother into doing them for you if you can; take them down and dump them in the laundromat if you wish. I don't care *what* you do, just as long as (1) all the dishes get sparkling clean; (2) none of them get broken; and (3) the job is completed not later than 11:00 P.M."

Again, mother would get her dishes done sparkling clean and on time, but *not necessarily in any single prescribed fashion.* Utter chaos might reign in the process of getting them done, but the end result would be accomplished just the same. The only difference between the first approach and the second would be a matter of control and orderliness; the first approach would be by far the more restrictive, the second by far the more permissive.

Similarly, the conservation laws can be regarded in traditional fashion as restrictive directives according to which events must occur in detailed, orderly manner; or they can be regarded as permissive outer limits beyond which events may not deviate, but within which literally *anything* may happen. But doesn't this really add up to exactly the same thing in the long run? Not quite. As we have seen, physicists recently have discovered a multitude of hitherto unsuspected elementary particles engaging in an enormous variety of different kinds of interactions. The conservation laws have certainly been useful guides in helping us to understand what interactions were occurring and how—but if they are considered merely as outer limits confining broad general areas of phenomena in which multitudes of possible interactions may take place, rather than as railroad tracks forcing interactions to follow one detailed course of events and one course only, then this in essence is the same as saying that *any one of a variety* of things *may* be happening during any given interaction between elementary particles, just as long as no single thing that might be happening exceeds the rather permissive outer limits of the conservation laws.

In fact, it is saying that as long as some phenomenon doesn't *ultimately* violate the conservation laws in the end, then there is no other limitation whatever upon *how* that phenomenon may occur in any given case at any given time. And this is a far cry from the old concept of the conservation laws as fixed and rigid templates or guidelines. It opens to nuclear physicists a whole new universe of possible subatomic events, all possibly occurring within the limits of the laws of conservation, but with no one event necessarily obliged to occur any more than any other event, under the given circumstances.

Philosophically the idea is disturbing. The old "guideline" concept fitted in perfectly with the philosophical idea of cause and effect, the neat and comfortable idea that everything happens predictably as the result of something else that happened before, which itself happened because of something else that happened before that. But in a part of the universe where probability enters the picture, as it does in the case of the micro-universe of nuclear physics, how can anyone tie in the continuous uncertainty of probability with the old rigid guideline concept of the conservation laws?

According to the concept of cause and effect, the universe and all of its functions first and foremost are orderly, and everything that occurs follows inevitably and predictably as a result of this orderliness, with no possible alternatives available. But the newer view of the conservation laws as nothing more than fairly broad limits beyond which no events can take place suggests a different kind of universe: a basically *disorderly* universe, a universe that would be basically chaotic, without order, except for a certain limited degree of order which is imposed upon it by the limitations of the conservation laws. In other words, these laws merely squeeze *some* degree of order out of chaos, and it is for this reason alone that we observe any degree of order whatever in the universe and in the events occurring in it.

Which view is correct? Nobody really knows, for sure, so far. Some workers in physics suspect that this is an area that lies outside the scope of science altogether. It is a concept, nothing more. It has intrigued modern physicists; it has opened up some new ways of looking at phenomena that have been observed, and has suggested some new directions in which to search further. In this sense, at least, it is a useful concept even if it cannot be demonstrated to be true by any techniques of science.

But on the other hand, it is quite possible that investigation of the concept of the conservation laws as limits, not guidelines, may well lie within the scope of science. It is even possible that this concept of the conservation laws may prove in the next few decades to be every bit as revolutionary and richly productive of new understanding of the way our universe works as the special relativity theory of Einstein proved to be half a century ago.

Obviously, the study of nuclear physics in the last fifty years has not produced nearly as many answers as it has produced new questions. What is more, almost all of the new questions have proved to be far more difficult for scientists to answer than the old ones they succeeded in answering. Certainly what once appeared to be a fairly simple and straightforward structure of matter in the micro-universe has proved to be far from simple. Indeed, it has proved to be extraordinarily complex and seems to become even more complex with each new discovery along the way.

Nevertheless, the search proceeds today with increasing vigor, and with no end in sight. Some physicists today have doubts that the questions they are now asking are ever likely to find answers; but most are confident that with time, with new directions of thought, and with the new techniques of investigation that are constantly being developed, answers will indeed be found. What is understood only vaguely today will be understood more clearly tomorrow, but, as day follows night, new and even more brain-wrenching questions will also be found.

Whether satisfactory answers to all possible questions ultimately will be found or not is a moot question; but there can be no doubt that the work being done today in the micro-universe of elementary particles is sharply challenging the most basic questions that have ever been asked about the true nature of matter and energy, and about the true underlying structure of the universe.

# Part VI

*Practicalities and Promises:*
*The Impact of Modern Physics*

# CHAPTER 28

## *Lasers, Transistors, and Other Practicalities*

In the earlier sections of this book we have discussed in greater or lesser detail quite a sobering variety of esoteric, complicated, and sometimes downright confusing aspects of the exciting work that is being done in modern physics today. We have devoted even more attention to the long historical march of events that forms the background of present-day work in physics. It may seem in some respects that we have been obsessed with history at the expense of discussion of current events, but if this is so, it has not been without purpose. At the very best, "current events" in modern physics are confusing and difficult for the nonscientist to comprehend. Without a clear understanding of the historic succession of discoveries which have ultimately led to our present-day concepts of the universe and how it works, we would long since have been trapped in the quagmire, fighting desperately to get one foot out while plunging the other foot in a couple of feet deeper.

So far, inevitably, we have concentrated our attention on the work of *theoretical* physics, the areas of pure scientific research where basic principles are sought that explain the multitudinous phenomena of nature. But in doing so we have consistently sought to avoid plunging ourselves into quantities of technical detail if we could find any way to avoid it. To physicists such detail is the breath of life; if it were not, they would not be physicists. But to most readers such technical detail is not important. What *is* important is to be aware of the basic ideas that have arisen from centuries of research in physics, and to see as clearly as possible why those ideas have been important and what new ideas they have led to.

We acknowledged from the first that such an approach would inevitably mean sacrificing a certain degree of scientific precision and accuracy. As a result, although we have discussed some areas of modern physics in considerable detail, we have scarcely broken the surface of other areas. It is quite impossible to discuss de Broglie's concept of the wavelike nature of electrons, for example, or the baffling complexity of the recent studies of the interaction of elementary particles in depth and with careful scientific accuracy without having recourse to mathematical analyses that are beyond the depth of most readers of this book, to say nothing of its author. Even so, we have attempted to show the *ideas underlying the mathematics* and to discuss the vitally important concepts that have arisen from both the theoretical and experimental work that the mathematical analyses of Heisenberg and others have engendered.

The really important question we must face, of course, is simply, *What does it matter, anyway*?

Much that we have been discussing might well be interesting, perhaps enlightening, possibly even fascinating, but would still be of little practical value to us if we could not see clearly where even the most esoteric work of modern physics is taking us. We might indeed be fascinated by the conflicting theories regarding the shape and form of the mega-universe and the birth and death of galaxies, but how does the knowledge that has been accumulated by astronomers and cosmologists affect *our* daily lives? The story of the exploration of the nucleus of atoms may make entrancing reading, but how does knowledge of this mysterious micro-universe affect *us*? Obviously the sun rises and sets without our having this knowledge, and always has. What good does all this knowledge do anyone but physicists?

In our modern world we pay taxes that go to support all sorts and varieties of front-line research in physics; but only a small proportion of that research is directly related to matters of our country's defense and national well-being. What does the rest accomplish, other than keeping physicists employed and off the streets? What does the rest of the world profit from this ever-growing concentration of effort, brain power, and money channeled into the exploration of areas of the universe that we can barely comprehend at all? It there really *any* way the rest of the world profits from all this?

Of course the answer is yes—although not always an unqualified yes in all cases. We have certainly discussed some areas of research in physics that seem to promise very little in concrete terms to the world in general under any imaginable circumstances. In many other areas where there *is* a practical connection between certain lines of basic research and ultimate benefit to the world at large, that line of connection is often so tenuous, so convoluted and obscure that no one can clearly define it, not even the physicists. Yet the overriding fact is that we are living today in the midst of a scientific and technological revolution which is in fact affecting each one of us more profoundly and in more everyday aspects of our lives than anything else that has happened to the human race in the course of the past five hundred years. This scientific and technological revolution of today is as profound in its implications for our lives now and in the future as the great intellectual revolution of the Renaissance was for the society of the sixteenth and seventeenth centuries. This revolution has already begun to transform our lives in the past few decades. It is going to transform them even more radically in the course of the next five years, or the next ten, or twenty, or a hundred years. It is going to alter our lives quite literally beyond recognition. Yet this revolution has its roots deeply buried in the great wave of discovery in modern physics which we have been discussing in such detail in the foregoing pages.

The truth is that this scientific and technological revolution has already altered our lives so widely, in so many ways, and at so many levels, that we would need another book half again as long as this one if we wished to begin to explore the alteration in detail. We will not even attempt an outline here. But it is entirely appropriate, before closing this book, for us to consider at least a few striking areas in which the esoteric work of theoretical and experimental physics has already touched our lives profoundly—to see a few exemplary areas in which the relationship between basic research and practical benefit to the world (or at least staggering impact upon it) is clearly evident.

One of these areas involves a discovery in which the practical implications of quantum theory and quantum mechanics were carried out to a logical extreme to provide a remarkable and incredibly useful device already being used to the benefit of mankind in a dozen different ways. Another area of applied research has led to a top-to-bottom revolution in the technology of electronics, and in the course of fifteen years or so has reached into the homes and lives of virtually everyone in the country. Still another area of research and experimental physics has led to the discovery of a whole new and bizarre assortment of physical properties of various substances under conditions of extremely low temperatures—properties so very strange that the search for practical applications today has barely begun as physicists scramble in their efforts to find out what is happening, and to learn how what is happening can be used. These few examples of the practicality and promise of research in modern physics will barely scratch the surface, but they will at least serve as dramatic examples of the impact on our lives that this complex field of scientific study has already had and will have in the future.

## MASERS AND LASERS: THE FUTURE OF WAVE AMPLIFICATION

In our world today the training of young people in the physical sciences has taken enormous and startling steps forward. If you want to know just what is going on in the frontier areas of physics today, a good place to start asking is among the students in any high school science laboratory.

Examples of the involvement of our young people in the complexities of physics keep turning up in the public press in remarkable fashion. For example, during the summer of 1964 two high school boys in a Western city began searching for a suitable project to enter in the National Science Fair, an annual exhibit sponsored by high-school science clubs all over the country, and decided that for their project they would build a laser. At first their instructors were skeptical, but the boys dug into the scientific and technical research necessary, and then with characteristic enterprise and zeal began gathering necessary materials and equipment together. By one means or another (no one is quite sure how) they talked a local industrial firm into selling them a $400 ruby rod for a mere $75, making and selling doughnuts to their classmates in order to foot the bill. Other industrial firms lent them critical and expensive electrical equipment, while still others sold them materials that they needed at prices far below cost.

The boys then began devoting their time after school—some 1,000 hours of it—to their project, and to the bemusement of their school counselors and the public press alike, began building in their high-school laboratory a simple model of one of the most fascinating and sophisticated devices ever to arise from research in modern physics. By the time of the competition they had assembled and tested a functioning laser unit which was every bit as operable as the ones developed by graduate research physicists working in their laboratories at Stanford University.

What was more, their laser *worked*. Those boys have a collection of stainless steel razor blades with neat holes punched through them to prove it. When the newspapers published stories of their achievement, adults in the community did not know whether to be proud or merely embarrassed; their own children, it seemed, were building from the ground up a gadget that most people had barely heard of, at best, and understood not in the slightest.

As for the boys, they were of course not interested in collecting razor blades with holes punched through them. Neither were they even remotely interested in using their laser as some sort of a "death ray" to bring down enemy planes in flight (conceivably possible, but not very practical) or even as a device with which to cut James Bond in two (still less practical). In fact, *they* were amazed at the number of ridiculous and improbable things that their friends and classmates thought that lasers could do. Far from putting their device to one or another practical use, the boys were busy complaining about its inadequacies of size and capability, and were busily planning how they might go on to build a bigger and better one. As a first step, they were working hard trying to *synthesize a larger ruby rod themselves* so that they could build a bigger and technically more perfect laser without having to spend all that money for a commercial synthetic ruby! Their earlier attempts at synthesizing rubies had not worked too well: the small ruby rods they had produced were of uneven quality, with too many flaws to be useful in a laser. But the way they figured it, with a little more work, experiment, and experience they could iron the bugs out of their ruby-synthesizing technique and make themselves just as big and perfect a ruby rod as they needed.

By now everyone has read about lasers. The popular press has been full of bizzare ideas about them. But very few nonscientists really know exactly what a laser is, how it works, or what it can do. The laser was a discovery which evolved from frontier research in physics during the last fifteen years, and although the function of a laser is deeply rooted in the basic principles of quantum mechanics, the device itself is not terribly difficult to understand. These boys who built their own laser had no advanced training in physics, but they understood how their device worked and why. Their curiosity and enthusiasm had simply led them into close acquaintance with a device that happens to have perfectly amazing potential for human benefit in the world today, and even greater potential in the future.

What is a laser? More properly, we should ask what is an L.A.S.E.R., because "laser" is a made-up word derived from the initials of a simple description of the device's basic principle: *l*ight *a*mplification by *s*timulated *e*mission of *r*adiation. Actually, the laser is really the unexpected offspring of an earlier device known as the maser (*m*icrowave *a*mplification by *s*timulated *e*mission of *r*adiation) which was first developed independently and simultaneously by a number of experimental physicists in the early 1950s.

For many years before then—indeed, since the earliest radio receivers and transmitters had been invented—physicists, technologists, and electrical engineers had been trying to find ways to generate microwaves of a completely "pure" or "coherent" wavelength. Most microwave transmitters generated waves of a variety of wavelengths at the same time, some "out of step" with each other or "incoherent," so that radio signals carried by waves of one wavelength met with constant interference or "static" sometimes spoken of as radio noise. Workers had also been searching for some way to amplify or magnify the intensity of microwaves, whether those being generated by transmitter or those reaching a receiver from a distant source, without at the same time creating so much "radio noise" in the form of undesired microwaves of outer wavelengths that the desired microwave signal would be drowned out by static.*

The problem was twofold. First, any generator of electromagnetic waves put out microwaves of a variety of different wavelengths all at once; some would interefere with others, some would complement others, and all would scatter in all directions from the source, so that aiming or "beaming" them at a distant target receiver was difficult. When it came to amplifying incoming microwaves at the receiving end, the procedure depended upon the use of vacuum tubes (like the old clumsy tubes you remember from pre-transistor radios) which themselves produced unwanted microwaves in the course of their function, often to such a degree that the desirable waves that were being amplified would be largely smothered in radio noise.

A possible solution to the problem occurred to several scientists at about the same time in the early 1950s, among them C. H. Townes and James P.

* Microwaves, we recall, are electromagnetic waves of very short wavelength and high energy, a part of a vast spectrum of electromagnetic waves that also includes very long radio waves, somewhat shorter waves of visible light, still shorter waves of ultraviolet light, very short radio waves of micro-wavelength and *extremely* short-wavelength high-energy radiations such as X-rays and gamma rays.

Gordon, at Columbia University; N. G. Basov and A. M. Prokhorov, of the U.S.S.R.; and J. Weber, at the University of Maryland. Rather than using vacuum tubes (which created streams of noise-generating electrons) for amplifying microwaves, these men began exploring the natural oscillation of molecules that was known to occur in a variety of substances including ammonia gas and various kinds of crystals. It was found that by "exciting" a great many molecules in a substance to a high-energy state and then exposing these excited molecules to a stream of microwave photons, these excited molecules would themselves emit more photons of exactly the same wavelength as the photons striking them, as they returned to an "unexcited" low-energy state again.

The result was that a weak stream of microwaves entering the area of the excited molecules would generate a much stronger or "amplified" stream of microwaves of the same wavelength, leaving unexcited molecules behind. Furthermore, by trapping the growing stream of microwaves within a closed cylindrical tube or chamber with mirrors facing each other at either end, a wave of new microwaves would build up as the remaining excited molecules gave up photons, and part of the stream of microwaves oscillating back and forth between the mirrors would be *transmitted through* one of the mirrors that was only partially silvered, traveling off through space as an intense surge of microwaves moving in a single tightly controlled direction. Only about one-tenth of 1 per cent of the microwaves trapped between the mirrors in the tube would seep through partially silvered mirror—a given microwave would be reflected 999 times out of 1,000, and would be transmitted through the mirror only once out of 1,000 reflections—but even so the transmitted surge of microwaves would be vastly amplified compared to the original stream that excited the molecules in the tube, and that amplified surge of waves would all be transmitted in the same straight-line direction.

In 1958 Townes and Gordon succeeded in building such a "microwave amplifier," using a chamber containing ammonia gas molecules at extremely low temperatures, and found that it worked exactly as predicted. In this device, which was called a maser, incoming microwaves of a given wavelength stimulated excited molecules of ammonia to emit more photons of the same micro-wavelength, so that the input stream of microwaves was markedly increased or amplified by the time the sudden crescendo surge of photons passed through the partially silvered mirror. Later on masers were built utilizing solid crystals of ruby, quartz, or even glass as the microwave-amplifying chamber instead of ammonia gas.

Crude as they were, even these early masers had practical applications. For one thing, they permitted tightly constricted beams of amplified microwaves to be transmitted from one place to another with remarkable accuracy and with very little unwanted "radio noise" interference. When attached to radio telescopes, masers were able to amplify very weak radio signals reaching the telescope from radio stars, whereas the old vacuum tube amplifiers tended to drown out such weak signals with the noise they themselves created. Thus masers could extend the effective "reach" of radio telescopes.

Again, the extremely pure or coherent beams of microwaves generated by the maser provided a remarkable opportunity for physicists to reconfirm the negative results of the Michelson-Morley experiment designed to detect the presence of an "ether wind" with far greater delicacy and accuracy than ever before. Such experiments conducted in 1958 produced precisely the same negative results that Michelson and Morley obtained, leaving physicists even more confident than ever that light and other electromagnetic waves travel at a constant velocity everywhere in space.

But the real value of the maser principle lay in quite a different direction. The first of these microwave amplifiers had hardly been built before other physicists, notably Arthur L. Shawlow, T. H. Maiman, and P. H. Manners, were wondering if the same principle of microwave amplification could not equally well be applied to amplify electromagnetic waves of longer wavelength—of the wavelength of *visible light,* for example. By the middle of 1960 several men had succeeded in building "optical masers" capable of remarkable amplification of visible light, the devices we know today as lasers.

## THE BEAM OF RUBY-RED LIGHT

In its basic construction, the most common type of laser is so simple as to seem almost ridiculous. Essentially it consists of little more than a ruby rod some two to three inches long and a quarter of an inch in diameter, with the two ends of the rod optically ground until they are perfectly parallel, and then coated with silver to form two mirrors facing each other through the length of the rod. The silvering on one end is lighter than on the other in order to form just a "partial mirror," so that a small amount—perhaps one-tenth of 1 per cent—of any light striking it will be *transmitted through it* rather than be reflected back by it.

This ruby rod is then mounted, as shown in Figure 54, within a coil of an evacuated glass tube which is capable of producing a dazzling flash of white light when connected up to a sufficiently high-voltage condenser to cause a spark to leap from one end to the other. When the flashtube circuit is closed and the flashtube flashes, momentarily flooding the ruby rod with white light, an incredibly intense beam of pure *red* light surges through the partial mirror at one end of the ruby rod—a beam of light so intense that, when properly focused, it can burn a hole through an eighth-inch steel plate a thousand times more readily than sunlight focused with a strong magnifying glass can burn a hole through a piece of paper.

Where does this intense beam of red light come from? Simple as the laser may be in its construction, its operation is a bit more complex. In order to understand the laser principle we must review for a moment what we learned about the way electrons behave, moving in their orbits around the nuclei of atoms. As we saw in an earlier chapter, the electrons surrounding an atom's nucleus, although sometimes called planetary electrons, do not follow the classical laws of Newtonian mechanics in their motion, but rather the laws of quantum mechanics worked out by Max Planck, Niels Bohr, Wolfgang Pauli, and other physicists in the early part of this century. Unlike planets around a sun, which might occupy orbits at virtually any distance from the sun, electrons are restricted to occupying certain specific orbital levels around the atom's nucleus and no others, and each such "orbital shell" which can contain electrons is associated with a certain energy level.

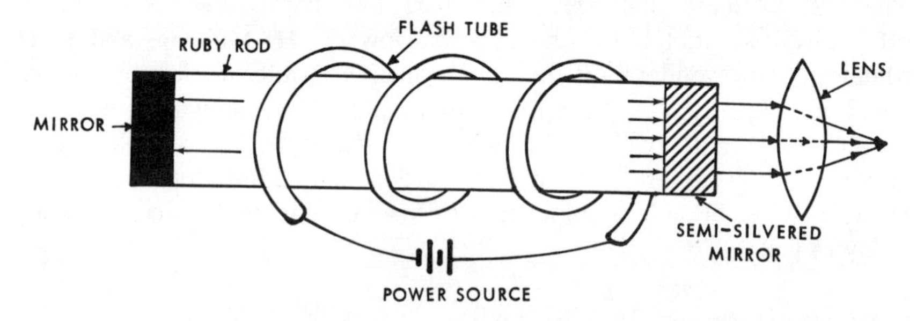


*Fig. 54* Basic structure of a laser.

Electrons in outer orbital shells possess high energy levels, while those occupying the orbital level closest to the nucleus have very low energy levels. The higher the energy level of a given electron the more unstable it is in its orbit—that is, the more "eager" it is to hop down to a lower energy level orbit—but as we saw, according to Wolfgang Pauli's exclusion principle, each energy level can be occupied only by a fixed or limited number of electrons and no more; so if all the low-energy positions are full, then none of the electrons with high-energy positions can move down.

We also saw that the only way that an electron can move from a low-energy level to a higher energy level is by absorbing a photon of radiant energy, as might occur if a light photon of precisely the right quantum size happened to pass by and collide with a low-energy electron. Conversely, a high-energy electron can jump down to fill an empty space in a lower energy level only by *emitting* a photon of radiant energy of precisely the right quantum size. Given an empty space to jump to, an electron might emit such a photon of its "excess energy" spontaneously and jump down to a lower energy state. But it also might not; such an event occurs on a basis of probability, not certainty. The probability of this occurring, however, is greatly increased if a light photon happens by to collide with the electron in a high-energy orbit; when this occurs, the electron emits *another* photon of light of precisely the same wavelength as the incoming photon, and simultaneously jumps down to a lower energy level.

We might conveniently think of electrons in low-energy-level orbits close to the nucleus as rather sluggish, indifferent, unexcited electrons which are not really interested in doing much but wandering placidly around in their orbits unless photons of radiant energy come along to strike them and kick up their energy level, literally "pumping" them up into a higher energy level. We could think of high-energy-level electrons as being highly nervous, twitchy, and excited electrons, eager to get rid of some of their excess energy at the drop of a hat by emitting photons of radiant energy of the right wavelength and settling back down to a lower energy level where they could rest a bit.

Let us regard electron behavior in this way, and let us suppose we had a group of atoms with a number of orbital shells of electrons located one outside the other like layers of an onion, and with a few electrons in the outermost shell in a highly excited state but unable to do anything about it because all the lower-energy-level positions were filled. Suppose, then, that there were some way that we could artificially bombard this collection of atoms with all sorts of light photons so that a great many electrons all at once would absorb photons and leap to higher energy levels, leaving low-energy spaces available for occupation by excited electrons in the outer levels—*lots and lots* of low-energy-level spaces vacant all at once.

Such a process of artificially "pumping" energy into electrons and "exciting" them to higher energy states would be much like cocking a revolver. Sooner or later one of the most excited electrons would emit a photon of light of some wavelength and leap down. That emitted photon might collide with another excited electron in another atom and cause that electron to emit a photon of the same wavelength of light so that there would be two photons instead of one. Each of those two might then collide with an excited electron causing emission of two more photons of the same wavelength of light, making four; and so on. If enough electrons had been excited by our artificial pumping maneuver, a chain reaction of photon emission might take place in that substance.

Of course many of the emitted photons might *not* collide with other excited electrons. They might simply fly out of the substance into space. But suppose we enclosed the substance on two sides with mirrors, so that emitted photons striking the mirrors would bounce back through the substance again, hit the far mirror, and bounce back through the substance again, etc. With each bounce the photon so "trapped" between two mirrors would have another chance to strike another excited electron, and as more and more photons were emitted and became trapped between the mirrors, more and more collisions and photon emissions would take place.

In such a case, the chain reaction of photon emission could grow by leaps and bounds, building up an enormously intense pulse of photons within a brief fraction of a second, all of the same wavelength and all traveling parallel to each other back and forth between the mirrors. If one of the mirrors trapping this sudden surge of oscillating photons was only partially silvered, a small percentage of the photons would be transmitted *through* the mirror in a rising wave as the chain reaction of photon emission within the crystal reached its height. Such a surge of transmitted or "escaping" photons would then travel on in space, acting as a pulse of light of a pure wavelength traveling in a very tight beam because all the light waves would be "in step" and moving in the same direction.

What we have been describing here is precisely what happens in the ruby crystal of the most common form of laser. We have already seen that in some kinds of atoms electrons in the high-energy outer orbits possess an "overflow" of energy, and will jump down to a lower-energy-level orbit, given a chance, by emitting a quantum of radiant energy. Under some circumstances this radiant energy may be in the form of microwaves; in other circumstances the electrons may emit a quantum of light of a certain wavelength. Precisely what size quantum (and therefore what wavelength) of radiant energy a given electron in a given atom may emit in order to make such a quantum jump to a lower energy level is a property of the particular atom and of the degree of "excitation" of the electron. In the right substance, the emitted quantum of radiant energy may be in the form of visible light.

Rubies are a crystalline form of aluminum oxide which contains a certain number of chromium atoms as contaminant. The red color of the ruby depends upon the quantity of the chromium impurity; natural rubies may vary in color all the way from a faint lavender-pink of poor stones (containing only 2 or 3 per cent chromium atoms) to the deep blood-red of very fine stones (containing 20 or 30 per cent chromium). Natural rubies are rarely large enough for practical use in a laser and are inordinately expensive as well, but in recent years techniques have been devised to synthesize rubies with carefully controlled percentages of impurity; so it is possible to manufacture good-sized rods of ruby containing precisely 15 or 20 per cent of chromium atoms.

Now the chromium atoms in ruby have a singular characteristic: When certain wavelengths of electromagnetic waves strike ruby, large numbers of electrons in low-energy-level orbits will absorb photons and jump to higher energy levels; in other words, great numbers of these electrons can be excited by the exposure to incoming radiation. This excitation can be achieved on a mass scale simply by exposing the ruby to a sudden blast of white light created by sending a high-voltage spark through a flashtube. Once these electrons are excited, however, with many low-energy-level positions suddenly available, they almost immediately emit photons and jump *part way* back down to medium-energy orbits. But then, from this "rest

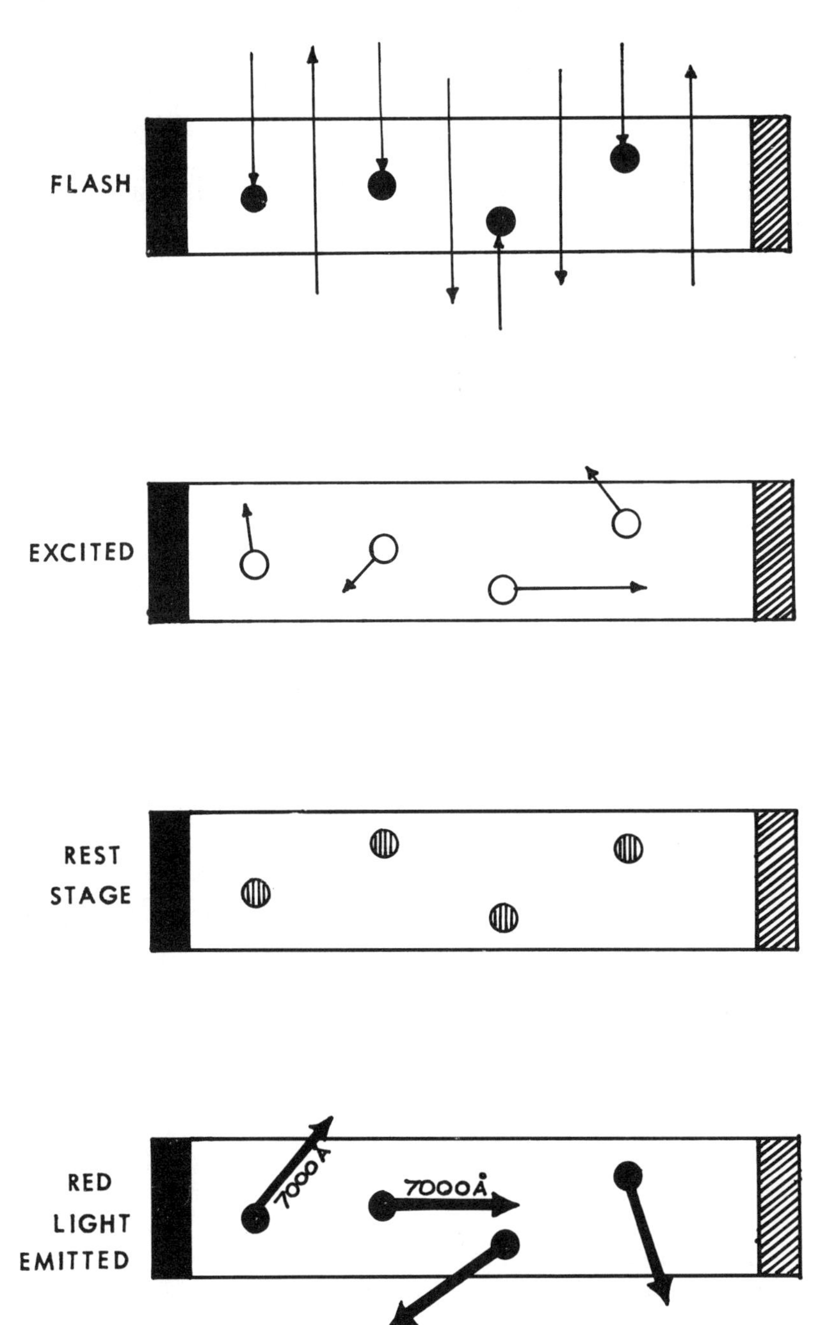


*Fig. 55 a & b*
*55a:* Function of a laser: the source of red light. White light from *flash* tube strikes chromium atoms in unexcited state, photons are absorbed, and atoms become *excited,* emitting photons and subsiding to a semi-excited *rest stage.* Then rest-stage atoms begin emitting more energy in form of a cascade of photons of *red light.*

stage" in which the electrons are still excited, they one by one begin emitting additional photons and jumping back down to their original, placid low energy levels again. The radiant energy these excited chromium atom electrons emit in this second "quantum jump" is always in the form of light quanta of a wavelength of approximately 7,000 angstrom units—the moderately long wavelength of pure red light.

In Figure 55a we see this sequence of events in a diagram. As we see, what is happening, essentially, is that the ruby rod of the laser is exposed to a burst of white, incoherent light of many wavelengths, and then, thanks to the peculiar properties of the chromium atoms in the ruby which are excited by this flash, *converts* that white light into pure red light which is *controlled* in its direction of motion by the mirrors and is immensely *amplified*. Finally it is released from one end of the rod as a sudden burst of intense red light (Fig. 55b). Modern lasers employ a carefully sychronized shutter at the partial-mirror end of the ruby rod so as to allow more chromium atoms to be triggered before the mirror effect takes place, resulting in even greater amplification of red light to produce a "giant pulse" of red light coming from the laser when the shutter is opened. What is more, if the ruby rod is repeatedly excited by flashes of white light, it will produce a succession of bursts of the intense red light, and modern lasers built with other substances having slightly different properties from chromium atoms, can now produce a virtually continuous beam of intense light.

It is important to understand that this "laser effect" is not a strange, quixotic quality of one particular substance. Many kinds of substances can be used in lasers, including neon gas, for example. The ruby laser is the most common form today simply because experiments have demonstrated that the chromium atoms in ruby can be excited reliably by the brilliant flash of white light produced in a flashtube, and will then almost instantaneously produce a much intensified burst of red light from the end of the ruby as all the "excited" electrons cascade back home again. Here was something that had never been achieved before: a source of pure light of a single wavelength which could be created by the artificial stimulation of electrons in the chromium atoms of a red ruby rod, and the emission of that red light in pure coherent form. The whole process takes a matter of a fraction of a second from the moment the flashtube is triggered to the moment the pulse of red light bursts forth from the laser.

And the result is a burst of radiant energy of magnificent proportions, a highly amplified, hyperintensified pulse of red light which will travel at the speed of light for enormous distances in an almost perfectly straight line without any significant divergence at all. Even the best-focused and most powerful searchlight known throws a beam of light which diverges from the source quite rapidly. A laser beam remains tightly controlled.

We see from this that the laser beam created in this fashion is not really a "beam" of light at all; rather it is a pulse of light. We might say that it is a bundle of identical photons emitted all at one instant and packed together to form a single large "bullet" of light. If this intensified burst of light is passed through the proper lenses, it behaves just as any other light

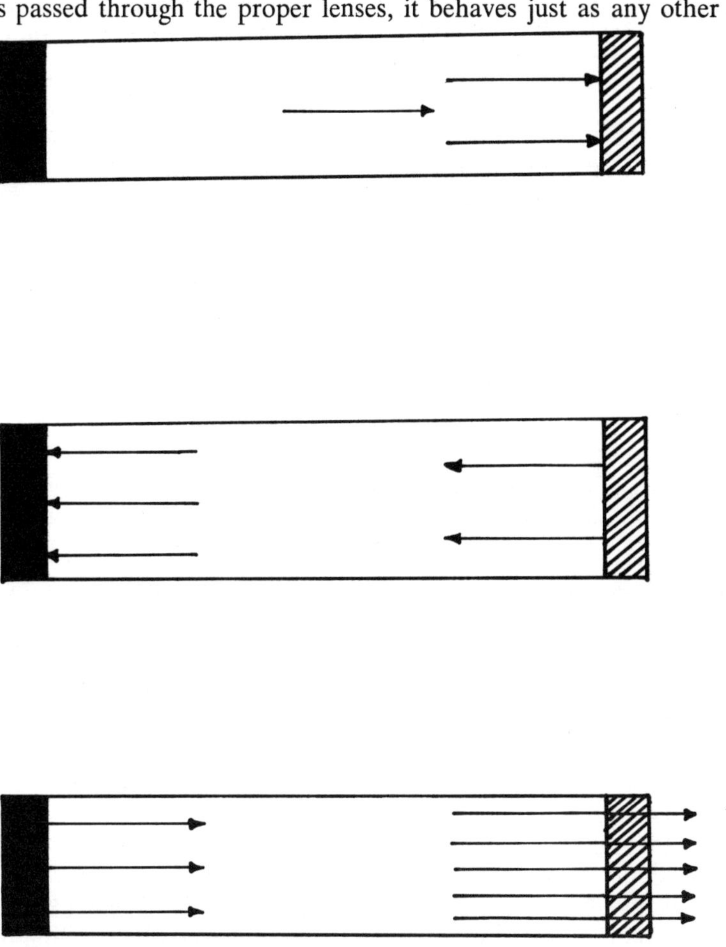

*55b:* Function of a laser: the control of red light. Photons of red light emitted as in 55a move in all directions, but some are trapped between mirrors at ends of laser tube, move back and forth between mirrors triggering more and more rest-stage atoms to emit more red-light photons. Finally intense, amplified, controlled beam of parallel red light waves bursts through semi-silvered mirror at right end of tube to become *laser beam.*

behaves: It can be focused to a pinpoint, if desired, or spread out in a spray. It is not, as many people erroneously imagine, a "heat ray," but when the laser burst is properly focused by means of lenses, for example, a perfectly staggering amount of radiant energy can be brought to bear upon a very small target area—sufficient radiant energy to vaporize virtually any substance that it strikes, from razor steel to cornstarch pudding.

A rapid sequence of laser bursts, created by a rapid sequence of stimulating flashtube flashes, could theoretically simulate a beam of light in the same way that a firing machine gun simulates a beam of lead, but because some of the energy absorbed by the ruby rod is dissipated into heat, the laser would heat up enormously if it were operated in this fashion. Thus, to achieve a continuously operating laser it was necessary to build one utilizing a "base substance" containing atoms that have more than one "middle-energy-level" rest stage so that some electrons can be pumped up to excitation levels at the same time that others are emitting their radiation, in order to form a steady laser beam. One such continuously acting laser, employing helium and neon gas as the base substance, has existed for several years; it can operate continuously and steadily for days and even months.

Already the laser has found some extremely important applications, and offers so many other exciting possible applications that history may ultimately judge it to be one of the most important practical discoveries of all time. For example, almost as soon as the first successful laser had been built, space technologists and astrophysicists were planning uses for it as a means of communication from one point to another in space. A laser burst from earth can be focused on a remarkably small target on the moon's surface, and serve either as a radio beam to guide and control landing space craft or as a means of direct communication. Since the light of a laser beam is traveling at light velocity and is perfectly coherent, it can be used to carry all kinds of electromagnetic signals, ranging from telephone messages to television channels. In fact, it is estimated that the ruby laser with laser beams of about 7,000 angstrom units could carry tens of thousands of separate television channels all at once. Of course a laser beam would be blocked by fog, rain, or intervening mountains just like any other light beam, and thus could not be effective as an open-air transmission system for television signals (although it could be used to increase greatly the number of television channels that could be handled by an Echo satellite, for example). But the laser beam is so readily controllable that a small hollow tube, equipped with mirrors and prisms to keep a laser beam moving in a series of straight-line jumps, could be run around the world, over mountains, under the ocean, anywhere, and television signals could be carried by a laser beam passed through such a "transmission tube." The practicability of such an idea is so great that it is now virtually certain that within a few years lasers will be used in this fashion to bring us television broadcasts from London, Moscow, Tokyo, or New Delhi with reception as clear and reliable as if the television signals were originating from the house next door. Lasers are destined to revolutionize the world of communication as thoroughly and radically as Marconi's first wireless radio.

Again, in the few short years since lasers were first developed, they have found surprisingly wide and beneficial application in medicine, especially in certain exceedingly delicate types of surgery. With the help of a laser a surgeon can concentrate and focus a split-second burst of radiant energy on a specific pinpoint target, perhaps involving only a few critical cells. One of the first surgical uses of the laser was in treating a type of disorder known as detachment of the retina, a condition in which the light-sensitive lining on the inside of the back of the eye begins detaching and peeling away, ultimately resulting in blindness. Previous to the development of the laser, retinal detachment was one of the most difficult of all eye disorders to treat. Now the process can be stopped almost as soon as it is discovered by use of a pinpoint laser burst focused to "tack down" a detaching retina, literally riveting it into place before detachment can progress to a serious degree. Since the laser burst lasts but a tiny fraction of a second, this procedure can be done without anesthetic and without any pain whatever to the patient; small lasers can be built into the ophthalmoscope with which the physician examines a patient's retina, and retinal detachment can be treated in any ophthalomologist's office.

In another area of medicine, pathologists at the Boston University Medical Center have used lasers focused through a microscope to perform fast and simple assays of the atomic make-up of biological specimens, a technique that may make "instant biopsies" of diseased tissues possible without either the pain or convalescence now necessary in order to manage extensive surgical biopsies. Finally, as surgeons explore deeper and deeper into the realm of delicate microsurgery—surgical manipulation of tiny blood vessels or nerve pathways—the laser's usefulness is certain to increase. Although it is hard to imagine surgeons working with microscopes and laser devices rather than with scalpel and clamps, the laser may open up whole new areas of exploration to the surgeon which he could never previously even have hoped to approach.

Even with existing equipment a laser beam can be focused and controlled so perfectly that it can be used to "erase" a pencil mark on a piece of paper without either injuring the paper or leaving a trace of the pencil mark. With such delicate control available, we may well see electric typewriters in the future with lasers attached, so that the harried typist can erase a mistyped letter with a flick of the finger.

For all the applications that have already been found, lasers are still in their infancy; work is progressing on a broad front both to develop more practical and efficient laser devices and to explore ways that laser beams can be applied. As Dr. Arthur L. Shawlow, now professor of physics at Stanford University, has said, "The laser today has the unexplored potential that the airplane had in 1910." And certainly the developments in just the last five years suggest that this may be a most conservative prognostication!

## THE STRANGE BEHAVIOR OF SEMICONDUCTORS

The laser is perhaps the most striking recent example of a brilliantly conceived practical application of basic research in physics. Men like Niels Bohr and Wolfgang Pauli were theorists; they were not particularly concerned with what *use* might be made of their development of quantum theory and their study of the way in which electrons behaved according to quantum mechanics. Yet the development of the laser was a direct result of other scientists' thinking out of the practical implications of quantum mechanics and then using this esoteric bit of knowledge about atomic structure to achieve a practical goal.

But even before attempts to manipulate the energy levels of electrons led to the development of the laser, physicists and engineers working in the field of *electronics* had discovered another quite different device with such widespread potential for practical application that it has literally altered the way we live in a multitude of ways. Practically from the time the existence of electrons was first discovered, physicists and engineers have concerned themselves with the development of a variety of electronic devices, tools, and instruments, all employing electromagnetic fields to control the movement of electrons and electrical currents. Devices such as radios, television sets, tape recorders, record players, and electronic computers have become so common in our lives that it is quite accurate to say that we are living in the "electronic age."

All of these devices depend upon control of the flow of electrons from one place to another by means of electrical and magnetic fields. Until the early 1950s such control of the flow of electrons was possible only by use of large and clumsy vacuum tubes or radio tubes, bulky, expensive, unreliable, and energy-consuming devices which greatly limited the potential of a number of electronic machines upon which the whole world had come to depend.

Then, in the early 1950s, a spectacularly useful new electronic device was invented, capable of replacing vacuum tubes—a device known as the *transistor.* Unbelievably simple in its operation, the discovery of the transistor came as a major breakthrough in electronic physics; yet it could have been discovered years earlier than it was. To understand what the ubiquitous transistor is and how it functions, we must understand something about a rather peculiar electrical property of a few common substances that have been known for years in the world of physics as *semiconductors.*

As we saw in an earlier section, the electrical current that supplies our household reading lamps, our washing machines, our vacuum cleaners, tape recorders, or electric typewriters is nothing more than a stream of electrons flowing from some source across an area of potential difference, an area where the electrical charge at one side is different from the electrical charge at the other side. Any time excess negative electrical charges pile up at one side of an area while positive electrical charges are in excess at the other side, electrons bearing negative charges will flow from the area of excess negative charge to the area of excess positive charge.

Of course, this movement of electrons is accomplished most readily if the electrons can move through a copper wire or some other kind of electrical conductor, but if no conductor is present to carry the electrons, and if the potential difference is great enough, the electrons may even leap a gap of empty space to complete the circuit. In either event, the flow of electrons constitutes an electrical current.

The electrons making up the electrical current may come from a chemical source such as a dry cell or storage battery. They may be generated by

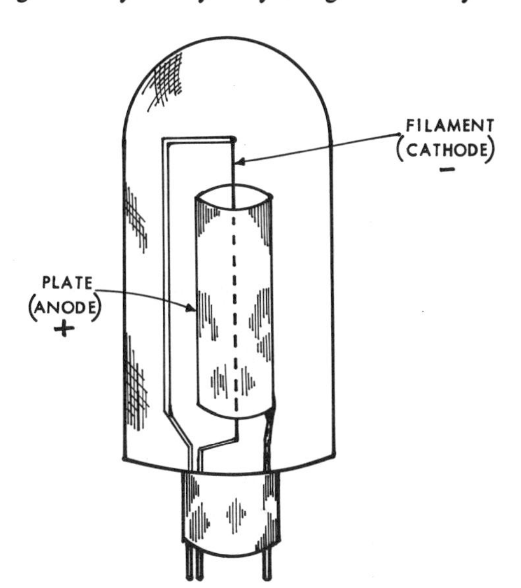


*Fig. 56* The basic structure of a vacuum tube. Disadvantages include size, ease of breakage, burnout, and need to maintain vacuum within tube.

means of conversion of mechanical energy to electrical energy in a generator or hydroelectric turbine. In fact, as we have seen, there are any number of ways in which heat, mechanical energy, radiant energy, or other kinds of energy may be converted into electrical energy. Thus when a piece of metal is heated red hot, the electrons from the atoms of that metal literally "boil off," flying free in space, and if that piece of metal is enclosed in an evacuated glass tube and connected to an electric circuit the electrons boiling off will "jump the gap" from the heated metal source (the cathode) to a metallic receiver (the anode) on the other side of the tube, creating a stream of electrons across the space in the vacuum tube (see Fig. 56).

This means of creating a stream of electrons across an enclosed space proved to be singularly useful, since it provided a means of controlling the *direction of flow* of an electrical current. Since all electrons bear negative charges, the stream of electrons flowing across a vacuum tube from cathode to anode can be manipulated very easily by forcing the stream of electrons to pass through electromagnetic fields. For example, simply by placing a grid carrying an electrical charge in the path of a stream of electrons moving from cathode to anode in a vacuum tube, and then increasing or decreasing the amount of charge on the grid, it is possible to slow down the stream of electrons, to stop the stream completely, or to make it fluctuate up and down according to the rising and falling of the current in the grid. It was essentially this simple form of manipulation of electrical currents by means of vacuum tubes that made radios, television sets, record players, or even electronic computers possible at all.

It had been known for centuries, of course, that certain kinds of materials seemed to conduct the flow of electrons or permit the passage of an electrical current very readily, while other substances stubbornly refused to carry an electrical current at all. Those substances that do permit the passage of electric current, including most metals, are known as "conductors" of electricity; substances such as glass, rubber, or pure distilled water, which tend to obstruct the flow of electric current are called "nonconductors." Some substances, such as rubber, are so strongly nonconductive that they obstruct the passage of any electrical current at all, and are thus sometimes called "insulators."

What is it that determines whether a given substance will be a good conductor of electricity or a nonconductor? The answer lies in the freedom with which electrons of the component atoms of the substance are themselves able to move. In the process of electrolysis a compound capable of breaking into positive and negative ions is mixed into pure water; then when the container is connected to an electrical circuit a potential difference is created from one side of the solution to the other. Negatively charged ions are drawn to the positive pole and the positively charged ions to the negative pole. Essentially this means that the solution conducts an electric current through it because ionized particles or compounds carrying an excess of negative charge (that is, carrying an excess of electrons) are drawn to the positive pole or anode. We could also say that those ionized particles bearing a deficit of negative charge (in other words, carrying a positive charge) are inexorably drawn to the negative pole or cathode.

Thus the passage of electrical current in the process of electroylsis involves the movement of whole charged atoms or ions, not merely electrons. But the conduction of electricity through a metal wire involves only the movement of electrons in the direction of a positively charged side of a potential difference. Good metallic conductors of electricity, such as copper or silver, are composed of atoms that have one, two, or more "extra" electrons in their outer electron shells. Although the atoms themselves may be held tightly bound in a rigid crystal structure, these extra or "unoccupied" electrons can move about from the outer orbit of one atom to the outer orbit of another. When the piece of metal is connected into an electrical circuit so that a potential difference exists across it, these extra electrons flow quite freely through the metal, so that the current passes through the metal with little or no resistance of any sort.

In the case of nonconductors, however, the opposite situation prevails. In the case of strong nonconductors, the component atoms have a *deficit* of electrons in their outer orbits. Such atoms, are, so to speak, "looking for additional electrons" to fill in their outer orbits, and can be thought of as "electron acceptors," whereas atoms of the conducting metals with excess electrons in their outer orbits can be thought of as "electron donors." Nonconducting materials, therefore, have few electrons that are free to move from one atom to the next. Thus, even when a potential difference exists across the substance there are no electrons free to move from atom to atom; electrons piling up at the negative pole or cathode are grabbed by the "electron acceptor" atoms and little or no electric current passes through the substance.

But not all materials qualify either as good conductors or good insulators; some, in fact, cannot properly be classified as either. These substances are generally the compounds of a group of elements known as "metalloids," including such elements as silicon, boron, germanium, selenium, and arsenic. Compounds of these elements do not have the rich supply of extra or unoccupied electrons that metals such as copper or silver have, but they do have *some* electrons which are capable of moving from atom to atom. Thus these materials are slightly conductive—capable of conducting a trickle of electric current when connected to a circuit—and are spoken of as "semiconductors," mostly to indicate that they are neither good conductors nor good insulators.

At first glance, it might seem that these semiconducting substances would be pretty useless for the manipulation of electrical currents. They are neither fish nor fowl, incapable of carrying a strong current, equally incapable of really insulating or preventing the conduction of a current. But certain semiconductors were discovered to have an additional rather singular property. It was found that if certain kinds of impurities were present in the crystalline structure of certain semiconductors, those impure crystals could become much more conductive than normal.

Suppose, for example, that a crystal of a silicon compound happened to contain a few arsenic atoms as a contaminant. The silicon atom is comfortably satisfied electronically; its outer shell contains a full complement of electrons, so that it is not eager either to donate one of its electrons or to accept an extra one. An arsenic atom, on the other hand, carries one "extra" electron which is free to move from atom to atom if it can find an "electron acceptor" anywhere. Thus, the arsenic atom can act as an "electron donor" (see Fig. 57a). This means that a crystal of silicon contaminated with arsenic atoms can allow a trickle of electric current to pass through it because of the extra negatively charged electrons of the arsenic atoms, and is spoken of as an N-type or negative-type semiconductor because the current can be carried by the negatively charged extra electrons.

Exactly the opposite situation prevails, however, when a crystal of a silicon compound is contaminated by a few atoms of boron, an element which has a *deficit* of one electron in its outer electron shell and thus acts as an electron acceptor. Here again such a substance can conduct an electric current if it is placed in contact with some electron source, since it has a deficit of negative charges (the equivalent of an excess of positive charges).

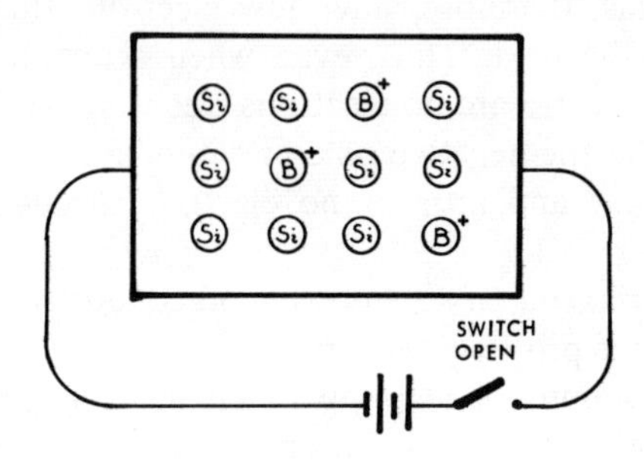


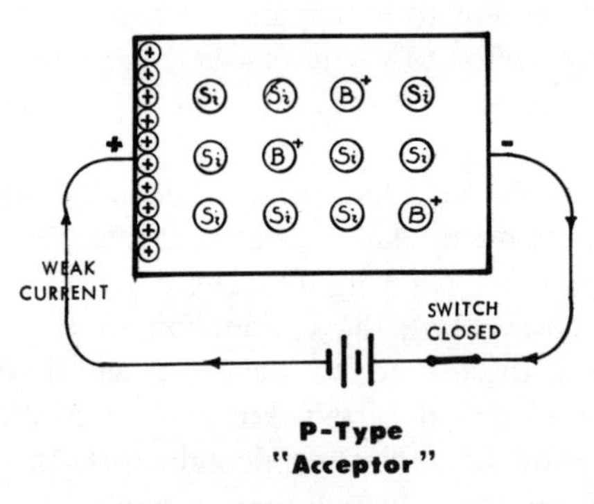


**P-Type**
**"Acceptor"**

*Fig. 57a*

Electron-acceptor–type semiconductors are called P-type or positive-type semiconductors for this reason (Fig. 57b). Precisely *how* electrically conductive a P-type semiconductor will be depends upon the number of electron deficits present—the amount of the contaminating boron in the crystal—while just *how* electrically conductive an N-type semiconductor will be depends upon the number of free electrons available from the arsenic electron donors and upon the ease with which those electrons can move through the crystal.

We can see from the diagrams that either an N-type or a P-type crystal, connected by itself into an electrical circuit, would be of little use in conducting a current; at best only a trickle of electrons would move through the crystal in either case. We can also see that if an N-type crystal and a P-type crystal were placed side by side there would be a certain amount of motion of electrons back and forth from one crystal to the other: the excess electrons in the N-type crystal would tend to diffuse over and fill the "electron holes" present in the P-type crystal. This diffusion, of course, could only happen at the interface between the two crystals, perhaps involving just a few atomic layers of the crystal on either side of the interface.

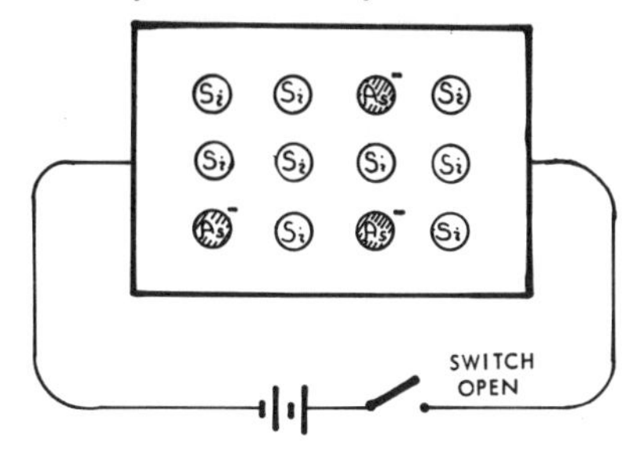


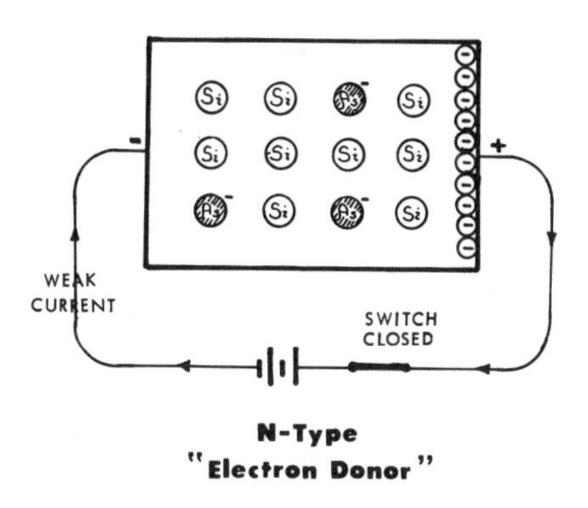


*Fig. 57b*

But suppose the two crystals placed together in this fashion were then connected to an electrical circuit with a dry cell battery, so that a potential difference existed across the crystal pair. If the battery were connected so that its positive pole were in contact with the P-type crystal bearing its deficit of negative charges and the negative pole were connected to the N-type crystal with its excess of negative charges, electrostatic force would then drive the electrons out of the N-type crystal and straight through the P-type crystal to reach the positive pole, so that a strong electric current would move through the paired system of crystals where only a trickle would move through either one or the other alone. On the other hand, if we re-

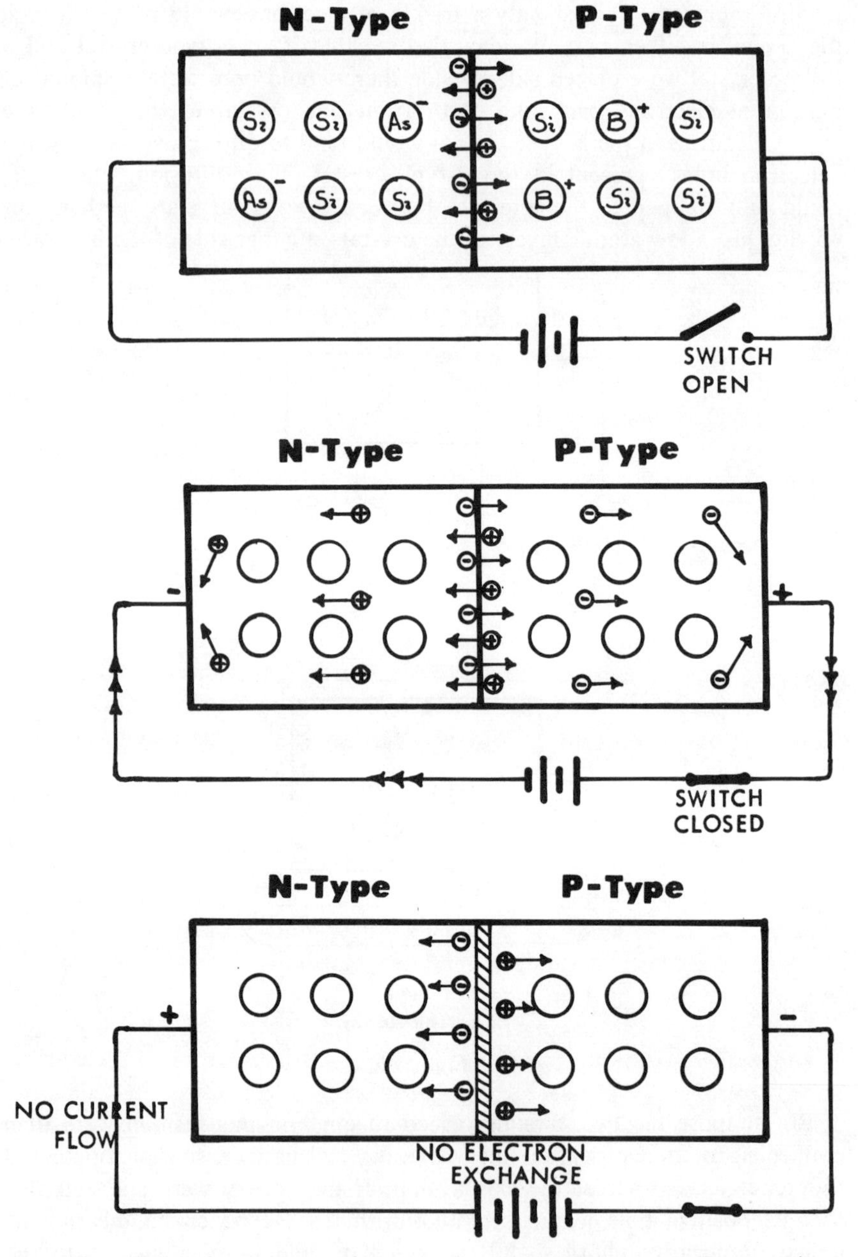


*Fig. 57c* The basis of the transistor. At the top, N-Type and P-Type semiconductors in contact allow surface diffusion of electrons only. In center, the N-P pair are connected to battery with negative pole attached to N-Type crystal, positive pole to P-Type, with vigorous current flow. But when poles of battery are *reversed,* as seen at bottom, no electron exchange is possible and current cannot flow at all.

versed the poles of the battery so that its positive pole was connected to the N-type crystal and its negative pole to the P-type crystal, virtually all electron movement from one crystal to the other would stop dead as though a solid insulating wall had been built between them. In Figure 57c we see diagrammatically how this would work.

Thus whenever an N-type semiconductor and a P-type semiconductor are placed side by side, we have an interesting situation. Such a crystal pair will allow a strong electric current to flow almost unimpeded through the paired crystals in one direction, but will prevent any current flow whatever in the opposite direction. Now suppose that such an N–P crystal pair were connected into an ordinary alternating current circuit, in which the electric current oscillates, reversing its direction sixty times each second. What would happen? Obviously the current flowing in one direction would be passed freely along, while the current flowing in the other direction would be stopped. In other words, such an N–P crystal pair would act to "rectify" or convert an incoming alternating current into a one-way direct current—a process which otherwise could be achieved only by using a far more complicated, fragile, and expensive electronic tube or vacuum tube.

Such N–P crystal pairs acting as rectifiers of alternating current were highly useful in themselves, but in recent years an additional refinement produced an enormously more valuable device. By placing another P-type crystal in contact with an N–P crystal pair in such a way that the N-type crystal was *sandwiched between* two P-type crystals, it was possible to make two separate electric currents flow through this sandwich arrangement of semiconductors in such a way that current A acted as a valve to control current B; slight adjustments in current A could cause current B to vary by many orders of magnitude. In short, such an N–P–N or P–N–P arrangement of semiconductors could carry out perfectly well all the functions of electron stream manipulation that had formerly been possible only with the use of vacum tubes!

Such a sandwich arrangement of N- and P-type crystals was first invented in 1948 by J. Bardeen and W. H. Brattain at Bell Telephone Laboratories, and was called a *transistor*. Later it was refined in collaboration with W. Shockley. With the use of these transistors, which could be constructed to occupy only a very tiny amount of space, it was possible to control the flow of electrons in a circuit just as effectively as with the old-fashioned bulky vacuum tubes—but with some enormous advantages thrown into the bargain.

For one thing, this control of electron flow could be accomplished entirely within a tiny bit of solid matter. With a transistor there was no vacuum to maintain, no glass tubes to overheat or to break, no filaments to burn out. A transistor half the size of a peanut could do the job of a vacuum tube the size of an orange; it could do the same job far more efficiently and could do it for a far, far longer period of time before repair or replacement was necessary. What was more, the transistor could do the job far more economically as well; since the flow of electrons did not have to come from a heated metal filament, much less electrical current was necessary to produce perfectly controlled streams of electrons than was the case with vacuum tubes, so that many electronic devices previously requiring many vacuum tubes for their operation could now not only be built in greatly miniaturized form with their electronic components made of solid crystalline material (so-called solid-state devices) that could not break or wear out, but they could be operated with far less energy input than before. A device previously requiring 110-volt current and depending upon vacuum tubes which might burn out at any time after the first few weeks of use could now be operated for months without trouble on the power provided by a single tiny 1½-volt battery!

The advantages of this simple and homely device were so great and so immediately obvious that within the space of less than twenty years the transistor has completely revolutionized the construction of all the multitudes of electronic devices that we use constantly in our daily lives. Pretransistor radios, for example, had to be bulky even in so-called portable models because they had to contain half a dozen vacuum tubes in order to operate. Those tubes were constantly vulnerable to shattering or burning out. The radio required so much power to operate it that it either had to be plugged into house current or supplied with large and heavy dry cell batteries which quickly ran down. By using transistors instead of vacuum tubes, it was possible to miniaturize radios to the point where extremely efficient radios can now be made smaller than the size of a cigarette pack. Such tiny radios have all the efficiency and range of the finest pretransistor monster, yet operate for months or even years on a tiny trickle of current from a battery half an inch wide and an inch long. They are so sturdy that they can be dropped, shaken, rattled, or even hurled against a wall without much ill effect. In addition to all these advantages, transistorized radios can be produced and sold at a fraction of the cost of the smallest tube-type radios.

Nor are radios the only instruments in which transistors are now used to the exclusion of vacuum tubes. Before transistors were available, music-playing systems needed bulky and expensive components operating with vacuum tubes, requiring ventilation because of the heat those tubes threw out, and built with huge speakers in order to reproduce sound with any degree of fidelity. Such music players, clumsy and delicate as they were, and vulnerable as they were to burned-out tubes, were also extremely costly, even if one built the components oneself. Today solid-state record players and speakers are capable of reproducing sound with higher fidelity than the old systems, yet are far more compact, even portable, more trouble-free, and can be made and sold profitably at a fifth the cost.

Transistors also permitted the miniaturization of electronic components of rocket ships, missile guidance systems, servo mechanisms, and multitudes of other devices used in our everyday lives, all the way from tape recorders (which can now be made so small that you can carry one in your pocket yet record over forty-five minutes of continuous interview or dictation on a single miniature tape) to television sets or adding machines. The list is almost endless. Perhaps of even more far-reaching importance to the future of our society, transistors have permitted the construction of electronic computers on a miniaturized basis. Before transistors, the biggest computers were truly giants, taking up whole warehouses of space and requiring thousands upon thousands of vacuum tubes for their operation. Since those tubes were constantly burning out, duplication and triplication of circuits was necessary to obtain reliable performance from such computers, and they required such vast amounts of power in order to operate at all that their use was very costly. Today computers are even larger than in pretransistor days, and require even more power to run; but with the use of miniaturized transistors instead of vacuum tubes, modern computers can achieve many times the volume of data storage in a tenth the space, and can operate with far greater reliability for far longer without requiring repair than the older vacuum tube computers. Thanks to transistors we get far more computer capability for the space, power, and money than could possibly be obtained without them.

## SEMICONDUCTORS AND SOLAR BATTERIES

Needless to say, research and development in this new and virgin field of solid-state physics did not end with the invention of the transistor, useful as that device was. The curious electron-controlling properties of semiconductors suggested an ever-growing variety of other possible applications, and physicists and engineers with experience in solid-state physics soon became key personnel in the research and development laboratories of computer manufacturers, electronics firms, aerospace industries, and communications companies. Probably there is no other single property of matter which has found such widespread practical application in so very brief a time; to try to list the areas of modern industry in which semiconductors play a role would be impossible here. But we can see at least one other good example of the versatility of these strange substances in the work that has been done in the Bell Telephone Laboratories during the last decade toward the development of an efficient and workable solar battery.

The world in which we live today literally turns upon electrical energy, and for centuries scientists have sought new, cheap, and efficient ways of converting other forms of energy into electrical energy without too much loss of energy to heat, friction, and other such wastage. For centuries physicists have longed for some way of tapping the greatest source of energy in the entire solar system, the sun itself, in order to convert even a tiny fraction of the incredible amounts of radiant energy it pours forth onto the earth's surface into some more usable form. Unfortunately, until the photoelectric effect was discovered in the early 1900s, no really efficient way of harnessing radiant energy from the sun and converting it to more useful forms had been found, although workers had gone to such extremes as using huge magnifying glasses to focus the sun's rays on a steam boiler in order to bring the water within it to a boil and operate a steam engine, to cite a single example. Inevitably these crude attempts proved far too clumsy to be practicable; yet there was all this energy pouring out of the sun and no way to harness it.

With the discovery of the photoelectric effect and the development of the concept of energy quanta, a possible new approach to harnessing the sun's radiant energy presented itself. It became clear that photons of light striking certain kinds of atoms under certain circumstances had the power to knock electrons loose from those atoms and send them flying, but there seemed no way to put this phenomenon to practical use. Then, as the function of semiconductors became better understood, physicists began taking a second look at the power of light photons to knock electrons out of atoms. Light rays from the sun striking a semiconductor crystal, for example, could knock loose some of the electrons from the contaminating atoms to which they belonged, thereby increasing the number of electron-accepting atoms available in the substance. It was found that by taking a silicon crystal containing arsenic contamination—as we saw, an electron donor or N-type crystal—and coating it with a thin layer of P-type crystal containing electron-acceptor boron atoms and then exposing this crystal combination to sunlight, the action of the sun knocking electrons free of the arsenic atoms could create a flow of electricity from the thin P-type layer to the thicker N-type layer, as diagramed in Figure 58. This current could then be tapped and stored. Such a simple device was, in fact, nothing more nor less than a battery—a source of electrical current—which converted radiant energy from the sun directly into usable electrical energy simply by allowing the sun to beat down on a pair of semiconductor crystals placed side by side.

Obviously such a "solar battery" would cost only as much as the cost of the semiconducting crystals and of some sort of reservoir for storing the electricity produced by it. The radiant energy from the sun itself is free. Of course the operation of such a solar battery depends upon direct exposure to the sun, and produces current in direct proportion to the area of semiconductor surface exposed. It cannot operate at all at night, or on gray days when the sun is hidden. At its present stage of development, such a solar battery really requires too much radiation-absorbing surface of semiconductor for it to be practicable. A ten-foot-square surface exposed to bright sunlight for eight hours a day would evolve only enough electricity to power a 100-watt light bulb during eight hours of darkness. Since semiconductors are not cheap to manufacture, and since the average household uses a good deal more than one 100-watt light bulb, such a device so far cannot offer us much relief from electric bills.

Indeed, solar batteries of this type may well never find practical household use, particularly in areas such as Seattle or London where clouds obscure the sun a significant part of the year, or in densely populated cities. In desert countries, however, the chance of practicability is much improved; in fact, if some way could be found for extremely inexpensive mass production of semiconducting materials, so that great areas of desert

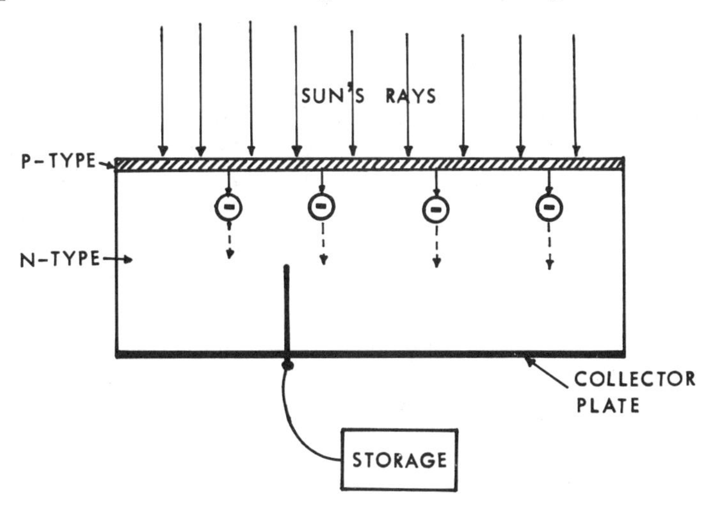


*Fig. 58* A simple solar battery.

or tropical ocean could be covered with permanently installed solar battery surface, electrical energy in significant and useful quantities could be stored and transported to where it is needed.

Such applications of this principle are still in the future, but already solar batteries utilizing semiconductors have proved invaluable in producing a continuing source of electric power in one area where sunlight is always reliably available and where other sources of electric power are either impractical or impossible—in the open reaches of outer space. Already artificial satellites, Apollo spacecraft, and lunar landing probes have been equipped with solar batteries to supply the electricity needed to operate their many instruments. What is more, under outer space conditions the extraordinarily long life of such a battery (which would continue operating just as long as sunlight continued to strike its surface) makes it ideal as a source of power for the instruments, guidance system, and other mechanical devices used on more prolonged space ventures such as the Mariner flights to Venus, or the unmanned and manned flights to Mars that are on the drawing boards for the future. Furthermore, since solar batteries can also operate from the radiant energy pouring out of radioactive materials, these devices are certain to play an important role in the future development of nuclear energy as an increasingly critical everyday power source.

One thing is certain: In the case of semiconductors, as in the case of lasers, the surface has barely been scratched; the future potential of solid-state physics, rooted firmly in the knowledge gained from basic research into quantum mechanics, is without foreseeable limits. Here again is an example of practical application of basic theoretical research which has already altered our way of life dramatically and can be counted on to alter it even more dramatically in the future.

But there is another area of frontier research and development in physics which also promises to change our lives, in this case, an area which has quite unexpectedly unearthed a series of the most baffling and peculiar phenomena ever before encountered on earth. Always exploring the outer limits of physical phenomena, it was inevitable that sooner or later physicists would begin to explore the regions of intense cold, the kind of phenomena that occur at temperatures so very low that almost all molecular motion comes to a standstill. Following their noses into this area of study, physicists suddenly found themselves with a bear by the tail, for they discovered that in this dim half-world of physics things were happening which could not possibly happen, according to any laws of physics known to man, and which no one is yet quite sure how to explain.

## THE UNIVERSE OF THE INCONCEIVABLY COLD

Throughout this book we have seen how physicists through the ages have worked to discover everything they could about the basic physical properties of matter—a long sequence of exploration and discovery that has gone on for better than three thousand years of human history. We might imagine that by now that exploration would have been completed, that everything there was to discover about the properties of matter would long since have been discovered already.

But if we were foolish enough to think that, we would be wrong. As in so many other areas in which physicists of the late nineteenth century had begun thinking they had gotten everything tied up into a neat and tidy bundle, some startling and totally unexpected discoveries in the early 1900s altered the situation radically. Physicists have certainly explored and studied the physical properties of matter diligently, and learned a great deal over the centuries; yet in recent years investigators studying the effects of extremely low temperatures on various forms of matter have discovered to their chagrin that certain substances, when extremely and intensely cold, begin exhibiting some singular properties—properties so strange and incredible that no one previously would have imagined them to be possible at all. These bizarre properties of intensely cold matter are not merely variations from the normal properties of matter; they are downright weird, and when first discovered they seemed to elude any imaginable explanation. Yet in the last ten years these properties have suddenly taken on a vast importance in our modern world of technological progress.

Just how "intensely cold" are the temperatures we are talking about, and how were they achieved? All of us are aware that under normal climatic conditions in our everyday lives the matter surrounding us may be found in three physical forms or "states": solid, liquid, and gaseous. In an earlier chapter we discussed some of the properties of these three familiar states of matter in some detail. We saw, in the case of every element or compound, that the physical state in which it might be found is always at least partly a function of its temperature, and that for each element or compound there are fixed temperature transition points at which a substance in the solid state becomes liquid, or a substance in the liquid state becomes gaseous. These transition temperatures, of course, differ widely from one substance to another; ice melts into liquid water at zero degrees Centigrade, and liquid water boils into steam at 100 degrees Centigrade, while solid iron must be heated to 1,530 degrees Centigrade before it melts, and molten iron can only be vaporized if it is heated to 2,735 degrees Centigrade.

These temperature transition points are unique for each separate substance, and so characteristic that they constitute important identifying physical properties of various elements and compounds. But we saw that, as a rule of thumb, a substance in the solid state would be associated with a relatively low temperature, in the liquid state with a higher temperature, and in the gaseous state with higher temperatures yet. This shifting of matter from one physical state to another at varying temperatures was characteristic of so many commonly known substances that it long was assumed that *any* element or compound could theoretically be changed to any of the three physical states that one desired provided one could attain high enough or low enough temperatures. Liquid mercury could be frozen if it were cooled to a low enough temperature, and could be made to boil if it were heated hot enough. Gold could be melted and then theoretically boiled into gold vapor, providing enough heat could be supplied in some container that would not itself melt before the gold boiled. Even gases such as oxygen or hydrogen could be liquefied by cooling and then theoretically could be frozen if cooled still further.

Why do these differing states of matter exist? As we saw earlier, the state of matter in which a given substance may be found is a function of the motion of the atoms or molecules of which that substance is made up. In all substances, whether in solid, liquid, or gaseous state, the atoms or molecules are in a continuous state of motion. This state of motion or "thermal energy" of atoms is at a minimum at low temperatures; the atoms or molecules of a substance become locked together in a crystalline structure which does not permit them to rove about freely, so that the best they can do is to vibrate in place like soldiers marking time or Eskimos shivering. The lower the temperature of the substance, the less of this "thermal energy" is exhibited by its molecules or atoms, down to a theoretical limiting point: a point at which even the very faintest molecular vibration comes to a halt.

No one has yet succeeded in cooling any substance down quite to that miraculous "lowest possible temperature," although, as we shall see, it has been approached very closely. The point at which molecular motion should theoretically stop altogether is spoken of as *absolute zero* or *zero degrees on the Kelvin scale.**

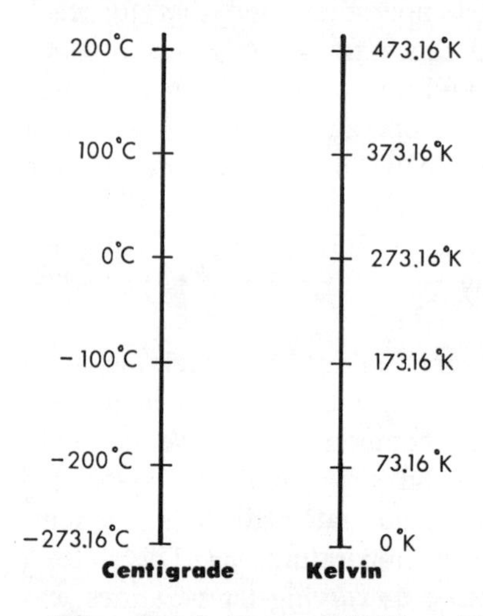


*Fig. 59* Comparison of Degrees Centigrade with Degrees Kelvin.

* So called in honor of William Thompson, Lord Kelvin, who first proposed an "absolute scale" of temperatures in 1848. The Kelvin scale differed from the Centigrade scale that had been used since 1742 only in the arbitrary "zero degree" point that was chosen. The Swedish astronomer Anders Celsius who proposed the Centigrade scale of temperatures picked the freezing point of water at sea level as an arbitrary zero point, and the boiling point of water as 100 degrees. Lord Kelvin argued that the "natural" zero point on a temperature scale should be the point at which all molecular motion ceased, in other words, the lowest possible temperature that could be obtained at all. This "lowest possible temperature" is now known to be 273.16 Centigrade degrees lower than zero on the Centigrade scale or minus 273.16 degrees Centigrade. The two comparative temperature scales, Centigrade and Kelvin, are compared in Figure 59. Low-temperature physicists customarily use the Kelvin scale in their work, and we will use it here, but since the magnitude of 1 degree Kelvin is exactly equivalent to the magnitude of 1 degree Centigrade, we can keep in mind as a rule of thumb that any temperature measured on the Kelvin scale will be roughly 273 degrees higher than it would be measured on the Centigrade scale (absolute zero can also be expressed as −459.6 degrees Fahrenheit).

Even today absolute zero (zero degrees Kelvin or −273.16 degrees Centigrade) remains a theoretical ideal rather than a practical and obtainable temperature. Although substances have been cooled to temperatures as low as a tiny fraction of one degree Kelvin, true absolute zero has never actually been achieved in the laboratory and in all probability never will be. The reason is simple to see: In order to measure the temperature of any substance, there must be some interchange of thermal energy between the substance being measured and the measuring instrument. But by definition absolute zero is the point at which molecular motion in a substance has ceased—in other words, the point at which it has no thermal energy whatever. Thus there could be no energy interchange with a measuring instrument unless the instrument contributed energy to the substance, in which case that substance's temperature would no longer be at absolute zero.

Just the same, very low temperatures have been obtained. Oxygen at comfortable room temperature of about 300 degrees Kelvin (27 degrees Centigrade) can be cooled down and liquefied at approximately 90 degrees Kelvin (−183 degrees Centigrade). Hydrogen has been cooled down to a liquid at 20 degrees Kelvin. Helium gas, which becomes liquid at 4.2 degrees Kelvin (−269.1 degrees Centigrade) has been chilled down still further to 0.01 or even 0.001 degrees Kelvin. It was in the course of attempting to cool substances as close to absolute zero as possible that physicists first discovered that certain ordinary mundane substances—mercury and lead, for example, or ordinary inert helium gas—began to take on some very bizarre special properties when they reach temperatures as low as a few degrees from absolute zero. The most striking of these strange properties are the phenomena of "superfluidity" and "superconductivity."

## FLUIDITY BEYOND REASON

The existence of a property now known as superfluidity was first discovered by low-temperature physicists as a result of their attempts to establish a freezing point for the one and only element which has never yet been satisfactorily frozen, the inert gas helium.

Helium was the last of the gases to be reduced to a liquid form, a feat accomplished by Dutch physicist Kamerlingh Onnes at Leiden University in 1908, by cooling helium gas down to 4.2 degrees above absolute zero. Other gases, such as oxygen, nitrogen, and hydrogen, once thought to be "permanent" gases because they could not be liquefied by means of primitive refrigerating techniques even under great pressure, had by 1908 not only each been liquefied but also cooled down to the freezing point. Naturally, once helium was liquefied the attempt was made to lower its temperature still further in order to solidify it. But the results were not as expected. As liquid helium was cooled from 4.2 degrees Kelvin down to

2.19 degrees Kelvin or even lower, it did not become solid as the other liquefied gases had done. Instead, it underwent a totally different and startling kind of change. At 2.19 degrees Kelvin, it suddenly became incredibly *more fluid* than any other liquid known to man.

How can a fluid become "more fluid"? We know from our everyday experience that some liquids are very viscous while others flow far more freely. For purposes of comparison, we might take water as a standard of fluidity, since we know that it pours readily, and spreads and covers surfaces it comes in contact with. Against such a standard, gasoline or paint thinner, for example, are more fluid than water: They pour in a thinner stream, and spread out and cover surfaces more quickly and more readily with a thinner layer than water does. On the other hand, 30-weight motor oil, although a liquid, is considerably less fluid than water, pouring more sluggishly and flowing and covering surfaces less readily at normal temperatures. Honey, also a liquid, is still less fluid than motor oil.

Physicists recognize that the fluidity of any liquid depends upon the freedom of movement of the molecules in that liquid past and around each other. No matter how fluid a liquid may be under normal circumstances, molecules of that fluid moving about encounter *some* resistance, banging into other molecules as they move. We speak of the "viscosity" of a liquid as a measure of the resistance the liquid presents to the motion of some foreign object through it; if we try to stir highly viscous (that is, highly resistant) honey with a spoon we feel marked resistance to the spoon's motion, whereas we would feel far less resistance in stirring water with the same spoon and even less resistance using it to stir gasoline.

Gases such as oxygen, hydrogen, or helium, when liquefied at very low temperatures, also presented *some* resistance to the free motion of their molecules at whatever low temperature they became liquid. Oxygen or hydrogen, when cooled enough more, froze solid. But liquid helium, when cooled further, far from freezing solid, seemed quite suddenly to *lose all of its fluid resistance completely* at a temperature of 2.19 degrees Kelvin. At that temperature liquid helium was found to become the most fluid fluid ever known, flowing readily through such narrow cracks that it seemed virtually impossible to seal it into any container. This "superfluid" helium (which came to be called helium II as distinct from liquid helium I above 2.19 degrees Kelvin) spread out to incredibly thin sheets on any surface it contacted, with no fluid resistance hindering its flow in any way.

This led to some truly bizarre behavior of helium II, behavior that almost has to be seen to be believed. For example, if an empty glass flask is dipped into a container of superfluid helium II, the liquid helium will creep up the *outside* of the flask, quite of its own accord, run over the edge, and proceed to fill the flask until the level of helium II inside is the same as the level outside. If the flask is then partially lifted out of the helium II bath, so that the level of helium II inside it is higher than that outside, the superfluid helium will reverse the process, creeping up the inside of the flask and sliding down the outside. If the flask containing helium II is lifted out altogether, the helium in the flask will climb out, run down the outside, and drip from the bottom until the flask literally empties itself.

Helium II presents other curious properties as well. It can conduct sound waves, for example, faster than any other known substance, and is capable of conducting two groups of sound waves of different frequencies simultaneously, each at a different speed. It also seems able, under certain circumstances, to flow past solid obstructions *without exerting any pressure upon them whatever*. The more physicists have studied this strange stuff, the more incredible its behavior seems.

What is this strange helium II that is formed when ordinary liquid helium I is cooled to a critical transition temperature of 2.19 degrees Kelvin? Certainly it isn't a solid—but equally certainly, it isn't the same liquid as helium I. Some physicists have referred to it as a "fourth state of matter"; others have described it as a "quantum fluid" made up of atoms with all their electrons lodged in the "ground state" or lowest possible energy level of electron orbits. Even to this day no one is entirely sure, and superfluid helium, so far, has no known practical use; but it has provided physicists with an unexpected and mysterious group of physical properties to study and a new approach to a deeper and more fundamental understanding of how quantum mechanics affects the behavior of matter at the level of the micro-universe.

## THE PUZZLE OF SUPERCONDUCTIVITY

*Superconductivity* is another strange property of certain substances at near-absolute-zero temperatures and, like superfluidity, it involves a sudden and unexpected disappearance of resistance in various substances at specific "transition temperatures"—in this case, resistance to the flow of an electrical current. But unlike superfluidity, superconductivity is a property that has very real and breathtaking potentials for practical usefulness in the world of technology.

Under normal temperature conditions virtually all substances, even excellent conductors of electricity like copper or silver, present at least *some* resistance to the passage of electric current through them. This resistance arises from the fact that the atoms or molecules of all substances have thermal energy and thus are vibrating slightly, even when locked into a rigid metallic crystalline structure. When an electric current—that is, a stream of electrons—flows through such a substance, some of the electrons inevitably strike atoms or molecules in their path, bounce and ricochet as they move, and thus are slightly impeded in moving from one end of the substance to another. This sort of bouncing and ricocheting also produces heat, so that the substance warms up slightly. Even a piece of copper wire, a splendid electrical conductor, will become warm when a current passes through it; if the wire is very narrow and is carrying a large amount of current, the heat arising from this resistance may be sufficient to make the wire glow red hot, or even burn out.

It has been known for years that the amount of resistance to passage of electric current presented by a given conductor is directly related to the temperature of the substance. The hotter a conducting wire becomes, the more resistance it presents to an orderly flow of a stream of electrons, since the atoms of the substance are vibrating more and more wildly the hotter the substance becomes. Similarly, if a substance is cooled down, its resistance decreases because of a decrease in the activity of its atoms or molecules.

In 1911 Kamerlingh Onnes, the first of the great low-temperature physicists, was measuring this decrease in resistance to the passage of electric current in solidified mercury at very low temperatures. As Onnes cooled the mercury down steadily in a bath of liquefied helium, its resistance decreased steadily, just as expected. At about 5 degrees above absolute zero its resistance had dropped to about 1/500 what it would normally be at zero degrees Centigrade. But as the temperature of the solidified mercury fell to 4.3 degrees Kelvin, a totally unexpected thing happened: Its resistance suddenly and abruptly dropped to the vanishing point. At that very low temperature, mercury suddenly and inexplicably became a *superconductor*, capable of carrying electric current of virtually any magnitude with no evidence of any resistance whatever.

During the next thirty years or so a number of other substances were found to become superconductors when cooled to excessively low temperatures. Many of these included metals that were ordinarily comparatively poor conductors of electricity—tin, lead, zinc, titanium, and cadmium, for example. Each individual substance seemed to have its own unique "transition point" at which it suddenly became superconductive. With mercury the transition temperature was 4.3 degrees Kelvin. In the case of tin it was 3.7 degrees Kelvin; for aluminum, 1.20 degrees Kelvin; and for lead, 7.2 degrees Kelvin. Attempts to make metallic alloys that became superconductive at somewhat more practicable higher temperatures were largely unsuccessful; to date the highest temperature at which any known substance becomes superconductive is about 18 degrees Kelvin—minus 255 degrees Centigrade!

Oddly enough, this property of superconductivity seemed to be extremely selective; only a comparatively few substances—perhaps three dozen or so to date—display this strange behavior, and even more oddly, most of the substances which do become superconductive at extremely low temperatures are substances which are comparatively *poor* conductors of electricity at normal temperature ranges. Metals such as copper and silver which are good conductors at normal temperatures do not become superconductive at all when cooled down to near-absolute-zero temperatures.

For many years the whole phenomenon of superconductivity was regarded more as a laboratory curiosity than as a phenomenon with any practical value, and research progressed very slowly, in part because exploration of low temperature ranges required such exceedingly costly and awkward equipment that only a very few laboratories could engage in this research. But as more and more was learned about this strange property of matter at low temperatures, more and more investigators became very curious about it, and more and more practical applications began to appear.

For example, it was possible to take a metal which was a relatively poor conductor of electricity, cool it down in a bath of liquid hydrogen or liquid helium to its transition-point temperature, and end up with a superconductive substance that would carry an electric current indefinitely *even after disconnecting the source of the current.* When a bar of lead, for example, is formed into a ring or doughnut and then cooled down below 7.2 degrees Kelvin (the transition point for lead) by immersing it in liquid helium, the lead ring becomes superconductive. When an electric current is started flowing in a circuit around such a superconductive ring, the current will continue flowing around and around without any appreciable loss in strength and without any further outside energy added to keep it going. You could disconnect the battery and walk off with it, and the current would keep on moving around and around the lead ring! In one laboratory such a current flowed in a frigid, superconductive ring of lead for over two years after the power source had been disconnected without any appreciable loss of strength.

This, in effect, seemed to be the age-old dream of the perpetual motion machine come true; and although physicists so far have no objective laboratory evidence that such a persisting current would actually continue flowing forever under its own steam in a superconductor, they do feel quite confident that it might persist for billions of years as long as the superconductor was kept at a temperature below its transition point.

This one characteristic of superconductors alone was enough to set the world of physics buzzing, for there are a number of ways in which such persisting currents could be extremely useful. We recall, for example, that an electric current flowing through a coil of copper wire wrapped around a bar of iron will induce magnetism in the iron, the basis of the electromagnet. But the ordinary electromagnet only retains its magnetism as long as a constant input of electrical energy keeps the current flowing through the wire coil. In a superconductor, the current can be made to persist perpetually without any additional energy input, and if channeled around an iron core, can create an electromagnet that operates forever once the current has been started.

Another place where persistent currents in superconductors have already found a useful application is in the tiny circuit components of high-speed computers. Complex printed circuits called "cryotrons" have been built into refrigerated areas of computer circuitry and require incredibly little energy input to maintain performance with previously unheard of reliability and, above all, speed. What is more, since a slight increase or decrease in a surrounding magnetic field can turn a superconducting cryotron off or on, causing persisting currents in the unit to dissipate instantly or start flowing again instantly, cryotrons in computers can be used as incredibly delicate on-off switches. Data can be stored in a sequence of such persisting circuits to form complex miniaturized "memory banks." This data can be kept available in the computer indefinitely just as long as superconductivity is maintained in the cryotrons, and then can be retrieved at will with fantastic speed simply by increasing the magnetic-field strength and thus destroying the cryotron's superconductivity. Just as swiftly, new data can be stored to replace that which has been retrieved by decreasing the magnetic field again and starting a new persisting circuit in the cryotron.

Still another remarkable and useful property of superconductors is the curious way they behave in the presence of ordinary magnetic fields. At normal temperature levels lead and tin, for example, have little or no effect on the lines of force of a magnetic field when magnets are placed near them. The magnetic fields seem to pass right through them unimpeded. But when cooled down to their respective transition points so that these metals become superconductors, they also become perfectly "diamagnetic"—that is, they seem to repel or exclude magnetic lines of force passing in their vicinity when they are placed in the field of a weak magnet. On the other hand, if a very powerful magnet is brought near a superconductor, the powerful magnetic field seems to be able to overcome the superconductor's ability to repel it—but when this happens and the magnetic field penetrates the substance, the superconductor abruptly loses its superconductive properties, even though its temperature still remains below its transition point, until the magnetic field is removed.

In other words, the property of superconductivity in a metal and the lines of force of a magnetic field seem to be mutually incompatible. Yet certain alloys of superconductive metals seem to be able to maintain their superconductivity (and thus maintain persisting electrical circuits) even in the presence of quite intense magnetic fields.

This strange behavior means that these superconductive materials can themselves be used to form the coils of electromagnets of considerable power. Furthermore, if electromagnets so constructed are bent into closed circles, they are found to maintain their full strength as magnets indefinitely *even when the electric current inducing the magnetism is cut off*. Now this is indeed extraordinary; ordinary conventional electromagnets with high magnetic-field intensity require a continuing input of hundreds or thousands of kilowatts of electrical energy to keep them working. But these strange superconductive magnets require *no energy whatever* to maintain their magnetism once magnetism is induced in them!

Needless to say, engineers and technologists, who love nothing more than the idea of getting something for nothing, have been discovering all sorts of practical applications for these "zero energy magnets." For example, we will see in the next chapter that huge and powerful electromagnets are ordinarily needed in hydrogen fusion experiments in order to pinch streams of high-energy hydrogen ions down into controllable paths while maintaining them at extremely high temperatures. But electromagnets powerful enough to do this job must be so very huge, require such staggering amounts of power to keep them operating, and such complex cooling systems to keep them cooled down while they are in operation that there is an upper practical limit to their size. By comparison, electromagnets built of superconductive material, when kept near the temperature of liquid helium, are literally mighty midgets. The working element of such a "cryogenic magnet," requiring *no* energy to keep it in operation, may be no larger than a pound of butter and still have as much magnetic power as a huge conventional electromagnet requiring millions of watts of power to operate it and massive equipment to keep it from melting from the heat it generates. Cryogenic magnets can even be made portable, so that they can easily be moved from place to place. And in properly designed equipment they may well help solve the problem of creating "magnetic bottles" in which to control the explosive violence of hydrogen fusion reactions.

One front-line researcher in this field of "cryogenics" or ultra-low-temperature physics is Stanford University's William M. Fairbank, who is highly optimistic that powerful cryogenic magnets may prove to be the vital key to controlling fusion reactions in the future. As he expresses it: "If so, we will have reached the ultimate in the absurdity of science: a 'bottle' kept at −269 degrees Centigrade used to contain a process involving millions of degrees of heat."

What accounts for the curious properties of superconductivity which allow certain metals and alloys to carry perpetual electric circuits without continuing energy input and makes such bizarre electromagnets possible? At this point no one knows for certain, any more than anyone can be sure that all the strange properties of matter at extremely cold temperatures have yet been discovered. Theoretical physicists are confident that the answer to these strange properties of ultracold matter lies in the energy levels of electrons of the atoms of superconductive materials, and that the natural laws governing this strange behavior will be found in the realm of quantum mechanics. Already a number of hypotheses appear at least promising, and work is now progressing in this field at an ever-accelerating pace. But confirmation of such theories still must await a wealth of experimental work that is yet to be done. We can be certain of one thing, however: if answers are forthcoming in the next five, ten, or fifty years, they will be accompanied by still more new and inexplicable discoveries which will remain to be explained.

Certainly the development of semiconductors and transistors, and the discovery of such practical applications of our knowledge of quantum mechanics as the maser and the laser, have been achievements in modern physics that have altered our lives and promised to alter them even more in the future. In recent years the whole field of cryogenic physics, improbable and strange as it may seem, has blossomed into a major area of applied research which promises in the end to change our lives in the modern world even more. But one area of work in nuclear physics, which began in 1939 and has continued at a fiercely accelerating pace ever since, has already had more dreadful impact on the affairs of men than any other discovery in the course of human history and today may well hold the key to the entire future of the world. In tampering with nature, as scientists have always done in order to uncover her secrets, physicists in recent decades have discovered a source of unimaginable quantities of power—power that can be released so violently as to destroy all life on earth in a matter of a few hours, but which, if harnessed, could make men once and for all the masters of their own environment and open the way to the stars. This book would not be complete without a review of the most intense search that has ever taken place in the history of science and is still proceeding today: the search for control of thermonuclear reactions.

# CHAPTER 29

## *Hydrogen Fusion and Thermonuclear Energy*

Day after day our sun blazes in the sky, pouring out radiant energy in inconceivable quantities and consuming itself day by day in the process. Every life process on earth can ultimately be traced to that radiant energy; without it there would be no life, simple or complex. For some 5 billion years now, astronomers estimate, the sun has continued this outpouring of energy without any significant change in its size, shape, or color, and it is expected that it will continue for still another 5 billion years or so without significant change. But even though we need not worry about it very much today, we know that that magnificent outpouring of energy ultimately will have to end, for it is based upon a simple fuel-consuming reaction: the fusing together of hydrogen ions into inert helium nuclei under conditions of incredibly high temperatures and intense pressures, with the resultant conversion or transformation of some of the mass of the hydrogen ions into radiant energy.

Albert Einstein was the first to recognize that matter and energy were two faces of the same coin, inseparable one from the other and theoretically interchangeable or intertransformable. It was he who calculated the ratio of equivalence between matter and energy, indicating that an incredibly tiny microquantity of matter would be equivalent to a staggeringly huge amount of energy, if some way could be found to transform it. But actual experimental laboratory evidence of this disproportionate equivalence ratio between matter and energy did not appear until nuclear physicists in the 1930s began splitting apart the nuclei of atoms by bombarding them with high-energy "atomic bullets."

### URANIUM FISSION, ENERGY, AND BOMBS

As we have seen, Germany's Otto Hahn was the first to demonstrate in the laboratory in 1939 that the nuclei of heavy radioactive atoms could literally be split in half under the impact of neutron bombardment, with the production of additional "fission neutrons" and the release of great quantities of radiant energy. The amount of energy released in uranium fission, and the possibility that a chain reaction could be set up in which the extra neutrons released in one fission would go on to produce fission in more atoms, producing more neutrons to cause fission in still more atoms, immediately suggested the basis for an extremely powerful bomb.

It was soon found that any uranium nucleus could be broken up if subjected to neutron bombardment, but the end products of the breakup differed sharply according to which isotope of uranium was used. The splitting of a single uranium atom released energy, but not enough energy to be used in any practical way. Such a single fission event, we must remember, occurred on a microcosmic scale. What was needed to create a bomb was a self-sustaining crescendo chain reaction of fission. But the nuclei of different isotopes of uranium behaved differently when they were bombarded with neutrons. Uranium of atomic weight 238, for example, would release energy when it was split apart, but it would not release any additional neutrons. An atom of uranium 235, on the other hand, would release three or four extra neutrons in the process of fission. These extra neutrons could then proceed to bombard other uranium 235 nuclei and thus start an expanding chain reaction of uranium fission.

Beyond question the discovery of uranium fission came at an inopportune time, in the midst of a bloody world war, and the employment of the fission reaction as the basis of a superbomb took priority over all other considerations. It might seem that the problem was simple: Gather a mass of uranium 235 together, bombard it with neutrons, and duck. In practice it was not all that simple. In natural uranium ore, virtually all of the uranium was in the form of uranium 238, with only the barest sprinkling of uranium 235 isotopes present. Any attempt to create an explosive chain reaction in $U^{238}$ seemed nearly hopeless; indeed, $U^{238}$ was primarily a nuisance, acting to dilute fissionable $U^{235}$ so that a huge pile had to be built for a relatively meager energy output.

Thus, when the famous Manhattan Project was begun, the first order of business was a search for ingenious ways of refining uranium in order to separate the usable isotope from the unusable one so that a sufficient mass of $U^{235}$ could be gotten together to sustain an expanding chain reaction—or even better, to accumulate enough of the even more fissionable man-made element *plutonium* to sustain such a chain reaction. The other problem was to find a way to accumulate such a "critical mass" of fissionable material under such conditions of control that the whole pile didn't blow to smithereens the instant a sufficient mass of the critical element had been collected in one place.

By now the story of that achievement is well known. Enrico Fermi had no way at first to separate $U^{235}$ from useless $U^{238}$ in the naturally occurring metal, so he constructed that first atomic pile with such a carefully planned geometrical arrangement as to maximize the chain reaction in the little $U^{235}$ present. Graphite blocks were placed between the chunks of uranium to slow down the velocity of free neutrons emitted by the uranium to the optimum, relatively slow velocity needed for neutrons to cause $U^{235}$ nuclei to split. To control the number of free neutrons in the pile, movable cadmium rods were inserted, since cadmium was a substance with a remarkable affinity for neutrons, avariciously absorbing any in its vicinity. Thus, as the first atomic pile was built in Chicago under Fermi's direction, a continuing fission chain reaction in the uranium 235 was finally achieved under perfect control, with the reaction either slowed down or speeded up at will by moving the cadmium rods farther into the pile or farther out of it.

Of course, in building a bomb out of fissionable material, exactly the opposite goal was sought: Rather than achieve a controlled chain reaction, it was necessary to find a way to trigger a chain reaction of explosive violence at will, but with some means of preventing the thing from going off before it was on target. To achieve this, the total "critical mass" of fissionable material needed to create an explosive chain reaction had to be separated and held apart into two aggregates which could then be pushed suddenly together at the time and place explosion was desired.

The development of such a device was carried on by a group of physicists and engineers working in strictest secrecy at Los Alamos, New Mexico, in 1942 and 1943. $U^{235}$ was still extremely difficult to obtain in significant quantities, but this problem was cleverly by-passed when it was discovered that an artificial radioactive element known as plutonium could be created in great quantity by bombarding uranium 238 with neutrons in a reactor. Plutonium was a new element, never before found in nature, with an atomic number of 94, two higher than uranium, and with an atomic weight of 239. This artificial element was not only easier to obtain than $U^{235}$, but was also much more fissionable, creating far more free neutrons from each fission and thus making possible a "hotter," faster explosive chain reaction. Once the Fermi uranium pile had been built, the major efforts of the physicists in the Manhattan Project were directed to producing, refining, and purifying plutonium and then using it to create the sort of explosive device that could be controlled until the moment of ground zero was reached.

There have been many exciting accounts written about the work of the Manhattan Project, culminating in the dramatically successful predawn test explosion of the first fission bomb, supported on a tower out in the New Mexican desert on July 16, 1945. The point that is often missed in such accounts of those dreadful days of creative work directed toward mass destruction is the simple fact that before a practical fission bomb could be made at all, it was first necessary to build a functioning *controlled* fission reactor capable of providing a steady flow of nuclear energy, a flow of energy that could be converted into useful power to serve mankind's multitudinous peaceful needs.

The bomb did its destructive job in spades, when it was employed, but long before Hiroshima, physicists had already found ways of controlling nuclear energy release—a power source which only today is really beginning to be exploited, but which has enormous potential for the future. Since the end of the Second World War the possibilities for peaceful use of nuclear energy have been steadily explored. Already atomic engines supply the power for the largest and longest-range submarines ever built. Nuclear-powered steamships will soon be crossing oceans in record times and in ever-growing numbers, and research moves forward to develop practical and economical nuclear engines for jet aircraft, freight trains, and far-ranging rocket ships.

Fission reactions can be put to other peaceful uses as well. A number of nuclear reactors are already in operation in England for the generation of electricity, and even among nations that do not have their own nuclear weapons, no less than forty-one possessed nuclear reactors capable of peaceful power generation in 1968, including such technological byways as Venezuela, Thailand, and the Congo. Thus reactors are certain to become a major source of light and electric power all over the world within a decade. Indeed, it is predicted that virtually *all* new power stations to be built after 1970 will be nuclear powered. In a world which can foresee the total exhaustion of all known petroleum reserves within the course of the next century, with most high-grade coal reserves already badly depleted, and with the growth of both populations and industrial needs fast outstripping all potential sources of hydroelectric power, the need for vast quantities of nuclear energy is clearly in the cards; the research going forward today to resolve the technical and engineering problems of producing enough nuclear energy for mass consumption cheaply enough to make it practicable is paving the way to fulfill a need that will be pressing upon us within ten years.

But if the achievement of controlled release of nuclear energy was a necessary forerunner to the use of nuclear fission in a bomb, the same cannot be said for the hydrogen *fusion* reaction basic to the vastly more powerful so-called thermonuclear devices or hydrogen bombs. As we discussed in an earlier chapter, it is possible under certain circumstances to force hydrogen nuclei together in such a way that the electrostatic repulsion of their positive electrical charges is overcome. Under these special circumstances, hydrogen nuclei can fuse together to form nuclei of atoms of a heavier element, the inert gas helium. But in the process a small portion of the mass of the hydrogen nuclei involved is converted into pure radiant energy.

This "thermonuclear reaction" or "fusion reaction" does not occur readily in the case of ordinary hydrogen nuclei containing only one proton; it is far more readily accomplished using the nuclei of a heavy isotope of hydrogen—atoms of so-called heavy hydrogen or deuterium—with nuclei containing one proton and one neutron each. But even with the use of these special hydrogen atoms, the fusion of such nuclei requires very special circumstances in which the heavy-hydrogen atoms are excited into violent thermal activity, heated to excessively high temperatures.

These special physical conditions needed to make hydrogen fusion reactions possible are known to exist in nature in multitudes of places throughout the universe, yet such special conditions are virtually unknown on our own planet. The attempt to find ways to create the special conditions has led modern physicists into one of the most exciting, fascinating, and incredibly difficult areas of search and discovery in the entire history of physics, into an area of challenge which many physicists today consider totally beyond the reach of human ingenuity. Yet if science could conquer and master the violent physical conditions necessary to control hydrogen fusion, mankind could be provided with such unthinkably huge quantities of power that men could literally hurl planets out of their orbits and mold and engineer the shape of solar systems to their will. The ability to control hydrogen fusion reactions and channel this power into peaceful use could in a few decades remold our lives as fundamentally as human lives were remolded by the control of fire or the discovery of the wheel. Such is the staggering challenge facing the physicists today who are working in the weird and wonderful world of *plasma physics*.

## THE FOURTH STATE OF MATTER

In the last chapter we saw that low-temperature physicists seeking to cool down liquid helium close enough to absolute zero that it would solidify discovered instead that at the exceedingly low temperature of 2.9 degrees Kelvin the frigid liquid underwent a striking change into a bizarre superfluid state, a state of fluidity so very different from the fluidity of normal liquids that investigators were tempted to call it "the fourth state of matter."

In more recent years physicists investigating reactions at the far extreme of the temperature scale—at temperatures of tens of thousands of degrees above zero Centigrade or Fahrenheit—discovered still another "fourth state of matter": an excessively hot gas which was not normal gas at all, but was composed of the stripped-down nuclei of hydrogen atoms bereft of their electrons, violently charged with thermal energy and violently reactive. This high-temperature "fourth state of matter," which has far more practical significance in the world today than the strangely superfluid helium II, is a form of matter commonly spoken of as a "plasma."

To understand what a plasma is, and why it is so important in modern physics, we must remember that the terms "heat" and "temperature" are terms that are often used rather loosely in the everyday world, but which have more precise and specific meaning in physics than we usually assign to them. Ordinarily we think of the temperature of a substance as a measure of its heat, and we regard the heat of a substance as a sort of palpable hotness, a degree of warmth or an ability to cause a burn. But strictly speaking, the temperature of any substance is not necessarily a measure of its apparent warmth or coldness. Temperature *is* a reference to the thermal energy of the atoms or molecules of a given substance, an indirect measure of the vigor of molecular motion within that substance.

This is not to say that a substance at a high temperature will never feel hot; if you hold your hand in a jet of steam coming out of a teapot, you will most assuredly feel an uncomfortable subjective sensation that we speak of as heat, you will probably sustain a burn, and you might well have a certain amount of skin and underlying tissue destroyed as a result of the exposure. Your hand is being exposed, in effect, to a dense concentration of water molecules possessing very high thermal energy, and in such a concentration you feel the thermal energy of the molecules subjectively as heat.

But if we could imagine a few water molecules in a box under extremely rarefied conditions—say only a few molecules in a box one foot square—and then if we could find some way to endow those few water molecules with very high thermal energy so that, few as they were, they were racing wildly around the inside of the container, you would feel no heat when you thrust your hand in and would not sustain a burn even though the temperature of the gas in that box might be far, far greater than the temperature of the steam coming out of the teapot spout. In ordinary life, of course, this distinction in terms does not have much significance, but as we shall soon see, it can have a very great deal of significance under certain special circumstances.

We also normally speak rather loosely of "heating up" a substance or of "adding heat" to it as a major means of increasing the thermal energy of its molecules and thus increasing its temperature. As we have seen, if we "add heat" to a solid and raise its temperature to a critical transition point, it will change from the solid state into a liquid state. If we continue "adding heat" to the liquid, presently it will reach a point where it will begin to boil and vaporize into a gas. In the case of ordinary solids and liquids this is a fairly acceptable statement of what happens; the most common way of increasing the temperature of an ordinary solid or liquid or even a gas is through heating it by some external means.

But the application of external heat is not by any means the *only* way of increasing the thermal energy of the molecules of a substance. That thermal energy can also be increased by applying external mechanical force in the form of pressure, thus shoving the molecules closer together. Of course solids or liquids, generally speaking, are relatively incompressible. Their molecules are already closely enough packed together that they resist being packed any tighter by external forces. But if, for example, *very great* pressure is applied to a block of iron, a tiny amount of compression can be brought about and the temperature of the iron will rise slightly.

Thus if you pound a four-inch spike vigorously with a hammer in order to straighten it, both the hammer and the spike will become noticeably warm to the touch after a while. Repeated blows have compressed the molecules of both the hammer and the spike together sufficiently that some of the kinetic energy of the hammer blows has been converted to heat energy and transmitted to the molecules of both the spike and the hammer. This is also the reason that an ice cube will melt if you squeeze it vigorously with a pair of pliers. But for all practical purposes, pressure is not really a significant factor in changing the thermal energy of molecules in either solids or liquids, and any diver who has ever belly-flopped in the water can testify that water's incompressibility makes it one of the hardest substances you could possibly find to fall upon.

In the case of gases, however, things are different. Changes in the pressure applied upon an enclosed quantity of gas can have a great deal of effect on the gas's temperature. As we have seen, the molecules of a gas are widely separated and move freely past each other. Gas molecules, if left unconfined, will diffuse steadily into the atmosphere and ultimately become randomly distributed throughout all the gas in the earth's gaseous envelope. When a gas is confined in a fixed container, the motion of its molecules is sharply curtailed. Confined gas molecules are continually dashing around, striking the walls that confine them and rebounding, so that a confined gas exerts an outward pressure on the walls of its container and, conversely, the walls of the container exert an inward pressure upon the gas. If such a confined quantity of gas is compressed into a smaller area, the molecules are pressed closer together, and have a shorter distance to travel from wall to wall. They also strike each other more frequently, since there are more of them there to run into per unit volume. Thus, the result of compression of a gas in a closed space is an increase in the thermal activity of all the molecules of the gas. Sudden reduction of pressure upon the gas by expanding the volume of the container, on the other hand, can sharply lower the temperature of the gas.

These relationships between temperature, pressure, and volume of quantities of gas in confined quarters were first carefully studied by a seventeenth-century chemist named Robert Boyle. Most of us have encountered Boyle's famous "gas laws" in high school chemistry or physics courses. Boyle's laws have proved to be of enormous practical value for centuries, permitting extremely accurate predictions to be made regarding changes in the temperature, pressure, or volume of enclosed quantities of gases.

Today these applications of Boyle's gas laws are so commonplace that we rarely think of them at all. It is the expansion of gas released from a closed compression area which cools down the coils in the ordinary household refrigerator or freezer, for example. In a physics laboratory, when the air in a cloud chamber is saturated with water vapor, and the pressure within the chamber is then suddenly reduced, conditions are created in which ionizing particles can trigger condensation of water droplets in the water vapor—the principle upon which the usefulness of the Wilson cloud chamber depends. The same principle operates in our atmosphere when a saturated mixture of water vapor, water droplets, dust, and other chemicals in a huge gray cloud reaches an area of lower barometric pressure. With the drop in pressure the water vapor begins condensing out to form more water droplets, the droplets coalesce, and a rainstorm occurs.

Again, when we force air from an air hose into an automobile tire, we cause a sharp rise in the temperature of the air confined within the tire and feel the tire become warm. Later when the gas has cooled down we find that the tire registers an air pressure somewhat lower than we had measured immediately after filling it, even though no air has escaped. But the release of compressed carbon dioxide from the nozzle of a high-pressure $CO_2$ fire extinguisher, on the other hand, creates a flow of gas which expands so rapidly that it supercools and solidifies on the spot, so that we see not a stream of gas but a shower of carbon dioxide snow when we use such a gadget to help us put out a fire in the mattress.

Thus if we have a confined quantity of gas, we can raise the temperature of that gas either by applying external heat to the container or by compressing the gas, reducing the volume in which it is confined. The higher we raise the temperature of the gas, the more violently the molecules fly about and the more violently they strike the walls of the container. Of course, as the thermal energy of the gas increases, more and more heat energy is lost to the air outside the container, conducted through the container wall and dissipated into the surrounding air by radiation. The higher the temperature and the faster the gas molecules fly about, the more rapidly heat is likely to be lost, but even so, it is perfectly possible to keep ahead of the heat loss by applying more external heat to the volume of gas or compressing it more and more, thus raising it to a higher and higher temperature.

But is there no limit to how hot a confined volume of gas can become? When a solid is heated up to a certain temperature, it melts. When a liquid is heated up to a certain temperature, it vaporizes. We recognize that the hotter we make a confined gas the more difficult it is to confine it and still sustain higher and higher temperatures; but what would happen if we were able to do just that? Suppose we could add external heat to a confined amount of gas steadily and in ever-increasing quantity. What would ultimately happen? Suppose we insulated the walls of the container so that little or no heat could escape. Could we continue heating that gas up indefinitely and end up merely with a hotter and hotter gas? Or would we end up with something else?

The answer is that we would end up with something else. Under such circumstances of superheating, with the temperature of the gas reaching five or six thousand degrees Fahrenheit, for example, the thermal energy and thermal activity of the gas's molecules would become extreme. They would begin thrashing about their enclosure so very violently and with such enormous energy that the gas atoms which would be electrically neutral under normal circumstances would begin smashing themselves into positively and negatively charged ions. Electrons would be stripped from their atoms and join the swarming gas as freely moving charged particles. The positively charged nuclei would begin racing about independent of the electrons normally associated with them. Every collision of a particle with a confining wall, or with another particle, would tend to dampen activity of the particles and "cool down" the furor; but if we continued to apply more and more heat, more and more of the gas would ionize itself in virtually the same situation that is present when an electrical discharge passes through a Crooke's tube, ionizing the gas within it.

What would happen to these "thermal ions" formed under such extreme conditions of temperature? At comparatively "low" temperatures of 6,000, 8,000, or 10,000 degrees Fahrenheit they would tend to recombine into neutral atoms as fast as they were ionized—there would be a continuous shuffling back and forth from ionized state to neutral state and back again. But if the temperature were raised to 60,000, or 70,000, or 100,000 degrees, virtually all the gas would become ionized into a chaotic swarm of charged particles hurling themselves about independently of each other with enormous thermal energy.

If such conditions could be achieved in the laboratory, we would have brought about a change of matter in the gaseous state into a totally new state of matter, a "fourth state of matter" quite different from normal atomic or molecular gas. As early as 1920 the first investigators of this strange and violent "fourth state of matter" began speaking of it not as a gas but as "plasma"; the name stuck, and remains in use today. Of course, it would be quite impossible to form plasma in the laboratory in the fashion that we have just described; any such "heating up" of a confined body of gas would have led to the complete vaporization of any possible containing vessel long before gas temperatures high enough for plasma formation could have been attained. In fact, here on earth, matter in the plasma state is all but unknown, existing in nature only in lightning bolts or within the aurora borealis, created artificially and momentarily only in electric arcs or in the leap of current between the plates of vacuum tubes.

Yet rare as plasma may be on earth, physicists today are quite certain that the overwhelming preponderance of all matter in the universe exists in the plasma state, with only an insignificant fraction of a per cent of all matter existing in what we consider "commonplace" gaseous, liquid, or solid states, and then only on such cold and remote backwaters of the universe as our planet earth. Except for the thin outermost layer of gas surrounding our sun, for example, virtually all the matter making up the sun is in the plasma state, a broiling mass of hot, furiously energetic, positively charged ions. Plasma is the raw material of which all the stars are made. Between the stars it fills all space with an exceedingly rarefied sprinkling of positive and negative ions, mostly hydrogen gas in the plasma state, so widely disseminated that any representative volume of interstellar space contains fewer atomic and subatomic particles of any sort than the most perfect vacuum ever achieved on earth.

And in the depths of the sun, under conditions of extreme pressure and extreme temperature, we know that hydrogen ions driven together at high velocities join together, fuse, and form helium nuclei. Four protons fused in this fashion create one helium nucleus, but the total weight of the helium nucleus thus formed is slightly less than the combined weight of the four fusing hydrogen nuclei. The remaining mass, although relatively insignificant in amount as *mass,* is converted in the course of the fusion process into radiant energy—a staggering 200 million electron volts of energy released for each and every single helium atom thus formed.

It is this enormous outpouring of energy that accounts, for the most part, for the "eternal fire" that keeps our sun alive. Even though our sun has been consuming huge amounts of hydrogen from its core every second and converting it to helium, with the resultant outpouring of radiant energy and light, for billions of years, astronomers estimate that the vast bulk of the matter in the interior of the sun is *still* composed of hydrogen plasma, so that this kind of hydrogen fusion reaction can be expected to continue without lagging for billions of years to come.

## MAN-MADE FUSION REACTIONS AND THE HYDROGEN BOMB

If it is true that the vast preponderance of all matter in the universe is in the plasma state, concentrated in the hearts of stars or scattered piecemeal throughout interstellar and intergalactic space in the form of excessively rarefied gas, one might think it would be a simple matter for physicists to get hold of some plasma, or artificially produce it in their laboratories, in order to bring about man-made hydrogen fusion reactions and take advantage of the huge quantities of energy that such reactions would release.

But physicists investigating plasma found this to be far from simple. Except in the rare circumstances we mentioned before, there is *no* matter existing on earth in the plasma state under natural conditions, and the creation of artificial conditions in which hydrogen plasma could be produced has presented physicists with virtually insuperable problems. How could it be possible to contain a body of gas superheated to such temperatures, for example, in any material container? How could it be possible to keep plasma particles at such a high state of thermal agitation that violent forceful collisions would take place—the conditions necessary in order to cause hydrogen ions to fuse into helium?

There were no ready answers to such questions. But in their study of the properties of plasma since the early 1920s, physicists have gradually learned a few things which have been of some help. For one thing, two heavy isotopes of hydrogen have been discovered which will enter into fusion reactions under far less extreme conditions than normal hydrogen nuclei. The temperature and pressure conditions necessary to make nuclei of ordinary atomic hydrogen—nothing more than individual raw protons with the electrons stripped away—enter into fusion reactions that were found to be so furiously extreme that some alternative had to be sought. Almost all hydrogen atoms found under normal conditions have a single proton and a single electron, but two other isotopes of hydrogen were discovered, one with the nucleus containing one proton and one neutron (an isotope spoken of as "deuterium") and the other containing one proton and two neutrons (commonly known a "tritium").

A very small percentage of atoms of hydrogen in the naturally occurring gas are atoms of deuterium, and these can be separated out. It is even possible to oxidize deuterium to form water molecules made of heavy-hydrogen atoms, so-called heavy water. Tritium atoms rarely occur in nature, but can be artificially produced. It was found that although nuclei of ordinary hydrogen atoms—in other words, single protons—would fuse together only with difficulty under such extreme conditions that no laboratory could manage them, both deuterium and tritium ions could be made to fuse far more readily at far lower temperatures. Furthermore, deuterium could be separated and purified by means of separating out the heavy-water molecules from ordinary water supplies, and tritium ions could be produced by bombarding nuclei of a heavier element, lithium (the lightest of the alkali metals, composed of three protons and three neutrons) with neutron projectiles produced by an atomic reactor.

The discovery that both deuterium and tritium atoms could readily be produced and could be made to fuse at relatively low temperatures immediately stimulated military interest in the possibility of building a so-called *thermonuclear device,* a bomb in which the enormous energy from an uncontrolled fusion reaction could be triggered in a gigantic destructive explosion. An urgent program of research and development was begun to produce such a bomb, and in 1952 the first true *fusion* bomb, commonly called a "hydrogen bomb" or more properly, a "thermonuclear explosive device" was detonated on a coral island called Elugelab in the Pacific Ocean. That bomb exploded with such incredible violence that when the fallout settled, the island could no longer be found; it had literally been blown off the map. Two years later, on Bikini Atoll, also in the Pacific Ocean, another thermonuclear bomb was detonated to enable scientists to study first hand the nature of the explosion (which occurred above the ground rather than on a solid base) and to learn more about the immense cloud of radioactive materials, so-called radioactive fallout, which was produced by the explosion.

Sad to say, the scientists working on the project found out more than they had bargained for. Much more radioactive material was produced than was expected, and an ill-fated shift of the wind carried the cloud of fallout materials far off its expected course and across the path of a Japanese fishing vessel that was working in an area expected to be perfectly safe. The radioactive exposure even to the fallout materials of this bomb was so intense that one crew member of the Japanese fishing boat died from radiation exposure, and the twenty-two others on board suffered intense radiation sickness. As an additional unfortunate bonus, over 200,000 pounds of fish caught by Japanese vessels in Pacific waters that year had to be condemned and destroyed because of the radioactivity that they had picked up from the blast.

There could have been no clearer warning of the two-edged blade presented to the world by that early hydrogen bomb and by its more refined and sophisticated relatives which were exploded later. Subsequent bomb tests have scattered enormous clouds of fallout material to the four winds, with a fallout cloud often sweeping a path across the entire earth high in the atmosphere for weeks or even months after the blast before it finally is dissipated. Fortunately, many of the radioactive fallout by-products had very short half-lives: they would decay into harmless nonradioactive substances in a matter of seconds, minutes, or days. But other by-products of these bombs were found to have far longer half-lives and to pose unique and unexpected threats to human and animal life.

For instance, one by-product of modern thermonuclear weapons is a radioactive isotope of strontium, strontium 90, with a particularly long half-life. An aggregation of strontium 90 will pour out deadly gamma rays for a period of decades before it has finally decayed into harmless nonradioactive stuff. This might not be so alarming, were it not for the fact that strontium is in the same chemical family as calcium, and when a radioactive dust containing strontium 90 settles to the ground it is likely to be eaten, along with grass and hay, by dairy cattle.

Now a cow is a delightful animal, specialized in its domestication to produce a protein-rich and calcium-rich milk. But biochemists have pointed out that strontium 90 can take the place of calcium in the milk of cows that have eaten it, and when such contaminated milk is drunk by humans the strontium 90 will be concentrated, along with the calcium, in the structure of human bones. Thus the radioactive fallout products from thermonuclear devices exploded high in the atmosphere can lead to an inordinate concentration of a long-term deadly radioactive substance in human bones—the part of the body in which white and red blood cells are normally produced. Theoretically, a large increase in the amount of strontium 90 floating around loose in the atmosphere will inevitably lead to a sharp increase in such diseases as leukemia and other bone cancers among people living in the areas over which such a deadly cloud has passed.

It was the recognition of the growing threat to all mankind of vast quantities of radioactive materials that were accumulating in the atmosphere as a result of multiple thermonuclear bomb tests that finally led the governments of the world to explore and develop treaties which hopefully will one day restrict *all* further testing of thermonuclear weapons to underground tests which cannot scatter poisonous radioactive debris far and wide over the earth. Meanwhile, physicists all over the world have been vigorously exploring ways to develop "clean" bombs—thermonuclear weapons which might explode with little or no long-lived radioactive fallout resulting from their explosion.

## THE ANATOMY OF HOLOCAUST

How does a hydrogen bomb actually work, and how can this kind of violent explosion be contained in safe storage, to be triggered only at the desired time and place? Obviously the technical details of building, storing, and triggering such weapons are closely guarded secrets; but even physicists who have had no connection whatever with this secret work can imagine the nuclear sequence of events that leads to a thermonuclear explosion, and can guess quite accurately what the general structure of such bombs must be.

Essentially, a hydrogen bomb in its present form consists of three major parts: a chunk of uranium 235 contained within an outer coating of lithium 6 and deuterium, with both these parts encased in an outer shell of uranium 238. Lithium is the lightest of all known metals, its atoms made up of three protons and three neutrons each. The fusion reaction causing the thermonuclear blast must occur in the segment of the bomb containing the lithium and deuterium atoms; sufficient neutrons must bombard the lithium to convert lithium nuclei into tritium ions to mix with the deuterium, and this mixture of hydrogen isotopes must be raised to the ultrahigh temperatures and pressures necessary to convert the tritium and deuterium into plasma. The plasma particles thus formed then undergo fusion to form helium 4 with the instantaneous liberation of energy.

But where do the neutrons come from to bombard the lithium to convert it into tritium, and how is the extremely high temperature to convert this mess into the plasma state achieved? As we have seen, one of the best sources of free neutrons is the nuclear *fission* reaction in uranium 235, and the explosion of the uranium *fission* bomb in the immediate vicinity of the lithium and deuterium container would also create the high temperatures necessary for a *fusion* reaction. Thus a hydrogen bomb uses a uranium fission bomb as its trigger, with the uranium bomb acting as a "percussion cap" to set off a massively more powerful explosion.

Once the uranium percussion cap has raised temperatures high enough and created enough neutrons to convert the lithium atoms to tritium, and once the tritium and deuterium atoms have been reduced to plasma and a fusion reaction takes place, the subsequent blast liberates still more neutrons which then bombard the outer casing of uranium 238, triggering a final fission reaction. Thus, a hydrogen bomb explosion is actually a three-stage explosion. First a small uranium 235 fission bomb is exploded. This converts lithium into tritium and triggers the fusion of tritium and deuterium into helium 4 in the thermonuclear blast, which finally sets off a uranium 238 fission explosion to cap the climax. Since all these stages are tightly enclosed in a confining container, it is estimated that this final fission explosion of uranium 238 accounts for from one-half to two-thirds of the bomb's entire energy release. It is also the radioactive by-products of this final stage of the explosion that produce the lethal fallout of radioactive dust.

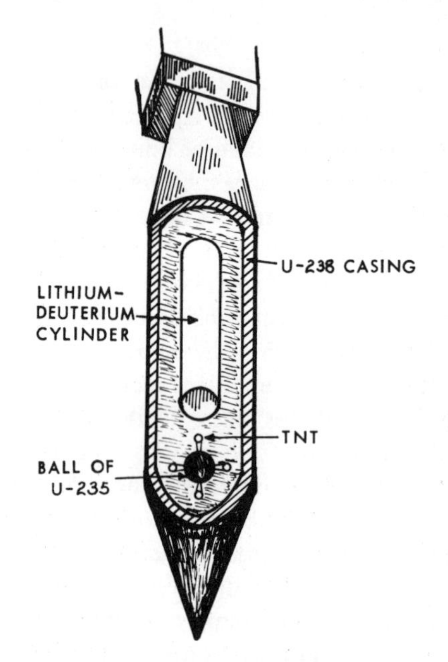


*Fig. 60* Although the actual construction of a hydrogen bomb is obviously classified, the basic elements it must contain are diagramed at left. TNT charges drive two chunks of $U^{235}$ together to create a critical mass, engendering a fission explosion. High temperature and release of free neutrons from this "percussion cap" convert lithium nuclei into tritium ions which mix with deuterium ions and temperature is raised high enough to convert this mixture into plasma and trigger an explosive fusion reaction. The subsequent blast releases more neutrons which bombard an outer casing of $U^{238}$ which sets off another fission reaction. This final explosion is probably responsible for much of the "dirty" radioactive fallout of such bombs, but also accounts for a large proportion of the bomb's total energy release, so that explosive power might have to be sacrificed if a so-called "cleaner" bomb were to be built.

One of the major problems of thermonuclear weaponry today is to find some alternative to uranium 238 for the last stage of the explosion, in order to reduce this radioactive fallout from today's "dirty" thermonuclear weapons. What is needed is a material that is fissionable, but which has fewer long-lived radioactive by-products, or some other kind of explosive material that does not produce radioactive by-products at all. Alternatively, researchers must find some way to filter the deadly fallout particles out of the atmosphere surrounding the bomb very quickly, so that they are not released to travel with the great air currents that circle the earth. Progress has not been rapid in this respect: The last atmospheric bomb tests conducted by the United States and Russia were still "dirty" bombs, and subsequent tests of bombs by the French and the Red Chinese have produced fallout which was just as "dirty" if not dirtier. In Figure 60 is a rough schematic diagram of how a hydrogen bomb might look, assuming the three-stage fission-fusion-fission reaction that we have been discussing. The steps in the thermonuclear reactions that occur in such a bomb are also outlined in the diagram.

## THE CHALLENGE OF CONTROLLED FUSION

Sometimes it seems to be the nature of human affairs that things are invariably done backwards, and it is a cruel irony that the major emphasis and the most intensive research into the behavior of plasma and fusion reactions have been carried out under the heat of military urgency. In the case of the uranium bomb, as we saw, physicists at least had to learn how to produce a controlled nuclear chain reaction before proceeding to produce bombs that released uncontrollable violence. In the case of hydrogen fusion reactions, however, so much more violent than fission chain reaction, the same cannot be said. The conditions necessary to make hydrogen isotopes fuse into helium involve such high temperatures even to create the necessary plasma state for such a reaction to occur that the reaction can be brought about at all only under the most violent of circumstances, with no hope whatever of controlling it or containing it.

We can see a rough parallel in the problem of trying to harness and control the energy of the buckshot pellets hurled from a shotgun shell when it is fired. When the firing pin strikes the percussion cap, over two hundred lead pellets from a shotgun shell are hurled away with great energy in one grand explosion. This might be the greatest thing imaginable if we wanted to knock ducks out of the air, but suppose we wanted to harness and control the energy of each of those pellets in order to convert it into a different kind of useful work. Suppose we were searching, for example, for some way to release the energy of the gunpowder in the shell more slowly, in order to drive the buckshot pellets out in a steady, controlled stream, one after another, instead of hurling them all out in one single instantaneous cloud of lead. If we could control the buckshot pellets in that fashion, we could conceivably direct a stream of those pellets to strike the blades of a turbine, just as a stream of water might, to set the turbines spinning and perhaps generate electric power to operate the machinery necessary to reload the shotgun shells.

Unfortunately there is no way to achieve this ideal with shotgun shells. If the powder in the shell were to burn slowly, most of its energy would be dissipated in heat and very little if any of it would be converted into kinetic energy driving the lead pellets forward. Firing a shotgun shell is an all-or-nothing proposition; we either get all the pellets moving forward violently at the same time, or we don't get any of them to move at all. The same is true of a firecracker: with its powder tightly contained, it explodes with a bang; but when we try to control the speed of the burning of the powder by, say, breaking open the firecracker, all we get is a fizzle.

There is a vast difference between engendering a hydrogen fusion reaction in a thermonuclear bomb and firing off the black powder in a shotgun shell, granted; yet there are certain inescapable similarities. The problem of trying to produce a controlled hydrogen fusion reaction centers around two conditions, neither of which is yet attainable. First, it is necessary to find some way to attain sufficient temperature of a gas to convert it to plasma state so that fusion can take place. Second it is necessary to maintain such temperatures with a sufficient number of fusible particles in order to sustain a reaction for any period of time and to contain it within any imaginable kind of laboratory apparatus.

In the hydrogen bomb, it was easy. The necessary high temperatures could be achieved by means of a uranium fission explosion. This is not very practical in the laboratory. What is more, it is all but impossible to keep hydrogen gas heated to the plasma state within any kind of closed container without having the relatively frigid temperature of the container itself constantly slowing down the plasma particles, or without increasing the temperature of plasma *and* container to such a point that the container itself is vaporized. How do you slow down the firing of a shotgun shell so that each of the lead pellets can individually be controlled? No one knows; there seems to be no way to do it. How do you slow down the violently explosive reaction of a hydrogen bomb without having an explosion, and in such a way that the energy can be tapped and drained away a little at a time? Nobody knows, so far.

The problem has been studied for a dozen years, and has still not been resolved. At first, work in this area was top secret, but gradually there has been more and more exchange of information between the scientists of different countries. It is not hard to see why. The successful discovery of a way to control and contain the conversion of mass into energy would provide the key to virtually limitless power to serve all mankind. This kind of problem supersedes political and national boundaries. It has provided, in fact, the incredible opportunity for scientists throughout the whole world to join hands to discover techniques which could not conceivably hurt anyone but which could ultimately benefit every human being alive on earth today.

So far, the most promising approach to such control has been to try to develop a "container" *which itself has no material substance* in which to hold and heat the plasma particles. Any material bottle or container would have solid walls of matter with which the plasma particles could interact, so that the plasma would inevitably be cooled down to a temperature that would make fusion reactions impossible. Thus the ideal container for plasma particles cannot have walls made of solid matter; it must be composed of electric and magnetic fields.

Such a container sounds bizarre, but it makes sense when we consider that the plasma particles we are speaking of are positively and negatively charged ions, vulnerable to the effects of electrical and magnetic fields. If a stream of charged plasma particles could somehow be held in a tight "pinched down" column within a tube formed by magnetic and electrical lines of force, that stream of particles might be kept from ever touching the material walls of the tube and could be agitated or "heated" to higher and higher temperatures by means of the driving force of high-voltage electric current. Furthermore, instead of using collections of plasma particles of high density such as are used in hydrogen bombs which, when entering into fusion reactions create such violent quantities of radiant energy that no walls of any kind could contain them, physicists have worked to develop streams of plasma at extremely low gas densities. The densities are so low, in fact, that the plasma particles contained in the sort of "magnetic hobbles" we have been talking about would be so rarefied that there would be fewer particles per unit of volume of a controlled stream of plasma particles than we would find in a high vacuum.

Consider what this means. We know that deuterium gas in the plasma state within an enclosed tube will consist entirely of positively charged deuterium nuclei or so-called deuterons, together with negatively charged electrons. When electrically charged particles move through a magnetic field, they experience a force perpendicular to the direction of the motion of that magnetic field. Thus, if powerful electromagnets surround a tube containing such a plasma, the individual particles will spiral along in the direction of the magnetic lines. If in addition a strong electric current is discharged through the tube the plasma particles will tend to be squeezed together, drawn toward each other in a narrow column at the center of the tube. Such an electrical discharge through the tube also speeds the motion of the particles. In other words, it increases their temperature. Theoretically, if such a "pinch effect" could be brought to bear upon a stream of plasma in a tube, and if such a pinch effect could be strong enough and maintained long enough, the deuterons making up part of the plasma particles would be driven together with sufficient energy for them to fuse, forming helium nuclei and releasing incredible amounts of radiant energy and large numbers of neutrons.

Such reactions can be achieved and *have been achieved,* at least momentarily, in many physics laboratories around the world. The equipment required for this kind of experiment is massive, complex, and costly. The electric power required to achieve a momentary fusion reaction is simply staggering. At the Plasma Physics Laboratory at Princeton University, for example, such a "magnetic bottle" has been built in the form of a hollow tube twisted into the shape of a huge figure 8 or "torus," shaped rather like an elongated doughnut which has been twisted upon itself (see Fig. 61). This huge gadget, which its manufacturers call a "Stellarator," fills a room the size of a barn. It cost $35 million to build, and requires a whole warehouse full of generators in order to develop the electric voltages required to

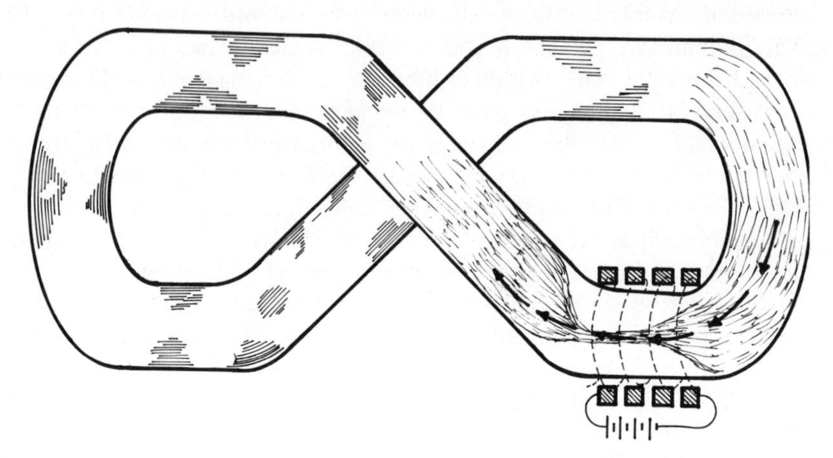

*Fig. 61* A diagramatic representation of a "magnetic bottle" in the shape of a torus, designed to contain high-energy plasma particles, accelerate them, and "pinch" them away from the sides of the tube by the force of electromagnetic fields, so that the particles might be driven into fusion reactions under controlled conditions. Successful controlled and sustained fusion reactions have not yet been achieved, but "fusion reactors" of the future may well contain such "magnetic bottles" as their vital components.

operate its electromagnets. For all of that, those generators can only be activated for brief periods during the hours between midnight and dawn; so much electric power is required to operate them that if they were activated at times when available current in the area was being used for normal household needs, the generator would simply black out the city of Princeton.

With such huge and expensive toys to work with, modern plasma physicists have learned more and more about the properties and behavior of plasma particles, about their containment, and about ways of improving the shapes of the tubes to hold them, about ways to achieve higher and higher temperatures, always moving toward the time that some kind of controlled and sustained fusion reaction may be possible. So far that utopian goal has not been reached. To date nothing even approaching sufficient controlled fusion to have any practical value has been achieved. Many of the men working in the field of plasma physics hold considerable doubt that they will ever find a way to contain plasma effectively and at the same time raise its temperature high enough for long enough periods at a time to sustain a controlled fusion conversion of mass to energy.

Other workers are more optimistic. They point to the long succession of almost unbelievable achievements that modern physicists have marked up, one by one. They point to the long history of solutions of apparently insoluble problems which modern physicists have achieved. Almost all physicists working in plasma physics recognize that the answer is not likely to be found merely by building bigger and better Stellarators, or by working out technical improvements in existing equipment, important as this may be. The basic problem is that there are still simply far too many gaping holes in our knowledge of this strange fourth state of matter called plasma. The more that is learned about plasma and its properties, the more likely it is that we will encounter the major breakthroughs necessary to understand plasma or devise ways of making it behave the way we want it to. These investigators also realize that a deeper and more fundamental knowledge of the true elemental particles making up the nuclei of atoms is necessary. Other workers are constantly probing deeper and deeper into the heart of nuclei, and providing more and more new knowledge of its structure. They may provide the necessary new knowledge and insight into the creation and maintenance of plasma, and as this basic research continues, a rich intellectual fallout is certain to follow.

And although the main effort in plasma research is concentrated upon achieving energy-rich fusion reactions, the plasma state of matter may well be put to other uses as well. One side branch of the basic research and technical engineering that has arisen from the study of plasma properties is a field of work spoken of as "magnetohydrodynamics,"—MHD, for short. This is the study of a number of useful effects that arise when magnetic fields are applied to various fluids (including plasmas) that are capable of conducting electric currents. Already much work is being done in the exploration of possible uses of MHD principles in building more efficient rocket engines for propelling spacecraft, for example, or for creating powerful braking fields around the spaceship to help slow and support the ship as it plunges through our atmosphere toward the earth.

Thus, modern physicists stand on the threshold of discoveries of techniques that can literally alter our lives within a few years or decades. The first step is the control of fusion reactions in some fashion—but this may be only the crudest beginning to an era of human history marked by the control of perfectly unthinkable amounts of virtually free energy. Even the most powerful hydrogen bomb taps only a tiny percentage of the *potential* energy that lies in the hearts of atoms. Only a trifle of the total mass of two deuterium nuclei is converted into energy when those nuclei fuse into a helium nucleus. What if the *total mass* of these reacting particles could be converted into energy under control and direction? This is precisely what is believed to happen when a "normal" particle and its respective antiparticle collide, as in the collision of an electron and a positron, for example, or the collision of a proton and its antiproton, or of a neutron and an antineutron. The energy released in the collision with and mutual annihilation of a single particle with its respective antiparticle exceeds the energy release that occurs when two deuterium nuclei join to form one helium nucleus by a factor of ten thousand.

So consider the possibilities. Electrically charged plasma particles such as deuterium ions and electrons can be contained and controlled in "magnetic bottles" so that they never come in contact with the cold matter composing the walls of that bottle. Why then could it not be possible to devise a source of electrically charged antiparticles—antielectrons (positrons) and antideuterons, for example—and contain them within magnetic and electric fields so that they could not come into annihilative contact with the "normal" particles forming their containers? They could be brought into contact at will with their respective antiparticles in controlled streams so that sustained annihilative reactions between particles and antiparticles could be achieved and the fantastic outpouring of energy tapped as it was being steadily produced.

What would happen, if that could be achieved, would simply be that mankind could have control over such fantastic and limitless quantities of energy that he could literally turn the motion of the planets to his will and control the structure of solar systems. Here would be the energy that would permit true planetary engineering, total control of matter in any form desired, the opportunity for altering climates here on earth, the opportunity for moving hot Venus out away from the sun and then providing it with an artificial breathable atmosphere, a capability of bringing cold Mars in closer to the sun and converting it from an essentially airless chunk of desiccated desert rock into a warm, moist, habitable planet.

This kind of reaction would provide mankind with the energy to grow virtually limitless quantities of food, to extract vast quantities of mineral riches from our sea water at incredibly low cost, to gather materials from other parts of our solar system, even from the sun itself, and to build more and better living space for our burgeoning populations. Today, of course, these are little more than science fiction daydreams. If the work of modern physics were to come to an end today, they would never be achieved. But there is no reason to imagine that the work of modern physics is about to end. Already it has found a way for mankind to destroy itself virtually overnight—but so far mankind has not taken up that option. It has found reason to hope that mankind can find the limitless and incredible quantities of power needed for the benefit of all mankind. This power is not yet available. Will it become available? The answer must be found in the long hard course of work that lies before the modern physicist today.

# CHAPTER 30

## *The Endless Investigation*

In the past few decades we have seen that a radical change has taken place in the over-all complexion of scientific research and in the manner in which the average man in the street regards the scientist and his work.

For centuries the scientist was a lone wolf, carrying on his work, making his discoveries, and communicating his advances in a sort of intellectual vacuum. The medieval alchemists were regarded by their contemporaries, probably quite reasonably, as fools or magicians or perhaps even as scholars and wise men, but generally people kept out of their way. Men such as Galileo, Copernicus, Isaac Newton, Michael Faraday, and the like lived in their own private worlds. The societies around them could pretty well take them or leave them alone, and generally they elected to leave them alone. Ordinary nonscientific people were largely indifferent to the advancement of science and to the work of scientists, except in those rare cases in which scientists were unwise or unwitting enough to step on the corns of kings or popes.

Even as late as the 1920s and 1930s, when the pioneering work in relativity theory, quantum mechanics, and nuclear physics was building up in a rising tide, physicists were still more likely to be regarded by the general public as the subject matter for jokes than as significant contributors to the welfare of society. The men working in these great frontier areas of physics were, for the most part, outside the mainstream of everyday life. Few nonscientists had any comprehension of what it was that the physicists were doing, or any recognition that their work might have any relationship to life in the everyday world. Physicists were regarded then as slightly eccentric but essentially harmless kooks working on the far-out edge of a strange intellectual world no one but their own colleagues could understand.

The Manhattan Project changed all that, once and for all. People became aware, with a shock, that the discoveries and the developments of modern physics were suddenly posing a real, live threat to the very survival of men and nations on earth. Then, as further work in physics began leading to a gathering wave of technological advances of enormous benefit to society, people began realizing that while physicists were not harmless eccentrics they also were not all mad scientists either. They were simply engaged in a work which meant change in the everyday life of society—change for the better in some respects, change for the worst in others. Of course all this did not happen at once, but has been gathering over the past three or four decades until today we are in the midst of a vast technological revolution which has arisen directly from the increasing understanding of the nature of matter and energy and the manner in which the universe is put together.

There is today no plausible reason to imagine that research in physics is going to slow down to any degree as we move ahead into the future. On the contrary, there is ample evidence on all sides that that work will continue to expand in the decades ahead. There is evidence to suggest that the major work of physics—the really tough, really important, work—has actually just barely begun. Already this work touches on our lives, in multitudes of ways, wherever we turn. In the decades to come it will affect our lives even more, and in ways we cannot even imagine today.

Not all physicists are quite so sanguine about the future of their work. Many feel that a major break point has been reached in the search of the outer reaches of the cosmos and the exploration of the micro-universe of nuclear physics. Some feel that we have reached a point where there is relatively little remaining to be penetrated that human minds are capable of penetrating. These men are not by any means contending that physicists have already discovered everything there is to know. They merely contend that many of the major problems faced by physicists today are likely to remain insoluble, beyond the reach of human capacity to understand them any better than they are already understood.

The trouble with championing this attitude, however, is that every time somebody publicizes this view the printer's ink is barely dry before some new and quite unexpected observation or discovery is made. Over a decade ago Dr. George Gamow, one of the ablest of all apologists for physicists and their work, remarked that while there was still plenty of interesting and exciting work to be done in the physical sciences, there would probably be no great undiscovered areas of physics suddenly appearing for investigation. In other words, he felt that we might be exploring known countries in greater and greater detail, but that no new countries would be found to explore. As a cosmologist, he spoke gloomily of scientists having already expanded man's knowledge to a limit beyond which nothing basically new would be likely to be found. In his book *Matter, Earth and Sky** Gamow said: "Even the 200-inch telescope of the Palomar Mountain Observatory permits us to see only more of the same galaxies than were seen through the 100 inch telescope on Mt. Wilson, and a 300 inch, 400 inch, etc. telescope will probably see just more of the same. Thus, it seems, that we now have a rather complete general picture of the universe around us . . . both of the vast expanses of the macrocosm and of the vanishing small structures of the microcosm."

* New York, Prentice-Hall, 1959.

Gamow freely admitted at the time that this opinion was strongly disputed by many of his colleagues, and well it might have been. Within three years of the time it was expressed, the first quasar was discovered and a "new area of geography" of startling proportions suddenly opened up for investigation. By now that one discovery has quite thoroughly revolutionized the outlook of astronomers and cosmologists the world over.

In addition, during that same interval of time, equally strange and revolutionary realms opened up in the areas of nuclear physics, plasma physics, the physics of semiconductors, cryogenic physics, and computer technology. Few physicists today would be so incautious as to predict that all possible areas of the universe have been discovered and explored. Most would predict that the next five, ten, fifty, or one hundred years will bring new observations, new, unexplained phenomena, and new "insoluble" problems to the attention of science even as the old "insoluble" problems are one by one resolved. Meanwhile, the effects upon our everyday lives of discoveries already made continue to multiply day by day.

Today, of course, all thinking people are preoccupied with the problem of war, with the existence of thermonuclear weapons, with the threat of radioactive contamination of our planet, and with the very real possibility that paranoid leaders of psychotic nations could unleash such destructive furies on mankind that *all* human achievement could be wiped out permanently, once and for all. But in a longer view, these grim preoccupations may well be obscuring from our thoughts the truly great and significant advances by which history will eventually judge our era. Not long ago science fiction writer Robert A. Heinlein, who for years has pointed out that scientific and technological advancement has consistently outstripped the most wild-eyed speculative predictions, made a startling observation. He suggested that for all of Einstein's work with relativity principles, for all of the recent discoveries about the form and structure of atoms and their nuclei, for all the fission bombs and hydrogen bombs that have been built, and for all of the space exploration this century has already witnessed so far, it is still very likely that our twentieth century, in the long view of history, will be remembered primarily as the century *in which people on earth learned to read*—a quiet by-product of our enormous technological advances, which may have far greater long-range effect on human history than all the hydrogen bombs ever made.

By the same token, the twenty-first century may well go down in history as the century in which men achieved full understanding and total control of physical matter on earth, and control of the unthinkable quantities of energy held captive in the nuclei of atoms. Control of that energy could open the door to achievements of magnitude and benefit that men today can only dream of. The advances of the last sixty-eight years would have staggered the imagination of the most optimistic physicists of the year 1900, had they suddenly learned what their work would be leading to in that short space of time. If we are to believe the lessons of history, our most sanguine predictions of further advancements likely to occur in the next sixty-five years will probably prove ridiculously conservative. If there is one thing we *can* be sure of, it is that the limits of human capability and ingenuity have not yet even been approached in our time, and there is no reason to think that those limits are likely to be approached at any time in the foreseeable future.

But when it comes to understanding the scientific background and basis for the changes that have come about in our lives as a result of the work of modern physics, we find that it becomes increasingly difficult. From one point of view, this work has made the over-all behavior of the physical universe (as far as we have been able to observe it) seem simpler and simpler to understand. Today it appears that that universe is controlled by nothing more than a few types of interactions between elementary particles which can take place within the surprisingly broad limits of a few conservation laws.

Yet it seems that the simpler the basic underlying natural laws become, the more complex and convoluted the implications of those laws seem to become. We are reaching a point at which language cannot describe what is actually happening; the "language" of modern physics has become the language of mathematical analysis and nonverbal symbols. The physicists working on the frontiers of research literally cannot describe, explain, or even discuss their work in any language of words. Books such as this one can only describe this work in terms of coarse generalities and inaccurate, inadequate, or sometimes misleading analogies. And it seems probable that the more the physicists' work impinges on our lives, the more difficult it will become for us even to grasp in terms of generalities what that work is all about.

If it is difficult for the nonscientist to grope his way through the mystery and confusion, he can be cheered to know that modern physicists and mathematicians are not in much better shape themselves. The most brilliant of them find themselves working at the very limit of their comprehension. Furthermore, there are areas of physics which are just as much a mystery to physicists now as they were two hundred years ago. So far, for example, physics has made very little headway toward explaining how gravitational force is transmitted throughout the universe. There is hope that some clue to the mystery of gravity may be found in the quantum theory, but so far the idea of actually controlling gravity or building practical antigravity machines remains in the realm of science fiction.

Similarly, physics has made little progress in understanding the part that the flow of time plays in the infinite scheme of things. We are totally unable to control the movement of time in any way, and there is no reason to think that any sudden enlightenment in this area is going to come about. Thus, time travel must also be considered the stuff of fantasy, and the likelihood that men will find any practical means to control the flow of time is slender indeed.

There are some areas in physics, so far barely explored at all, which offer more promise than this. One such area is the field of biophysics, the application of knowledge of electronic and nuclear physics to an understanding of the behavior of atoms and molecules in living cells. In particular, biophysicists are concerned with the whole question of how electrical impulses are transmitted through living neuron circuits, and face the staggering problem of understanding the operation of the master computer of all time, the human mind.

Quite aside from understanding the normal, recognized functioning of human minds, there is also the question of certain possible *extrasensory* functions of the mind—such functions as clairvoyance, telepathy, psychokinesis, etc. So far, no one is entirely certain whether such functions actually do even exist as true phenomena or not. If they do, there is certainly no explanation yet available of any physical mechanisms which might possibly account for such phenomena. Many physicists and other scientists become very upset even at the mention of such things; they insist that this is pure speculative nonsense, impossible and flatly ruled out by the physical nature of things.

Yet those same scientists have long recognized the existence of other "forces acting at a distance"—forces such as gravity or the force of electromagnetic fields—which cannot be regarded as "impossible" just because no one understands what they are or how they work. Conceivably human minds may generate a totally unsuspected type of electromagnetic field capable of acting upon other minds or even upon physical objects at a distance, capable of influencing the motion of objects backward in time as well as forward. At this point all that can be said is that nobody knows, but the future work of physicists exploring in these areas may well hold some remarkable surprises.

Indeed, when we consider the comparatively brief period of time that men have been grappling with the nature of the universe at all—a mere two or three thousand years—and then regard the enormous progress that has been made in just the last few decades in probing the depths of space and the mysterious realm of the nuclei of atoms, we must be a little cautious about setting any arbitrary limits whatever on what we think may be achieved in a few more years or a few more generations. At the same time, we must be increasingly concerned to find ways to understand what *is* being done in the physical sciences, and to see what *meaning* this work has in our everyday life.

In this book we have attempted, sometimes well, sometimes poorly, to present a coherent picture of the many worlds of physics. We have tried to present the historical background of work in science that has led to the exciting work that is going on today, and to understand better what the work of modern physics means and where it is going. At the same time, we have tried to convey some small impression of the grandeur, the excitement, and the drama of the search so far, and of the breathtaking vistas that lie ahead. For it is clear beyond doubt that the better we understand these things today and in the decades to come, the better we will be equipped to deal with the changes in our lives being wrought by the work of modern physics.

# Index

69 70 71 72 73 8 7 6 5 4 3 2 1